Association for Women in Mathematics Series

Volume 4

More information about this series at http://www.springer.com/series/13764

Association for Women in Mathematics Series

Focusing on the groundbreaking work of women in mathematics past, present, and future, Springer's Association for Women in Mathematics Series presents the latest research and proceedings of conferences worldwide organized by the Association for Women in Mathematics (AWM). All works are peer-reviewed to meet the highest standards of scientific literature, while presenting topics at the cutting edge of pure and appliedmathematics. Since its inception in 1971, The Association for Women in Mathematics has been a non-profit organization designed to help encourage women and girls to study and pursue active careers in mathematics and the mathematical sciences and to promote equal opportunity and equal treatment of women and girls in the mathematical sciences. Currently, the organization represents more than 3000 members and 200 institutions constituting a broad spectrum of the mathematical community, in the United States and around the world.

María Cristina Pereyra • Stefania Marcantognini
Alexander M. Stokolos • Wilfredo Urbina
Editors

Harmonic Analysis, Partial Differential Equations, Complex Analysis, Banach Spaces, and Operator Theory (Volume 1)

Celebrating Cora Sadosky's life

Springer

Editors
María Cristina Pereyra
Department of Mathematics and Statistics
University of New Mexico
Albuquerque, New Mexico, USA

Stefania Marcantognini
Department of Mathematics
Venezuelan Institute for Scientific Research
Caracas, Venezuela

Alexander M. Stokolos
Department of Mathematical Sciences
Georgia Southern University
Statesboro, Georgia, USA

Wilfredo Urbina
Department of Mathematics and Actuarial Science
Roosevelt University
Chicago, Illinois, USA

ISSN 2364-5733 ISSN 2364-5741 (electronic)
Association for Women in Mathematics Series
ISBN 978-3-319-30959-0 ISBN 978-3-319-30961-3 (eBook)
DOI 10.1007/978-3-319-30961-3

Library of Congress Control Number: 2016946423

Printed on acid-free paper

This Springer imprint is published by Springer Nature
The registered company is Springer International Publishing AG Switzerland

Photograph by Margaret Randall, February 2, 2004

Preface for Volume 1

On April 4, 2014, we celebrated Cora Sadosky's life with an *afternoon in her honor*, preceded by the 13th New Mexico Analysis Seminar[1] on April 3–4, 2014, and followed by the Western Sectional Meeting of the AMS on April 5–6, 2014, all held in Albuquerque, New Mexico, USA. It was a mathematical feast, gathering more than a 100 analysts—fledgling, junior, and senior, from all over the USA and the world such as: Canada, India, Mexico, Sweden, the UK, South Korea, Brazil, Israel, Hungary, Finland, Australia, and Spain—to remember her outspokenness, her uncompromising ways, her sharp sense of humor, her erudition, and above all her profound love for mathematics.

Many speakers talked about how their mathematical lives were influenced by Cora's magnetic personality and her mentoring early in their careers and as they grew into independent mathematicians. Particularly felt was her influence among young Argentinian and Venezuelan mathematicians. Rodolfo Torres, in a splendid lecture about Cora and her mathematics, transported us through the years from Buenos Aires to Chicago and then back to Buenos Aires, from Caracas to the USA and then back to Buenos Aires, and from Washington, D.C., to California. He reminded us of Cora always standing up for human rights, Cora as president of the Association for Women in Mathematics (AWM), and Cora always encouraging and fighting for what she thought was right.

---

[1]The 13th New Mexico Analysis Seminar and *An Afternoon in Honor of Cora Sadosky* were sponsored by National Science Foundation (NSF) grant DMS-140042, the Simons Foundation, and the Efroymson Foundation, and the events were done *in cooperation* with the Association for Women in Mathematics (AWM). See the conference's websites:

www.math.unm.edu/conferences/13thAnalysis

people.math.umass.edu/~nahmod/CoraSadosky.html

An Afternoon in Honor of Cora Sadosky was organized by Andrea Nahmod, Cristina Pereyra, and Wilfredo Urbina. The 13th New Mexico Analysis Seminar organizers were Matt Blair, Cristina Pereyra Anna Skripka, Maxim Zinchenko from University of New Mexico, and Nick Michalowsky from New Mexico State University.

Cora was born in Buenos Aires, Argentina, on May 23, 1940, and died on December 3, 2010, in Long Beach, CA. Cora got her PhD in 1965 at the University of Chicago under the supervision of both Alberto Calderón and Antoni Zygmund, the grandparents of the now known Calderón-Zygmund school. Shortly after her return from Chicago, she married Daniel J. Goldstein, her lifelong companion who sadly passed away on March 13, 2014, a few weeks before the Albuquerque gathering. Daniel and Cora had a daughter, Cora Sol, who is now a political science professor at California State University in Long Beach, and a granddaughter, Sasha Malena, who brightened their last years. During her life, Cora wrote more than 50 research papers, and a graduate textbook on *Interpolation of Operators and Singular Integrals: An Introduction to Harmonic Analysis* (Marcel Dekker, 1979), and she edited two volumes: one celebrating Mischa Cotlar's 70th birthday (*Analysis and Partial Differential Equations: A Collection of Papers Dedicated to Mischa Cotlar*, CRC Press, 1989) and one celebrating Alberto Calderón 75th birthday (*Harmonic Analysis and Partial Differential Equations: Essays in Honor of Alberto Calderón*, edited with M. Christ and C. Kenig, The University of Chicago Press, 1999). We have included a list as complete as possible of her scholarly work. Notable are her contributions to harmonic analysis and operator theory, in particular her lifelong very fruitful collaboration with Mischa Cotlar.

When news of Cora's passing spread like wildfire in December 2010, many people were struck. The mathematical community quickly reacted. The AWM organized an impromptu memorial at the 2011 Joint Mathematical Meeting (JMM), as reported by Jill Pipher, at the time the AWM president:

> Many people wrote to express their sadness and to send remembrances. The AWM business meeting on Thursday, January 6 at the 2011 JMM was largely devoted to a remembrance of Cora.

This appeared in the March–April issue of the AWM Newsletter[2] which was entirely dedicated to the memory of Cora Sadosky.

An obituary by Allyn Jackson for Cora Sadosky appeared in Notices of the American Mathematical Society in April 2011.[3]

In June 2011, Cathy O'Neal wrote in her blog mathbabe[4] a beautiful remembrance for Cora:

> [...] Cora, whom I met when I was 21, was the person that made me realize there is a community of women mathematicians, and that I was also welcome to that world. [...] And I felt honored to have met Cora, whose obvious passion for mathematics was absolutely awe-inspiring. She was the person who first explained to me that, as women mathematicians, we will keep growing, keep writing, and keep getting better at math as we grow older [...]. When I googled her this morning, I found out she'd died about 6 months ago. You can read

[2]*President's Report*, AWM Newsletter, Vol. 41, No. 2, March–April 2011, p. 1. This issue was dedicated to the memory of Cora Sadosky.

[3]Notices AMS, Vol. 58, Number 4, April 2011, pp. 613–614.

[4]http://mathbabe.org/2011/06/29/cora-sadosky/.

> about her difficult and inspiring mathematical career in this biography.[5] It made me cry and made me think about how much the world needs role models like Cora.

In 2013, the Association for Women in Mathematics established the biennial AWM-Sadosky Prize in Analysis,[6] to be awarded every other year starting in 2014. The purpose of the award is to highlight exceptional research in analysis by a woman early in her career. Svitlana Mayboroda was the first recipient of the AWM-Sadosky Research Prize in Analysis awarded in January 2014. Mayboroda is contributing a survey paper joint with Ariel Barton to this volume. We include the press release issued by the AWM on May 15, 2013, and the citation and Mayboroda's response that appeared in the March–April 2014 issue of the AWM Newsletter.[7] As this volume goes into press, the second recipient of the award, the 2016 AWM-Sadosky Prize, was announced: Daniela da Silva, from Columbia University.

In 2015, Kristin Lauter, president of the AWM, started her report in the May–June issue of the AWM Newsletter,[8] with a couple of paragraphs remembering Cora:

> I remember very clearly the day I met Cora Sadosky at an AWM event shortly after I got my PhD, and, knowing very little about me, she said unabashedly that she didn't see any reason that I should not be a professor at Harvard someday. I remember being shocked by this idea, and pleased that anyone would express such confidence in my potential, and impressed at the audacity of her ideas and confidence of her convictions.
> Now I know how she felt: when I see the incredibly talented and passionate young female researchers in my field of mathematics, I think to myself that there is no reason on this earth that some of them should not be professors at Harvard. But we are not there yet ... and there still remain many barriers to the advancement and equal treatment of women in our profession and much work to be done.

In these two volumes, friends, colleagues, and/or mentees have contributed research papers, surveys, and/or short remembrances about Cora. The remembrances were sometimes weaved into the article submitted (either at the begining or the end), and we have respected the format each author chose. Many of the authors gave talks in *the 13th New Mexico Analysis Seminar*, in *An Afternoon in Honor of Cora Sadosky*, and/or in the special sessions of the AMS; others could not attend these events but did not think twice when given the opportunity to contribute to this homage.

The mathematical contributions naturally align with Cora's mathematical interests: harmonic analysis and PDEs, weighted norm inequalities, Banach spaces and BMO, operator theory, complex analysis, and classical Fourier theory.

Volume 1 contains articles about Cora, her mathematics and mentorship, remembrances by colleagues and friends, her bibliography according to MathSciNet, and

[5]Biographies of Women in Mathematics: Cora Sadosky http://www.agnesscott.edu/lriddle/women/corasadosky.htm.

[6]More details in the AWM-Sadosky Research Prize in Analysis webpage: https://sites.google.com/site/awmmath/programs/sadosky-prize.

[7]AWM Newsletter, Volume 44, Number 2, March–April 2014.

[8]President's Report. AWM Newsletter, Vol. 45, No. 3, May–June, p. 1.

survey and research articles on harmonic analysis and partial differential equations, BMO, Banach and metric spaces, and complex and classical Fourier analysis.

Last year (2014) saw the resolution of the two-weight problem for the Hilbert transform à la Muckenhoupt by Michael Lacey, Eric Sawyer, Chun-Yen Shen, and Ignacio Uriarte-Tuero, a problem that had been open for 40 years. This problem was solved à la Helson-Szegö by Cora Sadosky and Mischa Cotlar in the early 1980s using complex analysis and operator theory methods. In the last 15 years, a number of techniques have been developed and refined to yield this result, including stopping time arguments, Bellman functions, Lerner's median approach, and bumped approach.

Volume 2 contains survey and research articles on weighted norm inequalities, operator theory, and dyadic harmonic analysis. The articles illustrate some of the recent techniques developed to understand weighted inequalities and more, including a survey of the two-weight problem for the Hilbert transform by Michael Lacey.

Contents of Volume 1

We now describe in more detail the contents of this first volume. Volume 1 consists of two parts, the first one devoted to remembering Cora in all her facets, and the second to mathematics which is, as we well know, a fundamental part of who Cora was.

Part I of Volume 1 contains articles about Cora, her mathematics and mentorship, as well as some remembrances by colleagues and friends. Chapter "Cora Sadosky: Her Mathematics, Mentorship, and Professional Contributions" is a written rendering of Rodolfo Torres' Albuquerque lecture *Cora Sadosky: her mathematics, mentorship, and professional contributions.* This should help us all not to forget this amazing and strong-willed mathematician and woman and the new generations to learn about her vibrant personality. Chapter "Cora's Scholarly Work: Publications According to MathSciNet" contains Cora's scholarly work according to MathSciNet. Chapter "Remembering Cora Sadosky" contains remembrances from friends and colleagues, such as Steven Krantz, María Dolores (Loló) Morán, Guido Weiss, and Mike Wilson, and reproduces the article *Remembering Cora Sadosky* with contributions from Georgia Benkart, Judy Green, Richard Bourgin, and Daniel Szyld and a remembrance written collectively by Estela Gavosto, Andrea Nahmod, Cristina Pereyra, Gustavo Ponce, Rodolfo Torres, and Wilfredo Urbina, published in the AWM Newsletter Volume 41, Number 2, March–April 2011.

Part II in Volume 1 contains a survey and research articles submitted by an array of mathematicians representing one or several of the mathematical themes close to Cora's heart. In chapter "Higher-Order Elliptic Equations in Non-Smooth Domains: A Partial Survey", Ariel Barton and Svitlana Mayboroda present an extensive survey dealing with the theory of higher-order elliptic operators in non-smooth settings. The first section of the paper deals with mostly constant coefficient operators, in arbitrary domains. It explains the Miranda-Agmon maximum principle and related regularity estimates, as well as the interesting extension to this setting of the Wiener test. The results explained are deep and important, and it is very useful to have the history and development detailed here. The second section deals with L^p boundary value problems for higher-order elliptic operators on Lipschitz domains. This is mainly discussed again in the constant coefficient case. This is a very rich subject, where, in spite of decades of steady progress, important problems remain open. The final section deals with variable coefficient higher-order operators. Two natural classes of such operators are discussed, and recent works by the authors are described. This is a part of the theory that is at its very beginning and where much remains to be done. The authors' recent pioneering works are explained, and also many areas of investigation that remain wide open are discussed.

Chapters "Victor Shapiro and the Theory of Uniqueness for Multiple Trigonometric Series", "A Last Conversation with Cora", and "Fourier Multipliers of the Homogeneous Sobolev Space $\dot{W}^{1,1}$" are from Cora's academic sibling Marchall Ash and dear friend and colleague Aline Bonami. Marshall Ash's paper focuses on aspects of Cantor's uniqueness theorem which states that if a trigonometric series converges to zero pointless everywhere, then all of its coefficients must be zero. The paper is written in the form of a survey and contains outlines of proofs, ideas and discussions, and personal experiences. Aline Bonami starts with a remembrance she titled *A last conversation with Cora*, followed by an article that studies the class of Fourier multipliers on the homogeneous Sobolev space $\dot{W}^{1,1}$ which is meant to be part of this final conversation.

In chapter "A Note on Nonhomogenous Weighted Div-Curl Lemmas", Galia Dafni, Der-Chen Chang, and Hong Yue present new results concerning local versions of the div-curl lemma in $\mathbb{R}^n$ in the context of weighted Lebesgue spaces and weighted localized Hardy spaces. Dafni and Chang were Cora's coauthors.

In chapter "A Remark on Bilinear Square Functions", Loukas Grafakos in an interesting note opens up a discussion regarding a bilinear version of a classical theorem due to Rubio de Francia on L^p-estimates for square functions based on disjoint and arbitrary (in particular, not necessarily dyadic) intervals on the real line.

In chapter "Unique Continuation for the Elasticity System and a Counterexample for Second-Order Elliptic Systems", Carlos Kenig and Jenn-Nan Wang present an interesting discussion of unique continuation for second-order elliptic systems in the plane. They discuss positive results for a class of systems of elasticity where coefficients are bounded and measurable (both isotropic and anisotropic systems) and where coefficients are Lipschitz (anisotropic system). They present an example which shows that unique continuation may fail for elliptic systems with bounded measurable coefficients in interesting contrast to the case of single equations where unique continuation results are known to hold.

In chapter "Hardy Spaces of Holomorphic Functions for Domains in $\mathbb{C}^n$ with Minimal Smoothness", Loredana Lanzani and Eli Stein continue their program analyzing Hardy spaces on domains in several complex variables with minimal boundary smoothness. As the reviewer said, "These will be of interest to a broad cross-section of analysts. And I must note that Cora Sadosky would have been quite interested in this paper."

In chapter "On the Preservation of Eccentricities of Monge–Ampère Sections", Diego Maldonado studies how the eccentricity of sections to solutions u to the Monge-Ampère equation behave when the right-hand side of the equation satisfies a Dini-type condition on sections. This is applied to obtain estimates of second derivatives of u in terms of the eccentricity. In addition, he uses these results to show existence of quasi-conformal solutions to Jacobian equations and also to improve recent estimates for second derivatives of solutions to the linearized Monge-Ampère equation.

In chapter "BMO: Oscillations, Self-Improvement, Gagliardo Coordinate Spaces, and Reverse Hardy Inequalities", Mario Milman writes about BMO touching on many topics including oscillations, self improvement, Gagliardo coordinate spaces, and reverse Hardy inequalities. As the author himself says about Cora in the last section, "I know that the space BMO had a very special place in her mathematical interests and, indeed, BMO spaces appear in many considerations throughout her works. For this very reason, and whatever the merits of my small contribution, I have chosen to dedicate this note on BMO inequalities to her memory."

In chapter "Besov Spaces, Symbolic Calculus, and Boundedness of Bilinear Pseudodifferential Operators", Virginia Naibo and Jodi Herbert continue their work on L^p boundedness properties for bilinear pseudodifferential operators with symbols in certain Besov spaces of product type. It is useful to notice that the classes of symbols considered are extensions of some particular instance of the nowadays well-understood bilinear Hörmander classes of symbols.

In chapter "Metric Characterizations of Some Classes of Banach Spaces", Mikhail Ostrowskii writes about metric characterization of some classes of Banach spaces; this topic is currently central in nonlinear analysis.

In chapter "On the IVP for the k-Generalized Benjamin–Ono Equation", Gustavo Ponce surveys recent developments regarding local and global well-posedness and special properties (such as decay and regularity) of solutions of the initial value problem (IVP) associated to the Benjamin-Ono equation and k-generalized Benjamin-Ono equation.

Mayboroda, Ponce, and Torres were invited speakers to "An Afternoon in Honor of Cora Sadosky." Dafni and Naibo gave talks in the AMS meeting in Albuquerque in April and honored there as they are doing here the life and work of Cora Sadosky. Other authors could not make it to the conference but were more than happy to contribute to this volume.

Acknowledgments

These volumes would not have been possible without the contributions from all the authors. We are grateful for the time and care you placed into crafting your manuscripts, and for this, we thank you!

All the articles were peer reviewed, and we are indebted to our dedicated referees, who timely and often enthusiastically pitched in to help make these volumes a reality. We used your well-placed comments and words to describe in the preface the articles in these volumes, and for this, we thank you!

We would like to thank Cora Sol Goldstein, Cora's daughter, who blessed the project and gave us a selection of beautiful photos for us to choose and use. When she first heard about the volumes, she said, "My father would had been so happy to know about these books."

Carlos Kenig, Cristian Gutierrez, and Mike Christ, Cora's academic siblings, helped us identify all the other siblings in the "Calderón family photo," a nontrivial task! Barry Mazur helped us localize Cathy O'Neil who in turn identified and secured permission from the author of the photo showing them both and Cora. Ramón Bruzual from Venezuela gave us permission to reproduce the photos he took in the 1994 Cotlar conference in Caracas. Margaret Randall, writer and long time friend of Corita and her parents, shared her memories and provided the opening photo for the book after a serendipitous encounter in Albuquerque. Thank you!

Our editor at Springer, Jay Popham, was very accommodating and patient, and so was editor Marc Strauss who came on board later as the editor of the AWM-Springer Series.

We thank Kristin Lauter, AWM president, for embracing this project and the AWM staff for helping us in the final laps. In particular, we thank Anne Leggett Macdonald, AWM Newsletter editor, who provided guidance and moral support.

We cannot help but to think that Cora's spirit was around helping us finish this project. Cora's legacy is strong and will continue inspiring many more mathematicians!

Albuquerque, New Mexico, USA	María Cristina Pereyra
Caracas, Venezuela	Stefania Marcantognini
Statesboro, Georgia, USA	Alexander M. Stokolos
Chicago, Illinois, USA	Wilfredo Urbina

Contents

Part I
Cora

Cora Sadosky: Her Mathematics, Mentorship, and Professional Contributions

Rodolfo H. Torres

Abstract We present some snapshots of Cora Sadosky's career focusing on her intertwined roles as mathematician, mentor, and leader in the profession. We recount some of her contributions to specific areas of mathematics as well as her broader impact on the mathematical profession.

Introduction

The content of this article is essentially that of the talk I presented at *"An Afternoon in Honor of Cora Sadosky,"* New Mexico Analysis Seminar, Albuquerque, New Mexico, April 4, 2014. I want to express my gratitude again to the organizers of the event and the editors of this volume for the opportunity to present that talk and write this article. In particular, I want to thank Cristina Pereyra for the very nice idea of having the mini conference to honor Cora and Andrea Nahmod for encouraging me to give the talk, which I consider a big honor and big responsibility. Special thanks go to Estela Gavosto too, for the enormous help she gave me preparing the materials for my presentation. Finally I also want to thank Cora Sol Goldstein for kindly reading a draft of this article, making some suggestions, and providing some photographs.

It is difficult to summarize in a few pages Cora's contributions to mathematics and the lives of many individuals. Shortly after she passed away on December 3, 2010, we wrote with several colleagues a brief note for the Newsletter of the Association for Women in Mathematics (AWM) [14][1] and, in doing so, we came across by chance with the poem by Emma Lazarus *"The New Colossus,"* which is engraved on the Statue of Liberty. We used the words from that poem:

> "Give me your tired, your poor,

[1]The article in the AWM Newsletter is reproduced elsewhere in this volume too.

R.H. Torres (✉)
Department of Mathematics, University of Kansas, Lawrence, KS 66045-7523, USA
e-mail: torres@ku.edu

M.C. Pereyra et al. (eds.), *Harmonic Analysis, Partial Differential Equations, Complex Analysis, Banach Spaces, and Operator Theory (Volume 1)*, Association for Women in Mathematics Series 4, DOI 10.1007/978-3-319-30961-3_1

> Your huddled masses yearning to breathe free..."

as an opening quote in that note. At a personal level, these words make many of us who knew Cora think a lot about her. Indeed, as we wrote in the AWM note,

> "Cora was a vibrant, strong minded and outspoken woman, who fought all her life for human rights, who helped uncountable young mathematicians without expecting anything in return. Cora was a phenomenal mathematician, in a time when it was not easy for women, and she championed this cause all her life, becoming president of the AWM in the early 90s."

Cora used to tell us many stories about her time as graduate student in Chicago. In particular, she told us that

> "Antoni Zygmund taught her to judge mathematicians by their theorems; that when someone spoke highly of a mathematician, Zygmund would always like to see the theorems such mathematician had proved. She did the same. If you said to her that you like mathematician X she would ask you to explain his/her mathematics" [14].

So, as we speak highly of her, we will have to remind the reader of some of her theorems. Unfortunately, we will only cover a small part of her mathematical accomplishments, but we will also try to provide glimpses of her personality through quotes of hers. Cora wrote a lot about mathematics and other topics and whenever possible we will use her own words to describe different aspects of both her work and professional life. Certainly we will not provide an exhaustive biographical account, but rather highlight a few aspects of her life, which, in our view, have left a mark.

Cora Sadosky.
Photo courtesy of Cora Sol Goldstein.

Cora always expressed her views in a very straightforward manner. We remember her saying, when she did not believe or agree with some statement made by someone in a "questionable authoritatively way,"

> Quién te dijo eso? (Who told you that?) or Y este señor/a quién se cree que es? (Who this sir/madam thinks he/she is?)

Cora would transparently let you know when she was not in harmony of opinion with someone. However, as we witnessed many times,

> "she [Cora] never let disagreements she may have with a mathematics colleague on other issues affect her appreciation of his/her mathematics" [14].

It is our hope that she is not looking at us saying: *Y este señor quién se cree que es?…* as we attempt to write about Cora in the way some of us knew and remember her. We shall select to present some of her mathematical contributions based in part on the impact we think they had and our affinity with some areas. In doing so, we may have unintentionally omitted some other results that some experts may have found more important to be presented. If so, that is simply the ignorance of this author.

Some Brief Bibliographical Notes

To put Cora's life in perspective we need to say a few words about her parents. She was the daughter of two mathematicians and outstanding individuals, Cora Ratto de Sadosky (1912–1981) and Manuel Sadosky (1914–2005). Like many others in Argentina at that time, they were the children of immigrant families. Upon coming to Argentina, many middle class and even lower income families of immigrants aspired for their children to get a university education and worked very hard to do so. They both got PhD's.

Cora Ratto was from a family of Italian origin. She showed her social commitment from an early age as a student leader at the University of Buenos Aires (UBA). Over her life she was a tireless activist in several organizations which fought against racism and discrimination. She was a founder member of the International Women's Union. She studied in Paris with M. Frechet but actually got her PhD in Mathematics in her 40s, back in Buenos Aires working with M. Cotlar. With him, she authored one of the first modern books in Linear Algebra in Spanish and also edited a prestigious publication series at the UBA (with authors like L. Schwartz and A.P. Calderón). She created the Albert Einstein Foundation to provide financial support to talented students. A prize in her name was established in Vietnam in 1996. It is awarded to young women participating in the math Olympics.[2]

[2]Historical information taken in part from: *Complexities: Women in Mathematics*, edited by B.A. Case and A.M. Leggett, Princeton University Press, 2005.

Cora Ratto de Sadosky (1912-1981).
Photo courtesy of Cora Sol Goldstein.

Manuel Sadosky (1914-2005).
Photo courtesy of *Libros Del Zorzal*, Argentina.

Manuel Sadosky was born in Buenos Aires in a family of Jewish Russian immigrants. He got his PhD in math and physics at the UBA and then studied in France and Italy. Back in Buenos Aires, in 1949 he was first denied a position at the UBA for political reasons, but eventually became professor and later associate dean. He founded the Computational Institute where he brought to Argentina the first computer for research (*Clementina*, an eighteen-meter long monster machine with only

5 KB RAM![3]). He and Cora Ratto were instrumental in creating a modern School of Sciences with a tremendous impact in the scientific landscape of Argentina. But in 1966 the university was taken over by the military (losing its autonomy), causing the resignation of 400 scientists from their positions, and eventually a life in exile for Cora's family in Venezuela, Spain, and the USA. He returned to Argentina in the 1980s (after Cora Ratto has passed away), in the midst of a new democratic government and became Secretary of Science and Technology. Manuel was elected "Illustrious Citizen" of the city of Buenos Aires in 2003.[4]

Cora (*"Corita,"* as those who knew her from a young age called her) was born on May 23, 1940, and was only 6 years old when she moved to France and Italy with her parents and attended many different schools. She entered college at age 15 and received her Licenciatura in Mathematics in 1960, almost at the same time her mother got her PhD. She then did her PhD in Chicago working with A.P. Calderón and A. Zygmund. She graduated in 1965 and after that she returned to Argentina and married Daniel J. Goldstein, who sadly passed away in 2014 shortly before the *Afternoon in Honor of Cora Sadosky* and after a long battle with illness.

Like her mother, Cora resigned in protest from the UBA in 1966 after the military intervention. After several temporary positions, including one at Johns Hopkins University, and periods where she had to take on other type of jobs to make a living, Cora and family had to leave Argentina in 1974 because of political persecution. They all moved to Venezuela, where Cora continued her lifelong collaboration with M. Cotlar, which had started in Buenos Aires. In 1980 she moved to Howard University where she became a full professor in 1985.

Cora returned for a year to Argentina in 1984 and held a position at the UBA. She conducted then the rest of her mathematical career in the USA but kept her interest in Argentina and Venezuela where she helped many young mathematicians. Some of them pursued then, with Cora's assistance, their mathematical careers in the USA. After she retired from Howard University, Cora moved to California to live close to her daughter Cora Sol, son in law Tom, and granddaughter Sasha.[5]

---

[3]Information about Clementina taken from www.sobretiza.com.ar/2013/05/17/clementina-cumplio-52-anos/#axzz2xO23TGBJ.

[4]Historical information taken in part from: www.fundacionsadosky.org.ar/en/institucional/biografia-dr-sadosky.

[5]Historical information taken in part from *Notable Women in Mathematics: A Biographical Dictionary*, edited by Morrow and Perl [19].

Cora with husband Daniel, daughter Cora Sol and granddaugther Sasha, California 2008.
Photo courtesy of Cora Sol Goldstein.

In addition to the positions already mentioned, Cora held/received the following appointments/honors[6]:

- Member in residence of the Institute of Advanced Studies, Princeton, 1978–1979.
- Visiting Professorship for Women, NSF to spend a year at the Institute of Advanced Studies, Princeton, 1983–1984.
- Visiting Professor, University of Buenos Aires, 1984–1985.
- Career Advancement Award, NSF to spend a year at the Mathematical Science Research Institute, Berkeley, 1987–1988.
- Visiting Professorship for Women, NSF to spend a year at the University of California Berkeley, 1995–1996.
- Fellow of the American Association for the Advancement of Science, 1997.

Research Areas and Collaborators

Cora Sadosky wrote about 60 articles in several areas of Harmonic Analysis and Operator Theory. According to her MathSciNet author profile[7] these were her areas and coauthors

[6]Some information taken from the Institute of Advanced Studies, https://www.ias.edu/people/cos/users/5624.

[7]http://www.ams.org/mathscinet/search/author.html?mrauthid=199038.

Publications (by number in area)

Fourier analysis · Functional analysis · Functions of a complex variable · History and biography · Integral equations · Integral transforms, operational calculus · Operator theory · Probability theory and stochastic processes · Real functions · Systems theory; control

Co-authors (by number of collaborations)

Alpay, Daniel · Arocena, Rodrigo · Ball, Joseph A. · Bolotnikov, Vladimir · Chang, Der-Chen E. · Christ, Michael · Cotlar, Mischa · Dafni, Galia · Dijksma, Aad · Fabes, Eugene Barry · Ferguson, Sarah H. · Kenig, Carlos E. · Pott, Sandra · Rovnyak, James · Vinnikov, Victor · Weiss, Guido L. · Wheeden, Richard L.

The profile clearly shows Cora's interest in Fourier Analysis and Operator Theory and Mischa Collar as her main collaborator. A better idea about her specific research topics is given by the following Wordle cloud of titles of a selected collection of her articles.

We can see clearly Hankel and Toeplitz operators and the Hilbert transform as the focus of many of Cora's works as well as the space *BMO*, weighted norm inequalities and other topics related to singular integrals, scattering, and lifting techniques.

We will present a few samples of her work in three areas: Parabolic Singular Integrals, The Helson–Szegö theorem (major collaboration with M. Cotlar), and Multiparameter Analysis, the last two topics being very much interrelated.

Some Samples of Cora Sadosky's Mathematics

Parabolic Singular Integral Operators

Cora did her thesis in this subject under the direction of Calderón and Zygmund. In her own words:

> "I was obsessed with parabolic singular integrals, which seemed the natural object to study after Calderón's success with elliptic and hyperbolic PDEs. Calderón encouraged me in that interest, and, as the problem was in the air, very soon afterwards a first paper on the subject appeared by B. Frank Jones. This did not discourage me, since I came up with a notion of principal value for the integral through a nonisotropic distance, an idea which Calderón thought was the right one"'

and she added

> "...through C-Z correspondence, we found out that Zygmund had assigned one of his students, Eugene Fabes, a problem close to mine and that we had both proved the pointwise convergence of parabolic singular integrals (by different methods)! Panic struck; Calderón defended my priority on the problem, but all was solved amicably, and upon my return Gene [Fabes] and I wrote our first result as a joint paper" [5].

In 1964, Jones [18] introduced singular integrals of the type

$$\lim_{\epsilon\to 0} T_\epsilon f(x,t) = \lim_{\epsilon\to 0} \int_0^{t-\epsilon} \int_{\mathbb{R}^n} K(x-y, t-s) f(y,s)\, dyds$$

where $K(x,t) = 0$ for $t < 0$,

$$K(\lambda x, \lambda^m t) = \lambda^{-n-m} K(x,t)$$

for some $m > 0$,

$$\int_{\mathbb{R}^n} K(x,1)\, dx = 0$$

and K has some appropriate smoothness. Typically

$$K(x,t) = \frac{1}{t^{d/m+1}} \Omega\left(\frac{x}{t^{1/m}}\right), \qquad \Omega(x) = K(x,1)$$

These singular integrals arise in parabolic differential equations of the form

$$\frac{\partial u}{\partial t}(x,t) = (-1)^m P(D) u(x,t), \quad (x,t) \in \mathbb{R}^d \times (0,\infty),$$

for P and appropriate homogeneous polynomial of degree m.

If Γ is the fundamental solution of one such equation, then

$$u(x,t) = \int_0^t \int_{\mathbb{R}^d} \Gamma(x-y, t-s) f(y,s)\, dyds$$

solves

$$\frac{\partial u}{\partial t} - (-1)^m P(D)u = f.$$

Taking α derivatives, with $|\alpha| = m$, in the integral representation of u, leads to the singular integrals considered. For example, if $d = 1$ and we consider the heat equation (m=2), taking two derivatives of the fundamental solution leads to integrals with kernels of the form

$$K(x,t) \approx \frac{1}{t^{3/2}} \left(\frac{x^2}{8t} - \frac{1}{4} \right) e^{\frac{-x^2}{4t}}.$$

Jones proved the convergence in L^p-sense of the principal valued parabolic singular integrals and posed the question about their almost everywhere convergence in $\mathbb{R}^n \times (0, \infty)$.

In 1966, Fabes and Sadosky [11] proved such convergence by obtaining the boundedness of the associated maximal truncated singular integrals; a method which is nowadays commonly used in proving pointwise convergence results.

Theorem (Fabes–Sadosky). *If* $f \in L^p$ *for* $1 < p < \infty$ *and*

$$T^* f(x,t) = \sup_{\epsilon > 0} |T_\epsilon f(x,t)|,$$

then

$$\|T^* f\|_p \lesssim \|f\|_p.$$

More generally Sadosky [21] studied operators given by kernels with homogeneities of the form

$$K(\lambda^{\alpha_1} x_1, \ldots, \lambda^{\alpha_n} x_n) = \lambda^{-(\alpha_1 + \cdots + \alpha_n)} K(x_1, \ldots, x_n)$$

as principal valued singular integrals with respect to a proper pseudodistance ρ, which matches the homogeneity of the problem.

For example, for appropriate m,

$$\rho(x,y) = \left(\sum_{j=1}^{n} |x_j - y_j|^{m/\alpha_j} \right)^{1/m}.$$

Sadosky also considered fractional singular integral versions of the above operators, and they led to work with Cotlar in 1967, [30], on *"quasi-homogeneous" Bessel potential spaces* (which is probably her only work jointly with Cotlar outside their main area of research in Operator Theory). They extended to the parabolic setting work of N. Aronszajn, K.T. Smith, A.P. Calderón, and others at a time when potential spaces where still being properly understood.

All this, of course, preceded the development in the 1970s of the Coifman–Weiss [6] theory of spaces of homogeneous type; the parabolic setting was one of the motivating examples for such theory.

The theory of parabolic singular integrals was then carried out much further by B.F. Jones; E. Fabes and N. Rivière; and many others; and successfully applied to parabolic differential equations and boundary value problems. Cora was very fond of this area of research but, due to isolation, she had to switch to other areas of mathematics.

The Helson–Szegö Theorem

It is hard to overstate the extent and relevance of the work that Cora did with Mischa Cotlar; they wrote together more than 30 articles. We cannot talk about Cora's work without briefly remembering a few facts about Mischa too.

An incredibly talented mathematician whose only formal education was his PhD from Chicago, Mischa Cotlar was also a devoted pacifists and, not surprisingly, a close friend of Cora's family. The 2007 New Mexico Seminar had a special afternoon dedicated to Mischa in which Cora also participated. The website of the conference in his honor portraits Cora's mathematical partner very well and reads:

> "Mischa Cotlar was an exceptional mathematician and human being. Generations of mathematicians in Venezuela, Argentina, and other Latin American countries grew under his guidance. He was one of the world experts in harmonic analysis and operator theory."

Laurent Schwartz, Mischa Cotlar, Concepción Ballester, and Cora Sadosky.
Conference in Honor of M. Cotlar, Caracas, Venezuela, 1994.
Photo courtesy of Ramón Bruzual.

Speaking of their joint work, Cora wrote in [28]:

> "In Caracas, Mischa and I began to collaborate in earnest and together we established an ambitious research program. Mathematically, our Caracas exile was extraordinarily productive. Although Mischa was part of the Zygmund school, he had an astonishing intellectual affinity with the Ukrainian school of Mathematics lead by Professors Krein and

> Gohberg, the leaders of the extraordinarily original and fertile school of operator theory. In spite of my analytic upbringing, I could not resist Mischa's daring approaches to operator theory."

Indeed, the ambitious Cotlar–Sadosky program brought to light beautiful connections between different areas in the study of moments theory and lifting theorems for measures, Toeplitz forms, Hankel operators, and scattering systems. All this was done both in a very general setting, but also on concrete applications in harmonic analysis involving weighted norm inequalities and the ubiquitous role of BMO in the intersection of many topics in analysis. We will briefly discuss some of their achievements related to the Helson–Szegö theorem, highlighting only a couple of remarks about different characterizations as described in their own works.

Helson and Szegö [15] proved in 1960 that for $\omega \geq 0$ and $d\mu = \omega dx$,

$$\int_{\mathbb{T}} |Hf|^2\, d\mu \lesssim \int_{\mathbb{T}} |f|^2\, d\mu$$

if and only if

$$\omega = e^{u+Hv}, \qquad u, v \in L^\infty, \qquad \|v\|_\infty < \pi/2. \tag{1}$$

Here $Hf = -if_1 + if_2$, and f_1 and f_2 are the analytic and anti-analytic parts of f, so the analytic projector is $P = 1/2(I + iH)$.

Helson and Szegö were working on prediction theory. According to Cora [27]:

> "When A. Zygmund read the manuscript of the paper [by Helson-Szegö] he remarked to the authors that they had given the necessary and sufficient condition for the boundedness of the Hilbert transform in L^2, solving an important problem. The consequences of this result continue to be of interest in a variety of analysis settings."

Of course there is also the Hunt–Muckenhoupt–Wheeden theorem [16] and one also has that H is bounded if and only if $\omega \in A_2$, i.e.,

$$\sup_I \left(\frac{1}{|I|}\int_I \omega\, dx\right)\left(\frac{1}{|I|}\int_I \omega^{-1}\, dx\right) < \infty. \tag{2}$$

The condition in (1) is very useful to construct weights but it is almost useless to verify that something is a weight; while for the condition in (2) the situation is the opposite.

Cora was always very much puzzled by the relationship between the two characterizations. They are, at some level, connected via *BMO* and together the two theorems provide a new proof of a characterization of such space. If $f \in BMO$, then by the John–Nirenberg inequality $e^{cf} \in A_2$ for some appropriate c. It follows from the Hunt–Muckenhoupt–Wheeden theorem that H is bounded on $L^2(e^{cf})$ and by the Helson and Szegö theorem

$$cf = u + Hv,$$

so that $f \in L^\infty + HL^\infty$, and so $BMO \subset L^\infty + HL^\infty$.

Since the other inclusion $L^\infty + HL^\infty \subset BMO$ is clear, we have another "proof" of

$$BMO = L^\infty + HL^\infty.$$

However, as Cora observed:

> "The trick of the proof sketched above is to apply the necessary condition of one theorem and the sufficient condition of the other. Those happen to be the hard part of both theorems..." [27].

A *moment problem* related to the Helson–Szegö theorem is that of characterizing the Fourier coefficients of the weights ω. It is interesting how Cotlar and Sadosky saw the moment problem and generalized Toeplitz operators as natural substitutes for *Cotlar's Lemma* on almost orthogonality. We can read in [23]:

> "Cotlar's Lemma showed that the Hilbert transform basic property on boundedness on L^2 is related to a more general property of operators admitting certain type of decomposition. Since the Lemma was not suited to characterize the positive measures μ such that H is bounded in $L^2(\mu)$ it was natural to think that such characterization could also be related to a general property of certain operators."
> In [23] the moment problem is defined as follows: "to characterize the Fourier transform $\hat{\mu}$ of such μ, i.e. those sequences $s : \mathbb{Z} \to \mathbb{C}$ such that there exist a measure $\mu \in R_2$ with $\hat{\mu} = s(n)$."

Such measures give rise to some building block operators, *generalized Toeplitz kernels* (GTK), i.e., $K : \mathbb{Z} \times \mathbb{Z} \to \mathbb{C}$ such that for some $M > 1$,

$$K(m,n) = \begin{cases} (M-1)s(m-n) & \text{if } sign(m) = sign(n) \\ (M+1)s(m-n) & \text{if } sign(m) \neq sign(n) \end{cases}$$

is positive definite. Note that K is not translation invariant but its restriction to each quadrant of $\mathbb{Z} \times \mathbb{Z}$ is translation invariant or Toeplitz, hence the name. Cora further wrote in [23],

> "Since, by the Bochner theorem, every positive definite Toeplitz kernel is the Fourier transform of a positive measure, Cotlar and Sadosky set to extend this result to GTKs, ..."

and they used the results then in applications to the estimates for the Hilbert transform.

The power of these ideas is that they work also for the *two-weight problem* for the Hilbert transform by relating matrices of bilinear forms with certain invariances to "positive" measures. The building blocks of these matrices are the invariant (under shifts in $\mathbb{T}$) transformations of the form

$$B(f,g) = \int_{\mathbb{T}} f\bar{g}\, d\mu.$$

Let V be the space of trigonometric polynomials, and write $V = W_1 + W_2$, where $W_1 = P(V)$. Consider now the problem of finding a pair of positive measures μ and ν on $\mathbb{T}$ such that

$$\int_{\mathbb{T}} |Hf|^2 \, d\mu \leq M^2 \int_{\mathbb{T}} |f|^2 \, d\nu.$$

Writing as before $f = f_1 + f_2$ and $Hf = -if_1 + if_2$, this is equivalent to

$$\int_{\mathbb{T}} f_1 \bar{f}_1 \, d\mu_{11} + \int_{\mathbb{T}} f_1 \bar{f}_2 \, d\mu_{12} + \int_{\mathbb{T}} f_2 \bar{f}_1 \, d\mu_{21} + \int_{\mathbb{T}} f_2 \bar{f}_2 \, d\mu_{22} \geq 0,$$

where

$$\mu_{11} = \mu_{22} = M^2 \nu - \mu \qquad \mu_{12} = \mu_{21} = M^2 \nu + \mu.$$

One can also verify,

$$\left| \int_{\mathbb{T}} f_1 \bar{f}_2 \, d\mu_{12} \right| \leq \left(\int |f_1|^2 \, d\mu_{11} \right)^{1/2} \left(\int |f_2|^2 \, d\mu_{22} \right)^{1/2}. \tag{3}$$

Cotlar–Sadosky showed that this can be *lifted* to a (complex) measure ρ differing form μ_{12} by an analytic function $h \in H^1(\mathbb{T})$ so that (3) holds for all $(f_1, f_2) \in V \times V$ and, hence,

$$|\rho(D)| \leq \mu_{11}(D)^{1/2} \mu_{22}(D)^{1/2}$$

holds for all Borel sets D. They obtain the following result:

Theorem (Cotlar–Sadosky). *A pair of positive measures μ and ν on $\mathbb{T}$ satisfy*

$$\int_{\mathbb{T}} |Hf|^2 \, d\mu \leq M^2 \int_{\mathbb{T}} |f|^2 \, d\nu$$

if and only if

$$|(M^2 \nu + \mu - h dx)(D)| \leq (M^2 \nu - \mu)(D)$$

for some $h \in H^1(\mathbb{T})$ and all Borel sets $D \subset \mathbb{T}$.

One can further show that in the case $\mu = \nu$ one recovers the Helson–Szegö theorem from the above. Together with R. Arocena, Cotlar and Sadosky also developed, among other things, a unified approach for the Helson–Szegö theorem in $\mathbb{T}$ and $\mathbb{R}$ [1]. In addition, Cotlar–Sadosky gave other characterizations using the notion of u-bounded operators on Banach lattices and they also obtained versions of the Helson–Szegö theorem in L^p. We refer the reader to [1, 8, 9, 27], and the survey article [29] and the reference therein for more details. The characterization of two-weight norm inequalities for more general singular integrals in $\mathbb{R}^n$ has received tremendous attention in recent years. The progress in the area will be described in other articles in this volume.

Cotlar and Sadosky also obtained versions of the Helson–Szegö theorem for the product Hilbert transform. In the higher dimensional setting the conditions on the weight are still *logarithmic BMO*. Here we just state the $\mathbb{T}^2$ case. See [9] and [10].

Theorem (Cotlar–Sadosky). *A weight ω on $\mathbb{T}^2$ satisfies*

$$\int_{\mathbb{T}^2} |Hf|^2\,\omega \le M^2 \int_{\mathbb{T}^2} |f|^2\,\omega,$$

where $H = H_1H_2$ and H_j is the Hilbert transform in the variable j, if and only if $\log\omega \in bmo(\mathbb{T}^2)$ *with*

$$\log\omega = u_1 + H_1 v_1 = u_2 + H_2 v_2$$

for u_j, v_j real valued in $\mathbb{T}^2$, $u_j \in L^\infty$ and $\|v_j\|_\infty \le \pi/2 - \epsilon_M$.

The *bmo* space in the theorem (see [10]) is precisely the space of all functions Φ so that for some $f_j, g_j \in L^\infty(\mathbb{T}^2)$

$$\Phi = f_1 + H_1 g_1 = f_2 + H_2 g_2,$$

with norm given by

$$\|\Phi\|_{bmo} = \inf\{\max_{j=1,2}\{\|f_j\|_\infty, \|g_j\|_\infty\} \text{ over all decompositions}\}.$$

This space *bmo* is also characterized by bounded mean oscillations on rectangles,

$$\|\Phi\|_{bmo} = \sup_{I,J\subset\mathbb{T}} \frac{1}{|I||J|}\int_{I\times J} |\Phi(x_1,x_2) - \Phi_{IJ}|dx_1dx_2. \tag{4}$$

This gives origin to the name "little BMO," since clearly this space is smaller than $BMO(\mathbb{T}^2)$, which as usual is defined as in (4) but with averages on cubes (squares) in $\mathbb{T}^2$.

Multiparameter Analysis

Other *BMO* spaces surfaced in Cotlar–Sadosky's work on the product Hilbert transform and the study of Hankel operators, as well as in further collaborations of Cora with Ferguson [12] and with Pott [20]. All such spaces coincide in 1-d but they become different substitutes for *BMO* in product dimensions. Moreover different characterizations in the one parameter case produce different spaces in the multiparameter setting.

Another natural *BMO*-like space is BMO_{rec}. It is defined to be the space of functions Φ such that

$$\sup_{I,J\subset\mathbb{T}} \frac{1}{|I||J|}\int_{I\times J} |\Phi(x_1,x_2) - \Phi_I(x_2) - \Phi_J(x_1) + \Phi_{IJ}|dx_1dx_2 < \infty. \tag{5}$$

It was shown by Carleson [2] that the norm defined by (5) does not characterize the dual of the Hardy space $H^1_{Re}(\mathbb{T}^2)$. On the other hand, such space can be characterized by an appropriate Carleson measure condition on open sets and not just rectangles (we will not need such definition here). The space in question is denoted by BMO_{prod}. One has $BMO_{prod} \subset BMO_{rec}$. A detailed survey on the subject is provided by Chang and Fefferman in [4].

Ferguson and Sadosky characterized the spaces *bmo* and BMO_{rec} in terms of (big) Hankel and little Hankel operators [12]. We shall only state the characterization of *bmo*.

Let $P : L^2 \to \mathcal{H}^2$ be the projection operator and $P^{\perp} = I - P$. Define the Hankel operator Γ_{Φ} with symbol Φ, to be $\Gamma_{\Phi}(f) = P^{\perp}(\Phi f)$ and let $M_{\Phi}f = \Phi f$.

Theorem (Ferguson–Sadosky). *The following are equivalent.*

- *The function* $\Phi \in bmo$.
- Γ_{Φ} *and* $\Gamma_{\bar{\Phi}}$ *are both bounded on* $\mathcal{H}^2$.
- *The commutators* $[M_{\Phi}, H_j] = M_{\Phi} H_j - H_j M_{\Phi}$ *are bounded on* L^2.
- *The commutator* $[M_{\Phi}, H_1 H_2] = M_{\Phi} H_1 H_2 - H_1 H_2 M_{\Phi}$ *is bounded on* L^2.

Notice that the third condition is a biparameter version of the Coifman–Rochberg–Weiss theorem [7] about the commutator of the Hilbert transform and pointwise multiplication. Moreover, it is also shown in [12] that if $\Phi \in BMO_{prod}$ then iterated commutator $[[M_{\Phi}, H_1], H_2]$ satisfies

$$\|[[M_{\Phi}, H_1], H_2]\|_{L^2 \to L^2} \lesssim \|\Phi\|_{prod}.$$

The converse was proved by Ferguson–Lacey [13]. This gave a dual formulation and solution to the so-called weak factorization problem in the context biholomorphic Hardy spaces in two variables. That is, the problem of being able to write $h \in \mathcal{H}^1$

$$h = \sum_j f_j g_j$$

with $f_j, g_j \in \mathcal{H}^2$ and

$$\sum_j \|f_j\|_2 \|g_j\|_2 \leq c\|h\|_1.$$

Cora Sadosky's Impact on the Profession

According to the Mathematics Genealogy Project, Cora had officially one PhD student: Sandra Farrier, Howard University, 2005. However, Cora played a tremendously influential role in the life of many young mathematicians (and some not so young too). We take the following from [14].

> "In particular she [Cora] interacted with many students in Venezuela and in Argentina. In her years in Caracas, Cora was very influential on a group of Venezuelan mathematicians, María Dolores Morán, Ramón Bruzual, Ma-risela Domínguez and Stefania Marcantognini, among others, and a Uru-guayan mathematician, Rodrigo Arocena, that got their PhD degrees at the Universidad Central de Venezuela (UCV). She also aided others pursue their doctorates in the US, including Gustavo Ponce who went to Courant Institute in 1978 and later Cristina Pereyra who went to Yale University in 1987.
>
> Likewise, during her sabbatical year in Buenos Aires in 1984–1985 she helped many Argentinian mathematicians come to the US for their doctoral degrees; among them, José Zero who went to University of Pennsylvania, Estela Gavosto and Rodolfo Torres who went to Washington University, and Andrea Nahmod and Lucas Monzón who went to Yale University."

Estela Gavosto (now my wife) and I met Cora at an annual meeting of the Argentine Mathematical Union when she was temporarily back in the country. We were introduced to her by two of our professors at the Universidad Nacional de Rosario, Pedro Aranda and Enrique Cattaneo. They have themselves studied at some point with Mischa Cotlar in Buenos Aires and they gave us a very strong formation in analysis in our undergraduate degree in Rosario. Cora was immediately eager to help us to continue our studies. We visited her many times in Buenos Aires and in one of those occasions we met also Cotlar. I remember he was giving a talk about *BMO* and we were very enthusiast about attending a lecture by such famous mathematician. It was the first time I heard about "bounded mean oscillations," and of course I did not understand anything. However, Cora took us aside after the talk and explained things again for us. That was our first mathematical interaction. Many more followed and Cora remained always interested in knowing what we were working on. She helped us pursue our mathematical dreams and became a great mentor for us over the years.

Many other students benefited from courses Cora taught in several countries. These lectures notes gave rise to her monograph *Interpolation of Operators and Singular Integrals: An Introduction to Harmonic Analysis* [22]. The book was conceived as a very accessible introduction for students to materials that were only covered in textbooks at the time by the famous treatises of E. Stein, *Singular Integrals and Differentiability Properties of Functions* [31], and E. Stein and G. Weiss *Introduction to Fourier Analysis on Euclidean Spaces* [32], which trained generations of harmonic analysts. In the preface of her book Cora wrote:

> "I hope that the book will be accessible to a wide audience that includes graduate students first approaching the subject" "The initial inspiration on the treatment of this subject comes from magnificent courses given by E. M. Stein, G. Weiss, and A. P. Calderón, which I attended as a graduate student a the University of Chicago and the University of Buenos Aires. The overall influence is that of Professor A. Zygmund who taught me how beautiful singular integrals are and induced the will to try to share with others the pleasure of their beauty."

Many of us were helped by Cora in other countless ways. In particular Estela and I benefited, among many other things, from the way she understood the "two-body problem" (I am referring of course to the couples' problem not the classical mechanics one) and brought awareness about this issue in our profession when few were paying attention. From the time we applied to graduate school to the last time we looked for a job, Cora was always advising us and trying to help solving the two-body problem, which we were extremely fortunate to do several times with her critical help. She not only mentored us as scientists, but she also gave us advise about other professional responsibilities and roles we held in our lives. As we wrote in [14],

> "Cora was also concerned with many other aspects of the academic life and she often told us to learn about them and encouraged us to be proactive. She taught us that for any change to take place people really need to get involved."

Guiding with the example, she did get involved. These are some of the many positions in which she served the profession in the USA.[8]

- Member of the Human Resources Advisory Committee, Mathematical Sciences Research Institute, Berkeley, 2002–2005.[9]
- Member of the Nominating Committee, AMS, 2001–2003.
- Member of the Council of the AMS, 1987–1988 and 1995–1998.
- Member of the Committee on Science Policy, AMS, 1996–1998.
- Member of the Committee on the Profession, ASM, 1995–1996.
- Member of the Committee on Human Rights of Mathematicians, AMS, 1990–1996.
- President of the AWM, 1993–1995.
- Member of the Committee on Cooperation with Latin American Mathematicians, AMS, 1990–1992.

As described in [33] some of Cora's accomplishments as president of the AWM included:

> "...the move of AWM headquarters to the University of Maryland and the concurrent staff changes. She increased AWM's international connections and involvement in science policy, in particular initiating (in coordination with other organizations) the first Emmy Noether Lecture at an ICM in 1994 and representing AWM at the International Congress of Mathematics Education in 1993."

We quote again from [14],

> "Cora Sadosky fought many battles and in many fronts. She had tremendous convictions against many injustices, gender inequality, and discrimination in our society. She often took the flag of the underrepresented, underserved, and underestimated. She chose to fight many of her battles from within the mathematical community, sometimes even risking her own mathematical career. She showed a lot of courage in this sense and never worried about the consequences for her. She was never afraid to express her views."

...and expressing her views indeed she did. She wrote several articles and was often quoted in others about issues on women and young mathematicians, immigration, and discrimination. She had a vision of inclusiveness. For example, speaking of immigration policies and the profession, she once wrote in the *Notices* [26]

[8]Some information taken from *Cora Sadosky (1940–2010)*, by Jackson [17].

[9]Exact dates could not be verified.

> "In the mathematics departments of many U.S. universities a substantial percentage of professors-in particular, of mathematicians-are foreign born. This would be unthinkable in Europe! It is a sign of how more socially open this country is with respect to the rest of the world. We should be collectively proud of the U.S. openness, which underlines an extraordinary social dynamism. The enlightened acceptance foreigners receive at our universities ought to be promoted as an example to follow."

And she pointed out later on the same article to some issues that remain of concern today:

> "Many young mathematicians are trying to develop research careers in difficult circumstances. Research mathematicians in nonresearch environments, including those at nondoctoral institutions, need support. For ages many women have faced these difficulties without help, and some have survived as mathematicians. Their experiences could help others, and their losses should not be repeated. Now some see promising young mathematicians take positions at non-research institutions, and they cry foul. Instead, we should help devise support systems to make small-college positions compatible with research."

Finally adding in her presentation:

> "Finding ways to support an active research population in the U.S. will become increasingly difficult in an era of globalization of the world economy, where the competition is fierce and international."

A believer in equal opportunity and affirmative action properly done, Cora organized in 1994 an AWM panel on *"Are Women Getting All the Jobs"*. She spoke often to dispel the notion that women were taking a disproportional percentage of the jobs in the 1990s (which AMS data at the time also showed it was not the case) with spirited statements like:

> "We strongly believe that this is false and dangerous, that pitting one group of under/unemployed mathematicians against another is just the old tactic of dividing people with similar interests in order to exploit them all" [33].

In a public lecture at the Meeting of the Canadian Mathematical Society in 1995 Cora stated [25] (quoted also in [3], p. 116–120),

> "We have achieved much. But we are striving for nothing less than the right of all people to do mathematics. For that we have to work together women and men, so that the mathematics community no longer needs constant reminders of the existence of women in its midst."

And further reflecting on the AWM, Cora wrote, as quoted in [33]:

> "Our Association really makes an impact on the situation of women in mathematics. ... Still, women continue to face formidable problems in their development as mathematicians ... To successfully confront these problems, we need the ideas and the work, the enthusiasm and the commitment of all–students and teachers and researchers and industrial mathematicians–of every woman and every man who stands for 'women's right to mathematics'."

Sometimes Cora preferred instead to rely on irony and humor to convey a similar message, as in [25] (also quoted in [3], p. 118):

> "Years ago, a friendly colleague told me his department was considering hiring a junior person in our field and asked me for a top candidate. After some thought I mentioned one

of the best junior researchers in the field. His answer was 'But we already have a woman!' and mine to him, 'So, would you hire a man for the job? I assume your faculty already has at least one man!'"

Some Final Memories

The two pictures below represent very special moment in my life in relation to Cora. Although in the first one Cora is not present, it is a picture taken when several of us met, through Cora, for the first time. These individuals will become some of my best colleagues and friends in the profession. The second picture is from the last time I saw Cora and I had a chance to introduce to her several of the young people with whom I have had the fortune to work.

Gustavo Ponce, Estela Gavosto, Wilfredo Urbina, Andrea Nahmod and Carlos Perez, Special Year in Harmonic Analysis and PDE, MSRI, Berkeley, 1988.
Photo by R.H. Torres.

Kabe Moen, Virginia Naibo, Diego Maldonado, Cora Sadosky, Rodolfo Torres, and Arpád Bényi, New Mexico Analysis Seminar, Albuquerque, 2009.
Photo courtesy of Erika Ward.

At the end of the day, for Cora it was (is) all about Mathematics, as she once wrote [24]:

> "Strange as it may seem at the end of this long report, I started out with the impression I had little to say. That impression stemmed from my doing something it may be unwise for a current AWM president to do: I plunged into mathematics head on for three full weeks. I came out of that stint dazed, vaguely guilty and deeply happy. It is very clear that I enjoy like crazy doing mathematics! "
> "But what remains with me is the sense of elation. I did not prove the Riemann Conjecture. My work was modest, but it gave me so much pleasure to do it! Thus I close this conversation with a wish to each of you for this summer: do some of the mathematics you want to do and do it with great pleasure!"

I have to say that I feel sort of the same way. When I started first to prepare the talk about Cora and then this article, I did not know exactly what to say or whether I will be able to make justice to her memory and contributions. Like Cora in the above quote, I think my work here has been *"modest"* but I hope I have accurately portrayed at least some of her multiple professional facets. But certainly, also like her, I ended up enjoying this adventure, which gave me *"much pleasure"* to relate one more time to Cora through her beloved mathematics. I will definitely continue to follow her advise for the summer.

We borrow a few more words from [14],

> "Cora taught all of us by example how to mentor, how to help younger mathematicians pursue their dreams. Cora touched our lives in many ways but she never wanted to be thanked for her good deeds; she believed instead in paying it forward to others. Helping students reach their potential may be the best way to honor her memory. We will deeply miss her'."

I would like to conclude by recognizing Cora Sadosky in another line from the poem by Emma Lazarus that we quoted at the beginning,

> "A mighty woman with a torch, ..."

References

1. R. Arocena, M. Cotlar, C. Sadosky, Weighted inequalities in L^2 and lifting properties. *Mathematical analysis and applications, Part A*. Advances in Mathematics: Supplementary Studies, vol. 7a (Academic Press, New York/London, 1981), pp. 95–128
2. L. Carleson, A counterexample for measures bounded on H^p for the bi-disc. Mittag-Leffler Report, 1974
3. B.A. Case, A.M. Leggett, *Complexities: Women in Mathematics* (Princeton University Press, Princeton, 2005)
4. S-Y.A. Chang, R. Fefferman, Robert a continuous version of duality of H1 with BMO on the bidisc. Ann. Math. (2) **112**(1), 179–201 (1980)
5. M. Christ, C.E. Kenig, C. Sadosky, G. Weiss, Alberto Pedro Calderón (1920–1998). Not. Am. Math. Soc. **45**(9), 1148–1153 (1998)
6. R.R. Coifman, G. Weiss, Extensions of Hardy spaces and their use in analysis. Bull. Am. Math. Soc. **83**(4), 569–645 (1977)
7. R.R. Coifman, R. Rochberg, G. Weiss, Factorization theorems for Hardy spaces in several variables. Ann. Math. **103**, 611–635 (1976)
8. M. Cotlar, C. Sadosky, On the Helson-Szegö theorem and a related class of modified Toeplitz kernels. *Harmonic Analysis in Euclidean Spaces (Proceedings of Symposium in Pure Mathematics, Williams College, Massachusetts, 1978), Part 1*. Proceedings of Symposia in Pure Mathematics, Part XXXV (American Mathematical Society, Providence, 1979), pp. 383–407
9. M. Cotlar, C. Sadosky, The Helson-Szegö theorem in L^p of the bidimensional torus. *Harmonic Analysis and Partial Differential Equations (Boca Raton, FL, 1988)*. Contemporary Mathematics, vol. 107 (American Mathematical Society, Providence, 1990), pp. 19–37
10. M. Cotlar, C. Sadosky, Two distinguished subspaces of product BMO and Nehari-AAK theory for Hankel operators on the torus. Integr. Equ. Oper. Theory **26**(3), 273–304 (1996)
11. E.B. Fabes, C. Sadosky, Pointwise convergence for parabolic singular integrals. Stud. Math. **26**, 225–232 (1966)
12. S. Ferguson, C. Sadosky, Characterizations of bounded mean oscillation on the polydisk in terms of Hankel operators and Carleson measures. J. Anal. Math. **81**, 239–267 (2000)
13. S. Ferguson, M. Lacey, A characterization of product BMO by commutators. Acta Math. **189**(2), 143–160 (2002)
14. E.A. Gavosto, A. Nahmod, C. Pereyra, G. Ponce, R.H. Torres, W. Urbina, Remembering Cora Sadosky. AWM Newsl. **41**(2), 10–14 (2011)
15. H. Helson, G. Szegö, A problem in prediction theory. Ann. Mat. Pura Appl. **51**(4), 107–138 (1960)
16. R. Hunt, B. Muckenhoupt, R. Wheeden, Weighted norm inequalities for the conjugate function and Hilbert transform. Trans. Am. Math. Soc. **176**, 227–251 (1973)
17. A. Jackson, Cora Sadosky (1940–2010). Not. Am. Math. Soc. **58**, 613–614 (2011)
18. B.F. Jones, A class of singular integrals. Am. J. Math. **86**, 441–462 (1964)
19. C. Morrow, T. Perl, *Notable Women in Mathematics: A Biographical Dictionary* (Greenwood Press, New York, 1988)
20. S. Pott, C. Sadosky, Bounded mean oscillation on the bidisk and operator BMO. J. Funct. Anal. **189**(2), 475–495 (2002)
21. C. Sadosky, On some properties of a class of singular integrals. Stud. Math. **27**, 73–86 (1966)
22. C. Sadosky, *Interpolation of Operators and Singular Integrals: An Introduction to Harmonic Analysis* (Marcel Dekker, Inc., New York, 1979)
23. C. Sadosky, The mathematical contributions of Mischa Cotlar since 1955. Lect. Notes Pure Appl. Math. **122**, 715–742 (1990)
24. C. Sadosky, President's report. AWM Newsl. **24**(4), 1–5 (1994)
25. C. Sadosky, Affirmative action: what is it and what should it be? AWM Newsl. **25**(5), 22–24 (1995)

26. C. Sadosky, On issues of immigration and employment for mathematicians. Not. Am. Math. Soc. **44**, 1474–1477 (1997)
27. C. Sadosky, A unified view of disparate results from scattering systems. Contemp. Math. **445**, 297–311 (2007)
28. C. Sadosky, On the life and work of Mischa Cotlar. Rev. Un. Mat. Argent. **49**(2), i–iv (2008)
29. C. Sadosky, The Hilbert transform and scattering. Rev. Un. Mat. Argent. **49**(2), 133–142 (2009)
30. C. Sadosky, M. Cotlar, On quasi-homogeneous Bessel potential operators. *Singular Integrals (Proceedings of Symposia in Pure Mathematics, Chicago, Illinois, 1966)* (American Mathematical Society, Providence, 1967), pp. 275–287
31. E. Stein, *Singular Integrals and Differentiability Properties of Functions* (Princeton University Press, Princeton, 1970)
32. E. Stein, G. Weiss, *Introduction to Fourier Analysis on Euclidean Spaces* (Princeton University Press, Princeton, 1971)
33. J.E. Taylor, S.M. Wiegand, AWM in the 1990s: a recent history of the association for women in mathematics. Not. Am. Math. Soc. **46**, 27–38 (1999)

Cora's Scholarly Work: Publications According to MathSciNet

Cora's Bibliography According to MathSciNET

- Sadosky Cora, *On some properties of a class of singular integrals.* Studia Math. 27 (1966), 73–86.
- Sadosky Cora; Cotlar, Mischa, *On quasi-homogeneous Bessel potential operators.* Singular integrals (Proc. Sympos. Pure Math., Chicago, Ill., 1966), pp. 275–287. Amer. Math. Soc., Providence, R.I., 1967.
- Cotlar, Mischa; Sadosky, Cora, *A moment theory approach to the Riesz theorem on the conjugate function with general measures.* Studia Math. 53 (1975), no. 1, 75–101.
- Cotlar, Mischa; Sadosky, Cora, *Transformée de Hilbert, théoréme de Bochner et le probléme des moments.* (French) C. R. Acad. Sci. Paris Sér. A-B 285 (1977), no. 6, A433–A436.
- Sadosky, Cora, *The problem of moments and the Hilbert transform.* (Spanish) Acta Cient. Venezolana 29 (1978), no. 2, 69–72.
- Sadosky, Cora, *Interpolation of operators and singular integrals. An introduction to harmonic analysis.* Monographs and Textbooks in Pure and Applied Math., 53. Marcel Dekker, Inc., New York, 1979. xii+375 pp.
- Cotlar, Mischa; Sadosky, Cora, *Characterization of two measures satisfying the Riesz inequality for the Hilbert transform in L^2.* Acta Cient. Venezolana 30 (1979), no. 4, 346–348.
- Cotlar, Mischa; Sadosky, Cora, *On the Helson-Szegö theorem and a related class of modified Toeplitz kernels.* Harmonic analysis in Euclidean spaces (Proc. Sympos. Pure Math., Williams Coll., Williamstown, Mass., 1978), Part 1, pp. 383–407, Proc. Sympos. Pure Math., XXXV, Part, Amer. Math. Soc., Providence, R.I., 1979.

M.C. Pereyra et al. (eds.), *Harmonic Analysis, Partial Differential Equations, Complex Analysis, Banach Spaces, and Operator Theory (Volume 1)*, Association for Women in Mathematics Series 4, DOI 10.1007/978-3-319-30961-3_2

- Arocena, Rodrigo; Cotlar, Mischa; Sadosky, Cora, *Weighted inequalities in L^2 and lifting properties.* Mathematical analysis and applications, Part A, pp. 95–128, Adv. in Math. Suppl. Stud., 7a, Academic Press, New York-London, 1981.
- Cotlar, Mischa; Sadosky, Cora, *Majorized Toeplitz forms and weighted inequalities with general norms.* Harmonic analysis (Minneapolis, Minn., 1981), pp. 139–168, Lecture Notes in Math., 908, Springer, Berlin-New York, 1982.
- Cotlar, Mischa; Sadosky, Cora, *Vector-valued inequalities of Marcinkiewicz-Zygmund and Grothendieck type for Toeplitz forms.* Harmonic analysis (Cortona, 1982), 278–308, Lecture Notes in Math., 992, Springer, Berlin, 1983.
- Sadosky, Cora, *Some applications of majorized Toeplitz kernels.* Topics in modern harmonic analysis, Vol. I, II (Turin/Milan, 1982), 581–626, Ist. Naz Alta Mat. Francesco Severi, Rome, 1983.
- Cotlar, Mischa; Sadosky, Cora, *Inégalités à poids pour les coefficients lacunaires de certaines fonctions analytiques.* (French) [Weighted inequalities for lacunary coefficients of analytic functions] C. R. Acad. Sci. Paris Sér. I Math. 299 (1984), no. 13, 591–594.
- Cotlar, Mischa; Sadosky, Cora, *Generalized Toeplitz kernels, stationarity and harmonizability.* J. Analyse Math. 44 (1984/85), 117–133.
- Sadosky Cora; Wheeden, Richard L. *Some weighted norm inequalities for the Fourier transform of functions with vanishing moments.* Trans. Amer. Math. Soc. 300 (1987), no. 2, 521–533.
- Cotlar, Mischa; Sadosky, Cora, *Prolongements des formes de Hankel gnralises en formes de Toeplitz.* (French) [Toeplitz extensions of generalized Hankel forms: the discrete and continuous cases] C. R. Acad. Sci. Paris Sér. I Math. 305 (1987), no. 5, 167–170.
- Cotlar, Mischa; Sadosky, Cora, *Toeplitz liftings of Hankel forms.* Function spaces and applications (Lund, 1986), 22–43, Lecture Notes in Math., 1302, Springer, Berlin, 1988.
- Cotlar, Mischa; Sadosky, Cora, *Integral representations of bounded Hankel forms defined in scattering systems with a multiparametric evolution group.* Contributions to operator theory and its applications (Mesa, AZ, 1987), 357–375, Oper. Theory Adv. Appl., 35, Birkhäuser, Basel, 1988.
- Cotlar, Mischa; Sadosky, Cora, *Generalized Bochner theorem in algebraic scattering systems.* Analysis at Urbana, Vol. II (Urbana, IL, 1986–1987), 144–169, London Math. Soc. Lecture Note Ser., 138, Cambridge Univ. Press, Cambridge, 1989.
- Cotlar, Mischa; Sadosky, Cora, *Nonlinear lifting theorems, integral representations and stationary processes in algebraic scattering systems.* The Gohberg anniversary collection, Vol. II (Calgary AB, 1988), 97–123, Oper. Theory Adv. Appl., 41, Birkhäuser, Basel, 1989.
- Sadosky, Cora, *The mathematical contributions of Mischa Cotlar since 1955.* Analysis and partial differential equations, 715–742, Lecture Notes in Pure and Appl. Math., 122, Dekker, New York, 1990.

- Cotlar, Mischa; Sadosky, Cora, *The Helson-Szegö theorem in L^p of the bidimensional torus.* Harmonic analysis and partial differential equations (Boca Raton, FL, 1988), 19–37, Contemp. Math., 107, Amer. Math. Soc., Providence, RI, 1990.
- Cotlar, Mischa; Sadosky, Cora, *Two-parameter lifting theorems and double Hilbert transforms in commutative and noncommutative settings.* J. Math. Anal. Appl. 150 (1990), no. 2, 439–480.
- Cotlar, Mischa; Sadosky, Cora, *Toeplitz and Hankel forms related to unitary representations of the symplectic plane.* Colloq. Math. 60/61 (1990), no. 2, 693–708.
- Cotlar, Mischa; Sadosky, Cora, *Toeplitz liftings of Hankel forms bounded by non-Toeplitz norms.* Integral Equations Operator Theory 14 (1991), no. 4, 501–532.
- Cotlar, Mischa; Sadosky, Cora, *Weakly positive matrix measures, generalized Toeplitz forms, and their applications to Hankel and Hilbert transform operators.* Continuous and discrete Fourier transforms, extension problems and Wiener-Hopf equations, 93–120, Oper. Theory Adv. Appl., 58, Birkhäuser, Basel, 1992.
- Cotlar, Mischa; Sadosky, Cora, *Transference of metrics induced by unitary couplings, a Sarason theorem for the bidimensional torus, and a Sz.-Nagy-Foias theorem for two pairs of dilations.* J. Funct. Anal. 111 (1993), no. 2, 473–488.
- Cotlar, Mischa; Sadosky, Cora, *Abstract, weighted, and multidimensional Adamjan-Arov-Krein theorems, and the singular numbers of Sarason commutants.* Integral Equations Operator Theory 17 (1993), no. 2, 169–201.
- Cotlar, Mischa; Sadosky, Cora, *Nehari and Nevanlinna-Pick problems and holomorphic extensions in the polydisk in terms of restricted BMO.* J. Funct. Anal. 124 (1994), no. 1, 205–210.
- Cotlar, Mischa; Sadosky, Cora, *The Adamjan-Arov-Krein theorem in general and regular representations of $\mathbb{R}^2$ and the symplectic plane.* Toeplitz operators and related topics (Santa Cruz, CA, 1992), 54–78, Oper. Theory Adv. Appl., 71, Birkhäuser, Basel, 1994.
- Cotlar, Mischa; Sadosky, Cora, *Two distinguished subspaces of product BMO and Nehari-AAK theory for Hankel operators on the torus.* Integral Equations Operator Theory 26 (1996), no. 3, 273–304.
- Sadosky, Cora, *On issues of immigration and employment for mathematicians.* Notices Amer. Math. Soc. 44 (1997), no. 11, 1474–1477.
- Cotlar, Mischa; Sadosky, Cora, *A polydisk version of Beurling's characterization for invariant subspaces of finite multi-codimension.* Operator theory for complex and hypercomplex analysis (Mexico City 1994), 51–56, Contemp. Math., 212, Amer. Math. Soc., Providence, RI, 1998.
- Sadosky, Cora, *Liftings of kernels shift-invariant in scattering systems.* Holomorphic spaces (Berkeley CA, 1995), 303–336, Math. Sci. Res. Inst. Publ., 33, Cambridge Univ. Press, Cambridge, 1998.
- Christ, Michael; Kenig, Carlos E.; Sadosky, Cora; Weiss, Guido, *Alberto Pedro Calderón (19201998).* Notices Amer. Math. Soc. 45 (1998), no. 9, 1148–1153.
- Alpay Daniel; Bolotnikov, Vladimir; Dijksma, Aad; Rovnyak, James; Sadosky, Cora, *Espaces de Hilbert inclus contractivement dans l'espace de Hardy du*

bi-disque. (French) [Hilbert spaces contractively included in the Hardy space of the bidisk] C. R. Acad. Sci. Paris Sér. I Math. 326 (1998), no. 12, 1365–1370.

- Christ, Michael; Kenig, Carlos E.; Sadosky, Cora, *Introduction.* Harmonic analysis and partial differential equations (Chicago, IL, 1996), ix–xv, Chicago Lectures in Math., Univ. Chicago Press, Chicago, IL, 1999.
- Cotlar, Mischa; Sadosky, Cora, *Hankel forms and operators in Hardy spaces with two Szegö weights.* Operator theory and interpolation (Bloomington, IN, 1996), 145–162, Oper. Theory Adv. Appl., 115, Birkhäuser, Basel, 2000.
- Ferguson, Sarah H.; Sadosky, Cora, *Characterizations of bounded mean oscillation on the polydisk in terms of Hankel operators and Carleson measures.* J. Anal. Math. 81 (2000), 239–267.
- Sadosky Cora, *Look again at Vilnius: "Vilnius between the wars" [Math. Intelligencer 22 (2000), no. 4, 47–50; 1796765] by S. Domoradzki and Z. Pawlikowska-Brożek.* Math. Intelligencer 23 (2001), no. 1, 5–6.
- Pott, Sandra; Sadosky, Cora, *Bounded mean oscillation on the bidisk and operator BMO* J. Funct. Anal. 189 (2002), no. 2, 475–495.
- Ball, Joseph A.; Sadosky, Cora; Vinnikov, Victor, *Conservative linear systems, unitary colligations and Lax-Phillips scattering: multidimensional generalizations.* Internat. J. Control 77 (2004), no. 9, 802–811.
- Ball, Joseph A.; Sadosky, Cora; Vinnikov, Victor, *Conservative input-state-output systems with evolution on a multidimensional integer lattice.* Multidimens. Syst. Signal Process. 16 (2005), no. 2, 133–198.
- Chang, Der-Chen; Dafni, Galia; Sadosky, Cora, *A div-curl lemma in BMO on a domain.* Harmonic analysis, signal processing, and complexity 55–65, Progr. Math., 238, Birkhäuser Boston, Boston, MA, 2005.
- Ball, Joseph A.; Sadosky, Cora; Vinnikov, Victor, *Scattering systems with several evolutions and multidimensional input/state/output systems.* Integral Equations Operator Theory 52 (2005), no. 3, 323–393.
- Chang, Der-Chen; Sadosky, Cora, *Functions of bounded mean oscillation.* Taiwanese J. Math. 10 (2006), no. 3, 573–601.
- Sadosky Cora, *The BMO extended family in product spaces.* Harmonic analysis, 63–78, Contemp. Math., 411, Amer. Math. Soc., Providence, RI, 2006.
- Sadosky Cora, *A unified view of disparate results from scattering systems.* Interpolation theory and applications, 297–311, Contemp. Math., 445, Amer. Math. Soc., Providence, RI, 2007.
- Christ, Michael; Kenig, Carlos E.; Sadosky, Cora, *Alberto P. Calderón the mathematician, his life and works*. Selected papers of Alberto P Calderón, xv–xx, Amer. Math. Soc., Providence, RI, 2008.
- Sadosky, Cora, *On the life and work of Mischa Cotlar.* Rev. Un. Mat. Argentina 49 (2008), no. 2, i–iv.
- Sadosky, Cora, *The Hilbert transform and scattering.* Rev. Un. Mat. Argentina 49 (2009), no. 2, 133–142.

Remembering Cora Sadosky

M.C. Pereyra et al. (eds.), *Harmonic Analysis, Partial Differential Equations, Complex Analysis, Banach Spaces, and Operator Theory (Volume 1)*, Association for Women in Mathematics Series 4, DOI 10.1007/978-3-319-30961-3_3

Remembering Cora Sadosky

Introduction by Georgia Benkart

It was with deep sadness that AWM learned of the death of AWM's eleventh president Cora Sadosky on December 3, 2010. Cora was born in Buenos Aires, Argentina, on May 23, 1940. As a young child, she accompanied her parents while they pursued their mathematical studies in several European countries. Her mother Cora Ratto de Sadosky was a mathematician and political activist who founded La Junta de la Victoria (The Victory Union), a women's organization in Argentina of over 50,000 members devoted to furthering the anti-Nazi war effort. In 1945, as representative of her organization, Cora Ratto was a founding member of the International Women's Union at its first meeting in Paris. Cora's father Manuel Sadosky was one of Latin America's first computer scientists and later served as vice dean of the University of Buenos Aires.

Cora entered the university at the age of fifteen with the intention of majoring in physics but switched to mathematics after her first semester. During her undergraduate years, she had the great fortune to study with University of Chicago professors Alberto Calderón and Antoni Zygmund when they visited the University of Buenos Aires. She received her Licendiada degree in 1960, just two years after her mother received her Ph.D. in mathematics from the same university. In 1965, Cora earned her Ph.D. from the University of Chicago with Calderón as her adviser, but also supervised by Zygmund.

Following graduate school, Cora returned to Argentina and married Daniel Goldstein, an Argentinean physician. She joined the faculty of the University of Buenos Aires as an assistant professor of mathematics but, like many of her fellow faculty members, resigned in protest after a brutal assault by the police on the School of Sciences. After one semester of teaching at the Uruguay National University, she was appointed an assistant professor at Johns Hopkins University, where Daniel held a postdoctoral position.

When Cora and Daniel returned to Argentina in 1968, there were no academic positions available for Cora, and she was forced to abandon mathematics for several years. Their daughter Cora Sol was born in 1971. Cora's thirty-year research collaboration with Mischa Cotlar began two years later. Cotlar had been her mother's Ph.D. adviser and coauthored with Cora Ratto a highly acclaimed and widely used text *Introducción al Algebra: Nociones de Álgebra Lineal.* In

Judy Green, Cora Sadosky, Carol Wood, and Lenore Blum

1974, Cora, her husband, daughter, and parents, along with Mischa Cotlar and his wife Yanny, were forced to flee Argentina by social and political unrest. They settled in Caracas, Venezuela, where Cora joined the faculty of the Mathematics Department at Universidad Central de Venezuela. In her tribute to Cotlar ("On the life and work of Mischa Cotlar." *Rev. Un. Mat. Argentina* 49 (2008), no. 2, i–iv) Cora remarked, "In Caracas, Mischa and I began to collaborate in earnest and together established an ambitious research program. Mathematically, our Caracas exile was extraordinarily productive." During this time, the seven exiles enjoyed the great friendship and hospitality of Concepcíon Ballester, who had left Argentina in 1966 to accept a faculty position in the Mathematics Department at UCV. Ballester is the mother of University of New Mexico mathematician Cristina Pereyra, who is a contributor to this *Newsletter* tribute. In 1980, Cora accepted an appointment as an associate professor of mathematics at Howard University in Washington, D.C., and she remained on the faculty there until her retirement.

As President of AWM from 1993 to 1995, Cora Sadosky organized AWM's move to the University of Maryland. She was instrumental in the establishment of the Emmy Noether Lecture at the International Congress of Mathematicians, which was given for the first time in 1994. In "Affirmative Action: What it is and What Should it Be?" which appears in the volume *Complexities* by Bettye Anne Case and Anne Leggett, and which is based on an invited address to the Canadian Mathematical Society upon its fiftieth anniversary and on an earlier article in the *AWM Newsletter* 25(5) (1995), 22–24, Cora recounts the following episode:

> Years ago, a friendly colleague told me his department was considering hiring a junior person in our field and asked me for a top candidate. After some thought I

continued on page 6

Cora Sadosky *continued from page 5*

mentioned one of the best junior researchers in the field. His answer was, "But we already have a woman!" and mine to him, "So would you hire a man for the job? I assume your faculty already has at least one man!"

Throughout her career Cora Sadosky remained a strong advocate for women in mathematics and an active proponent of the greater participation of African Americans in mathematics. Twice she was elected to the Council of the American Mathematical Society, and she was a fellow of the American Association for Advancement of Science.

A reorganization of the AWM business meeting at the Joint Mathematics Meetings in New Orleans in January enabled us to devote a portion of that meeting to a remembrance of Cora. Daniel Szyld of Temple University spoke about her early life in Argentina and her exile in Uruguay and Venezuela. The text of his talk appears below. Abdul-Aziz Yakubu, chair of the Mathematics Department at Howard University, recalled Cora's contributions as a much-treasured colleague and friend. James Donaldson, a mathematics colleague of Cora's, now Dean of the College of Arts and Sciences at Howard, travelled to New Orleans just to attend the remembrance. We were very grateful for his presence. Richard Bourgin has kindly shared his remembrances of Cora, his colleague at Howard, and of his thirty-year friendship with Cora and Daniel. Cora influenced an entire generation of young Argentinean mathematicians. Their memories of Cora appear below and were captured so capably by Reuben Hersh at the ceremony.

Announcement of Cora's death precipitated a flurry of e-mails among AWM's Executive Committee members and past officers. Judy Green, AWM treasurer during Cora's presidential term, describes below their years of AWM service and their enduring friendship. During the business meeting remembrance, Carol Wood echoed words she had shared in an earlier e-mails, "One of Cora's strengths, which I saw firsthand repeatedly during her/my terms as AWM presidents, was her profound interest in the youngest members of our community. The young women sensed her warmth and caring, and were accordingly drawn to her. She came to this role honestly through her mother, who influenced the lives and careers of budding mathematicians in Argentina. It sometimes felt as if Cora wanted to adopt the Schafer Prize winners, every one! Cora was an original, and it was a privilege and a delight for me to get to know her. Jill [Mesirov] made me cry when she mentioned the get-together with photo) of the three of us—my presidency was sandwiched between Jill's and Cora's, and their friendships alone would have made the work worthwhile."

Cora Sadosky, Carol Wood, and Jill Mesirov

Terri Edwards, Cora Sadosky, and Sylvia Bozeman

Linda Keen shared these thoughts: "I was the nominating committee member who convinced Cora to run. She always teased me about it saying that I hadn't told her how hard it really was. We got to work together during her presidency and became friends. Her spunk and her warmth were apparent to everyone she encountered."

Linda Rothschild wrote, "As I read all your kind messages about Cora, I felt that it was unfair that she should die so young. I have proudly displayed her graduate text on the bookshelf in my office for many years and have referred many people to it. Her gentle style has helped many young mathematicians understand the difficult (and sometimes obscure) points in the 'classical' texts on Fourier analysis. Cora was one of a kind, a unique person, and special to AWM."

At the remembrance I read a brief quote sent by Chandler Davis, a longtime supporter of AWM: "Cora Sadosky was a force of nature. We valued her tempestuous laughter; her incisive wit; her total impatience with fools, phoneys, and bigots; and her total patience in service of mathematics and justice. We will miss her but draw strength from her memory."

We extend our sympathy to Daniel, to Cora Sol Goldstein and her husband Thomas Johnson, to Cora's beloved granddaughter Sasha, and to all of Cora's many friends.

Judy Green

Cora Sadosky was an amazing person and I am grateful to AWM for having put me in a position that allowed me to become aware of just how amazing she was. In 1992 Cora and I, both of whom lived in the Washington, D.C., area, became president-elect and treasurer of AWM, and the AWM office moved to the nearby University of Maryland. Since we both were very involved in the move, Cora and I became close friends and over the years, together with our husbands, we spent many happy times, including holidays and vacations, together.

In the course of our friendship of almost twenty years, I learned a lot about Cora and her life. She and her husband, Daniel Goldstein, were political exiles. Because of their protest of a repressive military dictatorship, they were forced to flee Argentina to save their lives, leaving all their possessions behind. Despite having sacrificed the possibility of a career in her native Argentina, Cora was able to maintain an enthusiasm for involving new, young, people in doing what she loved so much, mathematics. I think her own words rather than my own will better let you see this. Those of you who knew Cora, or read her AWM president's reports, will, I hope, find the following typical.

Six AWM presidents at the MSRI twentieth-fifth anniversary celebration: Cora Sadosky, Lenore Blum, Bhama Srinivasan, Carol Wood, Georgia Benkart, and Cathy Kessel

continued on page 8

CALL FOR NOMINATIONS

2012 M. Gweneth Humphreys Award

The Executive Committee of the Association for Women in Mathematics has established a prize in memory of M. Gweneth Humphreys to recognize outstanding mentorship activities. This prize will be awarded annually to a mathematics teacher (female or male) who has encouraged female undergraduate students to pursue mathematical careers and/or the study of mathematics at the graduate level. The recipient will receive a cash prize and honorary plaque and will be featured in an article in the AWM *Newsletter.* The award is open to all regardless of nationality and citizenship. Nominees must be living at the time of their nomination.

The award is named for M. Gweneth Humphreys (1911–2006). Professor Humphreys graduated with honors in mathematics from the University of British Columbia in 1932, earning the prestigious Governor General's Gold Medal at graduation. After receiving her master's degree from Smith College in 1933, Humphreys earned her Ph.D. at age 23 from the University of Chicago in 1935. She taught mathematics to women for her entire career, first at Mount St. Scholastica College, then for several years at Sophie Newcomb College, and finally for over thirty years at Randolph Macon Woman's College. This award, funded by contributions from her former students and colleagues at Randolph-Macon Woman's College, recognizes her commitment to and her profound influence on undergraduate students of mathematics.

The nomination documents should include: a nomination cover sheet (available at http://sites.google.com/site/awmmath/programs/humphreys-award); a letter of nomination explaining why the nominee qualifies for the award; the nominee's vita; a list of female students mentored by the nominee during their undergraduate years, with a brief account of their post-baccalaureate mathematical careers and/or graduate study in the mathematical sciences; supporting letters from colleagues and/or students; at least one letter from a current or former student of the candidate must be included.

Nomination materials for this award should be sent to awm@awm-math.org. Nominations must be received by **April 30, 2011** and will be kept active for three years at the request of the nominator. For more information, phone (703) 934-0163, email awm@awm-math.org or visit www.awm-math.org/humphreysaward.html.

Cora Sadosky *continued from page 7*

For the July/August 1994 *AWM Newsletter* (24 no 4: 5) she wrote:

> Strange as it may seem at the end of this long report, I started out with the impression I had very little to say. That impression stemmed from my doing something it may be unwise for a current AWM president to do: I plunged into mathematics head on for three full weeks. I came out of that stint dazed, vaguely guilty and deeply happy. It is very clear that I enjoy like crazy doing mathematics! ... I did not prove the Riemann Conjecture. My work was modest, but it gave me so much pleasure to do it!

Six months later, at the end of her term as AWM president she wrote (25 no 1: 6–7):

> I had never met so many women mathematicians, so many math teachers, so many graduate and undergraduate math students. How wonderful it has been. The brightest of all: those precious, unique marvels that are the Schafer Prize awardees. And all the young people that participate in the workshops. And all the people at meetings, scientific conferences, and funding agencies, who collaborate and contribute in so many ways in paving the roads for women to travel into mathematics.
>
> To hear our multiple voices, to perceive our diversity, to see it in the flesh and in the papers—what a privilege.
>
> A couple of days ago I had to answer on our behalf a questionnaire on AWM's commitment to minority inclusion into mathematics. I was so proud to write that our work for the right of women to mathematics is intertwined with our total commitment to the struggle for right of all people to mathematics.
>
> As a woman, as a Latin American, as a mathematician, I am so proud of our struggle. Much has been gained, it is true, and we should be happy about it. What women mathematicians in the U.S. have conquered in our quest for equity is unmatched in the world. Still, so much remains to be done. Until all deserving people find opportunities that correspond to their abilities and their contributions.
>
> We must not tire. We cannot cease to care. To go to a classroom and be in touch with the bright young people striving to do mathematics is enough reminder of our duty to open doors for them....
>
> Contrary to the despicable joke about women mathematicians, Sofia Kovaleskaia and Emmy Noether were both women and mathematicians. We cannot be like them by mere will. But we can empower others to be like them. And better still. Let's do it.

As I copied these words, I heard them in Cora's voice and knew that both her mathematics and the inspiration she gave to young mathematicians will live on. What I hope will also live on is that her work was done despite the sacrifices she and her husband made for their beliefs. What Cora wrote of her mentor and long-time collaborator, Mischa Cotlar, applies to her and Daniel as well:

> Mischa's experiences in a country corrupted by decades of authoritarianism made him a staunch defender of human rights and civil liberties... For Mischa, social and ethical issues were not marginal problems. (On the Life and Work of Mischa Cotlar, *Rev. Un. Mat. Argentina* 49 no. 2 (2008): iii).

Cora, too, remained committed to making the world a better place to live for all people. We will miss her.

Cora and Schafer winners

Richard Bourgin

Cora and I were friends and colleagues for 30 years. We briefly met before classes started in August 1980 at Howard University, but I didn't get much of an impression of her then. Some days later her mother (who had been visiting her in Washington, D.C., and was also a mathematician and human rights activist) died suddenly. Despite that personal tragedy, Cora entered department life fully from the start. During those first months she made a number of suggestions or substantive changes in our undergraduate and graduate program offerings, most of which were later implemented. In the faculty discussions concerning curricular matters she argued cogently and strongly for her positions. Cora believed that

Daniel Goldstein and Cora Sadosky. Photo © Azia Yakubu

often less is more: students (especially at the undergraduate level) should be challenged by studying core ideas without obfuscating embellishments, enhanced by a few well chosen, easily understood examples. While her dedication to educational matters at college and graduate levels was steadfast, at times her ideas met strong resistance from some of her colleagues. She handled conflict well, neither seeking nor shying away from it. Cora and I saw eye to eye on most professional matters (and on many others), but as often as not we found ourselves in the minority, sometimes a minority of two. On those occasions in which we did prevail, her ability to speak clearly and forcefully to the issue at hand was often a key factor. She also understood when not to speak, which was just as important.

The lingua franca of many of our colleagues in the department is French. Cora matched their preferred language, switching seamlessly between French and English, often in rapid succession. In fact, she was trilingual and could move in and out of Spanish equally as well. She was very warm, quick to pick up social cues, urbane, and always civil. At the same time, it was important to Cora that people take full responsibility for their actions and she didn't care for those who, in her view, did not. She could be intense, sometimes dominating conversations when she felt strongly about what was being discussed.

On the other hand, Cora had great empathy and compassion for people who she believed were being treated unfairly. She provided succor and counsel to several such individuals, sometimes for an extended period. Over the years Cora helped a number of graduate students and young faculty (at Howard and elsewhere), new to the United States, who were products of the French educational system. Several ran into difficulties due to the unspoken differences in perceptions and assumptions in these two systems. Cora understood both extremely well and was able to guide them through problems of all sorts—from visas to tenure. She worked with many graduate students, watching over their progress with great care, advising and helping them, including giving financial help in the form of travel money to conferences and summer support. Their continued loyalty to her is notable. In fact, she had a Ph.D. student in the 1980s who still considers her his second mother and refers to her that way sometimes when we talk.

Cora and I became friends soon after her arrival at Howard. We socialized both on and off campus, often with her husband Daniel. Once Cora helped me pick out something for my future wife. As I remember it she was at least as happy with the choice as I was. She had wonderful taste and for many years thereafter I consulted her when shopping for my wife. Dinner at Cora and Daniel's was always delightful. They had a tiny kitchen with Daniel its undisputed master. Every meal I had there was memorable—the food was magnificent, and the conversation good. Before she went off to college we were usually joined by their daughter Cora Sol, to whom they were devoted. Cora and Daniel had wide ranging interests and areas of expertise, including a deep understanding of the causes and effects of oppression. One day I made an egregious (though innocent) mistake which caused hard feelings among some of my graduate students. When I later described to Cora what I thought had happened, she invited me to her home where she and Daniel placed that event in a much broader context. I finally understood what had really occurred and what misunderstandings I needed to address. It was a very important moment for me.

From time to time Cora said in passing that she would never see eighty. I always thought she was joking. She was so solid, such a presence, I thought she would outlast us all. I join the long list of family, friends and colleagues who will always miss her.

continued on page 10

Cora Sadosky and Howard students. Photo © Aziz Yakubu

Cora Sadosky *continued from page 9*

Daniel B. Szyld, Temple University

I was asked to say a few words about Cora's life in Buenos Aires. I am honored to be able to do so.

I would like to start by saying that Cora was a quintessential Buenos Aires intellectual: widely-read and extremely well-informed. She understood politics, and how politics influences all areas of our life including education and research. She was an activist, and by that I mean a passionate socially responsible individual, caring about many issues, including of course opportunities for women and African-Americans in mathematics.

Many people are aware that there was a military government in Argentina during the 1976–1983 period, during which thousands of people were disappeared, tortured and killed. There was also an earlier military government from 1966 to 1973. In July 1966, barely a month after the Junta had taken power, the infamous "Noche de los bastones largos" (or night of the long batons) took place. Students and faculty had taken over several schools of the Universidad de Buenos Aires in protest against the military coup and their stated goal of reversing the 1918 law granting the University autonomy, or self-government. On that night, police assaulted the schools, hitting the occupiers with their long batons and destroying libraries and laboratories. Four hundred people were detained. Later, many professors were fired, and many more resigned in protest. Cora was a young assistant professor at that time, and she was one of those who resigned. I should note that her father and mother also resigned at that time. He was a vice-dean, and she was an associate professor.

After two years teaching in Montevideo and at Johns Hopkins, Cora was back in Buenos Aires, still under the dictatorship. She was unable to have an academic position, and she worked for a literary press for a while. Thus, for a few years she left mathematics. She told me once that one of the hardest things in her life was the process of going back to doing mathematics after that hiatus. But eventually back she was, starting her collaboration with Mischa Cotlar, a collaboration which would last over thirty years.

But then the second coup came in 1973, at a time when a paramilitary group was active, threatening those who they perceived as leftists. The group known as the "triple A" was the Argentine Anti-communist Alliance and had threatened Cora's family. If you watched the Argentine movie, *The Secret in Her Eyes,* which won last year's Oscar, you would recognize this climate of terror. Cora's parents, Manuel Sadosky and Cora Ratto de Sadosky, were well-known intellectuals and political activists. Cora Ratto had been a leader of the student union at the University in the 1930s. Later she created a women's organization devoted to helping the anti-Nazi war effort. The group eventually had fifty thousand members. She also co-edited a magazine called *Columna diez* (*Column Ten*) in the 1960s, devoted to the analysis of the impact of science and technology on international politics, and in particular it brought to light many issues related to the Vietnam War.

Cora Sadosky, her parents, her husband Daniel Goldstein, and daughter Cora Sol Goldstein, thus left Argentina again, this time for exile in Venezuela. Cotlar had taught in Venezuela before and had contacts there. Cotlar also went into exile at that time, and Cora and he worked together at the Universidad Central de Venezuela until Cora left Caracas for the U.S. I met her first at the Courant Institute in the late 1970s when she was at the Institute for Advanced Studies and would frequently visit Louis Niremberg at NYU. In 1980 she moved to Washington for her job at Howard University.

Cora's passion for social issues was only matched by her passion for mathematics, and her love of her immediate family, including in recent years her granddaughter Sasha Malena. Her intellect was rich, her intensity was penetrating. She left a big mark on many of us. There is a little bit of Cora in many of us.

Give me your tired, your poor,
Your huddled masses yearning to breathe free...[1]

Cora Sadosky fought many battles and on many fronts. She had tremendous convictions against many injustices, gender inequality, and discrimination in our society. She often took the flag of the underrepresented, underserved, and underestimated. She chose to fight many of her battles from within the mathematical community, sometimes even risking her own mathematical career. She showed a lot of courage in this sense and never worried about the consequences for her. She was never afraid to express her views.

We are deeply saddened for her passing, but through her teachings we celebrate her life. To remember her we choose to focus on personal experiences that are not so well-known, particularly about her mentorship role and her unselfish devotion to help young mathematicians.[2]

---

[1] from *The New Colossus* by Emma Lazarus, the poem engraved on the Statue of Liberty

[2] A biographical sketch of Cora Sadosky and some of the historical events surrounding her life in different countries can be found for example at http://www.agnesscott.edu/lriddle/women/corasadosky.htm.

Cora always wanted to help students who were interested in following a research career in mathematics. In particular she interacted with many students in Venezuela and in Argentina. In her years in Caracas, Cora was very influential on a group of Venezuelan mathematicians, María Dolores Morán, Ramón Bruzual, Marisela Domínguez and Stefania Marcantognini, among others, and a Uruguayan mathematician, Rodrigo Arocena, who earned their Ph.D. degrees at the Universidad Central de Venezuela (UCV). She also aided others to pursue their doctorates in the US, including Gustavo Ponce who went to Courant Institute in 1978 and later Cristina Pereyra who went to Yale University in 1987. Likewise, during her sabbatical year in Buenos Aires in 1984–1985 she helped many Argentinian mathematicians come to the US for their doctoral degrees, among them, José Zero who went to the University of Pennsylvania, Estela Gavosto and Rodolfo Torres who went to Washington University, and Andrea Nahmod and Lucas Monzón who went to Yale University.

Over the last thirty years Cora conducted her professional life in the US, but she continued to be interested in mathematics in Argentina and Venezuela, where she often visited Mischa Cotlar. She was always trying to help colleagues and students in those countries in any way she could.

Three years ago Cora retired from Howard University where she had been a professor since 1980. With her husband, Daniel Goldstein, they moved to California to be closer to their daughter Cora Sol, son-in-law Tom, and beloved granddaughter Sasha.

We remember here particular moments and aspects of her mentorship.

Gustavo Ponce: In 1976, the Universidad Central de Venezuela started a graduate program in mathematics. At that time I was Cora's TA in a course on Advanced Calculus. She convinced me to take two more advanced courses, one with Mischa Cotlar in functional analysis and a seminar on harmonic analysis where Cora and Mischa alternated as lecturers. These courses shaped my view of mathematics and the enthusiasm of both lecturers was as inspiring as the content. In retrospect, it is amazing that the topics were and still are so fruitful in problems related to my research interests, the field of nonlinear partial differential equations.

Cora suggested that I should pursue a career as a mathematician and recommended that I apply to the Courant Institute. She even made a providential call to Louis Nirenberg asking for an answer to my application: it turned out that it had been lost. This was just one of the many favors for which I feel so much in debt to her. She has been a systematic source of inspiration and academic advice and a model of intellectual integrity since those years at UCV.

Estela Gavosto and Rodolfo Torres: We met Cora at an annual meeting of the Unión Matemática Argentina in the mid '80s and often visited her then in Buenos Aires (we were living and studying in Rosario, another city in Argentina). From the first time we met she showed a lot of interest in getting to know us, mathematically and otherwise, and listening to what we wanted to do in our lives, so she could help us maximize our options. She never pushed us in any direction, but she was the first to suggest that we study in the US. We still remember her words when we told her that it would be impossible to get fellowships for both of us at the same university. Her almost irritated response was one we heard from her many times over the years and in many different circumstances: "... and who told you that? If you plan and do the things right you can do it." Her attitude remained the same over the years. She was always positive and enthusiastic. Nothing ever seemed impossible to her. With her help we went to do our Ph.D. degrees at Washington University in St. Louis. She strongly encouraged us to go there and told us it would be a great place for us and a good match for many of our interests. She was right. Not only did we find a great math graduate program but also a terrific atmosphere where we met fantastic teachers and other students who have become long lasting friends and mathematical colleagues. Several years later, we also thought it would be impossible to get postdoctoral positions at the same place or nearby ones, and then impossible to get tenure-track jobs at the same university. "And who told you that?" Cora repeated time and time again. She was there always ready to give advice and help us in what she could. She understood the "two-body problem" better than anybody, brought awareness about this common issue in our profession, and worked hard to help people in this situation. She was tireless in trying to achieve any goal she set her mind to and always infected us with her optimism.

Cora was also concerned with many other aspects of the academic life and she often told us to learn about them and encouraged us to be proactive. She taught us that for any change to take place people really need to get involved. We will always be very thankful for all her help, mentorship and guiding example.

Andrea Nahmod: Cora Sadosky was my professor and undergraduate advisor in Buenos Aires. She was also a friend. In 1984–1985, during the last year of my Licenciatura en Matemáticas (a degree similar to a Master of Science), several research mathematicians living abroad during the military

continued on page 12

Cora Sadosky *continued from page 11*

regime returned to the Universidad de Buenos Aires for a year, to teach. Cora Sadosky was one of them. I knew of her before her arrival in Buenos Aires, where she spent her sabbatical year, through one of my uncles, Victor E. Nahmod. Victor, a biomedical researcher and physician, had been a clinical teacher of Daniel Goldstein, Cora's husband, at the Instituto de Investigaciones Medicas at the Universidad de Buenos Aires. Victor and Daniel then became scientific collaborators for many years. Of course, Victor knew Cora's parents and Mischa Cotlar. But I first met Cora at the campus of the Universidad de Buenos Aires at the beginning of Fall 1984 and like many others rushed to enroll in her topics course on "Singular Integral Operators and the Theory of A_p Weights." To say that this course opened up our heads would be an understatement. To this date, I marvel at the course notes and at how deep and useful was the mathematics she taught us back then. It was an easy decision to do my Licenciatura's thesis under her supervision. I recall going one morning to her office to discuss it and spending practically the whole day with her at the blackboard passionately explaining to me lots of "hard analysis." It was exhilarating. For my thesis, we settled on *Factorization of operators and the Nikishin-Stein theory.* Garcia Cuerva and Rubio de Francia were at that time writing their celebrated book, and Cora had drafts of some of its chapters. I recall leaving her office that day late in the afternoon with these, together with a pile of papers by Grothendieck, Maurey, Nikishin, Stein, Pisier and others. These were the first research papers I read, and I learned to do so with Cora.

When I decided later that year to pursue my Ph.D. in the US, Cora did much more than write letters of recommendation on my behalf. She truly guided me and helped me each step of the way. Alberto Calderón was also in Argentina at that time after his first retirement from the University of Chicago. He had been my professor in Functional Analysis, and thanks to Cora he was a member of my Licenciatura's thesis committee and after my defense wrote a letter of recommendation on my behalf. I graduated in November of 1985, and by then Cora had returned to the US. Since the earliest I could start graduate school in the US was

September 1986, Cora worried about how to, mathematically, make the most out of the 10 months in between. She thought it was fundamental I continued learning mathematics until I started my Ph.D. in the US. She suggested several topics and gave me her own copy of J.L. Journé's book, while asking C. Segovia to read it with me and make sure I learned the material inside out in the months ahead. Cora went out of her way also at a personal level, reassuring my parents I was not going far away alone, for she and Daniel would take care of me. And so they did. My first two weeks in the US and ever away from home were at her home in Washington, D.C. Cora later insisted I call her collect from New Haven often to her home in D.C. to tell her about my studies and life at Yale, to let her know I was alright. I visited Cora and stayed at her home many times afterwards, in Washington, D.C. and while she spent a year at MSRI in Berkeley during 1987–1988. She was always there for me as a mentor and as a friend throughout my graduate school years in New Haven.

Cora had an impeccable work ethic and very high standards; she was all about being the best you can be, as a mathematician and as a person. She inspired all of us to work harder and be better, each day, every day. Many of us would probably not be professional mathematicians had it not been for Cora. Our gratitude is infinite.

Cristina Pereyra: Corita, as we knew her, was a dear friend of my mother, Concepción Ballester, from the time before I was born; so were her parents Cora Ratto and Manuel. All of them, my mother included, were mathematicians who grew up in the golden era of mathematics in Argentina, the fifties and the sixties, a time when Laurent Schwartz and Antoni Zygmund would visit Buenos Aires because the US and Europe were interested in identifying and helping talented budding mathematicians. In one of those visits Zygmund recruited Alberto Calderón and Mischa Cotlar to the University of Chicago. In turn, Calderón and Zygmund would become Corita's advisors at Chicago, where she received her Ph.D. in 1965.

In 1964, my mother, my older brother and me joined my father in the US where he was getting his Ph.D. We never returned to Argentina; by 1967 many of my parents' former colleagues in Buenos Aires had been fired or had resigned from the university for political reasons. Instead we landed in Caracas, where my parents became professors at the Universidad Central de Venezuela. Things in Argentina only got worse, and in 1974, when Corita, her husband Daniel, young daughter Cora Sol and her parents came as political refugees to Venezuela, we all lived in the same building (San Bartolomé; in Caracas every building has a name instead of a number). We were in the 12th floor of one tower, Cora Ratto and Manuel on the first floor of the same tower, Corita, Daniel and Cora Sol in the 11th floor of the other tower, with magnificent views to El Avila, the mountain that separates Caracas from the Caribbean Sea. This was no coincidence, my mother must have arranged things for her dear friends.

Mischa Cotlar also arrived in Venezuela, escaping the asphyxiating political climate in Argentina. Before leaving Buenos Aires Corita had started her lifelong collaboration with Mischa, which spanned more than 30 years, with more than 30 influential joint papers in harmonic and functional analysis and in operator theory. After Corita left Venezuela in the early '80s, she kept visiting once or twice a year to work with Mischa; during those visits she would stay with my mother. Cora and Mischa would work like crazy for many hours each day, but Cora would return to my mother's house to relax and to enjoy her company.

I was too young to have Cora as my teacher in Caracas, when in turn I decided that after all mathematics was my call. However Cora was instrumental in helping me choose which graduate schools to apply to, and her support was fundamental to enter Yale. I remember before arriving in New Haven in 1987 spending a week in Washington warming up with Cora and Daniel, as Andrea Nahmod had done the year before, and as Lucas Monzón did the next year when he also came. The rest is history: Cora became part of my mathematical family, or more precisely, I became part of hers. Cora always was there through my professional career to offer support and advice.

In October 2007, Cora helped me and Wilfredo Urbina organize an afternoon in honor of Mischa Cotlar who had died that January. She came to Albuquerque and stayed with my mother together with María Dolores (Loló) Moran and Stefania Marcantognini who had come from Venezuela. We saw Cora last in April 2009 when she came to Albuquerque and stayed again with my mother for five days for the 12th New Mexico Analysis Seminar; she looked well. It was a shock to learn of her passing, hard to accept that such a vital woman and life-long dear friend had left us.

Wilfredo Urbina: I met Cora at the Universidad Central de Venezuela where I was doing my undergraduate studies in mathematics. Later I was her TA in an Advanced Calculus course and then in 1978, when I started my master's degree, I had the privilege to take her course Introduction to Harmonic Analysis. The notes of that course were the base of her famous book *Interpolation of Operators and Singular Integrals* published by Marcel Dekker a year later. It was a very tough course for me but I must say that it had a lasting influence on my career. Also I attended the analysis seminar that Mischa

continued on page 12

Cora Sadosky *continued from page 9*

and Cora organized and that is still running today at UCV. Later I went to get my Ph.D. at the University of Minnesota, in principle to study probability, but I met Gene Fabes so I went back to analysis, and then back to the things that I had learned from Mischa and Cora in Caracas.

Cora was a real enthusiast of mathematics, willing to help anyone who showed serious interest in studying it. Her solidarity and support for so many mathematicians from Latin America is very well known. Cora's help in the organization of an international conference to celebrate Mischa's 80th birthday in Caracas in January 1994 was invaluable, as was her help and input when we organized an afternoon in honor of Mischa Cotlar in Albuquerque in October 2007. When I came back to the States in 2004, due to the political situation in Venezuela, Cora was an important reference for me, and I got her support in the painful process of looking for jobs. We met several times in Albuquerque (in October 2007 and April 2009) and in Zacatecas during the AMS-SMM Joint Meeting in July 2007; on every occasion we had a wonderful time together, enjoying good mathematics and good food.

We all met each other through Cora. At a special year in harmonic analysis at MSRI in 1987–1988, she introduced Andrea, Gustavo, Wilfredo, Estela and Rodolfo to each other, and we had a great time together. Cora jump-started the beautiful friendships and professional relations we have kept among us over the years. Every time we talked to Cora, she asked us what we were up to in mathematics, and she wanted details! She also wanted to know how what we were working on, fitted into the big picture of mathematics. She was very critical about mathematics. She once told us that Antoni Zygmund taught her to judge mathematicians by their theorems, that when someone spoke highly of a mathematician, Zygmund would always like to see the theorems the mathematician had proved. She did the same. If you said to her that you like mathematician X she would ask you to explain his/her mathematics. Cora was a person who always expressed her views in a very direct way. But she never let disagreements she might have with a mathematics colleague on other issues affect her appreciation of his/her mathematics. When it came to mathematics, Cora indeed judged mathematicians by their theorems.

Cora was a vibrant, strong minded and outspoken woman, who fought all her life for human rights, who helped uncountable young mathematicians without expecting anything in return. Cora was a phenomenal mathematician, in a time when it was not easy for women, and she championed this cause all her life, becoming president of the AWM in the early '90s.

Cora taught all of us by example how to mentor, how to help younger mathematicians pursue their dreams. Cora touched our lives in many ways but she never wanted to be thanked for her good deeds; she believed instead in paying it forward to others. Helping students reach their potential may be the best way to honor her memory. We will deeply miss her.

Estela A. Gavosto (University of Kansas, Lawrence)
Andrea R. Nahmod (University of Massachusetts, Amherst)
María Cristina Pereyra (University of New Mexico, Albuquerque)
Gustavo Ponce (University of California, Santa Barbara)
Rodolfo H. Torres (University of Kansas, Lawrence)
Wilfredo Urbina (Roosevelt University, Chicago)

CALL FOR NOMINATIONS:

2012 Louise Hay Award

The Executive Committee of the Association for Women in Mathematics has established the Louise Hay Award for Contributions to Mathematics Education, to be awarded annually to a woman at the Joint Prize Session at the Joint Mathematics Meetings in January. The purpose of this award is to recognize outstanding achievements in any area of mathematics education, to be interpreted in the broadest possible sense. The annual presentation of this award is intended to highlight the importance of mathematics education and to evoke the memory of all that Hay exemplified as a teacher, scholar, administrator, and human being.

The nomination documents should include: a one to three page letter of nomination highlighting the exceptional contributions of the candidate to be recognized, a curriculum vitae of the candidate not to exceed three pages, and three letters supporting the nomination. It is strongly recommended that the letters represent a range of constituents affected by the nominee's work. Nomination materials for this award should be sent to awm@awm-math.org. Nominations must be received by **April 30, 2011** and will be kept active for three years. For more information, phone (703) 934-0163, email awm@awm-math.org or visit www.awm-math.org.

In Memory of Cora Sadosky

Steven Krantz, Washington University in St. Louis

Cora Sadosky was a student of Alberto Calderón and Antoni Zygmund. And she was very closely associated with Mischa Cotlar. It is easy to see that she inherited from these wonderful mentors her intense sense of curiosity about everything mathematical. Apart from her personal charms, Cora was an intense mathematician who wanted to learn about everything.

This passion also shows in the remarkable array of collaborators that she had. It is clear that she loved to develop mathematics, to communicate mathematics, and to write mathematics. She wrote mathematics easily and well, and her publication record attests to her success at mathematical communication.

I knew Cora Sadosky for almost my entire career. For me she was a friend, a colleague, and a role model. She was the paradigm of what a mathematician should be. She was always ready with friendly advice, with sympathy, with ideas, or whatever the situation called for. I always looked forward to seeing Cora at conferences or other gatherings.

Cora's mathematical interests were in harmonic analysis. She kept up with all the latest developments, and wrote about many aspects of harmonic analysis on Euclidean space. Her papers covered such diverse topics as singular integrals, moment problems, interpolation of operators Toeplitz operators, weighted norm inequalities, lifting theorems, HelsonSzego theorems, Hankel forms, matrix measures, functional analysis, operator theory, conservative systems, BMO, and scattering theory. Space prevents me from enumerating all the subjects that were in her arsenal. She was a talented and diverse mathematician with catholic interests.

Cora will be missed by her friends in the American Mathematical Society and also by her colleagues in the Association for Women in Mathematics. Cora's intense feelings for the role of women in mathematics were well known and much appreciated. She was a mentor for woman mathematicians and a role model for all.

— Steven G. Krantz

Cora

Maria Dolores (Loló) Morán

There are people that mark the lives of others, Cora definitely influenced mine.

I met Cora in 1975. I had started a computer science major and she taught a pre-calculus course at the Facultad de Ciencias of the Universidad Central de Venezuela that I took. Each and every day to my delight, Cora assigned me a problem. At the end of the course, she said, " I'm here for you, for math, whenever you want."

A year later, when I decided to change my major to Mathematics (it was not easy in my country), she helped me with all the bureaucracy.

Cora also taught my first graduate course, Functional Analysis. She had a job offer in the US, that fulfilled my dream of working with the first mathematician, feminist I had ever met, my mentor. Anyhow, she gave me one of my favorites books, Reed-Simon's *Functional Analysis*. Her continued generosity helped me also with my first publication.

The last time I saw her was at the Albuquerque meeting in honor of Misha Cotlar, I recognized her weariness. She was about to retire, so she would see her granddaughter grow.

Cora, Misha and Rodrigo were crazy southern analysts with a dream: a school of mathematics, they succeeded: Venezuela is now a reference in Operator Theory.

A Statement About Cora Sadosky

I am very happy to write a short note about Cora Sadosky. She was a close friend an excellent mathematician who did much to help women mathematicians. I met her often and discussed many topics Let me give you more information about us two.

In the first semester of 1960,[1] I was invited to give a course at the University of Buenos Aires. During a very long period that included this date Argentina was "occupied by its own army." It was badly governed sometimes brutally. There were a few brief reprieves, including this period in 1960, when there was a better rule There was a surprisingly good group of mathematicians associated with this university at this time, including A. P Calderon and Mischa Cotlar This group was heavily influenced by A. Zygmund Cora Sadosky was a graduate student in the class I taught. She was an excellent student and she and I became good friends during this course. The notes I wrote for this course were published in the series "Cursos y seminarios de matematica" Volume 9 "Analisis Armonico en Varias Variables, teoria de los Espacios H^P" Alberto Calderon who had just accepted a position at the University of Chicago, worked in an area very close to this one and I encouraged Cora to pursue a PhD under his direction. At about the same time, I accepted a position at Washington University in St Louis. I visited the University of Chicago often and our friendship continued After completing her PhD, Cora made many attempts to find a position for herself and her husband, Daniel Goldstein, in Argentina. Unable to do that, Cora ultimately took a faculty position at Howard University in Washington D.C At that time I had to attend several committee meetings in Washington and reconnected with her. By then, we had established a long-lasting friendship which included not only mathematics, but long discussions about politics and the place of women in mathematics. She became a leading activist and presided the Association for Women in Mathematics from 1993 to 1995. By this time she exuded enormous energy, confidence and warmth I was one of her strongest supporters.

August 28, 2015

Guido L. Weiss
Elinor Anheuser Professor
of Mathematics

[1]Washington University in St. Louis, Campus Box 1146, One Brookings Drive, St. Louis, MO 63130-4899 (314) 9356711, Fax: (314) 935-6839, Email: guido@math.wustl.edu.

Michael Wilson's Remembrance

I first met Cora in Montreal in 1987, at a workshop on weighted norm inequalities. Doug Kurtz had flown in. His luggage didn't make it.

After the 24 (or 48?) hour period passed and it still hadn't arrived, he went out to get new luggage and clothes at the airline's expense. A bunch of us traipsed along with him.

"What is it with shopping?" Cora asked "I don't understand this fascination with just… shopping."

"Me neither," I said. "Except for books"

"Oh, books!" she said "*Completely* different!"

Clearly, she knew what was important.

Mike Wilson

From left to right: Weissman, Cora, Manuel Sadosky, Rome 1948
Photo Courtesy of Cora Sol Goldstein, photographer unknown

Graduation Photo: Cora Sadosky, PhD University of Chicago, 1965
Photo Courtesy of Cora Sol Goldstein, photographer unknown

From left to right: [*unknown*], Zygmund, Cora Sadosky, Chicago, 1981
Photo Courtesy of Cora Sol Goldstein, photographer unknown

Group photo, taken in Caracas in January 1994, on the occasion of Mischa Cotlar's conference celebrating his 80th birthday.
Front row: Concepción Ballester (in a blue dress); to the right, Laurent Schwartz (in a black tie); Mischa Cotlar (with a black shirt and beige jacket); Cora Sadosky (in a black dress), her arm on Stefania Marcantognini's shoulder. Behind Laurent Schwartz to the left is Steve Hoffman; behind Schwartz to the right, Eli Stein. Between Stein and Cotlar is Fernando Soria (with a beard).
Photo reproduced with the kind permission of Ramón Bruzual

Attendees of the Cotlar Conference.
1) Ileana Iribarren, 2) Aline Bonami, 3) Zeljko Cuckovic 5) Concepción Ballester 6) Laurent Schwartz 7) Mischa Cotlar 8) Stefania Marcantognini 9) Cora Sadosky 10) Damir Arov 11) Ilya Spitkovskiy 12) Israel Gohberg 13) Marisela Dominguez 14) Alfredo Octavio 16) Richard Gundy 17) Gustavo Ponce 18) Wilfredo Urbina Romero 21) Dmitry Yakubovich 24) Vadim Adamjan 28) Gerardo Mendoza 29) James Rovnyak 34) Michael Dritschel 37) Steve Hoffman 40) Elias Stein 44) Fernando Soria 56) Mario Milman 59) Javier Duoandikoetxea 63) Norberto Salinas 64) Victor Vinnikov 67) Vladimir Peller 70) Daniel Alpay 72) Nikolai Vasilevski 75) Gustavo Corach 76) Pedro Alegría 77) Victor Padrón 80) Lázaro Recht 81) Carlos Segovia 82) Ventura Echandía 84) José Andrés 87) Cristina Cerruti [unknown attendees unlisted]

1996 University of Chicago Conference in honor of Calixto Calderón's 75th Birthday: Front row, seated (left to right): M. Christ, C. Sadosky, A.P. Calderon, M.A. Muschietti. First row, standing (left to right): C.E. Kenig, J. Alvarez Alonso, C. Gutierrez, E. Berkson, J. Neuwirth. Second row, standing (left to right): A. Torchinsky, J. Polking, S. Vagi, R.R. Reitano, E. Gatto, R. Seeley.

(The editors would like to thank Carlos Kenig for help identifying most of the people appearing in the photo, to Mike Christ who help trying to identify the missing character, and pointed us in the direction of Cristian Gutierrez, who identified Reitano as well as Jerome Neuwirth. Thanks, Cristian!) Photo courtesy of Cora Sol Goldstein, photographer unknown

Family photo: From left to right, Daniel Goldstein, Cora Sol Goldstein & Cora Sadosky, on the day of Cora Sol's PhD Graduation from University of Chicago in 2002.
Photo reproduced with the kind permission of Thomas Johnson. Photo courtesy of Cora Sol Goldstein

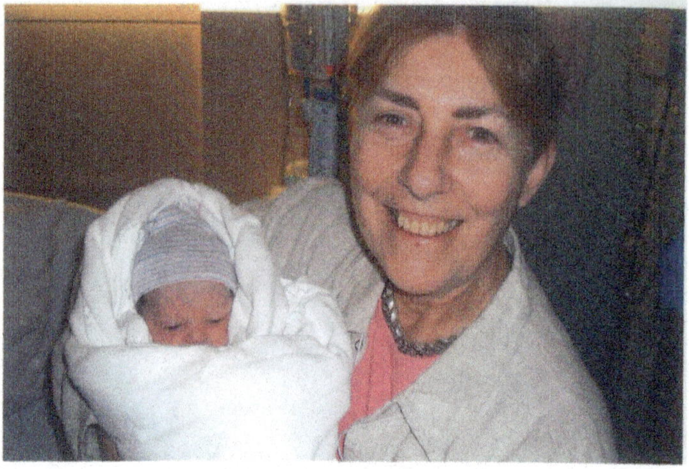

The next generation: Cora Sadosky and her newborn granddaughter, Sasha Malena, in 2006. *Photo reproduced with the kind permission of Thomas Johnson. Photo Courtesy of Cora Sol Goldstein*

Photo Courtesy of Cora Sol Goldstein, photographer unknown

Barry Mazur, Cathy O'Neil and Cora Sadosky: Cathy O'Neil being embraced and supported by Cora Sadosky on one side and Barry Mazur on the other. This picture was taken in 1993 in Vancouver, where O'Neil received the Alice T. Schafer prize. In O'Neil's words, "It was a critical moment for me, and both of those people have influenced me profoundly. Barry became my thesis advisor; part of the reason I went into number theory was to become his student [. . .] Cora became my mathematical role model and spiritual mother. [. . .] Well, Cora, whom I met when I was 21, was the person that made me realize there is a community of women mathematicians, and that I was also welcome to that world." Accessed June 29, 2011 at http://mathbabe.org/2011/06/29/cora-sadosky/

Photo reproduced with the kind permission of Aaron Abrams (Photo courtesy of Cathy O'Neil)

Author: George Bergman, Source: Archives of the Mathematishes Forchungsinstitute Oberwalfach

Mischa Cotlar, Cora Sadosky's mentor and long time friend and collaborator. *Author: Ramón Bruzual. Photo reproduced with the kind permission of Ramón Bruzual*

Part II
Harmonic and Complex Analysis, Banach and Metric Spaces, and Partial Differential Equations

Higher-Order Elliptic Equations in Non-Smooth Domains: a Partial Survey

Ariel Barton and Svitlana Mayboroda

Abstract Recent years have brought significant advances in the theory of higher-order elliptic equations in non-smooth domains. Sharp pointwise estimates on derivatives of polyharmonic functions in arbitrary domains were established, followed by the higher-order Wiener test. Certain boundary value problems for higher-order operators with variable non-smooth coefficients were addressed, both in divergence form and in composition form, the latter being adapted to the context of Lipschitz domains. These developments brought new estimates on the fundamental solutions and the Green function, allowing for the lack of smoothness of the boundary or of the coefficients of the equation. Building on our earlier account of history of the subject (published in *Concrete operators, spectral theory, operators in harmonic analysis and approximation*). Operator Theory: Advances and Applications, vol. 236, Birkhäuser/Springer, Basel, 2014, pp. 53–93), this survey presents the current state of the art, emphasizing the most recent results and emerging open problems.

1991 *Mathematics Subject Classification*: Primary 35–02; Secondary 35B50, 35B65, 35J40, 35J55

Introduction

The theory of boundary value problems for second-order elliptic operators on Lipschitz domains is a well-developed subject. It has received a great deal of study in the past decades and while some important open questions remain, well-posedness

---

A. Barton
Department of Mathematical Sciences, University of Arkansas, Fayetteville, AR 72701, USA
e-mail: aeb019@uark.edu

S. Mayboroda (✉)
Department of Mathematics, University of Minnesota, Minneapolis, MN 55455, USA
e-mail: svitlana@math.umn.edu

M.C. Pereyra et al. (eds.), *Harmonic Analysis, Partial Differential Equations, Complex Analysis, Banach Spaces, and Operator Theory (Volume 1)*, Association for Women in Mathematics Series 4, DOI 10.1007/978-3-319-30961-3_4

of the Dirichlet, Neumann, and regularity problems in L^p and other function spaces has been extensively studied in the full generality of divergence-form operators $-\operatorname{div} A\nabla$ with bounded measurable coefficients.

The corresponding theory for elliptic equations of order greater than two is much less well developed. Such equations are common in physics and in engineering design, with applications ranging from standard models of elasticity [101] to cutting-edge research of Bose–Einstein condensation in graphene and similar materials [124]. They naturally appear in many areas of mathematics too, including conformal geometry (Paneitz operator and Q-curvature [30, 31]), free boundary problems [1], and nonlinear elasticity [9, 32, 133].

It was realized very early in the study of higher-order equations that most of the methods developed for the second-order scenario break down. Further investigation brought challenging hypotheses and surprising counterexamples, and few general positive results. For instance, Hadamard's 1908 conjecture regarding positivity of the biharmonic Green function [56] was actually refuted in 1949 (see [46, 54, 127]), and later on the weak maximum principle was proved to fail as well, at least in high dimensions [96, 119]. Another curious feature is a paradox of passage to the limit for solutions under approximation of a smooth domain by polygons [19, 92].

For the sake of concreteness, we will mention that the prototypical example of a higher-order elliptic operator, well known from the theory of elasticity, is the bilaplacian $\Delta^2 = \Delta(\Delta)$; a more general example is the polyharmonic operator Δ^m, $m \geq 2$. The biharmonic problem in a domain $\Omega \subset \mathbb{R}^n$ with Dirichlet boundary data consists, roughly speaking, of finding a function u such that for given f, g, h,

$$\Delta^2 u = h \text{ in } \Omega, \quad u\big|_{\partial\Omega} = f, \quad \partial_\nu u\big|_{\partial\Omega} = g, \tag{1}$$

subject to appropriate estimates on u in terms of the data. To make it precise, as usual, one needs to properly interpret restriction of solution to the boundary $u\big|_{\partial\Omega}$ and its normal derivative $\partial_\nu u\big|_{\partial\Omega}$, as well as specify the desired estimates.

This survey concentrates on three directions in the study of the higher-order elliptic problems. First, we discuss the fundamental a priori estimates on solutions to biharmonic and other higher-order differential equations in arbitrary bounded domains. For the Laplacian, these properties are described by the maximum principle and by the 1924 Wiener criterion. The case of the polyharmonic operator has been only settled in 2014–2015 [83, 84], and is one of the main subjects of the present review. Then we turn to the known well-posedness results for higher-order boundary problems on Lipschitz domains with data in L^p, still largely restricted to the constant-coefficient operators and, in particular, to the polyharmonic case. Finally, we present some advancements of the past several years in the theory of variable-coefficient higher-order equations. In contrast to the second-order operators, here the discussion splits according to several forms of underlying operators. Let us now outline some details.

On smooth domains the study of higher-order differential equations went hand-in-hand with the second-order theory; in particular, the weak maximum principle was established in 1960 ([6]; see also [102, 103]). Roughly speaking, for a solution u to the equation $Lu = 0$ in Ω, where L is a differential operator of order $2m$ and $\Omega \subset \mathbb{R}^n$ is smooth, the maximum principle guarantees

$$\max_{|\alpha|\leq m-1} \|\partial^\alpha u\|_{L^\infty(\Omega)} \leq C \max_{|\beta|\leq m-1} \|\partial^\beta u\|_{L^\infty(\partial\Omega)}, \tag{2}$$

with the usual convention that the zeroth-order derivative of u is simply u itself. For the Laplacian ($m = 1$), this formula is a slightly weakened formulation of the maximum principle. In striking contrast with the case of harmonic functions, the maximum principle for an elliptic operator of order $2m \geq 4$ may fail, even in a Lipschitz domain. To be precise, in general, the derivatives of order $(m - 1)$ of a solution to an elliptic equation of order $2m$ need not be bounded. However, in the special case of three dimensions, (2) was proven for the m-Laplacian $(-\Delta)^m$ in domains with Lipschitz boundary ([117, 119]; see also [38, 116, 128, 130] for related work), and, by different methods, in three-dimensional domains diffeomorphic to a polyhedron [72, 96].

Quite recently, in 2014, the boundedness of the $(m - 1)$st derivatives of a solution to the polyharmonic equation $(-\Delta)^m u = 0$ was established in arbitrary three-dimensional domains [84]. Moreover, the authors derived sharp bounds on the k-th derivatives of solutions in higher dimensions, with k strictly less than $m - 1$ when the dimension is bigger than 3. These results were accompanied by pointwise estimates on the polyharmonic Green function, also optimal in the class of arbitrary domains. Furthermore, introducing the new notion of polyharmonic capacity, in [83] the authors established an analogue of the Wiener test. In parallel with the celebrated 1924 Wiener criterion for the Laplacian, the higher-order Wiener test describes necessary and sufficient capacitory conditions on the geometry of the domain corresponding to continuity of the derivatives of the solutions. Some earlier results were also available for boundedness and continuity of the solutions themselves (see, e.g., [86, 88, 90]).

We shall extensively describe all these developments and their historical context in section "Boundedness and Continuity of Derivatives of Solutions."

Going further, in sections "Boundary Value Problems with Constant Coefficients" and "Boundary Value Problems with Variable Coefficients" we consider boundary value problems in irregular media. Irregularity can manifest itself through lack of smoothness of the boundary of the domain and/or lack of smoothness of the coefficients of the underlying equation. Section "Boundary Value Problems with Constant Coefficients" largely concentrates on constant-coefficient higher-order operators, in particular, the polyharmonic equation, in domains with Lipschitz boundaries. Large parts of this section are taken verbatim from our earlier survey [24]; we have added some recent results of I. Mitrea and M. Mitrea. We have chosen to keep our description of the older results, for completeness, and also to provide background and motivation for study of the boundary problems for operators with variable coefficients—the subject of section "Boundary Value Problems with Variable Coefficients."

The simplest example is the L^p-Dirichlet problem for the bilaplacian

$$\Delta^2 u = 0 \quad \text{in } \Omega, \quad u\big|_{\partial\Omega} = f \in W_1^p(\partial\Omega), \quad \partial_\nu u\big|_{\partial\Omega} = g \in L^p(\partial\Omega), \tag{3}$$

in which case the expected sharp estimate on the solution is

$$\|N(\nabla u)\|_{L^p(\partial\Omega)} \leq C\|\nabla_\tau f\|_{L^p(\partial\Omega)} + C\|g\|_{L^p(\partial\Omega)}, \tag{4}$$

where N denotes the nontangential maximal function and $W_1^p(\partial\Omega)$ is the Sobolev space of functions with one tangential derivative in L^p (cf. section "Higher-Order Operators: Divergence Form and Composition Form" for precise definitions). In sections "The Dirichlet Problem: Definitions, Layer Potentials, and Some Well–Posedness Results"–"The Maximum Principle in Lipschitz Domains" we discuss (3) and (4), and more general higher-order homogeneous Dirichlet and regularity "Boundary Value Problems with Constant Coefficients" with boundary data in L^p. Section "Biharmonic Functions in Convex Domains" describes the specific case of convex domains. The Neumann problem for the bilaplacian is addressed in section "The Neumann Problem for the Biharmonic Equation." In section "Inhomogeneous Problems for the Biharmonic Equation," we discuss inhomogeneous boundary value problems (such as problem (1), with $h \neq 0$); for such problems it is natural to consider boundary data f, g in Besov spaces, which, in a sense, are intermediate between those with Dirichlet and regularity data.

Finally, in section "Boundary Value Problems with Variable Coefficients" we discuss higher-order operators with non-smooth coefficients. The results are still very scarce, but the developments of past several years promise to lay a foundation for a general theory.

To begin, let us mention that contrary to the second-order scenario, there are several natural generalizations of higher-order differential equations to the variable-coefficient context. Recall that the prototypical higher-order operator is the biharmonic operator Δ^2; there are two natural ways of writing the biharmonic operator, either as a *composition* $\Delta^2 u = \Delta(\Delta u)$ or in higher-order divergence form

$$\Delta^2 u = \sum_{j=1}^{n}\sum_{k=1}^{n} \partial_j\partial_k(\partial_j\partial_k u) = \sum_{|\alpha|=2} \frac{2}{\alpha!}\partial^\alpha(\partial^\alpha u).$$

If we regard Δ^2 as a composition of two copies of the Laplacian, then one generalization to variable coefficients is to replace each copy by a more general second-order variable-coefficient operator $L_2 = -\operatorname{div} A\nabla$ for some matrix A; this yields operators in *composition form*

$$Lu(X) = \operatorname{div} B(X)\nabla(a(X) \operatorname{div} A(X)\nabla u(X)) \tag{5}$$

for some scalar-valued function a and two matrices A and B. Conversely, if we regard Δ^2 as a divergence-form operator, we may generalize to a variable-coefficient operator in *divergence form*

$$Lu(X) = (-1)^m \operatorname{div}_m A(X)\nabla^m u(X) = (-1)^m \sum_{|\alpha|=|\beta|=m} \partial^\alpha (a_{\alpha\beta}(X)\partial^\beta u(X)). \qquad (6)$$

Both classes of operators will be defined more precisely in section "Higher-Order Operators: Divergence Form and Composition Form"; we will see that the composition form is closely connected to changes of variables, while operators in divergence form are directly associated with positive bilinear forms.

Sections "The Kato Problem and the Riesz Transforms" and "The Dirichlet Problem for Operators in Divergence Form" discuss higher-order operators in divergence form (6): these sections discuss, respectively, the Kato problem and the known well-posedness results for higher-order operators, all of which at present require boundary data in fractional smoothness spaces (cf. section "Inhomogeneous Problems for the Biharmonic Equation"). Section "The Dirichlet Problem for Operators in Composition Form" addresses well-posedness of the Dirichlet boundary value problem for a fourth-order operator in a composition form (5) with data in $L^p, p = 2$, in particular, generalizing the corresponding results for the biharmonic problem (3). To date, this is the only result addressing well-posedness with L^p boundary data for higher-order elliptic operators with bounded measurable coefficients, and its extensions to more general operators, other values of p, and other types of boundary data are still open. Returning to divergence-form operators (6), one is bound to start with the very foundations of the theory—the estimates on the fundamental solutions. This is a subject of section "The Fundamental Solution," and the emerging results are new even in the second-order case. Having those at hand, and relying on the Kato problem solution [16], we plan to pass to the study of the corresponding layer potentials and eventually, to the well-posedness of boundary value problems with data in L^p. In this context, an interesting new challenge, unparalleled in the second-order case, is a proper definition of "natural" Neumann boundary data. For higher-order equations the choice of Neumann data is not unique. Depending on peculiarities of the Neumann operator, one can be led to well-posed and ill-posed problems even for the bilaplacian, and more general operators give rise to new issues related to the coercivity of the underlying form. We extensively discuss these issues in the body of the paper and present a certain functional analytic approach to definitions in section "Formulation of Neumann Boundary Data." Finally, section "Open Questions and Preliminary Results" lays out open questions and some preliminary results.

To finish the introduction, let us point out a few directions of analysis of higher-order operators on non-smooth domains not covered in this survey. First, an excellent expository paper [89] by Vladimir Maz'ya on the topic of the Wiener criterion and pointwise estimates details the state of the art near the end of the previous century and provides a considerably more extended discussion of the questions we raised in section "Boundedness and Continuity of Derivatives of

Solutions," surrounding results and open problems. Here, we have concentrated on recent developments for the polyharmonic equation and their historical context. Secondly, we do not touch upon the methods and results of the part of elliptic theory studying the behavior of solutions in the domains with isolated singularities, conical points, cuspidal points, etc. For a good first exposure to that theory, one can consult, e.g., [72] and references therein. Instead, we have intentionally concentrated on the case of Lipschitz domains, which can display accumulating singularities—a feature drastically affecting both the available techniques and the actual properties of solutions.

Higher-Order Operators: Divergence Form and Composition Form

As we pointed out in the introduction, the prototypical higher-order elliptic equation is the biharmonic equation $\Delta^2 u = 0$, or, more generally, the polyharmonic equation $\Delta^m u = 0$ for some integer $m \geq 2$. It naturally arises in numerous applications in physics and in engineering, and in mathematics it is a basic model for a higher-order partial differential equation. For second-order differential equations, the natural generalization of the Laplacian is a divergence-form elliptic operator. However, it turns out that even *defining* a suitable general higher-order elliptic operator with variable coefficients is already a challenging problem with multiple *different* solutions, each of them important in its own right.

Recall that there are two important features possessed by the polyharmonic operator. First, it is a "divergence form" operator in the sense that there is an associated positive bilinear form, and this positive bilinear form can be used in a number of ways; in particular, it allows us to define weak solutions in appropriate Sobolev space. Secondly, it is a "composition operator," that is, it is defined by composition of several copies of the Laplacian. Moreover, if one considers the differential equation obtained from the polyharmonic equation by change of variables, the result would again be a composition of second-order operators. Hence, both generalizations are interesting and important for applications, albeit leading to *different* higher-order differential equations.

Let us discuss the details. To start, a general *constant-coefficient* elliptic operator is defined as follows.

Definition 2.1. Let L be an operator acting on functions $u : \mathbb{R}^n \mapsto \mathbb{C}^\ell$. Suppose that we may write

$$(Lu)_j = \sum_{k=1}^{\ell} \sum_{|\alpha|=|\beta|=m} \partial^\alpha a_{\alpha\beta}^{jk} \partial^\beta u_k \tag{7}$$

for some coefficients $a^{jk}_{\alpha\beta}$ defined for all $1 \leq j, k \leq \ell$, and all multiindices α, β of length n with $|\alpha| = |\beta| = m$. Then we say that L is a *differential operator of order* $2m$.

Suppose the coefficients $a^{jk}_{\alpha\beta}$ are constant and satisfy the Legendre–Hadamard ellipticity condition

$$\operatorname{Re} \sum_{j,k=1}^{\ell} \sum_{|\alpha|=|\beta|=m} a^{jk}_{\alpha\beta} \xi^{\alpha} \xi^{\beta} \zeta_j \bar{\zeta}_k \geq \lambda |\xi|^{2m} |\zeta|^2 \tag{8}$$

for all $\xi \in \mathbb{R}^n$ and all $\zeta \in \mathbb{C}^{\ell}$, where $\lambda > 0$ is a real constant. Then we say that L is an *elliptic operator of order* $2m$.

If $\ell = 1$, we say that L is a *scalar operator* and refer to the equation $Lu = 0$ as an elliptic equation; if $\ell > 1$, we refer to $Lu = 0$ as an elliptic system. If $a^{jk} = a^{kj}$, then we say the operator L is *symmetric*. If $a^{jk}_{\alpha\beta}$ is real for all α, β, j, and k, we say that L has real coefficients.

Here if α is a multiindex of length n, then $\partial^{\alpha} = \partial^{\alpha_1}_{x_1} \ldots \partial^{\alpha_n}_{x_n}$.

Now let us discuss the case of variable coefficients. A divergence-form higher-order elliptic operator is given by

$$(Lu)_j(X) = \sum_{k=1}^{\ell} \sum_{|\alpha|=|\beta|=m} \partial^{\alpha} (a^{jk}_{\alpha\beta}(X) \partial^{\beta} u_k(X)). \tag{9}$$

If the coefficients $a^{jk}_{\alpha\beta} : \mathbb{R}^n \to \mathbb{C}$ are sufficiently smooth, we may rewrite (9) in nondivergence form

$$(Lu)_j(X) = \sum_{k=1}^{\ell} \sum_{|\alpha|\leq 2m} a^{jk}_{\alpha}(X) \partial^{\alpha} u_k(X). \tag{10}$$

This form is particularly convenient when we allow equations with lower-order terms [note their appearance in (10)].

A simple criterion for ellipticity of the operators L of (10) is the condition that (8) holds with $a^{jk}_{\alpha\beta}$ replaced by $a^{jk}_{\alpha}(X)$ for any $X \in \mathbb{R}^n$, that is, that

$$\operatorname{Re} \sum_{j,k=1}^{\ell} \sum_{|\alpha|=2m} a^{jk}_{\alpha}(X) \xi^{\alpha} \zeta_j \bar{\zeta}_k \geq \lambda |\xi|^{2m} |\zeta|^2 \tag{11}$$

for any fixed $X \in \mathbb{R}^n$ and for all $\xi \in \mathbb{R}^n$, $\zeta \in \mathbb{C}^2$. This means in particular that ellipticity is only a property of the highest-order terms of (10); the value of a^{jk}_{α}, for $|\alpha| < m$, is not considered.

Returning to divergence-form operators (9), notice that in this case we have a notion of weak solution; we say that $Lu = h$ weakly if

$$\sum_{j=1}^{\ell} \int_\Omega \varphi_j \, h_j = \sum_{|\alpha|=|\beta|=m} \sum_{j,k=1}^{\ell} (-1)^m \int_\Omega \partial^\alpha \varphi_j \, a_{\alpha\beta}^{jk} \, \partial^\beta u_k \tag{12}$$

for any function $\varphi : \Omega \mapsto \mathbb{C}^\ell$ smooth and compactly supported. The right-hand side $\langle \varphi, Lu \rangle$ may be regarded as a bilinear form $A[\varphi, u]$. If L satisfies the ellipticity condition (11), then this bilinear form is positive definite. A more general ellipticity condition available in the divergence case is simply that $\langle \varphi, L\varphi \rangle \geq \lambda \|\nabla^m \varphi\|_{L^2}^2$ for all appropriate test functions φ (see formula (108) below); this condition is precisely that the form $A[\varphi, u]$ be positive definite.

As mentioned above, there is another important form of higher-order operators. Observe that second-order divergence-form equations arise from a change of variables as follows. If $\Delta u = h$, and $\tilde{u} = u \circ \rho$ for some change of variables ρ, then

$$a \operatorname{div} A \nabla \tilde{u} = \tilde{h},$$

where $a(X)$ is a real number and $A(X)$ is a real symmetric matrix (both depending only on ρ; see Fig. 1). In particular, if u is harmonic, then $\tilde{u}$ satisfies the divergence-form equation

$$\sum_{|\alpha|=|\beta|=1} \partial^\alpha (a_{\alpha\beta}(X) \, \partial^\beta \tilde{u}(X)) = 0$$

and so the study of divergence-form equations in simple domains (such as the upper half-space) encompasses the study of harmonic functions in more complicated, not necessarily smooth, domains (such as the domain above a Lipschitz graph).

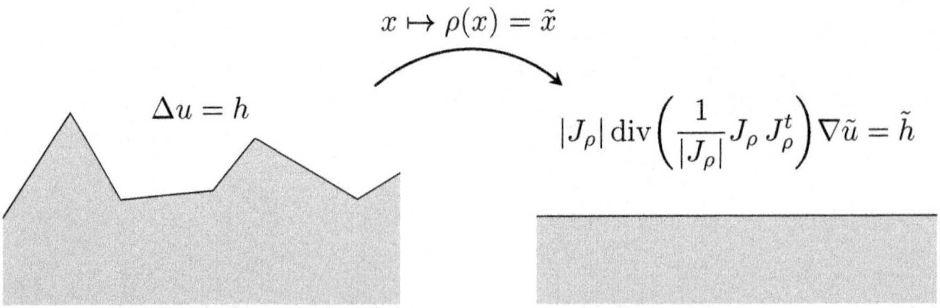

Fig. 1 The behavior of Laplace's equation after change of variables. Here $\tilde{u} = u \circ \rho$ and J_ρ is the Jacobean matrix for the change of variables ρ

If $\Delta^2 u = 0$, however, then after a change of variables $\tilde{u}$ does not satisfy a divergence-form equation [that is, an equation of the form (9)]. Instead, $\tilde{u}$ satisfies an equation of the following composition form:

$$\operatorname{div} A\nabla(a \operatorname{div} A\nabla\tilde{u}) = 0. \tag{13}$$

In section "The Dirichlet Problem for Operators in Composition Form" we shall discuss some new results pertaining to such operators.

Finally, let us mention that throughout we let C and ε denote positive constants whose value may change from line to line. We let $\fint$ denote the average integral, that is, $\fint_E f\,d\mu = \frac{1}{\mu(E)}\int_E f\,d\mu$. The only measures we will consider are the Lebesgue measure dX (on $\mathbb{R}^n$ or on domains in $\mathbb{R}^n$) or the surface measure $d\sigma$ (on the boundaries of domains).

Boundedness and Continuity of Derivatives of Solutions

Miranda–Agmon Maximum Principle and Related Geometric restrictions on the boundary

The maximum principle for harmonic functions is one of the fundamental results in the theory of elliptic equations. It holds in arbitrary domains and guarantees that every solution to the Dirichlet problem for the Laplace equation, with bounded data, is bounded. Moreover, it remains valid for all second-order divergence-form elliptic equations with real coefficients.

In the case of equations of higher order, the maximum principle has been established only in relatively nice domains. It was proven to hold for operators with smooth coefficients in smooth domains of dimension two in [102, 103], and of arbitrary dimension in [6]. In the early 1990s, it was extended to three-dimensional domains diffeomorphic to a polyhedron [72, 96] or having a Lipschitz boundary [117, 119]. However, in general domains, no direct analog of the maximum principle exist (see Problem 4.3, p. 275, in Nečas's book [113]). The increase of the order leads to the failure of the methods which work for second-order equations, and the properties of the solutions themselves become more involved.

To be more specific, the following theorem was proved by Agmon.

Theorem 3.1 ([6, Theorem 1]). *Let $m \geq 1$ be an integer. Suppose that Ω is domain with C^{2m} boundary. Let*

$$L = \sum_{|\alpha|\leq 2m} a_\alpha(X)\partial^\alpha$$

be a scalar operator of order $2m$*, where* $a_\alpha \in C^{|\alpha|}(\overline{\Omega})$*. Suppose that* L *is elliptic in the sense of* (11)*. Suppose further that solutions to the Dirichlet problem for* L *are unique.*

Then, for every $u \in C^{m-1}(\overline{\Omega}) \cap C^{2m}(\Omega)$ *that satisfies* $Lu = 0$ *in* Ω*, we have*

$$\max_{|\alpha| \leq m-1} \|\partial^\alpha u\|_{L^\infty(\Omega)} \leq C \max_{|\beta| \leq m-1} \|\partial^\beta u\|_{L^\infty(\partial\Omega)}. \tag{14}$$

We remark that the requirement that the Dirichlet problem have unique solutions is not automatically satisfied for elliptic equations with lower-order terms; for example, if λ is an eigenvalue of the Laplacian, then solutions to the Dirichlet problem for $\Delta u - \lambda u$ are not unique.

Equation (14) is called the *Agmon–Miranda maximum principle*. In [123], Šul'ce generalized this to systems of the form (10), elliptic in the sense of (11), that satisfy a positivity condition (strong enough to imply Agmon's requirement that solutions to the Dirichlet problem be unique).

Thus the Agmon–Miranda maximum principle holds for sufficiently smooth operators and domains. Moreover, for some operators, the maximum principle is valid even in domains with Lipschitz boundary, provided the dimension is small enough. We postpone a more detailed discussion of the Lipschitz case to section "The Maximum Principle in Lipschitz Domains"; here we simply state the main results. In [117, 119], Pipher and Verchota showed that the maximum principle holds for the biharmonic operator Δ^2, and more generally for the polyharmonic operator Δ^m, in bounded Lipschitz domains in $\mathbb{R}^2$ or $\mathbb{R}^3$. In [140, Sect. 8], Verchota extended this to symmetric, strongly elliptic systems with real constant coefficients in three-dimensional Lipschitz domains.

For Laplace's equation and more general second-order elliptic operators, the maximum principle continues to hold in *arbitrary* bounded domains. In contrast, the maximum principle for higher-order operators in rough domains generally *fails*.

In [98], Maz'ya et al. studied the Dirichlet problem (with zero boundary data) for constant-coefficient elliptic systems in cones. Counterexamples to (14) for systems of order $2m$ in dimension $n \geq 2m + 1$ immediately follow from their results. (See [98, formulas (1.3), (1.18), and (1.28)].) Furthermore, Pipher and Verchota constructed counterexamples to (14) for the biharmonic operator Δ^2 in dimension $n = 4$ in [116, Sect. 10], and for the polyharmonic equation $\Delta^m u = 0$ in dimension n, $4 \leq n < 2m + 1$, in [119, Theorem 2.1]. Independently Maz'ya and Rossmann showed that (14) fails in the exterior of a sufficiently thin cone in dimension n, $n \geq 4$, where L is any constant-coefficient elliptic scalar operator of order $2m \geq 4$ (without lower-order terms). See [97, Theorem 8 and Remark 3].

Moreover, with the exception of [97, Theorem 8], the aforementioned counterexamples actually provide a stronger negative result than simply the failure of the maximum principle: they show that the left-hand side of (14) may be infinite even if the data of the elliptic problem is as nice as possible, that is, smooth and compactly supported.

The counterexamples, however, pertain to high dimensions. This phenomenon raises two fundamental questions: whether the boundedness of the $(m-1)$st derivatives remains valid in dimensions $n \leq 3$, and whether there are some other, possibly lower-order, estimates that characterize the solutions when $n \geq 4$. This issue has been completely settled in [82, 84] for the polyharmonic equation in arbitrary domains.

Sharp Pointwise Estimates on the Derivatives of Solutions in Arbitrary Domains

The main results addressing pointwise bounds for solutions to the polyharmonic equation in arbitrary domains are as follows.

Theorem 3.2 ([84]). *Let Ω be a bounded domain in $\mathbb{R}^n$, $2 \leq n \leq 2m+1$, and*

$$(-\Delta)^m u = f \ in \ \Omega, \quad f \in C_0^\infty(\Omega), \quad u \in \mathring{W}_m^2(\Omega). \tag{15}$$

Then the solution to the boundary value problem (15) *satisfies*

$$\nabla^{m-n/2+1/2} u \in L^\infty(\Omega) \ when\ n\ is\ odd,\ \nabla^{m-n/2} u \in L^\infty(\Omega) \ when\ n\ is\ even. \tag{16}$$

In particular,

$$\nabla^{m-1} u \in L^\infty(\Omega) \ when\ n = 2, 3. \tag{17}$$

Here the space $\mathring{W}_m^2(\Omega)$ is, as usual, a completion of $C_0^\infty(\Omega)$ in the norm given by $\|u\|_{\mathring{W}_m^2(\Omega)} = \|\nabla^m u\|_{L^2(\Omega)}$. We note that $\mathring{W}_m^2(\Omega)$ embeds into $C^k(\Omega)$ only when k is strictly smaller than $m - \frac{n}{2}$, $n < 2m$. Thus, whether the dimension is even or odd, Theorem 3.2 gains one derivative over the outcome of Sobolev embedding.

The results of Theorem 3.2 are sharp, in the sense that the solutions do not exhibit higher smoothness than warranted by (16)–(17) in general domains. Indeed, assume that $n \in [3, 2m+1] \cap \mathbb{N}$ is odd and let $\Omega \subset \mathbb{R}^n$ be the punctured unit ball $B_1 \setminus \{O\}$, where $B_r = \{x \in \mathbb{R}^n : |x| < r\}$. Consider a function $\eta \in C_0^\infty(B_{1/2})$ such that $\eta = 1$ on $B_{1/4}$. Then let

$$u(x) := \eta(x)\, \partial_x^{m-\frac{n}{2}-\frac{1}{2}} (|x|^{2m-n}), \qquad x \in B_1 \setminus \{O\}, \tag{18}$$

where ∂_x stands for a derivative in the direction of x_i for some $i = 1, \ldots, n$. It is straightforward to check that $u \in \mathring{W}_m^2(\Omega)$ and $(-\Delta)^m u \in C_0^\infty(\Omega)$. While $\nabla^{m-\frac{n}{2}+\frac{1}{2}} u$ is bounded, the derivatives of order $m - \frac{n}{2} + \frac{3}{2}$ are not, and moreover, $\nabla^{m-\frac{n}{2}+\frac{1}{2}} u$ is not continuous at the origin. Therefore, the estimates (16)–(17) are optimal in general domains.

As for the case when n is even, the results in [72, Sect. 10.4] demonstrate that in the exterior of a ray there is an m-harmonic function behaving as $|x|^{m-\frac{n}{2}+\frac{1}{2}}$. Thus, upon truncation by the aforementioned cut-off η, one obtains a solution to (15) in $B_1\setminus\{x_1 = 0, \ldots, x_{n-1} = 0, 0 \leq x_n < 1\}$, whose derivatives of order $m-\frac{n}{2}+1$ are not bounded. More delicate examples can be obtained from our results for the Wiener test to be discussed in section "The Wiener Test: Continuity of Solutions." Those show that the derivatives of order $m - \frac{n}{2}$ need not be continuous in even dimensions. Therefore, in even dimensions (16) is a sharp property as well.

It is worth noting that the results above address also boundedness of solutions (rather than their derivatives) corresponding to the case when $m - \frac{n}{2} + \frac{1}{2} = 0$ in odd dimensions, or, respectively, $m - \frac{n}{2} = 0$ in the even case. In this respect, we would also like to mention higher dimensional results following from the Green function estimates in [89]. As will be discussed in the next section, one can show that, in addition to our results above, if $\Omega \subset \mathbb{R}^n$ is bounded for $n \leq 2m + 2$, and if u is a solution to the polyharmonic equation (15), then $u \in L^\infty(\Omega)$. This result also holds if $\Omega \subset \mathbb{R}^7$ and $m = 2$.

If $\Omega \subset \mathbb{R}^n$ is bounded and $n \geq 2m + 3$, or if $m = 2$ and $n \geq 8$, then the question of whether solutions u to (15) are bounded is open. In particular, it is not known whether solutions u to

$$\Delta^2 u = h \text{ in } \Omega, \quad u \in \mathring{W}_2^2(\Omega)$$

are bounded if $\Omega \subset \mathbb{R}^n$ for $n \geq 8$. However, there exist another fourth-order operator whose solutions are *not* bounded in higher dimensional domains. In [93], Maz'ya and Nazarov showed that if $n \geq 8$ and if $a > 0$ is large enough, then there exists an open cone $K \subset \mathbb{R}^n$ and a function $h \in C_0^\infty(\overline{K} \setminus \{0\})$ such that the solution u to

$$\Delta^2 u + a\,\partial_n^4 u = h \text{ in } K, \quad u \in \mathring{W}_2^2(K) \tag{19}$$

is unbounded near the origin.

Green Function Estimates

Theorem 3.2 has several quantitative manifestations, providing specific estimates on the solutions to (15). Most importantly, the authors established sharp pointwise estimates on Green's function of the polyharmonic operator and its derivatives, once again without any restrictions on the geometry of the domain.

To start, let us recall the definition of the fundamental solution for the polyharmonic equation (see, e.g., [10]). A fundamental solution for the m-Laplacian is a linear combination of the characteristic singular solution (defined below) and any m-harmonic function in $\mathbb{R}^n$. The characteristic singular solution is

$$C_{m,n}|x|^{2m-n}, \qquad \text{if } n \text{ is odd, or if } n \text{ is even with } n \geq 2m+2, \tag{20}$$

$$C_{m,n}|x|^{2m-n}\log|x|, \qquad \text{if } n \text{ is even with } n \leq 2m. \tag{21}$$

The exact expressions for constants $C_{m,n}$ can be found in [10], p. 8. Hereafter we will use the fundamental solution given by

$$\Gamma(x) = C_{m,n}\begin{cases} |x|^{2m-n}, & \text{if } n \text{ is odd}, \\ |x|^{2m-n}\log\frac{\operatorname{diam}\Omega}{|x|}, & \text{if } n \text{ is even and } n \leq 2m, \\ |x|^{2m-n}, & \text{if } n \text{ is even and } n \geq 2m+2. \end{cases} \tag{22}$$

As is customary, we denote the Green's function for the polyharmonic equation by $G(x,y)$, x, $y \in \Omega$, and its regular part by $S(x,y)$, that is, $S(x,y) = G(x,y) - \Gamma(x-y)$. By definition, for every fixed $y \in \Omega$ the function $G(\cdot, y)$ satisfies

$$(-\Delta_x)^m G(x,y) = \delta(x-y), \qquad x \in \Omega, \tag{23}$$

in the space $\mathring{W}^2_m(\Omega)$. Here Δ_x stands for the Laplacian in the x variable. Similarly, we use the notation Δ_y, ∇_y, ∇_x for the Laplacian and gradient in y, and gradient in x, respectively. By $d(x)$ we denote the distance from $x \in \Omega$ to $\partial\Omega$.

Theorem 3.3. *Let $\Omega \subset \mathbb{R}^n$ be an arbitrary bounded domain, $m \in \mathbb{N}$, $n \in [2, 2m+1] \cap \mathbb{N}$, and let*

$$\lambda = \begin{cases} m - n/2 + 1/2 & \text{when } n \text{ is odd}, \\ m - n/2 & \text{when } n \text{ is even}. \end{cases} \tag{24}$$

Fix any number $N \geq 25$. Then there exist a constant C depending only on m, n, N such that for every $x, y \in \Omega$ the following estimates hold.

If $n \in [3, 2m+1] \cap \mathbb{N}$ is odd, then

$$|\nabla_x^i \nabla_y^j G(x,y)| \leq C\frac{d(y)^{\lambda-j}}{|x-y|^{\lambda+n-2m+i}}, \quad \text{when } |x-y| \geq N\,d(y), \quad 0 \leq i,j \leq \lambda, \tag{25}$$

and

$$|\nabla_x^i \nabla_y^j G(x,y)| \leq C\frac{d(x)^{\lambda-i}}{|x-y|^{\lambda+n-2m+j}}, \quad \text{when } |x-y| \geq N\,d(x), \quad 0 \leq i,j \leq \lambda. \tag{26}$$

Next,

$$|\nabla_x^i \nabla_y^j G(x,y)| \leq \frac{C}{|x-y|^{n-2m+i+j}}, \tag{27}$$

when $|x-y| \leq N^{-1} \max\{d(x), d(y)\}$, *and* $i+j \geq 2m-n$, $0 \leq i,j \leq m-n/2+1/2$, *and*

$$|\nabla_x^i \nabla_y^j G(x,y)| \leq C \min\{d(x), d(y)\}^{2m-n-i-j}, \tag{28}$$

when $|x-y| \leq N^{-1} \max\{d(x), d(y)\}$, *and* $i+j \leq 2m-n$, $0 \leq i,j \leq m-n/2+1/2$. *Finally,*

$$\begin{aligned} |\nabla_x^i \nabla_y^j G(x,y)| &\leq \frac{C}{\min\{d(x), d(y), |x-y|\}^{n-2m+i+j}} \\ &\approx \frac{C}{\max\{d(x), d(y), |x-y|\}^{n-2m+i+j}}, \end{aligned} \tag{29}$$

when $N^{-1} d(x) \leq |x-y| \leq Nd(x)$ *and* $N^{-1} d(y) \leq |x-y| \leq Nd(y)$, $0 \leq i,j \leq \lambda$.

Furthermore, if $n \in [3, 2m+1] \cap \mathbb{N}$ *is odd, the estimates on the regular part of the Green function* S *are as follows:*

$$|\nabla_x^i \nabla_y^j S(x-y)| \leq \frac{C}{|x-y|^{n-2m+i+j}} \quad \text{when } |x-y| \geq N \min\{d(x), d(y)\}, \ 0 \leq i,j \leq \lambda. \tag{30}$$

Next,

$$|\nabla_x^i \nabla_y^j S(x,y)| \leq \frac{C}{\max\{d(x), d(y)\}^{n-2m+i+j}}, \tag{31}$$

when $|x-y| \leq N^{-1} \max\{d(x), d(y)\}$, *and* $i+j \geq 2m-n$, $0 \leq i,j \leq m-n/2+1/2$, *and*

$$|\nabla_x^i \nabla_y^j S(x,y)| \leq C \min\{d(x), d(y)\}^{2m-n-i-j}, \tag{32}$$

when $|x-y| \leq N^{-1} \max\{d(x), d(y)\}$, *and* $i+j \leq 2m-n$, $0 \leq i,j \leq m-n/2+1/2$. *Finally,*

$$\begin{aligned} |\nabla_x^i \nabla_y^j S(x,y)| &\leq \frac{C}{\min\{d(x), d(y), |x-y|\}^{n-2m+i+j}} \\ &\approx \frac{C}{\max\{d(x), d(y), |x-y|\}^{n-2m+i+j}}, \end{aligned} \tag{33}$$

when $N^{-1} d(x) \leq |x-y| \leq Nd(x)$ *and* $N^{-1} d(y) \leq |x-y| \leq Nd(y)$, $0 \leq i,j \leq m-n/2+1/2$.

If $n \in [2, 2m] \cap \mathbb{N}$ *is even, then* (25)–(26) *and* (29) *are valid with* $\lambda = m - \frac{n}{2}$, *and*

$$|\nabla_x^i \nabla_y^j G(x,y)| \leq C \min\{d(x), d(y)\}^{2m-n-i-j} \left(C' + \log \frac{\min\{d(x), d(y)\}}{|x-y|} \right), \tag{34}$$

when $|x-y| \leq N^{-1} \max\{d(x), d(y)\}$ *and* $0 \leq i,j \leq m-n/2$.

Furthermore, if $n \in [2,2m] \cap \mathbb{N}$ is even, the estimates on the regular part of the Green function S are as follows:

$$|\nabla_x^i \nabla_y^j S(x-y)| \leq C\,|x-y|^{-n+2m-i-j}\left(C' + \log\frac{\operatorname{diam}(\Omega)}{|x-y|}\right) \tag{35}$$

when $|x-y| \geq N \min\{d(x), d(y)\}$, $0 \leq i,j \leq m-n/2$. Next,

$$|\nabla_x^i \nabla_y^j S(x,y)| \leq C \min\{d(x),d(y)\}^{2m-n-i-j}\left(C' + \log\frac{\operatorname{diam}\Omega}{\max\{d(x),d(y)\}}\right), \tag{36}$$

when $|x-y| \leq N^{-1}\max\{d(x),d(y)\}$, $0 \leq i,j \leq m-n/2$. Finally,

$$|\nabla_x^i \nabla_y^j S(x,y)| \leq C \min\{d(x),d(y),|x-y|\}^{2m-n-i-j}\times$$
$$\times\left(C' + \log\frac{\operatorname{diam}\Omega}{\max\{d(x),d(y),|x-y|\}^{n-2m+i+j}}\right) \tag{37}$$

when $N^{-1}d(x) \leq |x-y| \leq Nd(x)$ and $N^{-1}d(y) \leq |x-y| \leq Nd(y)$, $0 \leq i$, $j \leq m-n/2$.

We would like to highlight the most important case of the estimates above, pertaining to the highest-order derivatives.

Corollary 3.4. *Let $\Omega \subset \mathbb{R}^n$ be an arbitrary bounded domain. If $n \in [3,2m+1] \cap \mathbb{N}$ is odd, then for all x, $y \in \Omega$,*

$$\left|\nabla_x^{m-\frac{n}{2}+\frac{1}{2}} \nabla_y^{m-\frac{n}{2}+\frac{1}{2}} (G(x,y) - \Gamma(x-y))\right| \leq \frac{C}{\max\{d(x),d(y),|x-y|\}}, \tag{38}$$

and, in particular,

$$\left|\nabla_x^{m-\frac{n}{2}+\frac{1}{2}} \nabla_y^{m-\frac{n}{2}+\frac{1}{2}} G(x,y)\right| \leq \frac{C}{|x-y|}. \tag{39}$$

If $n \in [2,2m] \cap \mathbb{N}$ is even, then for all x, $y \in \Omega$,

$$\left|\nabla_x^{m-\frac{n}{2}} \nabla_y^{m-\frac{n}{2}} (G(x,y) - \Gamma(x-y))\right|$$
$$\leq C\log\left(1 + \frac{\operatorname{diam}\Omega}{\max\{d(x),d(y),|x-y|\}}\right), \tag{40}$$

and

$$\left|\nabla_x^{m-\frac{n}{2}} \nabla_y^{m-\frac{n}{2}} G(x,y)\right| \leq C\log\left(1 + \frac{\min\{d(x),d(y)\}}{|x-y|}\right). \tag{41}$$

The constant C in (38)–(41) *depends on m and n only. In particular, it does not depend on the size or the geometry of the domain Ω.*

We mention that the pointwise bounds on the absolute value of Green's function itself have been treated previously in dimensions $2m + 1$ and $2m + 2$ for $m > 2$ and dimensions $5, 6, 7$ for $m = 2$ in [89, Sect. 10] (see also [86]). In particular, in [89, Sect. 10], Maz'ya showed that the Green's function $G_m(x, y)$ for Δ^m in an arbitrary bounded domain $\Omega \subset \mathbb{R}^n$ satisfies

$$|G_m(x, y)| \le \frac{C(n)}{|x - y|^{n-2m}} \tag{42}$$

if $n = 2m + 1$ or $n = 2m + 2$. If $m = 2$, then (42) also holds in dimension $n = 7 = 2m + 3$ (cf. [86]). Whether (42) holds in dimension $n \ge 8$ (for $m = 2$) or $n \ge 2m + 3$ (for $m > 2$) is an open problem; see [89, Problem 2]. Also, similarly to the case of general solutions discussed above, there exist results for Green functions in smooth domains [44, 73, 134, 135], in conical domains [72, 94], and in polyhedra [96].

Furthermore, using standard techniques, the Green's function estimates can be employed to establish the bounds on the solution to (15) for general classes of data f, such as L^p for a certain range of p, Lorentz spaces, etc. A sample statement to this effect is as follows.

Proposition 3.5. *Let $\Omega \subset \mathbb{R}^n$ be an arbitrary bounded domain, $m \in \mathbb{N}$, $n \in [2, 2m + 1] \cap \mathbb{N}$, and let λ retain the significance of* (24). *Consider the boundary value problem*

$$(-\Delta)^m u = \sum_{|\alpha| \le \lambda} c_\alpha \partial^\alpha f_\alpha, \quad u \in \mathring{W}^2_m(\Omega). \tag{43}$$

Then the solution satisfies the following estimates.

If $n \in [3, 2m + 1] \cap \mathbb{N}$ is odd, then for all $x \in \Omega$,

$$|\nabla^{m-\frac{n}{2}+\frac{1}{2}} u(x)| \le C_{m,n} \sum_{|\alpha| \le m-\frac{n}{2}+\frac{1}{2}} \int_\Omega \frac{d(y)^{m-\frac{n}{2}+\frac{1}{2}-|\alpha|}}{|x - y|} |f_\alpha(y)|\, dy, \tag{44}$$

whenever the integrals on the right-hand side of (44) *are finite. In particular,*

$$\|\nabla^{m-\frac{n}{2}+\frac{1}{2}} u\|_{L^\infty(\Omega)} \le C_{m,n,\Omega} \sum_{|\alpha| \le m-\frac{n}{2}+\frac{1}{2}} \|d(\cdot)^{m-\frac{n}{2}-\frac{1}{2}-|\alpha|} f_\alpha\|_{L^p(\Omega)}, \quad p > \frac{n}{n-1}, \tag{45}$$

provided that the norms on the right-hand side of (45) *are finite.*

If $n \in [2, 2m] \cap \mathbb{N}$ is even, then for all $x \in \Omega$,

$$|\nabla^{m-\frac{n}{2}} u(x)|$$

$$\leq C_{m,n} \sum_{|\alpha| \leq m-\frac{n}{2}} \int_\Omega d(y)^{m-\frac{n}{2}-|\alpha|} \log\left(1 + \frac{d(y)}{|x-y|}\right) |f_\alpha(y)|\, dy, \tag{46}$$

whenever the integrals on the right-hand side of (46) *are finite. In particular,*

$$\|\nabla^{m-\frac{n}{2}} u\|_{L^\infty(\Omega)} \leq C_{m,n,\Omega} \sum_{|\alpha| \leq m-\frac{n}{2}} \|d(\cdot)^{m-\frac{n}{2}-|\alpha|} f_\alpha\|_{L^p(\Omega)}, \ p > 1, \tag{47}$$

provided that the norms on the right-hand side of (47) *are finite.*

The constants $C_{m,n}$ above depend on m and n only, while the constants denoted by $C_{m,n,\Omega}$ depend on m, n, and the diameter of the domain Ω.

By the same token, if (42) holds, then solutions to (15) satisfy

$$\|u\|_{L^\infty(\Omega)} \leq C(m,n,p)\, \mathrm{diam}(\Omega)^{2m-n/p} \|f\|_{L^p(\partial\Omega)}$$

provided $p > n/2m$ (see, e.g., [89, Sect. 2]). Thus, e.g., if $\Omega \subset \mathbb{R}^n$ is bounded for $n = 2m+2$, and if u satisfies (15) for a reasonably nice function f, then $u \in L^\infty(\Omega)$. This result also holds if $\Omega \subset \mathbb{R}^7$ and $m = 2$. This complements the results in Theorem 3.2, as discussed in section "Sharp Pointwise Estimates on the Derivatives of Solutions in Arbitrary Domains."

To conclude our discussion of Green's functions, we mention two results from [106]; these results are restricted to relatively well-behaved domains. In [106], D. Mitrea and I. Mitrea showed that, if Ω is a bounded Lipschitz domain in $\mathbb{R}^3$, and G denotes the Green's function for the bilaplacian Δ^2, then the estimates

$$\nabla^2 G(x,\cdot) \in L^3(\Omega), \quad \mathrm{dist}(\cdot, \partial\Omega)^{-\alpha} \nabla G(x,\cdot) \in L^{3/\alpha,\infty}$$

hold, uniformly in $x \in \Omega$, for all $0 < \alpha \leq 1$.

Moreover, they considered more general elliptic systems. Suppose that L is an arbitrary elliptic operator of order $2m$ with constant coefficients, as defined by Definition 2.1, and that G denotes the Green's function for L. Suppose that $\Omega \subset \mathbb{R}^n$, for $n > m$, is a Lipschitz domain, and that the unit outward normal ν to Ω lies in the Sarason space $VMO(\partial\Omega)$ of functions of vanishing mean oscillations on $\partial\Omega$. Then the estimates

$$\nabla^m G(x,\cdot) \in L^{\frac{n}{n-m},\infty}(\Omega), \tag{48}$$

$$\mathrm{dist}(\cdot, \partial\Omega)^{-\alpha} \nabla^{m-1} G(x,\cdot) \in L^{\frac{n}{n-m-1+\alpha},\infty}(\Omega)$$

hold, uniformly in $x \in \Omega$, for any $0 \leq \alpha \leq 1$.

The Wiener Test: Continuity of Solutions

In this section, we discuss conditions that ensure that solutions (or appropriate gradients of solutions) are continuous up to the boundary. These conditions parallel the famous result of Wiener, who in 1924 formulated a criterion that ensured continuity of *harmonic* functions at boundary points [143]. Wiener's criterion has been extended to a variety of second-order elliptic and parabolic equations ([3, 41, 47, 50, 51, 74, 76, 78, 137]; see also the review papers [2, 87]). However, as with the maximum principle, extending this criterion to higher-order elliptic equations is a subtle matter, and many open questions remain.

We begin by stating the classical Wiener criterion for the Laplacian. If $\Omega \subset \mathbb{R}^n$ is a domain and $Q \in \partial\Omega$, then Q is called *regular* for the Laplacian if every solution u to

$$\Delta u = h \text{ in } \Omega, \quad u \in \mathring{W}_1^2(\Omega)$$

for $h \in C_0^\infty(\Omega)$ satisfies $\lim_{X\to Q} u(X) = 0$. According to Wiener's theorem [143], the boundary point $Q \in \partial\Omega$ is regular if and only if the equation

$$\int_0^1 \operatorname{cap}_2(\overline{B(Q,s)} \setminus \Omega) s^{1-n}\, ds = \infty \tag{49}$$

holds, where

$$\operatorname{cap}_2(K) = \inf\Big\{\|u\|^2_{L^2(\mathbb{R}^n)} + \|\nabla u\|^2_{L^2(\mathbb{R}^n)} : u \in C_0^\infty(\mathbb{R}^n),\ u \geq 1 \text{ on } K\Big\}. \tag{50}$$

For example, suppose Ω satisfies the exterior cone condition at Q. That is, suppose there is some open cone K with vertex at Q and some $\varepsilon > 0$ such that $K \cap B(Q,\varepsilon) \subset \Omega^C$. It is elementary to show that $\operatorname{cap}_2(\overline{B(Q,s)} \setminus \Omega) \geq C(K)s^{n-2}$ for all $0 < s < \varepsilon$, and so (49) holds and Q is regular. Regularity of such points was known prior to Wiener (see [75, 120, 144]) and provided inspiration for the formulation of the Wiener test.

By [76], if $L = -\operatorname{div} A\nabla$ is a second-order divergence-form operator, where the matrix $A(X)$ is bounded, measurable, real, symmetric, and elliptic, then $Q \in \partial\Omega$ is regular for L if and only if Q and Ω satisfy (49). In other words, $Q \in \partial\Omega$ is regular for the Laplacian if and only if it is regular for all such operators. Similar results hold for some other classes of second-order equations; see, for example, [41, 50], or [47].

One would like to consider the Wiener criterion for higher-order elliptic equations, and that immediately gives rise to the question of natural generalization of the concept of a regular point. The Wiener criterion for the second-order PDEs ensures, in particular, that weak $\mathring{W}_1^2$ solutions are *classical*. That is, the solution approaches its boundary values in the pointwise sense (continuously). From that point of view, one would extend the concept of regularity of a boundary point as continuity of derivatives of order $m-1$ of the solution to an equation of order $2m$ up to the boundary. On the other hand, as we discussed in the previous section, even the boundedness of solutions cannot be guaranteed in general, and thus, in some dimensions the study of the continuity up to the boundary for solutions themselves is also very natural. We begin with the latter question, as it is better understood.

Let us first define a regular point for an arbitrary differential operator L of order $2m$ analogously to the case of the Laplacian, by requiring that every solution u to

$$Lu = h \text{ in } \Omega, \quad u \in \mathring{W}^2_m(\Omega) \tag{51}$$

for $h \in C_0^\infty(\Omega)$ satisfies $\lim_{X\to Q} u(X) = 0$. Note that by the Sobolev embedding theorem, if $\Omega \subset \mathbb{R}^n$ for $n \leq 2m-1$, then every $u \in \mathring{W}^2_m(\Omega)$ is Hölder continuous on $\overline{\Omega}$ and so satisfies $\lim_{X\to Q} u(X) = 0$ at every point $Q \in \partial\Omega$. Thus, we are only interested in continuity of the solutions at the boundary when $n \geq 2m$.

In this context, the appropriate concept of capacity is the potential-theoretic Riesz capacity of order $2m$, given by

$$\operatorname{cap}_{2m}(K) = \inf\Big\{ \sum_{0\leq|\alpha|\leq m} \|\partial^\alpha u\|^2_{L^2(\mathbb{R}^n)} : u \in C_0^\infty(\mathbb{R}^n),\ u \geq 1 \text{ on } K\Big\}. \tag{52}$$

The following is known. If $m \geq 3$, and if $\Omega \subset \mathbb{R}^n$ for $n = 2m, 2m+1$, or $2m+2$, or if $m = 2$ and $n = 4, 5, 6$, or 7, then $Q \in \partial\Omega$ is regular for Δ^m if and only if

$$\int_0^1 \operatorname{cap}_{2m}(\overline{B(Q,s)} \setminus \Omega) s^{2m-n-1}\, ds = \infty. \tag{53}$$

The biharmonic case was treated in [85, 86], and the polyharmonic case for $m \geq 3$ in [88, 91].

Let us briefly discuss the method of the proof in order to explain the restrictions on the dimension. Let L be an arbitrary elliptic operator, and let F be the fundamental solution for L in $\mathbb{R}^n$ with pole at Q. We say that L is positive with weight F if, for all $u \in C_0^\infty(\mathbb{R}^n \setminus \{Q\})$, we have that

$$\int_{\mathbb{R}^n} Lu(X)\cdot u(X)\, F(X)\, dX \geq c \sum_{k=1}^m \int_{\mathbb{R}^n} \left|\nabla^k u(X)\right|^2 |X|^{2k-n}\, dX. \tag{54}$$

The biharmonic operator is positive with weight F in dimension n if $4 \leq n \leq 7$, and the polyharmonic operator Δ^m, $m \geq 3$, is positive with weight F in dimension $2m \leq n \leq 2m+2$. (The Laplacian Δ is positive with weight F in any dimension.) The biharmonic operator Δ^2 is not positive with weight F in dimensions $n \geq 8$, and Δ^m is not positive with weight F in dimension $n \geq 2m+3$. See [88, Propositions 1 and 2].

The proof of the Wiener criterion for the polyharmonic operator required positivity with weight F. In fact, it turns out that positivity with weight F suffices to provide a Wiener criterion for an *arbitrary* scalar elliptic operator with constant coefficients.

Theorem 3.6 ([90, Theorems 1 and 2]). *Suppose $\Omega \subset \mathbb{R}^n$ and that L is a scalar elliptic operator of order $2m$ with constant real coefficients, as defined by Definition 2.1.*

If $n = 2m$, then $Q \in \partial\Omega$ is regular for L if and only if (53) *holds.*

If $n \geq 2m+1$, and if the condition (54) *holds, then again $Q \in \partial\Omega$ is regular for L if and only if* (53) *holds.*

This theorem is also valid for certain variable-coefficient operators in divergence form; see the remark at the end of [88, Sect. 5].

Similar results have been proven for some second-order elliptic *systems*. In particular, for the Lamé system $Lu = \Delta u + \alpha \operatorname{grad} \operatorname{div} u$, $\alpha > -1$, positivity with weight F and Wiener criterion have been established for a range of α close to zero, that is, when the underlying operator is close to the Laplacian [77]. It was also shown that positivity with weight F may in general fail for the Lamé system. Since the present review is restricted to the higher-order operators, we shall not elaborate on this point and instead refer the reader to [77] for more detailed discussion.

In the absence of the positivity condition (54), the situation is much more involved. Let us point out first that the condition (54) is *not* necessary for regularity of a boundary point, that is, the continuity of the solutions. There exist fourth-order elliptic operators that are not positive with weight F whose solutions exhibit nice behavior near the boundary; there exist other such operators whose solutions exhibit very bad behavior near the boundary.

Specifically, recall that (54) fails for $L = \Delta^2$ in dimension $n \geq 8$. Nonetheless, solutions to $\Delta^2 u = h$ are often well-behaved near the boundary. By [95], the vertex of a cone is regular for the bilaplacian in any dimension. Furthermore, if the capacity condition (53) holds with $m = 2$, then by [90, Sect. 10], any solution u to

$$\Delta^2 u = h \text{ in } \Omega, \quad u \in \mathring{W}_2^2(\Omega)$$

for $h \in C_0^\infty(\Omega)$ satisfies $\lim_{X\to Q} u(X) = 0$ provided the limit is taken along a *nontangential* direction.

Conversely, if $n \geq 8$ and $L = \Delta^2 + a\partial_n^4$, then by [93], there exists a cone K and a function $h \in C_0^\infty(\overline{K} \setminus \{0\})$ such that the solution u to (19) is not only discontinuous but *unbounded* near the vertex of the cone. We remark that a careful examination of the proof in [93] implies that solutions to (19) are unbounded even along some nontangential directions.

Thus, conical points in dimension eight are regular for the bilaplacian and irregular for the operator $\Delta^2 + a\,\partial_n^4$. Hence, a relevant Wiener condition *must* use different capacities for these two operators. This is a striking contrast with the second-order case, where the same capacity condition implies regularity for all divergence-form operators, even with variable coefficients.

This concludes the discussion of regularity in terms of continuity of the solution. We now turn to regularity in terms of continuity of the $(m - 1)$st derivatives. Unfortunately, much less is known in this case.

The Higher-Order Wiener Test: Continuity of Derivatives of Polyharmonic Functions

The most natural generalization of the Wiener test to the higher-order scenario concerns the continuity of the derivatives of the solutions, rather than solutions themselves, as derivatives constitute part of the boundary data. However, a necessary prerequisite for such results is boundedness of the corresponding derivatives of the solutions—an extremely delicate matter in its own right as detailed in section "Sharp Pointwise Estimates on the Derivatives of Solutions in Arbitrary Domains." In the context of the polyharmonic equation, Theorem 3.2 has set the stage for an extensive investigation of the Wiener criterion and, following earlier results in [81], the second author of this paper and Maz'ya have recently obtained a full extension of the Wiener test to the polyharmonic context in [83]. One of the most intricate issues is the proper definition of the polyharmonic capacity, and we start by addressing it.

At this point Theorem 3.2 finally sets the stage for a discussion of the *Wiener test* for continuity of the corresponding derivatives of the solution, which brings us to the main results of the present paper.

Assume that $m \in \mathbb{N}$ and $n \in [2, 2m + 1] \cap \mathbb{N}$. Let us denote by Z the following set of indices:

$$Z = \{0, 1, \ldots, m - n/2 + 1/2\} \quad \text{if } n \text{ is odd,} \tag{55}$$

$$Z = \{-n/2 + 2, -n/2 + 4, \ldots, m - n/2 - 2, m - n/2\} \cap (\mathbb{N} \cup \{0\}) \quad \text{if } n \text{ is even, } m \text{ is even,} \tag{56}$$

$$Z = \{-n/2 + 1, -n/2 + 3, \ldots, m - n/2 - 2, m - n/2\} \cap (\mathbb{N} \cup \{0\}) \quad \text{if } n \text{ is even, } m \text{ is odd.} \tag{57}$$

Now let Π be the space of linear combinations of spherical harmonics

$$P(x) = \sum_{p \in Z} \sum_{l=-p}^{p} b_{pl} Y_l^p(x/|x|), \qquad b_{pl} \in \mathbb{R}, \quad x \in \mathbb{R}^n \setminus \{O\}, \tag{58}$$

with the norm

$$\|P\|_\Pi := \left(\sum_{p \in Z} \sum_{l=-p}^{p} b_{pl}^2 \right)^{\frac{1}{2}} \quad \text{and} \quad \Pi_1 := \{P \in \Pi : \|P\|_\Pi = 1\}. \tag{59}$$

Then, given $P \in \Pi_1$, an open set D in $\mathbb{R}^n$ such that $O \in \mathbb{R}^n \setminus D$, and a compactum K in D, we define

$$\begin{aligned} &\mathrm{Cap}_P\,(K, D) \\ &:= \inf \left\{ \int_D |\nabla^m u(x)|^2\,dx : \; u \in \mathring{W}^2_m(D), \; u{=}P \text{ in a neighborhood of } K \right\}, \end{aligned} \tag{60}$$

with

$$\mathrm{Cap}\,(K, D) := \inf_{P \in \Pi_1} \mathrm{Cap}_P\,(K, D). \tag{61}$$

In the context of the Wiener test, we will be working extensively with the capacity of the complement of a domain $\Omega \subset \mathbb{R}^n$ in the balls $B_{2^{-j}}$, $j \in \mathbb{N}$, and even more so, in dyadic annuli, $C_{2^{-j},2^{-j+2}}$, $j \in \mathbb{N}$, where $C_{s,as} := \{x \in \mathbb{R}^n : s < |x| < as\}$, s, $a > 0$. As is customary, we will drop the reference to the "ambient" set

$$\mathrm{Cap}_P\,(\overline{C_{2^{-j},2^{-j+2}}} \setminus \Omega) := \mathrm{Cap}_P\,(\overline{C_{2^{-j},2^{-j+2}}} \setminus \Omega, C_{2^{-j-2},2^{-j+4}}), \quad j \in \mathbb{N}, \tag{62}$$

and will drop the similar reference for Cap. In fact, it will be proven below that there are several equivalent definitions of capacity, in particular, for any $n \in [2, 2m+1]$ and for any $s > 0$, $a > 0$, $K \subset \overline{C_{s,as}}$, we have

$$\begin{aligned} &\mathrm{Cap}_P(K, C_{s/2,2as}) \\ &\qquad \approx \inf \left\{ \sum_{k=0}^m \int_{\mathbb{R}^n} \frac{|\nabla^k u(x)|^2}{|x|^{2m-2k}}\,dx : \; u \in \mathring{W}^2_m(\mathbb{R}^n \setminus \{O\}), \right. \\ &\qquad\qquad \left. u = P \text{ in a neighborhood of } K \right\}. \end{aligned} \tag{63}$$

In the case when the dimension is odd, also

$$\mathrm{Cap}_P\,(\overline{C_{s,as}} \setminus \Omega, C_{s/2,2as}) \approx \mathrm{Cap}_P\,(\overline{C_{s,as}} \setminus \Omega, \mathbb{R}^n \setminus \{O\}).$$

Thus, either of the above can be used in (62), as convenient.

Let Ω be a domain in $\mathbb{R}^n$, $n \geq 2$. The point $Q \in \partial\Omega$ is *k-regular* with respect to the domain Ω and the operator $(-\Delta)^m$, $m \in \mathbb{N}$, if the solution to the boundary problem

$$(-\Delta)^m u = f \text{ in } \Omega, \quad f \in C_0^\infty(\Omega), \quad u \in \mathring{W}^2_m(\Omega), \tag{64}$$

satisfies the condition

$$\nabla^k u(x) \to 0 \text{ as } x \to Q,\ x \in \Omega, \tag{65}$$

that is, all partial derivatives of u of order k are continuous. Otherwise, we say that $Q \in \partial\Omega$ is k-irregular.

Theorem 3.7 ([81, 83]). *Let Ω be an arbitrary open set in $\mathbb{R}^n$, $m \in \mathbb{N}$, $2 \le n \le 2m+1$. Let λ be given by*

$$\lambda = \begin{cases} m - n/2 + 1/2 & \text{when } n \text{ is odd,} \\ m - n/2 & \text{when } n \text{ is even.} \end{cases} \tag{66}$$

If

$$\sum_{j=0}^{\infty} 2^{-j(2m-n)} \inf_{P \in \Pi_1} \mathrm{Cap}_P\,(\overline{C_{2^{-j},2^{-j+2}}} \setminus \Omega) = +\infty, \quad \text{when } n \text{ is odd,} \tag{67}$$

and

$$\sum_{j=0}^{\infty} j\, 2^{-j(2m-n)} \inf_{P \in \Pi_1} \mathrm{Cap}_P\,(\overline{C_{2^{-j},2^{-j+2}}} \setminus \Omega) = +\infty, \quad \text{when } n \text{ is even,} \tag{68}$$

then the point O is λ-regular with respect to the domain Ω and the operator $(-\Delta)^m$.

Conversely, if the point $O \in \partial\Omega$ is λ-regular with respect to the domain Ω and the operator $(-\Delta)^m$, then

$$\inf_{P \in \Pi_1} \sum_{j=0}^{\infty} 2^{-j(2m-n)} \mathrm{Cap}_P\,(\overline{C_{2^{-j},2^{-j+2}}} \setminus \Omega) = +\infty, \quad \text{when } n \text{ is odd,} \tag{69}$$

and

$$\inf_{P \in \Pi_1} \sum_{j=0}^{\infty} j\, 2^{-j(2m-n)} \mathrm{Cap}_P\,(\overline{C_{2^{-j},2^{-j+2}}} \setminus \Omega) = +\infty, \quad \text{when } n \text{ is even.} \tag{70}$$

Here, as before, $C_{2^{-j},2^{-j+2}}$ is the annulus $\{x \in \mathbb{R}^n : 2^{-j} < |x| < 2^{-j+2}\}$, $j \in \mathbb{N} \cup \{0\}$.

Let us now discuss the results of Theorem 3.7 in more detail. This was the first treatment of the continuity of derivatives of an elliptic equation of order $m \ge 2$ at the boundary, and the first time the capacity (60) appeared in the literature. When applied to the case $m = 1$, $n = 3$, it yields the classical Wiener criterion for continuity of a harmonic function [cf. (49)]. Furthermore, as discussed in the previous section, continuity of the solution itself (rather than its derivatives) has been previously treated for the polyharmonic equation, and for $(-\Delta)^m$ the resulting criterion also follows from Theorem 3.7, in particular, when $m = 2n$, the new notion of capacity (55)–(59) coincides with the potential-theoretical Bessel capacity used in [90]. In the case $\lambda = 0$, covering both of the above, necessary and sufficient condition in Theorem 3.7 are trivially the same, as $P \equiv 1$ when $n = 2m$ in even dimensions and $n = 2m + 1$ in odd ones. For lower dimensions n the discrepancy is not artificial, for, e.g., (67) may fail to be necessary as was shown in [81].

It is not difficult to verify that we also recover known bounds in Lipschitz and in smooth domains, as the capacity of a cone, and hence capacity of an intersection with a complement of a Lipschitz domain, assures divergence of the series in (67)–(68). On the other hand, given Theorem 3.7 and following considerations traditional in this context (choosing sufficiently small balls in the consecutive annuli to constitute a complement of the domain), we can build a set with a convergent capacitory integral and, respectively, an irregular solution with discontinuous derivatives of order λ at the point O. Note that this yields further sharpness of the results of Theorem 3.2. In particular, in even dimensions, it is a stronger counterexample than that of a continuum (not only $m - n/2 + 1$ derivatives are not bounded, but $m - n/2$ derivatives might be discontinuous). We refer the reader back to section "Sharp Pointwise Estimates on the Derivatives of Solutions in Arbitrary Domains" for more details.

One of the most difficult aspects of proof of Theorem 3.7 is finding a correct notion of polyharmonic capacity and understanding its key properties. A peculiar choice of linear combinations of spherical harmonics [see (55)–(57) and (58)] is crucial at several stages of the argument, specific to the problem at hand, and no alterations would lead to reasonable necessary and sufficient conditions. At the same time, the new capacity and the notion of higher-order regularity sometimes exhibit surprising properties, such as sensitivity to the affine changes of coordinates [81], or the aforementioned fact that in sharp contrast with the second-order case [76], one does not expect the same geometric conditions to be responsible for regularity of solutions to all higher-order elliptic equations.

It is interesting to point out that despite fairly involved definitions, capacitory conditions may reduce to a simple and concise criterion, e.g., in a case of a graph. To be precise, let $\Omega \subset \mathbb{R}^3$ be a domain whose boundary is the graph of a function φ, and let ω be its modulus of continuity. If

$$\int_0^1 \frac{t\,dt}{\omega^2(t)} = \infty, \tag{71}$$

then every solution to the biharmonic equation satisfies $\nabla u \in C(\overline{\Omega})$. Conversely, for every ω such that the integral in (71) is convergent, there exists a $C^{0,\omega}$ domain and a solution u of the biharmonic equation such that $\nabla u \notin C(\overline{\Omega})$. In particular, as expected, the gradient of a solution to the biharmonic equation is always bounded in Lipschitz domains and is not necessarily bounded in a Hölder domain. Moreover, one can deduce from (71) that the gradient of a solution is always bounded, e.g., in a domain with $\omega(t) \approx t \log^{1/2} t$, which is not Lipschitz, and might fail to be bounded in a domain with $\omega(t) \approx t \log t$. More properties of the new capacity and examples can be found in [81, 83].

Boundary Value Problems in Lipschitz Domains for Elliptic Operators with Constant Coefficients

The maximum principle (14) provides estimates on solutions whose boundary data lies in L^∞. Recall that for second-order partial differential equations with real coefficients, the maximum principle is valid in arbitrary bounded domains. The corresponding sharp estimates for boundary data in L^p, $1 < p < \infty$, are much more delicate. They are *not* valid in arbitrary domains, even for harmonic functions, and they depend in a delicate way on the geometry of the boundary. At present, boundary value problems for the Laplacian and for general real symmetric elliptic operators of the second order are fairly well understood on Lipschitz domains. See, in particular, [65].

We consider biharmonic functions and more general higher-order elliptic equations. The question of estimates on biharmonic functions with data in L^p was raised by Rivière in the 1970s [28], and later Kenig redirected it towards Lipschitz domains in [64, 65]. The sharp range of well-posedness in L^p, even for biharmonic functions, remains an open problem (see [65, Problem 3.2.30]). In this section we shall review the current state of the art in the subject, the main techniques that have been successfully implemented, and their limitations in the higher-order case.

Most of the results we will discuss are valid in Lipschitz domains, defined as follows.

Definition 4.1. A domain $\Omega \subset \mathbb{R}^n$ is called a *Lipschitz domain* if, for every $Q \in \partial\Omega$, there is a number $r > 0$, a Lipschitz function $\varphi : \mathbb{R}^{n-1} \mapsto \mathbb{R}$ with $\|\nabla\varphi\|_{L^\infty} \leq M$, and a rectangular coordinate system for $\mathbb{R}^n$ such that

$$B(Q,r) \cap \Omega = \{(x,s) : x \in \mathbb{R}^{n-1},\ s \in \mathbb{R},\ |(x,s) - Q)| < r, \text{ and } s > \varphi(x)\}.$$

If we may take the functions φ to be C^k (that is, to possess k continuous derivatives), we say that Ω is a C^k *domain.*

The outward normal vector to Ω will be denoted ν. The surface measure will be denoted σ, and the tangential derivative along $\partial\Omega$ will be denoted ∇_τ.

In this paper, we will assume that all domains under consideration have connected boundary. Furthermore, if $\partial\Omega$ is unbounded, we assume that there is a single Lipschitz function φ and coordinate system that satisfies the conditions given above; that is, we assume that Ω is the domain above (in some coordinate system) the graph of a Lipschitz function.

In order to properly state boundary value problems on Lipschitz domains, we will need the notions of nontangential convergence and nontangential maximal function.

In this and subsequent sections we say that $u\big|_{\partial\Omega} = f$ if f is the *nontangential limit* of u, that is, if

$$\lim_{X\to Q,\, X\in\Gamma(Q)} u(X) = f(Q)$$

for almost every $(d\sigma)$ $Q \in \partial\Omega$, where $\Gamma(Q)$ is the *nontangential cone*

$$\Gamma(Q) = \{Y \in \Omega : \operatorname{dist}(Y, \partial\Omega) < (1+a)|X - Y|\}. \tag{72}$$

Here $a > 0$ is a positive parameter; the exact value of a is usually irrelevant to applications. The *nontangential maximal function* is given by

$$NF(Q) = \sup\{|F(X)| : X \in \Gamma(Q)\}. \tag{73}$$

The normal derivative of u of order m is defined as

$$\partial_\nu^m u(Q) = \sum_{|\alpha|=m} \nu(Q)^\alpha \frac{m!}{\alpha!} \partial^\alpha u(Q),$$

where $\partial^\alpha u(Q)$ is taken in the sense of nontangential limits as usual.

The Dirichlet Problem: Definitions, Layer Potentials, and Some Well-Posedness Results

We say that the L^p-Dirichlet problem for the biharmonic operator Δ^2 in a domain Ω is well-posed if there exist a constant $C > 0$ such that, for every $f \in W_1^p(\partial\Omega)$ and every $g \in L^p(\partial\Omega)$, there exist a unique function u that satisfies

$$\begin{cases} \Delta^2 u = 0 & \text{in } \Omega, \\ u = f & \text{on } \partial\Omega, \\ \partial_\nu u = g & \text{on } \partial\Omega, \\ \|N(\nabla u)\|_{L^p(\partial\Omega)} \le C\|g\|_{L^p(\partial\Omega)} + C\|\nabla_\tau f\|_{L^p(\partial\Omega)}. \end{cases} \tag{74}$$

The L^p-Dirichlet problem for the polyharmonic operator Δ^m is somewhat more involved, because the notion of boundary data is necessarily more subtle. We say that the L^p-Dirichlet problem for Δ^m in a domain Ω is well-posed if there exist a constant $C > 0$ such that, for every $g \in L^p(\partial\Omega)$ and every $\dot{f}$ in the Whitney–Sobolev space $WA_{m-1}^p(\partial\Omega)$, there exist a unique function u that satisfies

$$\begin{cases} \Delta^m u = 0 & \text{in } \Omega, \\ \partial^\alpha u\big|_{\partial\Omega} = f_\alpha & \text{for all } 0 \le |\alpha| \le m-2, \\ \partial_\nu^{m-1} u = g & \text{on } \partial\Omega, \\ \|N(\nabla^{m-1}u)\|_{L^p(\partial\Omega)} \le C\|g\|_{L^p(\partial\Omega} + C \sum\limits_{|\alpha|=m-2} \|\nabla_\tau f_\alpha\|_{L^p(\partial\Omega)}. \end{cases} \tag{75}$$

The space $WA_m^p(\partial\Omega)$ is defined as follows.

Definition 4.2. Suppose that $\Omega \subset \mathbb{R}^n$ is a Lipschitz domain, and consider arrays of functions $\dot{f} = \{f_\alpha : |\alpha| \le m-1\}$ indexed by multiindices α of length n, where $f_\alpha : \partial\Omega \mapsto \mathbb{C}$. We let $WA_m^p(\partial\Omega)$ be the completion of the set of arrays $\dot{\psi} = \{\partial^\alpha \psi : |\alpha| \le m-1\}$, for $\psi \in C_0^\infty(\mathbb{R}^n)$, under the norm

$$\sum_{|\alpha|\le m-1} \|\partial^\alpha \psi\|_{L^p(\partial\Omega)} + \sum_{|\alpha|=m-1} \|\nabla_\tau \partial^\alpha \psi\|_{L^p(\partial\Omega)}. \tag{76}$$

If we prescribe $\partial^\alpha u = f_\alpha$ on $\partial\Omega$ for some $\dot{f} \in WA_m^p(\partial\Omega)$, then we are prescribing the values of u, $\nabla u, \ldots, \nabla^{m-1}u$ on $\partial\Omega$, and requiring that (the prescribed part of) $\nabla^m u\big|_{\partial\Omega}$ lie in $L^p(\partial\Omega)$.

The study of these problems began with biharmonic functions in C^1 domains. In [125], Selvaggi and Sisto proved that, if Ω is the domain above the graph of a compactly supported C^1 function φ, with $\|\nabla\varphi\|_{L^\infty}$ small enough, then solutions to the Dirichlet problem exist provided $1 < p < \infty$. Their method used certain biharmonic layer potentials composed with the Riesz transforms.

In [33], Cohen and Gosselin proved that, if Ω is a bounded, simply connected C^1 domain contained in the plane $\mathbb{R}^2$, then the L^p-Dirichlet problem is well-posed in Ω for any $1 < p < \infty$. In [34], they extended this result to the complements of such domains. Their proof used multiple layer potentials introduced by Agmon in [5] in order to solve the Dirichlet problem with continuous boundary data. The general outline of their proof paralleled that of the proof of the corresponding result [49] for Laplace's equation. We remark that by [109, Theorem 6.30], we may weaken the condition that Ω be C^1 to the condition that the unit outward normal ν to Ω lies in $VMO(\partial\Omega)$. (Recall that this condition has been used in [106]; see formula (48) above and preceding remarks. This condition was also used in [100]; see section "The Dirichlet Problem for Operators in Divergence Form.")

As in the case of Laplace's equation, a result in Lipschitz domains soon followed. In [39], Dahlberg et al. showed that the L^p-Dirichlet problem for the biharmonic equation is well-posed in any bounded simply connected Lipschitz domain $\Omega \subset \mathbb{R}^n$, provided $2 - \varepsilon < p < 2 + \varepsilon$ for some $\varepsilon > 0$ depending on the domain Ω.

In [138], Verchota used the construction of [39] to extend Cohen and Gosselin's results from planar C^1 domains to C^1 domains of arbitrary dimension. Thus, the L^p-Dirichlet problem for the bilaplacian is well-posed for $1 < p < \infty$ in C^1 domains.

In [139], Verchota showed that the L^p-Dirichlet problem for the polyharmonic operator Δ^m could be solved for $2-\varepsilon < p < 2+\varepsilon$ in starlike Lipschitz domains by induction on the exponent m. He simultaneously proved results for the L^p-regularity problem in the same range; we will thus delay discussion of his methods to section "The Regularity Problem and the L^p-Dirichlet Problem."

All three of the papers [33, 39, 125] constructed biharmonic functions as potentials. However, the potentials used differ. Selvaggi and Sisto [125] constructed their solutions as

$$u(X) = \int_{\partial\Omega} \partial_n^2 F(X-Y) f(Y)\, d\sigma(Y) + \sum_{i=1}^{n-1} \int_{\partial\Omega} \partial_i \partial_n F(X-Y) R_i g(Y)\, d\sigma(Y) \qquad (77)$$

where R_i are the Riesz transforms. Here $F(X)$ is the fundamental solution to the biharmonic equation; thus, u is biharmonic in $\mathbb{R}^n \setminus \partial\Omega$. As in the case of Laplace's equation, well-posedness of the Dirichlet problem follows from the boundedness relation $\|N(\nabla u)\|_{L^p(\partial\Omega)} \leq C\|f\|_{L^p(\partial\Omega)} + C\|g\|_{L^p(\partial\Omega)}$ and from invertibility of the mapping $(f, g) \mapsto (u\big|_{\partial\Omega}, \partial_\nu u)$ on $L^p(\partial\Omega) \times L^p(\partial\Omega) \mapsto W_1^p(\partial\Omega) \times L^p(\partial\Omega)$.

The multiple layer potential of [33] is an operator of the form

$$\mathcal{L}\dot{f}(P) = \text{p.v.} \int_{\partial\Omega} \mathcal{L}(P,Q)\dot{f}(Q)\, d\sigma(Q) \qquad (78)$$

where $\mathcal{L}(P,Q)$ is a 3×3 matrix of kernels, also composed of derivatives of the fundamental solution to the biharmonic equation, and $\dot{f} = (f, f_x, f_y)$ is a "compatible triple" of boundary data, that is, an element of $W^{p_1}(\partial\Omega) \times L^p(\partial\Omega) \times L^p(\partial\Omega)$ that satisfies $\partial_\tau f = f_x \tau_x + f_y \tau_y$. Thus, the input is essentially a function and its gradient, rather than two functions, and the Riesz transforms are not involved.

The method of [39] is to compose two potentials. First, the function $f \in L^2(\partial\Omega)$ is mapped to its Poisson extension v. Next, u is taken to be the solution of the inhomogeneous equation $\Delta u(Y) = (n + 2Y \cdot \nabla)v(Y)$ with $u = 0$ on $\partial\Omega$. If $G(X, Y)$ is the Green's function for Δ in Ω and k^Y is the harmonic measure density at Y, we may write the map $f \mapsto u$ as

$$u(X) = \int_{\Omega} G(X,Y)(n + 2Y \cdot \nabla) \int_{\partial\Omega} k^Y(Q) f(Q)\, d\sigma(Q)\, dY. \qquad (79)$$

Since $(n + 2Y \cdot \nabla)v(Y)$ is harmonic, u is biharmonic, and so u solves the Dirichlet problem.

The L^p-Dirichlet Problem: The Summary of Known Results on Well-Posedness and Ill-Posedness

Recall that by [139], the L^p-Dirichlet problem is well-posed in Lipschitz domains provided $2 - \varepsilon < p < 2 + \varepsilon$. As in the case of Laplace's equation (see [48]), the range $p > 2-\varepsilon$ is sharp. That is, for any $p < 2$ and any integers $m \geq 2$, $n \geq 2$, there exist a bounded Lipschitz domain $\Omega \subset \mathbb{R}^n$ such that the L^p-Dirichlet problem for Δ^m is ill-posed in Ω. See [39, Sect. 5] for the case of the biharmonic operator Δ^2, and the proof of Theorem 2.1 in [119] for the polyharmonic operator Δ^m.

The range $p < 2 + \varepsilon$ is not sharp and has been studied extensively. Proving or disproving well-posedness of the L^p-Dirichlet problem for $p > 2$ in general Lipschitz domains has been an open question since [39], and was formally stated as such in [65, Problem 3.2.30]. (Earlier in [28, Question 7], the authors had posed the more general question of what classes of boundary data give existence and uniqueness of solutions.)

In [116, Theorem 10.7], Pipher and Verchota constructed Lipschitz domains Ω such that the L^p-Dirichlet problem for Δ^2 was ill-posed in Ω, for any given $p > 6$ (in four dimensions) or any given $p > 4$ (in five or more dimensions). Their counterexamples built on the study of solutions near a singular point, in particular upon [95, 98]. In [119], they provided other counterexamples to show that the L^p-Dirichlet problem for Δ^m is ill-posed, provided $p > 2(n-1)/(n-3)$ and $4 \leq n < 2m+1$. They remarked that if $n \geq 2m+1$, then ill-posedness follows from the results of [98] provided $p > 2m/(m-1)$.

The endpoint result at $p = \infty$ is the Agmon–Miranda maximum principle (14) discussed above. We remark that if $2 < p_0 \leq \infty$, and the L^{p_0}-Dirichlet problem is well-posed (or (14) holds) then by interpolation, the L^p-Dirichlet problem is well-posed for any $2 < p < p_0$.

We shall adopt the following definition (justified by the discussion above).

Definition 4.3. Suppose that $m \geq 2$ and $n \geq 4$. Then $p_{m,n}$ is defined to be the extended real number that satisfies the following properties. If $2 \leq p \leq p_{m,n}$, then the L^p-Dirichlet problem for Δ^m is well-posed in any bounded Lipschitz domain $\Omega \subset \mathbb{R}^n$. Conversely, if $p > p_{m,n}$, then there exist a bounded Lipschitz domain $\Omega \subset \mathbb{R}^n$ such that the L^p-Dirichlet problem for Δ^m is ill-posed in Ω. Here, well-posedness for $1 < p < \infty$ is meant in the sense of (75), and well-posedness for $p = \infty$ is meant in the sense of the maximum principle (see (91) below).

As in [39], we expect the range of solvability for any *particular* Lipschitz domain Ω to be $2-\varepsilon < p < p_{m,n} + \varepsilon$ for some ε depending on the Lipschitz character of Ω.

Let us summarize here the results currently known for $p_{m,n}$. More details will follow in section "The Regularity Problem and the L^p-Dirichlet Problem."

For any $m \geq 2$, we have that

- If $n = 2$ or $n = 3$, then the L^p-Dirichlet problem for Δ^m is well-posed in any Lipschitz domain Ω for any $2 \leq p < \infty$ [116, 119].
- If $4 \leq n \leq 2m+1$, then $p_{m,n} = 2(n-1)/(n-3)$ [119, 130].

- If $n = 2m + 2$, then $p_{m,n} = 2m/(m-1) = 2(n-2)/(n-4)$ [98, 128].
- If $n \geq 2m + 3$, then $2(n-1)/(n-3) \leq p_{m,n} \leq 2m/(m-1)$ [98, 130].

The value of $p_{m,n}$, for $n \geq 2m + 3$, is open.

In the special case of biharmonic functions ($m = 2$), more is known.

- $p_{2,4} = 6$, $p_{2,5} = 4$, $p_{2,6} = 4$, and $p_{2,7} = 4$ [128, 130].
- If $n \geq 8$, then

$$2 + \frac{4}{n - \lambda_n} < p_{2,n} \leq 4$$

  where

$$\lambda_n = \frac{n + 10 + 2\sqrt{2(n^2 - n + 2)}}{7}.$$

  [129]
- If Ω is a C^1 or convex domain of arbitrary dimension, then the L^p-Dirichlet problem for Δ^2 is well-posed in Ω for any $1 < p < \infty$ [70, 129, 139].

We comment on the nature of ill-posedness. The counterexamples of [39, 119] for $p < 2$ are failures of uniqueness. That is, those counterexamples are non-zero functions u, satisfying $\Delta^m u = 0$ in Ω, such that $\partial_\nu^k u = 0$ on $\partial\Omega$ for $0 \leq k \leq m-1$, and such that $N(\nabla^{m-1}u) \in L^p(\partial\Omega)$.

Observe that if Ω is bounded and $p > 2$, then $L^p(\partial\Omega) \subset L^2(\partial\Omega)$. Because the L^2-Dirichlet problem is well-posed, the failure of well-posedness for $p > 2$ can only be a failure of the optimal estimate $N(\nabla^{m-1}u) \in L^p(\partial\Omega)$. That is, if the L^p-Dirichlet problem for Δ^m is ill-posed in Ω, then for some Whitney array $\dot{f} \in WA^p_{m-1}(\partial\Omega)$ and some $g \in L^p(\partial\Omega)$, the unique function u that satisfies $\Delta^m u = 0$ in Ω, $\partial^\alpha u = f_\alpha$, $\partial_\nu^{m-1} u = g$ and $N(\nabla^{m-1}u) \in L^2(\partial\Omega)$ does not satisfy $N(\nabla^{m-1}u) \in L^p(\partial\Omega)$.

The Regularity Problem and the L^p-Dirichlet Problem

In this section we elaborate on some of the methods used to prove the Dirichlet well-posedness results listed above, as well as their historical context. This naturally brings up a consideration of a different boundary value problem, the L^q-regularity problem for higher-order operators.

Recall that for second-order equations the regularity problem corresponds to finding a solution with prescribed tangential gradient along the boundary. In analogy, we say that the L^q-regularity problem for Δ^m is well-posed in Ω if there exist a constant $C > 0$ such that, whenever $\dot{f} \in WA^q_m(\partial\Omega)$, there exist a unique function u that satisfies

$$\begin{cases} \Delta^m u = 0 & \text{in } \Omega, \\ \partial^\alpha u\big|_{\partial\Omega} = f_\alpha & \text{for all } 0 \le |\alpha| \le m-1, \\ \|N(\nabla^m u)\|_{L^q(\partial\Omega)} \le C \sum_{|\alpha|=m-1} \|\nabla_\tau f_\alpha\|_{L^q(\partial\Omega)}. \end{cases} \tag{80}$$

There is an important endpoint formulation at $q = 1$ for the regularity problem. We say that the H^1-regularity problem is well-posed if there exist a constant $C > 0$ such that, whenever $\dot{f}$ lies in the Whitney–Hardy space $H^1_m(\partial\Omega)$, there exist a unique function u that satisfies

$$\begin{cases} \Delta^m u = 0 & \text{in } \Omega, \\ \partial^\alpha u\big|_{\partial\Omega} = f_\alpha & \text{for all } 0 \le |\alpha| \le m-1, \\ \|N(\nabla^m u)\|_{L^1(\partial\Omega)} \le C \sum_{|\alpha|=m-1} \|\nabla_\tau f_\alpha\|_{H^1(\partial\Omega)}. \end{cases}$$

The space $H^1_m(\partial\Omega)$ is defined as follows.

Definition 4.4. We say that $\dot{a} \in WA^q_m(\partial\Omega)$ is a $H^1_m(\partial\Omega)$-L^q atom if $\dot{a}$ is supported in a ball $B(Q, r) \cap \partial\Omega$ and if

$$\sum_{|\alpha|=m-1} \|\nabla_\tau a_\alpha\|_{L^q(\partial\Omega)} \le \sigma(B(Q, r) \cap \partial\Omega)^{1/q-1}.$$

If $\dot{f} \in WA^1_m(\partial\Omega)$ and there are H^1_m-L^2 atoms $\dot{a}_k$ and constants $\lambda_k \in \mathbb{C}$ such that

$$\nabla_\tau f_\alpha = \sum_{k=1}^{\infty} \lambda_k \nabla_\tau (a_k)_\alpha \text{ for all } |\alpha| = m-1$$

and such that $\sum |\lambda_k| < \infty$, we say that $\dot{f} \in H^1_m(\partial\Omega)$, with $\|\dot{f}\|_{H^1_m(\partial\Omega)}$ being the smallest $\sum |\lambda_k|$ among all such representations.

In [139], Verchota proved well-posedness of the L^2-Dirichlet problem and the L^2-regularity problem for the polyharmonic operator Δ^m in any bounded starlike Lipschitz domain by simultaneous induction.

The base case $m = 1$ is valid in all bounded Lipschitz domains by [35, 62]. The inductive step is to show that well-posedness for the Dirichlet problem for Δ^{m+1} follows from well-posedness of the lower-order problems. In particular, solutions with $\partial^\alpha u = f_\alpha$ may be constructed using the regularity problem for Δ^m, and the boundary term $\partial^m_\nu u{=}g$, missing from the regularity data, may be attained using the *inhomogeneous* Dirichlet problem for Δ^m. On the other hand, it was shown that the well-posedness for the regularity problem for Δ^{m+1} follows from well-posedness of the lower-order problems and from the Dirichlet problem for Δ^{m+1}, in some sense, by realizing the solution to the regularity problem as an integral of the solution to the Dirichlet problem.

As regards a broader range of p and q, Pipher and Verchota showed in [116] that the L^p-Dirichlet and L^q-regularity problems for Δ^2 are well-posed in all bounded Lipschitz domains $\Omega \subset \mathbb{R}^3$, provided $2 \leq p < \infty$ and $1 < q \leq 2$. Their method relied on duality. Using potentials similar to those of [39], they constructed solutions to the L^2-Dirichlet problem in domains above Lipschitz graphs. The core of their proof was the invertibility on $L^2(\partial\Omega)$ of a certain potential operator T. They were able to show that the invertibility of its adjoint T^* on $L^2(\partial\Omega)$ implies that the L^2-regularity problem for Δ^2 is well-posed. Then, using the atomic decomposition of Hardy spaces, they analyzed the H^1-regularity problem. Applying interpolation and duality for T^* once again, now in the reverse regularity-to-Dirichlet direction, the full range for both regularity and Dirichlet problems was recovered in domains above graphs. Localization arguments then completed the argument in bounded Lipschitz domains.

In four or more dimensions, further progress relied on the following theorem of Shen.

Theorem 4.5 ([128]). *Suppose that $\Omega \subset \mathbb{R}^n$ is a Lipschitz domain. The following conditions are equivalent.*

- *The L^p-Dirichlet problem for L is well-posed, where L is a symmetric elliptic system of order $2m$ with real constant coefficients.*
- *There exists some constant $C > 0$ and some $p > 2$ such that*

$$\left(\fint_{B(Q,r)\cap\partial\Omega} N(\nabla^{m-1}u)^p \, d\sigma\right)^{1/p} \leq C\left(\fint_{B(Q,2r)\cap\partial\Omega} N(\nabla^{m-1}u)^2 \, d\sigma\right)^{1/2} \tag{81}$$

holds whenever u is a solution to the L^2-Dirichlet problem for L in Ω, with $\nabla u \equiv 0$ on $B(Q,3r) \cap \partial\Omega$.

For the polyharmonic operator Δ^m, this theorem was essentially proven in [130]. Furthermore, the reverse Hölder estimate (81) with $p = 2(n-1)/(n-3)$ was shown to follow from well-posedness of the L^2-regularity problem. Thus the L^p-Dirichlet problem is well-posed in bounded Lipschitz domains in $\mathbb{R}^n$ for $p = 2(n-1)/(n-3)$. By interpolation, and because reverse Hölder estimates have self-improving properties, well-posedness in the range $2 \leq p \leq 2(n-1)/(n-3) + \varepsilon$ for any particular Lipschitz domain follows automatically.

Using regularity estimates and square-function estimates, Shen was able to further improve this range of p. He showed that with $p = 2 + 4/(n-\lambda)$, $0 < \lambda < n$, the reverse Hölder estimate (81) is true, provided that

$$\int_{B(Q,r)\cap\Omega} |\nabla^{m-1}u|^2 \leq C\left(\frac{r}{R}\right)^{\lambda} \int_{B(Q,R)\cap\Omega} |\nabla^{m-1}u|^2 \tag{82}$$

holds whenever u is a solution to the L^2-Dirichlet problem in Ω with $N(\nabla^{m-1}u) \in L^2(\partial\Omega)$ and $\nabla^k u\big|_{B(Q,R)\cap\Omega} \equiv 0$ for all $0 \leq k \leq m-1$.

It is illuminating to observe that the estimates arising in connection with the pointwise bounds on the solutions in arbitrary domains (cf. section "Miranda–Agmon Maximum Principle and Related Geometric restrictions on the boundary") and the Wiener test (cf. section "The Wiener Test: Continuity of Solutions") take essentially the form (82). Thus, Theorem 4.5 and its relation to (82) provide a direct way to transform results regarding local boundary regularity of solutions, obtained via the methods underlined in sections "Miranda–Agmon Maximum Principle and Related Geometric restrictions on the boundary" and "The Wiener Test: Continuity of Solutions," into well-posedness of the L^p-Dirichlet problem.

In particular, consider [90, Lemma 5]. If u is a solution to $\Delta^m u = 0$ in $B(Q,R) \cap \Omega$, where Ω is a Lipschitz domain, then by [90, Lemma 5] there is some constant $\lambda_0 > 0$ such that

$$\sup_{B(Q,r)\cap\Omega} |u|^2 \leq \left(\frac{r}{R}\right)^{\lambda_0} \frac{C}{R^n} \int_{B(Q,R)\cap\Omega} |u(X)|^2\, dX \tag{83}$$

provided that r/R is small enough, that u has zero boundary data on $B(Q,R) \cap \partial\Omega$, and where $\Omega \subset \mathbb{R}^n$ has dimension $n = 2m+1$ or $n = 2m+2$, or where $m = 2$ and $n = 7 = 2m+3$. (The bound on dimension comes from the requirement that Δ^m be positive with weight F; see Eq. (54).)

It is not difficult to see (cf., e.g., [128, Theorem 2.6]) that (83) implies (82) for some $\lambda > n - 2m + 2$, and thus implies well-posedness of the L^p-Dirichlet problem for a certain range of p. This provides an improvement on the results of [130] in the case $m = 2$ and $n = 6$ or $n = 7$, and in the case $m \geq 3$ and $n = 2m+2$. Shen has stated this improvement in [128, Theorems 1.4 and 1.5]: the L^p-Dirichlet problem for Δ^2 is well-posed for $2 \leq p < 4 + \varepsilon$ in dimensions $n = 6$ or $n = 7$, and the L^p-Dirichlet problem for Δ^m is well-posed if $2 \leq p < 2m/(m-1) + \varepsilon$ in dimension $n = 2m+2$.

The method of weighted integral identities, related to positivity with weight F [cf. (54)], can be further finessed in a particular case of the biharmonic equation. Shen [129] uses this method (extending the ideas from [86]) to show that if $n \geq 8$, then (82) is valid for solutions to Δ^2 with $\lambda = \lambda_n$, where

$$\lambda_n = \frac{n + 10 + 2\sqrt{2(n^2 - n + 2)}}{7}. \tag{84}$$

We now return to the L^q-regularity problem. Recall that in [116], Pipher and Verchota showed that if $2 < p < \infty$ and $1/p + 1/q < 1$, then the L^p-Dirichlet problem and the L^q-regularity problem for Δ^2 are both well-posed in three-dimensional Lipschitz domains. They proved this by showing that, in the special case of a domain above a Lipschitz graph, there is duality between the L^p-Dirichlet and L^q-regularity problems. Such duality results are common. See [67, 68, 131] for duality results in the second-order case; although even in that case, duality is not always guaranteed. (See [79].) Many of the known results concerning the regularity problem for the polyharmonic operator Δ^m are results relating the L^p-Dirichlet problem to the L^q-regularity problem.

In [105], I. Mitrea and M. Mitrea showed that if $1 < p < \infty$ and $1/p + 1/q = 1$, and if the L^q-regularity problem for Δ^2 and the L^p-regularity problem for Δ were both well-posed in a particular bounded Lipschitz domain Ω, then the L^p-Dirichlet problem for Δ^2 was also well-posed in Ω. They proved this result (in arbitrary dimensions) using layer potentials and a Green representation formula for biharmonic equations. Observe that the extra requirement of well-posedness for the Laplacian is extremely unfortunate, since in bad domains it essentially restricts consideration to $p < 2 + \varepsilon$ and thus does not shed new light on well-posedness in the general class of Lipschitz domains. As will be discussed below, later Kilty and Shen established an optimal duality result for biharmonic Dirichlet and regularity problems.

Recall that the formula (81) provides a necessary and sufficient condition for well-posedness of the L^p-Dirichlet problem. In [71], Kilty and Shen provided a similar condition for the regularity problem. To be precise, they demonstrated that if $q > 2$ and L is a symmetric elliptic system of order $2m$ with real constant coefficients, then the L^q-regularity problem for L is well-posed if and only if the estimate

$$\left(\fint_{B(Q,r)\cap\Omega} N(\nabla^m u)^q \, d\sigma\right)^{1/q} \leq C\left(\fint_{B(Q,2r)\cap\Omega} N(\nabla^m u)^2 \, d\sigma\right)^{1/2} \tag{85}$$

holds for all points $Q \in \partial\Omega$, all $r > 0$ small enough, and all solutions u to the L^2-regularity problem with $\nabla^k u\big|_{B(Q,3r)\cap\partial\Omega} = 0$ for $0 \leq k \leq m-1$. Observe that (85) is identical to (81) with p replaced by q and $m-1$ replaced by m.

As a consequence, well-posedness of the L^q-regularity problem in Ω for certain values of q implies well-posedness of the L^p-Dirichlet problem for some values of p. Specifically, arguments using interior regularity and fractional integral estimates (given in [71, Sect. 5]) show that (85) implies (81) with $1/p = 1/q - 1/(n-1)$. But recall from [128] that (81) holds if and only if the L^p-Dirichlet problem for L is well-posed in Ω. Thus, if $2 < q < n-1$, and if the L^q-regularity problem for a symmetric elliptic system is well-posed in a Lipschitz domain Ω, then the L^p-Dirichlet problem for the same system and domain is also well-posed, provided $2 < p < p_0 + \varepsilon$ where $1/p_0 = 1/q - 1/(n-1)$.

For the bilaplacian, a full duality result is known. In [70], Kilty and Shen showed that, if $1 < p < \infty$ and $1/p + 1/q = 1$, then well-posedness of the L^p-Dirichlet problem for Δ^2 in a Lipschitz domain Ω and well-posedness of the L^q-regularity problem for Δ^2 in Ω were both equivalent to the bilinear estimate

$$\begin{aligned}\left|\int_\Omega \Delta u \, \Delta v\right| \leq C\Big(\|\nabla_\tau \nabla f\|_{L^p} + |\partial\Omega|^{-1/(n-1)}\|\nabla f\|_{L^p} + |\partial\Omega|^{-2/(n-1)}\|f\|_{L^p}\Big) \\ \times \Big(\|\nabla g\|_{L^q} + |\partial\Omega|^{-1/(n-1)}\|g\|_{L^q}\Big)\end{aligned} \tag{86}$$

for all $f, g \in C_0^\infty(\mathbb{R}^n)$, where u and v are solutions of the L^2-regularity problem with boundary data $\partial^\alpha u = \partial^\alpha f$ and $\partial^\alpha v = \partial^\alpha g$. Thus, if $\Omega \subset \mathbb{R}^n$ is a bounded Lipschitz domain, and if $1/p + 1/q = 1$, then the L^p-Dirichlet problem is well-posed in Ω if and only if the L^q-regularity problem is well-posed in Ω.

All in all, we see that the L^p-regularity problem for Δ^2 is well-posed in $\Omega \subset \mathbb{R}^n$ if

- Ω is C^1 or convex, and $1 < p < \infty$.
- $n = 2$ or $n = 3$ and $1 < p < 2 + \varepsilon$.
- $n = 4$ and $6/5 - \varepsilon < p < 2 + \varepsilon$.
- $n = 5, 6$, or 7, and $4/3 - \varepsilon < p < 2 + \varepsilon$.
- $n \geq 8$, and $2 - \frac{4}{4+n-\lambda_n} < p < 2 + \varepsilon$, where λ_n is given by (84). The above ranges of p are sharp, but this range is still open.

Higher-Order Elliptic Systems

The polyharmonic operator Δ^m is part of a larger class of elliptic higher-order operators. Some study has been made of boundary value problems for such operators and systems.

The L^p-Dirichlet problem for a strongly elliptic system L of order $2m$, as defined in Definition 2.1, is well-posed in Ω if there exist a constant C such that, for every $\dot{f} \in WA^p_{m-1}(\partial\Omega \mapsto \mathbb{C}^\ell)$ and every $\vec{g} \in L^p(\partial\Omega \mapsto \mathbb{C}^\ell)$, there exist a unique vector-valued function $\vec{u} : \Omega \mapsto \mathbb{C}^\ell$ such that

$$\begin{cases} (L\vec{u})_j = \sum_{k=1}^{\ell} \sum_{|\alpha|=|\beta|=m} \partial^\alpha a^{jk}_{\alpha\beta} \partial^\beta u_k = 0 & \text{in } \Omega \text{ for each } 1 \leq j \leq \ell, \\ \partial^\alpha \vec{u} = f_\alpha & \text{on } \partial\Omega \text{ for } |\alpha| \leq m-2, \\ \partial_\nu^{m-1} \vec{u} = \vec{g} & \text{on } \partial\Omega, \\ \|N(\nabla^{m-1} u)\|_{L^q(\partial\Omega)} \leq C \sum_{|\alpha|=m-2} \|\nabla_\tau f_\alpha\|_{L^q(\partial\Omega)} + C\|\vec{g}\|_{L^p(\partial\Omega)}. \end{cases} \tag{87}$$

The L^q-regularity problem is well-posed in Ω if there is some constant C such that, for every $\dot{f} \in WA^p_m(\partial\Omega \mapsto \mathbb{C}^\ell)$, there exist a unique $\vec{u}$ such that

$$\begin{cases} (L\vec{u})_j = \sum_{k=1}^{\ell} \sum_{|\alpha|=|\beta|=m} \partial^\alpha a^{jk}_{\alpha\beta} \partial^\beta u_k = 0 & \text{in } \Omega \text{ for each } 1 \leq j \leq \ell, \\ \partial^\alpha \vec{u} = f_\alpha & \text{on } \partial\Omega \text{ for } |\alpha| \leq m-1, \\ \|N(\nabla^m u)\|_{L^q(\partial\Omega)} \leq C \sum_{|\alpha|=m-1} \|\nabla_\tau f_\alpha\|_{L^q(\partial\Omega)}. \end{cases} \tag{88}$$

In [118], Pipher and Verchota showed that the L^p-Dirichlet and L^p-regularity problems were well-posed for $2 - \varepsilon < p < 2 + \varepsilon$, for any higher-order elliptic partial differential equation with real constant coefficients, in Lipschitz domains of arbitrary dimension. This was extended to symmetric elliptic systems in [140]. A key ingredient of the proof was the boundary Gårding inequality

$$\frac{\lambda}{4}\int_{\partial\Omega} |\nabla^m u|(-\nu_n)\,d\sigma \\ \leq \sum_{j,k=1}^{\ell} \sum_{|\alpha|=|\beta|=m} \int_{\partial\Omega} \partial^\alpha a_{\alpha\beta}^{jk} \partial^\beta u_k(-\nu_n)\,d\sigma + C\int_{\partial\Omega} \left|\nabla^{m-1}\partial_n u\right|^2 d\sigma$$

valid if $u \in C_0^\infty(\mathbb{R}^n)^\ell$, if $L = \partial^\alpha a_{\alpha\beta}^{jk} \partial^\beta$ is a symmetric elliptic system with real constant coefficients, and if Ω is the domain above the graph of a Lipschitz function. We observe that in this case, $(-\nu_n)$ is a positive number bounded from below. Pipher and Verchota then used this Gårding inequality and a Green's formula to construct the nontangential maximal estimate. See [119] and [140, Sects. 4 and 6].

As in the case of the polyharmonic operator Δ^m, this first result concerned the L^p-Dirichlet problem and L^q-regularity problem only for $2 - \varepsilon < p < 2 + \varepsilon$ and for $2 - \varepsilon < q < 2 + \varepsilon$. The polyharmonic operator Δ^m is an elliptic system, and so we cannot in general improve upon the requirement that $2 - \varepsilon < p$ for well-posedness of the L^p-Dirichlet problem.

However, we can improve on the requirement $p < 2+\varepsilon$. Recall that Theorem 4.5 from [128] and its equivalence to (82) were proven in the general case of strongly elliptic systems with real symmetric constant coefficients. As in the case of the polyharmonic operator Δ^m, (81) follows from well-posedness of the L^2-regularity problem provided $p = 2(n-1)/(n-3)$, and so if L is such a system, the L^p-Dirichlet problem for L is well-posed in Ω provided $2 - \varepsilon < p < 2(n-1)/(n-3) + \varepsilon$. This is [128, Corollary 1.3]. Again, by the counterexamples of [119], this range cannot be improved if $m \geq 2$ and $4 \leq n \leq 2m + 1$; the question of whether this range can be improved for general operators L if $n \geq 2m + 2$ is still open.

Little is known concerning the regularity problem in a broader range of p. Recall that (85) from [71] was proven in the general case of strongly elliptic systems with real symmetric constant coefficients. Thus, we known that for such systems, well-posedness of the L^q-regularity problem for $2 < q < n - 1$ implies well-posedness of the L^p-Dirichlet problem for appropriate p. The question of whether the reverse implication holds, or whether this result can be extended to a broader range of q, is open.

The Area Integral

One of major tools in the theory of second-order elliptic differential equations is the Lusin area integral, defined as follows. If w lies in $W^2_{1,loc}(\Omega)$ for some domain $\Omega \subset \mathbb{R}^n$, then the area integral (or square function) of w is defined for $Q \in \partial\Omega$ as

$$Sw(Q) = \left(\int_{\Gamma(Q)} |\nabla w(X)|^2 \operatorname{dist}(X, \partial\Omega)^{2-n} dX \right)^{1/2}.$$

In [36], Dahlberg showed that if u is harmonic in a bounded Lipschitz domain Ω, if $P_0 \in \Omega$ and $u(P_0) = 0$, then for any $0 < p < \infty$,

$$\frac{1}{C} \int_{\partial\Omega} (Su)^p \, d\sigma \le \int_{\partial\Omega} (Nu)^p \, d\sigma \le C \int_{\partial\Omega} (Su)^p \, d\sigma \tag{89}$$

for some constants C depending only on p, Ω, and P_0. Thus, the Lusin area integral bears deep connections to the L^p-Dirichlet problem. In [38], Dahlberg et al. generalized this result to solutions to second-order divergence-form elliptic equations with real coefficients for which the L^r-Dirichlet problem is well-posed for at least one r.

If L is an operator of order $2m$, then the appropriate estimate is

$$\frac{1}{C} \int_{\partial\Omega} N(\nabla u^{m-1})^p \, d\sigma \le \int_{\partial\Omega} S(\nabla u^{m-1})^p \, d\sigma \le C \int_{\partial\Omega} N(\nabla u^{m-1})^p \, d\sigma. \tag{90}$$

Before discussing their validity for particular operators, let us point out that such square-function estimates are very useful in the study of higher-order equations. In [128], Shen used (90) to prove the equivalence of (82) and (81), above. In [70], Kilty and Shen used (90) to prove that well-posedness of the L^p-Dirichlet problem for Δ^2 implies the bilinear estimate (86). The proof of the maximum principle (14) in [140, Sect. 8] (to be discussed in section "The Maximum Principle in Lipschitz Domains") also exploited (90). Estimates on square functions can be used to derive estimates on Besov space norms; see [4, Proposition S].

In [115], Pipher and Verchota proved that (90) (with $m = 2$) holds for solutions u to $\Delta^2 u = 0$, provided Ω is a bounded Lipschitz domain, $0 < p < \infty$, and $\nabla u(P_0) = 0$ for some fixed $P_0 \in \Omega$. Their proof was an adaptation of Dahlberg's proof [36] of the corresponding result for harmonic functions. They used the L^2-theory for the biharmonic operator [39], the representation formula (79), and the L^2-theory for harmonic functions to prove good-λ inequalities, which, in turn, imply L^p estimates for $0 < p < \infty$.

In [40], Dahlberg et al. proved that (90) held for solutions u to $Lu = 0$, for a symmetric elliptic system L of order $2m$ with real constant coefficients, provided as usual that Ω is a bounded Lipschitz domain, $0 < p < \infty$, and $\nabla^{m-1} u(P_0) = 0$ for some fixed $P_0 \in \Omega$. The argument is necessarily considerably more involved than the argument of [115] or [36]. In particular, the bound $\|S(\nabla^{m-1} u)\|_{L^2(\partial\Omega)} \le C\|N(\nabla^{m-1} u)\|_{L^2(\partial\Omega)}$ was proven in three steps.

The first step was to reduce from the elliptic system L of order $2m$ to the scalar elliptic operator $M = \det L$ of order $2\ell m$, where ℓ is as in formula (7). The second step was to reduce to elliptic equations of the form $\sum_{|\alpha|=m} a_\alpha \partial^{2\alpha} u = 0$, where $|a_\alpha| > 0$ for all $|\alpha| = m$. Finally, it was shown that for operators of this form

$$\sum_{|\alpha|=m} \int_\Omega a_\alpha\, \partial^\alpha u(X)^2 \operatorname{dist}(X, \partial\Omega)\, dX \le C \int_{\partial\Omega} N(\nabla^{m-1} u)^2\, d\sigma.$$

The passage to $0 < p < \infty$ in (90) was done, as usual, using good-λ inequalities. We remark that these arguments used the result of [118] that the L^2-Dirichlet problem is well-posed for such operators L in Lipschitz domains.

It is quite interesting that for second-order elliptic systems, the only currently known approach to the square-function estimate (89) is this reduction to a higher-order operator.

The Maximum Principle in Lipschitz Domains

We are now in a position to discuss the maximum principle (14) for higher-order equations in Lipschitz domains.

We say that the maximum principle for an operator L of order $2m$ holds in the bounded Lipschitz domain Ω if there exist a constant $C > 0$ such that, whenever $f \in WA^\infty_{m-1}(\partial\Omega) \subset WA^2_{m-1}(\partial\Omega)$ and $g \in L^\infty(\partial\Omega) \subset L^2(\partial\Omega)$, the solution u to the Dirichlet problem (87) with boundary data f and g satisfies

$$\|\nabla^{m-1} u\|_{L^\infty} \le C \|g\|_{L^\infty(\partial\Omega)} + C \sum_{|\alpha|=m-2} \|\nabla_\tau f_\alpha\|_{L^\infty(\partial\Omega)}. \tag{91}$$

The maximum principle (91) was proven to hold in three-dimensional Lipschitz domains by Pipher and Verchota in [117] (for biharmonic functions), in [119] (for polyharmonic functions), and by Verchota in [140, Sect. 8] (for solutions to symmetric systems with real constant coefficients). Pipher and Verchota also proved in [117] that the maximum principle was valid for biharmonic functions in C^1 domains of arbitrary dimension. In [70, Theorem 1.5], Kilty and Shen observed that the same technique gives validity of the maximum principle for biharmonic functions in convex domains of arbitrary dimension.

The proof of [117] uses the L^2-regularity problem in the domain Ω to construct the Green's function $G(X, Y)$ for Δ^2 in Ω. Then if u is biharmonic in Ω with $N(\nabla u) \in L^2(\partial\Omega)$, we have that

$$u(X) = \int_{\partial\Omega} u(Q)\, \partial_\nu \Delta G(X, Q)\, d\sigma(Q) + \int_{\partial\Omega} \partial_\nu u(Q)\, \Delta G(X, Q)\, d\sigma(Q) \tag{92}$$

where all derivatives of G are taken in the second variable Q. If the H^1-regularity problem is well-posed in appropriate subdomains of Ω, then $\nabla^2\nabla_X G(X,\cdot)$ is in $L^1(\partial\Omega)$ with L^1 norm independent of X, and so the second integral is at most $C\|\partial_\nu u\|_{L^\infty(\partial\Omega)}$. By taking Riesz transforms, the normal derivative $\partial_\nu \Delta G(X,Q)$ may be transformed to tangential derivatives $\nabla_\tau \Delta G(X,Q)$; integrating by parts transfers these derivatives to u. The square-function estimate (90) implies that the Riesz transforms of $\nabla_X \Delta_Q G(X,Q)$ are bounded on $L^1(\partial\Omega)$. This completes the proof of the maximum principle.

Similar arguments show that the maximum principle is valid for more general operators. See [119] for the polyharmonic operator, or [140, Sect. 8] for arbitrary symmetric operators with real constant coefficients.

An important transitional step is the well-posedness of the H^1-regularity problem. It was established in three-dimensional (or C^1) domains in [117, Theorem 4.2] and [119, Theorem 1.2] and discussed in [140, Sect. 7]. In each case, well-posedness was proven by analyzing solutions with atomic data $\dot{f}$ using a technique from [37]. A crucial ingredient in this technique is the well-posedness of the L^p-Dirichlet problem for some $p < (n-1)/(n-2)$; the latter is valid if $n = 3$ by [39], and (for Δ^2) in C^1 and convex domains by [70, 139], but fails in general Lipschitz domains for $n \geq 4$.

Biharmonic Functions in Convex Domains

We say that a domain Ω is *convex* if, whenever $X, Y \in \Omega$, the line segment connecting X and Y lies in Ω. Observe that all convex domains are necessarily Lipschitz domains but the converse does not hold. Moreover, while convex domains are in general no smoother than Lipschitz domains, the extra geometrical structure often allows for considerably stronger results.

Recall that in [81], the second author of this paper and Maz'ya showed that the gradient of a biharmonic function is bounded in a three-dimensional domain. This is a sharp property in dimension three, and in higher dimensional domains the solutions can be even less regular (cf. section "Miranda–Agmon Maximum Principle and Related Geometric restrictions on the boundary"). However, using some intricate linear combination of weighted integrals, the same authors showed in [80] that *second* derivatives to biharmonic functions were locally bounded when the domain was convex. To be precise, they showed that if Ω is convex, and $u \in \mathring{W}_2^2(\Omega)$ is a solution to $\Delta^2 u = h$ for some $h \in C_0^\infty(\Omega \setminus B(Q, 10R))$, $R > 0$, $Q \in \partial\Omega$, then

$$\sup_{B(Q,R/5)\cap\Omega} |\nabla^2 u| \leq \frac{C}{R^2}\left(\fint_{\Omega\cap B(Q,5R)\setminus B(Q,R/2)} |u|^2\right)^{1/2}. \tag{93}$$

In particular, not only are all boundary points of convex domains 1-regular, but the gradient ∇u is Lipschitz continuous near such points.

Kilty and Shen noted in [70] that (93) implies that (85) holds in convex domains for any q; thus, the L^q-regularity problem for the bilaplacian is well-posed for any $2 < q < \infty$ in a convex domain. Well-posedness of the L^p-Dirichlet problem for $2 < p < \infty$ has been established by Shen in [129]. By the duality result (86), again from [70], this implies that both the L^p-Dirichlet and L^q-regularity problems are well-posed, for any $1 < p < \infty$ and any $1 < q < \infty$, in a convex domain of arbitrary dimension. They also observed that, by the techniques of [117] (discussed in section "The Maximum Principle in Lipschitz Domains" above), the maximum principle (91) is valid in arbitrary convex domains.

It is interesting to note how, once again, the methods and results related to pointwise estimates, the Wiener criterion, and local regularity estimates near the boundary are intertwined with the well-posedness of boundary problems in L^p.

The Neumann Problem for the Biharmonic Equation

So far we have only discussed the Dirichlet and regularity problems for higher-order operators. Another common and important boundary value problem that arises in applications is the Neumann problem. Indeed, the principal physical motivation for the inhomogeneous biharmonic equation $\Delta^2 u = h$ is that it describes the equilibrium position of a thin elastic plate subject to a vertical force h. The Dirichlet problem $u\big|_{\partial\Omega} = f$, $\nabla u\big|_{\partial\Omega} = g$ describes an elastic plate whose edges are clamped, that is, held at a fixed position in a fixed orientation. The Neumann problem, on the other hand, corresponds to the case of a free boundary. Guido Sweers has written an excellent short paper [136] discussing the boundary conditions that correspond to these and other physical situations.

More precisely, if a thin two-dimensional plate is subject to a force h and the edges are free to move, then its displacement u satisfies the boundary value problem

$$\begin{cases} \Delta^2 u = h & \text{in } \Omega, \\ \rho\Delta u + (1-\rho)\partial_\nu^2 u = 0 & \text{on } \partial\Omega, \\ \partial_\nu \Delta u + (1-\rho)\partial_{\tau\tau\nu} u = 0 & \text{on } \partial\Omega. \end{cases}$$

Here ρ is a physical constant, called the Poisson ratio. This formulation goes back to Kirchhoff and is well known in the theory of elasticity; see, for example, Sect. 3.1 and Chap. 8 of the classic engineering text [112]. We remark that by [112, formula (8–10)],

$$\partial_\nu \Delta u + (1-\rho)\partial_{\tau\tau\nu} u = \partial_\nu \Delta u + (1-\rho)\partial_\tau\left(\partial_{\nu\tau} u\right).$$

This suggests the following homogeneous boundary value problem in a Lipschitz domain Ω of arbitrary dimension. We say that the L^p-Neumann problem is well-posed if there exist a constant $C > 0$ such that, for every $f_0 \in L^p(\partial\Omega)$ and $\Lambda_0 \in W^p_{-1}(\partial\Omega)$, there exist a function u such that

$$\begin{cases} \Delta^2 u = 0 & \text{in } \Omega, \\ M_\rho u := \rho \Delta u + (1-\rho)\partial_\nu^2 u = f_0 & \text{on } \partial\Omega, \\ K_\rho u := \partial_\nu \Delta u + (1-\rho)\dfrac{1}{2}\partial_{\tau_{ij}}\left(\partial_{\nu\tau_{ij}} u\right) = \Lambda_0 & \text{on } \partial\Omega, \\ \|N(\nabla^2 u)\|_{L^p(\partial\Omega)} \leq C\|f_0\|_{W^p_1(\partial\Omega)} + C\|\Lambda_0\|_{W^p_{-1}(\partial\Omega)}. \end{cases} \tag{94}$$

Here $\tau_{ij} = \nu_i \mathbf{e}_j - \nu_j \mathbf{e}_i$ is a vector orthogonal to the outward normal ν and lying in the $x_i x_j$-plane.

In addition to the connection to the theory of elasticity, this problem is of interest because it is in some sense adjoint to the Dirichlet problem (74). That is, if $\Delta^2 u = \Delta^2 w = 0$ in Ω, then $\int_{\partial\Omega} \partial_\nu w\, M_\rho u - w\, K_\rho u\, d\sigma = \int_{\partial\Omega} \partial_\nu u\, M_\rho w - u\, K_\rho w\, d\sigma$, where M_ρ and K_ρ are as in (94); this follows from the more general formula

$$\int_\Omega w\,\Delta^2 u = \int_\Omega \left(\rho \Delta u\, \Delta w + (1-\rho)\partial_{jk} u\, \partial_{jk} w\right) + \int_{\partial\Omega} w\, K_\rho u - \partial_\nu w\, M_\rho u\, d\sigma \tag{95}$$

valid for arbitrary smooth functions. This formula is analogous to the classical Green's identity for the Laplacian

$$\int_\Omega w\,\Delta u = -\int_\Omega \nabla u \cdot \nabla w + \int_{\partial\Omega} w\,\nu \cdot \nabla u\, d\sigma. \tag{96}$$

Observe that, contrary to the Laplacian or more general second-order operators, there is a *family* of relevant Neumann data for the biharmonic equation. Moreover, different values (or, rather, ranges) of ρ correspond to different natural physical situations. We refer the reader to [141] for a detailed discussion.

In [34], Cohen and Gosselin showed that the L^p-Neumann problem (94) was well-posed in C^1 domains contained in $\mathbb{R}^2$ for $1 < p < \infty$, provided in addition that $\rho = -1$. The method of proof was as follows. Recall from (78) that Cohen and Gosselin showed that the L^p-Dirichlet problem was well-posed by constructing a multiple layer potential $\mathcal{L}\dot{f}$ with boundary values $(I + \mathcal{K})\dot{f}$, and showing that $I + \mathcal{K}$ is invertible. We remark that because Cohen and Gosselin preferred to work with Dirichlet boundary data of the form $(u, \partial_x u, \partial_y u)\big|_{\partial\Omega}$ rather than of the form $(u, \partial_\nu u)\big|_{\partial\Omega}$, the notation of [34] is somewhat different from that of the present paper. In the notation of the present paper, the method of proof of [34] was to observe that $(I+\mathcal{K})^*\dot{\theta}$ is equivalent to $(K_{-1}v\dot{\theta}, M_{-1}v\dot{\theta})_{\partial\Omega^C}$, where v is another biharmonic layer potential and $(I + \mathcal{K})^*$ is the adjoint to $(I + \mathcal{K})$. Well-posedness of the Neumann problem then follows from invertibility of $I + \mathcal{K}$ on $\partial\Omega^C$.

In [141], Verchota investigated the Neumann problem (94) in full generality. He considered Lipschitz domains with compact, connected boundary contained in $\mathbb{R}^n$, $n \geq 2$. He showed that if $-1/(n-1) \leq \rho < 1$, then the Neumann problem is well-posed provided $2 - \varepsilon < p < 2 + \varepsilon$. That is, the solutions exist, satisfy the desired estimates, and are unique either modulo functions of an appropriate class or (in the case where Ω is unbounded) when subject to an appropriate growth condition. See [141, Theorems 13.2 and 15.4]. Verchota's proof also used boundedness and invertibility of certain potentials on $L^p(\partial\Omega)$; a crucial step was a coercivity estimate $\|\nabla^2 u\|_{L^2(\partial\Omega)} \leq C\|K_\rho u\|_{W^2_{-1}(\partial\Omega)} + C\|M_\rho u\|_{L^2(\partial\Omega)}$. (This estimate is valid provided u is biharmonic and satisfies some mean-value hypotheses; see [141, Theorem 7.6]).

More recently, in [132], Shen improved upon Verchota's results by extending the range on p (in bounded simply connected Lipschitz domains) to $2(n-1)/(n+1) - \varepsilon < p < 2 + \varepsilon$ if $n \geq 4$, and $1 < p < 2 + \varepsilon$ if $n = 2$ or $n = 3$. This result again was proven by inverting layer potentials. Observe that the L^p-regularity problem is also known to be well-posed for p in this range, and (if $n \geq 6$) in a broader range of p; see section "The Regularity Problem and the L^p-Dirichlet Problem." The question of the sharp range of p for which the L^p-Neumann problem is well-posed in a Lipschitz domain is still open.

Finally, in [109, Sect. 6.5], I. Mitrea and M. Mitrea showed that if $\Omega \subset \mathbb{R}^n$ is a simply connected domain whose unit outward normal ν lies in $VMO(\partial\Omega)$ (for example, if Ω is a C^1 domain), then the acceptable range of p is $1 < p < \infty$; this may be seen as a generalization of the result of Cohen and Gosselin to higher dimensions, to other values of ρ, and to slightly rougher domains.

It turns out that extending the well-posedness results for the Neumann problem beyond the case of the bilaplacian is an excruciatingly difficult problem, even if one considers only fourth-order operators with constant coefficients. Even defining Neumann boundary values for more general operators is a difficult problem (see section "Formulation of Neumann Boundary Data"), and while some progress has been made (see [7, 20, 25, 109], or section "Formulation of Neumann Boundary Data" below), at present there are no well-posedness results for the Neumann problem with L^p boundary data.

In analogy to (95) and (96), one can write

$$\int_\Omega w\, Lu = A[u, w] + \int_{\partial\Omega} w\, K_A u - \partial_\nu w\, M_A u\, d\sigma, \tag{97}$$

where $A[u, w] = \sum_{|\alpha|=|\beta|=2} a_{\alpha\beta} \int_\Omega D^\beta u\, D^\alpha w$ is an energy form associated with the operator $L = \sum_{|\alpha|=|\beta|=2} a_{\alpha\beta} D^\alpha D^\beta$. Note that in the context of fourth-order operators, the pair $(w, \partial_\nu w)$ constitutes the Dirichlet data for w on the boundary, and so one can say that the operators $K_A u$ and $M_A u$ define the Neumann data for u. One immediately faces the problem that the same higher-order operator L can be rewritten in many different ways and gives rise to different energy forms. The corresponding Neumann data will be different. (This is the reason why there is a family of Neumann data for the biharmonic operator.)

Furthermore, whatever the choice of the form, in order to establish well-posedness of the Neumann problem, one needs to be able to estimate all second derivatives of a solution on the boundary in terms of the Neumann data. In the analogous second-order case, such an estimate is provided by the Rellich identity, which shows that the tangential derivatives are equivalent to the normal derivative in L^2 for solutions of elliptic PDEs. In the higher-order scenario, such a result calls for certain coercivity estimates which are still rather poorly understood. We refer the reader to [142] for a detailed discussion of related results and problems.

Inhomogeneous Problems and Other Classes of Boundary Data

In [4], Adolfsson and Pipher investigated the inhomogeneous Dirichlet problem for the biharmonic equation with data in Besov and Sobolev spaces. While resting on the results for homogeneous boundary value problems discussed in sections "The Dirichlet Problem: Definitions, Layer Potentials, and Some Well–Posedness Results" and "The Regularity Problem and the L^p-Dirichlet Problem," such a framework presents a completely new setting, allowing for the inhomogeneous problem and for consideration of classes of boundary data which are, in some sense, intermediate between the Dirichlet and the regularity problems.

They showed that if $\dot{f} \in WA^p_{1+s}(\partial\Omega)$ and $h \in L^p_{s+1/p-3}(\Omega)$, then there exist a unique function u that satisfies

$$\begin{cases} \Delta^2 u = h & \text{in } \Omega, \\ \operatorname{Tr} \partial^\alpha u = f_\alpha, & \text{for } 0 \leq |\alpha| \leq 1 \end{cases} \tag{98}$$

subject to the estimate

$$\|u\|_{L^p_{s+1/p+1}(\Omega)} \leq C\|h\|_{L^p_{s+1/p-3}(\Omega)} + C\|\dot{f}\|_{WA^p_{1+s}(\partial\Omega)} \tag{99}$$

provided $2 - \varepsilon < p < 2 + \varepsilon$ and $0 < s < 1$. Here $\operatorname{Tr} w$ denotes the trace of w in the sense of Sobolev spaces; that these may be extended to functions $u \in L^p_{s+1+1/p}$, $s > 0$, was proven in [4, Theorem 1.12].

In Lipschitz domains contained in $\mathbb{R}^3$, they proved these results for a broader range of p and s, namely for $0 < s < 1$ and for

$$\max\left(1, \frac{2}{s+1+\varepsilon}\right) < p < \begin{cases} \infty, & s < \varepsilon, \\ \frac{2}{s-\varepsilon}, & \varepsilon \leq s < 1. \end{cases} \tag{100}$$

Finally, in C^1 domains, they proved these results for any p and s with $1 < p < \infty$ and $0 < s < 1$.

In [107], Mitrea et al. extended the three-dimensional results to $p = \infty$ (for $0 < s < \varepsilon$) or $2/(s+1+\varepsilon) < p \leq 1$ (for $1-\varepsilon < s < 1$). They also extended these results to data h and $\dot{f}$ in more general Besov or Triebel–Lizorkin spaces.

In [108], I. Mitrea and M. Mitrea extended the results of [4] to higher dimensions. That is, they showed that if $\Omega \subset \mathbb{R}^n$ is a Lipschitz domain and $n \geq 4$, then there is a unique solution to the problem (98) subject to the estimate (99) provided that $0 < s < 1$ and that

$$\max\left(1, \frac{n-1}{s+(n-1)/2+\varepsilon}\right) < p < \begin{cases} \infty, & (n-3)/2+s < \varepsilon, \\ \frac{n-1}{(n-3)/2+s-\varepsilon}, & \varepsilon \leq s < 1. \end{cases}$$

As in [107], their results extend to more general function spaces.

I. Mitrea and M. Mitrea also showed that, for the same values of p and s, there exist unique solutions to the inhomogeneous *Neumann* problem

$$\begin{cases} \Delta^2 u = h & \text{in } \Omega, \\ M_\rho u = f & \text{on } \partial\Omega, \\ K_\rho u = \Lambda & \text{on } \partial\Omega \end{cases}$$

where M_ρ and K_ρ are as in section "The Neumann Problem for the Biharmonic Equation," subject to the estimate

$$\|u\|_{L^p_{s+1/p+1}(\Omega)} \leq C\|h\|_{L^p_{s+1/p-3}(\Omega)} + C\|f\|_{B^{p,p}_{s-1}(\partial\Omega)} + C\|\Lambda\|_{B^{p,p}_{s-2}(\partial\Omega)}. \tag{101}$$

Finally, in [109, Sect. 6.4], I. Mitrea and M. Mitrea proved similar results for more general constant-coefficient elliptic operators. That is, if L is an operator given by formula (7) whose coefficients satisfy the ellipticity condition (8), and if $\Omega \subset \mathbb{R}^n$ is a bounded Lipschitz domain, then there exist a unique solution to the Dirichlet problem

$$\begin{cases} Lu = h & \text{in } \Omega, \\ \operatorname{Tr} \partial^\alpha u = f_\alpha, & \text{for } 0 \leq |\alpha| \leq 1 \end{cases} \tag{102}$$

subject to the estimate

$$\|u\|_{B^{p,q}_{m-1+s+1/p}(\Omega)} \leq C\|h\|_{B^{p,q}_{-m-1+s+1/p}(\Omega)} + C\|\dot{f}\|_{WA^{p,q}_{m-1+s}(\partial\Omega)} \tag{103}$$

provided $2-\varepsilon < p < 2+\varepsilon$, $2-\varepsilon < q < 2+\varepsilon$, and $1/2-\varepsilon < s < 1/2+\varepsilon$. Furthermore, with a slightly stronger ellipticity condition

$$\operatorname{Re} \sum_{|\alpha|=|\beta|=m} \sum_{j,k=1}^{\ell} a^{jk}_{\alpha\beta} \zeta^\alpha_j \overline{\zeta^\beta_k} \geq \lambda|\zeta|^2,$$

they established well-posedness of the inhomogeneous Neumann problem for arbitrary constant-coefficient operators. See section "Formulation of Neumann Boundary Data" below for a formulation of Neumann boundary data in the case of arbitrary operators. Finally, if L is self-adjoint and $n > 2m$, and if the unit outward normal ν to $\partial\Omega$ lies in $VMO(\partial\Omega)$, then the Dirichlet problem (102) has a unique solution satisfying the estimate (103) for any $0 < s < 1$ and any $1 < p < \infty$, $1 < q < \infty$.

Let us define the function spaces appearing above. $L^p_\alpha(\mathbb{R}^n)$ is defined to be $\{g : (I - \Delta)^{\alpha/2} g \in L^p(\mathbb{R}^n)\}$; we say $g \in L^p_\alpha(\Omega)$ if $g = h\big|_\Omega$ for some $h \in L^p_\alpha(\mathbb{R}^n)$. If k is a nonnegative integer, then $L^p_k = W^p_k$. If m is an integer and $0 < s < 1$, then the Whitney–Besov space $WA^p_{m-1+s} = WA^{p,p}_{m-1+s}$ or $WA^{p,q}_{m-1+s}$ is defined analogously to WA^p_m (see Definition 4.2), except that we take the completion with respect to the Whitney–Besov norm

$$\sum_{|\alpha|\leq m-1} \|\partial^\alpha \psi\|_{L^p(\partial\Omega)} + \sum_{|\alpha|=m-1} \|\partial^\alpha \psi\|_{B^{p,q}_s(\partial\Omega)} \tag{104}$$

rather than the Whitney–Sobolev norm

$$\sum_{|\alpha|\leq m-1} \|\partial^\alpha \psi\|_{L^p(\partial\Omega)} + \sum_{|\alpha|=m-1} \|\nabla_\tau \partial^\alpha \psi\|_{L^p(\partial\Omega)}.$$

In [4], the general problem (98) for Δ^2 was first reduced to the case $h = 0$ (that is, to a homogeneous problem) by means of trace/extension theorems, that is, subtracting $w(X) = \int_{\mathbb{R}^n} F(X,Y)\,\tilde{h}(Y)\,dY$, and showing that if $h \in L^p_{s+1/p-3}(\Omega)$ then $(\operatorname{Tr} w, \operatorname{Tr} \nabla w) \in WA^p_{1+s}(\partial\Omega)$. Next, the well-posedness of Dirichlet and regularity problems discussed in sections "The Dirichlet Problem: Definitions, Layer Potentials, and Some Well-Posedness Results" and "The Regularity Problem and the L^p-Dirichlet Problem" provide the endpoint cases $s = 0$ and $s = 1$, respectively. The core of the matter is to show that, if u is biharmonic, k is an integer, and $0 \leq \alpha \leq 1$, then $u \in L^p_{k+\alpha}(\Omega)$ if and only if

$$\int_\Omega |\nabla^{k+1} u(X)|^p \operatorname{dist}(X, \partial\Omega)^{p-p\alpha} + |\nabla^k u(X)|^p + |u(X)|^p \, dX < \infty, \tag{105}$$

(cf. [4, Proposition S]). With this at hand, one can use square-function estimates to justify the aforementioned endpoint results. Indeed, observe that for $p = 2$ the first integral on the left-hand side of (105) is exactly the L^2 norm of $S(\nabla^k u)$. The latter, by [115] (discussed in section "The Area Integral"), is equivalent to the L^2 norm of the corresponding nontangential maximal function, connecting the estimate (99) to the nontangential estimates in the Dirichlet problem (74) and the regularity problem (80). Finally, one can build an interpolation-type scheme to pass to well-posedness in intermediate Besov and Sobolev spaces.

The solution in [109] to the problem (102), at least in the case of general Lipschitz domains, was constructed in the opposite way, by first reducing to the case where the boundary data $\dot{f} = 0$. Using duality it is straightforward to establish well-posedness in the case $p = q = 2$, $s = 1/2$; perturbative results then suffice to extend to p, q near 2 and s near $1/2$.

Boundary Value Problems with Variable Coefficients

Results for higher-order differential equations with variable coefficients are very scarce. As we discussed in section "Higher-Order Operators: Divergence Form and Composition Form," there are two natural manifestations of higher-order operators with variable coefficients. Operators in divergence form arise via the weak formulation framework. Conversely, operators in composition form generalize the bilaplacian under a pull-back of a Lipschitz domain to the upper half-space.

Both classes of operators have been investigated. However, operators in divergence form have received somewhat more study; thus, we begin this section by reviewing the definition of divergence-form operator. A divergence-form operator L, acting on $W^2_{m,loc}(\Omega \mapsto \mathbb{C}^\ell)$, may be defined weakly via (12); we say that $Lu = h$ if

$$\sum_{j=1}^{\ell} \int_\Omega \varphi_j \, h_j = (-1)^m \sum_{j,k=1}^{\ell} \sum_{|\alpha|=|\beta|=m} \int_\Omega \partial^\alpha \varphi_j \, a^{jk}_{\alpha\beta} \, \partial^\beta u_k \tag{106}$$

for all φ smooth and compactly supported in Ω.

The Kato Problem and the Riesz Transforms

We begin with the Kato problem and the properties of the Riesz transform; this is an important topic in elliptic theory, which formally stands somewhat apart from the well-posedness issues.

Suppose that L is a variable-coefficient operator in divergence form, that is, an operator defined by (106). Suppose that L satisfies the bound

$$\left| \sum_{|\alpha|=|\beta|=m} \sum_{j,k=1}^{\ell} \int_{\mathbb{R}^n} a^{jk}_{\alpha\beta} \, \partial^\beta f_k \, \partial^\alpha g_j \right| \leq C \|\nabla^m f\|_{L^2(\mathbb{R}^n)} \|\nabla^m g\|_{L^2(\mathbb{R}^n)} \tag{107}$$

for all f and g in $\dot{W}^2_m(\mathbb{R}^n)$, and the ellipticity estimate

$$\operatorname{Re} \sum_{|\alpha|=|\beta|=m} \sum_{j,k=1}^{\ell} \int_\Omega a^{jk}_{\alpha\beta}(X) \partial^\beta \varphi_k(X) \partial^\alpha \overline{\varphi_j}(X)\, dX \geq \lambda \sum_{|\alpha|=m} \sum_{k=1}^{\ell} \int_\Omega |\partial^\alpha \varphi_k|^2 \tag{108}$$

for all functions $\varphi \in C^\infty_0(\Omega \mapsto \mathbb{C}^\ell)$. Notice that this is a weaker requirement than the pointwise ellipticity condition (11).

Auscher et al. [16] proved that under these conditions, the Kato estimate

$$\frac{1}{C}\|\nabla^m f\|_{L^2(\mathbb{R}^n)} \leq \|\sqrt{L} f\|_{L^2(\mathbb{R}^n)} \leq C\|\nabla^m f\|_{L^2(\mathbb{R}^n)} \tag{109}$$

is valid for some constant C. They also proved similar results for operators with lower-order terms.

It was later observed in [11] that by the methods of [15], if $1 \leq n \leq 2m$, then the bound on the Riesz transform $\nabla^m L^{-1/2}$ in L^p [that is, the first inequality in formula (109)] extends to the range $1 < p < 2 + \varepsilon$, and the reverse Riesz transform bound [that is, the second inequality in formula (109)] extends to the range $1 < p < \infty$. This also holds if the Schwartz kernel $W_t(X, Y)$ of the operator e^{-tL} satisfies certain pointwise bounds (e.g., if L is second order and the coefficients of A are real).

In the case where $n > 2m$, the inequality $\|\nabla^m L^{-1/2} f\|_{L^p(\mathbb{R}^n)} \leq C\|f\|_{L^p(\mathbb{R}^n)}$ holds for $2n/(n+2m) - \varepsilon < p \leq 2$; see [11, 26]. The reverse inequality holds for $\max(2n/(n+4m) - \varepsilon, 1) < p < 2$ by [11, Theorem 18], and for $2 < p < 2n/(n-2m) + \varepsilon$ by duality (see [12, Sect. 7.2]).

In the case of second-order operators, the Kato estimate implies well-posedness of boundary value problems with L^2 data in the upper half-space for certain coefficients in a special ("block") form. We conjecture that the same is true in the case of higher-order operators; see section "Open Questions and Preliminary Results."

The Dirichlet Problem for Operators in Divergence Form

In this section we discuss boundary value problems for divergence-form operators with variable coefficients. At the moment, well-posedness results for such operators are restricted in that the boundary problems treated fall *strictly* between the range of L^p-Dirichlet and L^p-regularity, in the sense of section "Inhomogeneous Problems for the Biharmonic Equation." That is, there are at present no well-posedness results for the L^p-Dirichlet, regularity, and Neumann problems on Lipschitz domains with the usual sharp estimates in terms of the nontangential maximal function for these divergence-form operators. (Such problems are now being considered; see section "Open Questions and Preliminary Results.")

To be more precise, recall from the discussion in section "Inhomogeneous Problems for the Biharmonic Equation" that the classical Dirichlet and regularity problems, with boundary data in L^p, can be viewed as the $s = 0, 1$ endpoints of the boundary problem studied in [4, 107, 108]

$$\Delta^2 u = h \text{ in } \Omega, \quad \partial^\alpha u\big|_{\partial\Omega} = f_\alpha \text{ for all } |\alpha| \le 1$$

with $\dot{f}$ lying in an *intermediate* smoothness space $WA^p_{1+s}(\partial\Omega)$, $0 \le s \le 1$. In the context of divergence-form higher-order operators with variable coefficients, essentially the known results pertain *only* to boundary data of intermediate smoothness.

In [7], Agranovich investigated the inhomogeneous Dirichlet problem, in Lipschitz domains, for such operators L that are elliptic [in the pointwise sense of (11), and not the more general condition (108)] and whose coefficients $a^{jk}_{\alpha\beta}$ are Lipschitz continuous in Ω.

He showed that if $h \in L^p_{-m-1+1/p+s}(\Omega)$ and $\dot{f} \in WA^p_{m-1+s}(\partial\Omega)$, for some $0 < s < 1$, and if $|p-2|$ is small enough, then the Dirichlet problem

$$\begin{cases} Lu = h & \text{in } \Omega, \\ \operatorname{Tr} \partial^\alpha u = f_\alpha & \text{for all } 0 \le |\alpha| \le m-1 \end{cases} \tag{110}$$

has a unique solution u that satisfies the estimate

$$\|u\|_{L^p_{m-1+s+1/p}(\Omega)} \le C\|h\|_{L^p_{-m-1+1/p+s}(\partial\Omega)} + C\|\dot{f}\|_{WA^p_{m-1+s}(\partial\Omega)}. \tag{111}$$

Agranovich also considered the Neumann problem for such operators. As we discussed in section "The Neumann Problem for the Biharmonic Equation," defining the Neumann problem is a delicate matter. In the context of zero boundary data, the situation is a little simpler as one can take a formal functional analytic point of view and avoid to some extent the discussion of estimates at the boundary. We say that u solves the Neumann problem for L, with homogeneous boundary data, if Eq. (106) in the weak formulation of L is valid for all test functions φ compactly supported in $\mathbb{R}^n$ (but not necessarily in Ω). Agranovich showed that, if $h \in \mathring{L}^p_{-m-1+1/p+s}(\Omega)$, then there exist a unique function $u \in L^p_{m-1+1/p+s}(\Omega)$ that solves this Neumann problem with homogeneous boundary data, under the same conditions on p, s, L as for his results for the Dirichlet problem. Here $h \in \mathring{L}^p_\alpha(\Omega)$ if $h = g\big|_\Omega$ for some $g \in L^p_\alpha(\mathbb{R}^n)$ that in addition is supported in $\bar{\Omega}$.

In [100], Maz'ya et al. considered the Dirichlet problem, again with boundary data in intermediate Besov spaces, for much rougher coefficients. They showed that if $f \in WA^p_{m-1+s}(\partial\Omega)$, for some $0 < s < 1$ and some $1 < p < \infty$, if h lies in an appropriate space, and if L is a divergence-form operator of order $2m$ [as defined by (106)], then under some conditions, there is a unique function u that satisfies the Dirichlet problem (110) subject to the estimate

$$\|u\|_{W^p_{m,1-s-1/p}} = \left(\sum_{|\alpha|\le m} \int_\Omega |\partial^\alpha u(X)|^p \operatorname{dist}(X,\partial\Omega)^{p-ps-1}\, dX\right)^{1/p} < \infty. \tag{112}$$

See [100, Theorem 8.1]. The inhomogeneous data h is required to lie in the space $V^p_{-m,1-s-1/p}(\Omega)$, the dual space to $V^q_{m,s+1/p-1}(\Omega)$, where

$$\|w\|_{V^p_{m,a}} = \left(\sum_{|\alpha|\leq m} \int_\Omega |\partial^\alpha u(X)|^p \operatorname{dist}(X,\partial\Omega)^{pa+p|\alpha|-pm}\, dX\right)^{1/p}. \tag{113}$$

Notice that $w \in V^p_{m,a}$ if and only if $w \in W^p_{m,a}$ and $\partial^\alpha w = 0$ on $\partial\Omega$ for all $0 \leq |\alpha| \leq m-1$.

The conditions are that the coefficients $a^{jk}_{\alpha\beta}$ satisfy the weak ellipticity condition (108) considered in the theory of the Kato problem, that Ω be a Lipschitz domain whose normal vector ν lies in $VMO(\partial\Omega)$, and that the coefficients $a^{ij}_{\alpha\beta}$ lie in $L^\infty(\mathbb{R}^n)$ and in $VMO(\mathbb{R}^n)$. Recall that this condition on Ω has also arisen in [106] [it ensures the validity of formula (48)]. Notice that the L^∞ bound on the coefficients is a stronger condition than the bound (107) of [16], and the requirement that the coefficients lie in $VMO(\mathbb{R}^n)$ is a regularity requirement that is weaker than the requirement of [7] that the coefficients be Lipschitz continuous.

In fact, [100] provides a more intricate result, allowing one to deduce a well-posedness range of s and p, given information about the oscillation of the coefficients $a^{jk}_{\alpha\beta}$ and the normal to the domain ν. In the extreme case, when the oscillations for both are vanishing, the allowable range expands to $0 < s < 1$, $1 < p < \infty$, as stated above.

The construction of solutions to the Dirichlet problem may be simplified using trace and extension theorems. In [100, Proposition 7.3], the authors showed that if $\dot f \in WA^p_{m-1+s}(\partial\Omega)$, then there exist a function $F \in W^p_{m,a}$ such that $\partial^\alpha F = f_\alpha$ on $\partial\Omega$. It is easy to see that if $F \in W^p_{m,a}$, and the coefficients $a^{jk}_{\alpha\beta}$ of L are bounded pointwise, then $LF \in V_{-m,1-s-1/p}$. Thus, the Dirichlet problem

$$\begin{cases} Lu = h \text{ in } \Omega, \\ \partial^\alpha u\big|_{\partial\Omega} = f_\alpha, \text{ for all } |\alpha| \leq m-1, \\ \|u\|_{W^p_{m,1-s-1/p}} \leq \|h\|_{V^p_{-m,1-s-1/p}} + \|\dot f\|_{WA^p_{m-1+s}(\partial\Omega)} \end{cases}$$

may be solved by solving the Dirichlet problem with homogeneous boundary data

$$\begin{cases} Lw = h - LF \text{ in } \Omega, \\ \partial^\alpha w\big|_{\partial\Omega} = 0 \text{ for all } |\alpha| \leq m-1, \\ \|w\|_{W^p_{m,1-s-1/p}} \leq \|h\|_{V^p_{-m,1-s-1/p}} + \|LF\|_{V^p_{-m,1-s-1/p}} \end{cases}$$

for some extension F and then letting $u = w + F$.

We comment on the estimate (112). First, by [4, Proposition S] [listed above as formula (105)], if u is biharmonic, then the estimate (112) is equivalent to the estimate (111) of [7]. Second, by (90), if the coefficients $a^{jk}_{\alpha\beta}$ are constant, one can draw connections between (112) for $s = 0, 1$ and the nontangential maximal estimates of the Dirichlet or regularity problems (87) or (88). However, as we pointed out earlier, this endpoint case, corresponding to the true L^p-Dirichlet and regularity problems, has not been achieved.

The Dirichlet Problem for Operators in Composition Form

Let us now discuss variable-coefficient fourth-order operators in composition form. Recall that this particular form arises naturally when considering the transformation of the bilaplacian under a pull-back from a Lipschitz domain [cf. (13)]. The authors of the present paper have shown the well-posedness, for a class of such operators, of the Dirichlet problem with boundary data in L^2, thus establishing the first results concerning the L^p-Dirichlet problem for variable-coefficient higher-order operators.

Consider the Dirichlet problem

$$\begin{cases} L^*(aLu) = 0 & \text{in } \Omega, \\ u = f & \text{on } \partial\Omega, \\ \nu \cdot A\nabla u = g & \text{on } \partial\Omega, \\ \|\tilde{N}(\nabla u)\|_{L^2(\partial\Omega)} \leq C\|\nabla f\|_{L^2(\partial\Omega)} + C\|g\|_{L^2(\partial\Omega)}. \end{cases} \tag{114}$$

Here L is a *second-order* divergence-form differential operator $L = -\operatorname{div} A(X)\nabla$, and a is a scalar-valued function. (For rough coefficients A, the exact weak definition of $L^*(aLu) = 0$ is somewhat delicate, and so we refer the reader to [23].) The domain Ω is taken to be the domain above a Lipschitz graph, that is, $\Omega = \{(x,t) : x \in \mathbb{R}^{n-1}, t > \varphi(x)\}$ for some function φ with $\nabla\varphi \in L^\infty(\mathbb{R}^{n-1})$. As pointed out above, the class of equations $L^*(aLu) = 0$ is preserved by a change of variables, and so well-posedness of the Dirichlet problem (114) in such domains follows from well-posedness in upper half-spaces $\mathbb{R}^n_+$. Hence, in the remainder of this section, $\Omega = \mathbb{R}^n_+$.

The appropriate ellipticity condition is then

$$\lambda \leq a(X) \leq \Lambda, \qquad \lambda|\eta|^2 \leq \operatorname{Re} \overline{\eta}^t A(X)\eta, \quad |A(X)| \leq \Lambda \tag{115}$$

for all $X \in \mathbb{R}^n$ and all $\eta \in \mathbb{C}^n$, for some constants $\Lambda > \lambda > 0$. The modified nontangential maximal function $\tilde{N}(\nabla u)$, defined by

$$\tilde{N}(\nabla u)(Q) = \sup_{X\in\Gamma(Q)} \left(\fint_{B(X,\operatorname{dist}(X,\partial\Omega)/2)} |\nabla u|^2 \right)^{1/2},$$

is taken from [67] and is fairly common in the study of variable-coefficient elliptic operators.

In this case, we say that $u\big|_{\partial\Omega} = f$ and $\nu \cdot A\nabla u = g$ if

$$\lim_{t\to 0+} \|u(\cdot\ + t\mathbf{e}) - f\|_{W_1^2(\partial\Omega)} = 0,$$

$$\lim_{t\to 0+} \|\nu \cdot A\nabla u(\cdot\ + t\mathbf{e}) - g\|_{L^2(\partial\Omega)} = 0$$

where $\mathbf{e} = \mathbf{e}_n$ is the unit vector in the vertical direction. Notice that by the restriction on the domain Ω, $\mathbf{e}$ is transverse to the boundary at all points. We usually refer to the vertical direction as the t-direction, and if some function depends only on the first $n - 1$ coordinates, we say that function is t-independent.

In [23], the authors of the present paper have shown that if $n \geq 3$, and if a and A satisfy (115) and are t-independent, then for every $f \in W_1^2(\partial\Omega)$ and every $g \in L^2(\partial\Omega)$, there exist a u that satisfies (114), provided that the second-order operator $L = \operatorname{div} A\nabla$ is good from the point of view of the second-order theory.

Without going into the details, we mention that there are certain restrictions on the coefficients A necessary to ensure the well-posedness even of the corresponding second-order boundary value problems; see [27]. The key issues are good behavior in the direction transverse to the boundary, and symmetry. See [61, 67] for results for symmetric t-independent coefficients, [58, 59, 68, 69, 122] for well-posedness results and important counterexamples for non-symmetric coefficients, and [8, 17, 18] for perturbation results for t-independent coefficients.

In particular, using the results of [8, 17, 18], we have established that the L^2-Dirichlet problem (114) in the upper half-space is well-posed, provided the coefficients a and A satisfy (115) and are t-independent, if in addition one of the following conditions holds:

1. The matrix A is real and symmetric,
2. The matrix A is constant,
3. The matrix A is in block form (see section "Open Questions and Preliminary Results") and the Schwartz kernel $W_t(X, Y)$ of the operator e^{-tL} satisfies certain pointwise bounds, or
4. There is some matrix A_0, satisfying (1), (2), or (3), that again satisfies(115) and is t-independent, such that $\|A - A_0\|_{L^\infty(\mathbb{R}^{n-1})}$ is small enough (depending only on the constants λ, Λ in (115)).

The solutions to (114) take the following form. Inspired by formula (79) (taken from [39]), and a similar representation in [116], we let

$$\mathcal{E}h = \int_\Omega F(X, Y)\frac{1}{a(Y)}\partial_n^2 \mathcal{S}_* h(Y)\, dY \tag{116}$$

for h defined on $\partial\Omega$, where $\mathcal{S}_*$ is the (second-order) single layer potential associated with L^* and F is the fundamental solution associated with L. Then $a(X)\, L(\mathcal{E}h)(X) = \partial_n^2 \mathcal{S}_* f(X)$ in Ω (and is zero in its complement); if A^* is t-independent, then $L^*(\partial_n^2 \mathcal{S}_* h) = \partial_n^2 L^*(\mathcal{S}_* h) = 0$. Thus $u = w + \mathcal{E}h$ is a solution to (114), for any solution w to $Lw = 0$. The estimate $\|\tilde{N}(\nabla\mathcal{E}h)\|_{L^2(\partial\Omega)} \leq \|h\|_{L^2(\partial\Omega)}$ must then be established. In the case of biharmonic functions (considered in [116]), this estimate follows from the boundedness of the Cauchy integral; in the case of (114), this is the most delicate part of the construction, as the operators involved are far from being Calderón–Zygmund kernels. Once this estimate has been established, it can be shown, by an argument that precisely parallels that of [116], that there exists a w and h such that $Lw = 0$ and $u = w + \mathcal{E}h$ solves (114).

The Fundamental Solution

A set of important tools, and interesting objects of study in their own right, are the fundamental solutions and Green's functions of differential operators in various domains. To mention some applications presented in this survey, recall from sections "Sharp Pointwise Estimates on the Derivatives of Solutions in Arbitrary Domains" and "Green Function Estimates" that bounds on Green's functions G are closely tied to maximum principle estimates, and from sections "The Wiener Test: Continuity of Solutions" and "The Higher-Order Wiener Test: Continuity of Derivatives of Polyharmonic Functions" that the fundamental solution F is used to establish regularity of boundary points (that is, the Wiener criterion). See in particular Theorem 3.6.

Furthermore, fundamental solutions and Green's functions are often crucial elements of the construction of solutions to boundary value problems. In the case of boundary value problems in divergence form, the fundamental solution or Green's function for the corresponding higher-order operators is often useful; see the constructions in [33, 100, 118, 119, 125, 140], or in formulas (77) and (78) above. In the case of operators in composition form, it is often more appropriate to use the fundamental solution for the lower-order components; see, for example, formulas (79), (92), and (116), or the paper [139], which makes extensive use of the Green's function for $(-\Delta)^m$ to solve boundary value problems for $(-\Delta)^{m+1}$.

We now discuss some constructions of the fundamental solution. In the case of the biharmonic equation, and more generally in the case of constant-coefficient equations, the fundamental solution may be found in a fairly straightforward fashion, for example, by use of the Fourier transform; see, for example, formulas (20) and (21) above, [60, 63, 110, 114, 126] (the relevant results of which are summarized as [109, Theorem 4.2]), or [42, 43]. In the case of variable-coefficient second-order operators, the fundamental solution has been constructed in [45, 53, 55, 57, 66, 76, 121] under progressively weaker assumptions on the operators. The most recent of these papers, [121], constructs the fundamental solution F for a second-order operator L under the assumption that if $Lu = 0$ in some ball $B(X, r)$, then we have the local boundedness estimate

$$|u(X)| \leq C\left(\frac{1}{r^n}\int_{B(X,r)} |u|^2\right)^{1/2} \tag{117}$$

for some constant C depending only on L and not on u, X, or r. This assumption is true if L is a scalar second-order operator with real coefficients (see [111]) but is not necessarily true for more general elliptic operators (see [52]).

If $Lu = 0$ in $B(X, r)$ for some elliptic operator of order $2m$, where $2m > n$, then the local boundedness estimate (117) follows from the Poincaré inequality, the Caccioppoli inequality

$$\int_{B(X,r/2)} |\nabla^m u|^2 \leq \frac{C}{r^{2m}} \int_{B(X,r)} |u|^2 \tag{118}$$

and Morrey's inequality

$$|u(X)| \leq C \sum_{j=0}^{N} r^j \left(\frac{1}{r^n} \int_{B(X,r/2)} |\nabla^j u|^2 \right)^{1/2} \quad \text{whenever } N/2 > n.$$

Weaker versions of the Caccioppoli inequality (118) (that is, bounds with higher-order derivatives appearing on the right-hand side) were established in [14, 29]. In [21], the first author of the present paper established the full Caccioppoli inequality (118), thus establishing that if $2m > n$ then solutions to $Lu = 0$ satisfy the estimate (117). (Compare the results of section "Sharp Pointwise Estimates on the Derivatives of Solutions in Arbitrary Domains," in which solutions to $(-\Delta)^m u = f$ are shown to be pointwise bounded only if $2m > n - 2$; as observed in that section, Morrey's inequality yields one fewer degree of smoothness but was still adequate for the purpose of [21].)

Working much as in the second-order papers listed above, Barton then constructed the fundamental solution for divergence-form differential operators L with order $2m > n$ and with bounded coefficients satisfying the ellipticity condition (108). In the case of operators L with $2m \leq n$, she then constructed an auxiliary operator $\widetilde{L}$ with $2m > n$ and used the fundamental solution for $\widetilde{L}$ to construct the fundamental solution for L; this technique was also used in [16] to pass from operators of high order to operators of arbitrary order, and for similar reasons (i.e., to exploit pointwise bounds present only in the case of operators of very high order).

This technique allowed the proof of the following theorem, the main result of [21].

Theorem 5.1. *Let L be a divergence-form operator of order $2m$, acting on functions defined on $\mathbb{R}^n$, that satisfies the ellipticity condition* (108) *and whose coefficients A are pointwise bounded. Then there exist an array of functions $F^L_{j,k}(x, y)$ with the following properties.*

Let q and s be two integers that satisfy $q + s < n$ and the bounds $0 \leq q \leq \min(m, n/2)$, $0 \leq s \leq \min(m, n/2)$.

Then there is some $\varepsilon > 0$ such that if $x_0 \in \mathbb{R}^n$, if $0 < 4r < R$, if $A(x_0, R) = B(x_0, 2R) \setminus B(x_0, R)$, and if $q < n/2$ then

$$\int_{y \in B(x_0,r)} \int_{x \in A(x_0,R)} |\nabla_x^{m-s} \nabla_y^{m-q} F^L(x,y)|^2 \, dx \, dy \leq C r^{2q} R^{2s} \left(\frac{r}{R}\right)^{\varepsilon}. \tag{119}$$

If $q = n/2$, then we instead have the bound

$$\int_{y \in B(x_0,r)} \int_{x \in A(x_0,R)} |\nabla_x^{m-s} \nabla_y^{m-q} F^L(x,y)|^2 \, dx \, dy \leq C(\delta)\, r^{2q} R^{2s} \left(\frac{R}{r}\right)^{\delta} \tag{120}$$

for all $\delta > 0$ and some constant $C(\delta)$ depending on δ.

We also have the symmetry property

$$\partial_x^\gamma \partial_y^\delta F_{j,k}^L(x,y) = \overline{\partial_x^\gamma \partial_y^\delta F_{k,j}^{L^*}(y,x)} \tag{121}$$

as locally L^2 functions, for all multiindices γ, δ with $|\gamma| = m - q$ and $|\delta| = m - s$.

If in addition $q + s > 0$, then for all p with $1 \leq p \leq 2$ and $p < n/(n - (q + s))$, we have that

$$\int_{B(x_0,r)} \int_{B(x_0,r)} |\nabla_x^{m-s} \nabla_y^{m-q} F^L(x,y)|^p \, dx \, dy \leq C(p)\, r^{2n+p(s+q-n)} \tag{122}$$

for all $x_0 \in \mathbb{R}^n$ and all $r > 0$.

Finally, there is some $\varepsilon > 0$ such that if $2 - \varepsilon < p < 2 + \varepsilon$ then $\nabla^m \Pi^L$ extends to a bounded operator $L^p(\mathbb{R}^n) \mapsto L^p(\mathbb{R}^n)$. If γ satisfies $m - n/p < |\gamma| \leq m - 1$ for some such p, then

$$\partial_x^\gamma \Pi_j^L \dot{h}(x) = \sum_{k=1}^{N} \sum_{|\beta|=m} \int_{\mathbb{R}^n} \partial_x^\gamma \partial_y^\beta F_{j,k}^L(x,y)\, h_{k,\beta}(y)\, dy \quad \text{for a.e. } x \in \mathbb{R}^n \tag{123}$$

for all $\dot{h} \in L^p(\mathbb{R}^n)$ that are also locally in $L^P(\mathbb{R}^n)$, for some $P > n/(m - |\gamma|)$. In the case of $|\alpha| = m$, we still have that

$$\partial^\alpha \Pi_j^L \dot{h}(x) = \sum_{k=1}^{N} \sum_{|\beta|=m} \int_{\mathbb{R}^n} \partial_x^\alpha \partial_y^\beta F_{j,k}^L(x,y)\, h_{k,\beta}(y)\, dy \quad \text{for a.e. } x \notin \operatorname{supp} \dot{h} \tag{124}$$

for all $\dot{h} \in L^2(\mathbb{R}^n)$ whose support is not all of $\mathbb{R}^n$.

Here, if $\dot{h} \in L^2(\mathbb{R}^n)$, then $\Pi^L \dot{h}$ is the unique function in $\dot{W}_m^2(\mathbb{R}^n)$ that satisfies

$$\sum_{j=1}^{\ell} \sum_{|\alpha|=m} \int_\Omega \partial^\alpha \varphi_j\, h_{j,\alpha} = (-1)^m \sum_{j,k=1}^{\ell} \sum_{|\alpha|=|\beta|=m} \int_\Omega \partial^\alpha \varphi_j\, a_{\alpha\beta}^{jk}\, \partial^\beta (\Pi^L \dot{h})_k$$

for all $\varphi \in \dot{W}^2_m(\mathbb{R}^n)$. That is, $u = \Pi^L \dot{h}$ is the solution to $Lu = \operatorname{div}_m \dot{h}$. The formulas (123) and (124) represent the statement that F^L is the fundamental solution for L, that is, that $L_x F^L(x,y) = \delta_y(x)$ in some sense.

Thus, [21] contains a construction of the fundamental solution for divergence-form operators L of arbitrary order, with no smoothness assumptions on the coefficients of L or on solutions to $Lu = 0$ beyond boundedness, measurability, and ellipticity. These results are new even in the second-order case, as there exist second-order operators $L = -\operatorname{div} A\nabla$ whose solutions do not satisfy the local boundedness estimate (117) (see [52, 99]) and thus whose fundamental solution cannot be constructed as in [121].

Formulation of Neumann Boundary Data

Recall from section "The Neumann Problem for the Biharmonic Equation" that even defining the Neumann problem is a delicate matter. In the case of higher-order divergence-form operators with variable coefficients, the Neumann problem has thus received little study.

As discussed in section "The Dirichlet Problem for Operators in Divergence Form," Agranovich has established some well-posedness results for the inhomogeneous problem $Lu = h$ with homogeneous Neumann boundary data. He has also provided a formulation of inhomogeneous Neumann boundary values; see [7, Sect. 5.2].

This formulation is as follows. Observe that if the test function φ does not have zero boundary data, then formula (106) becomes

$$\sum_{j=1}^{\ell} \int_\Omega (Lu)_j \, \varphi_j = (-1)^m \sum_{j,k=1}^{\ell} \sum_{|\alpha|=|\beta|=m} \int_\Omega \partial^\alpha \varphi_j(X)\, a^{jk}_{\alpha\beta}(X)\, \partial^\beta u_k(X)\, dX \qquad (125)$$

$$+ \sum_{i=0}^{m-1} \sum_{j=1}^{\ell} \int_{\partial\Omega} B^j_{m-1-i} u \, \partial^i_\nu \varphi_j \, d\sigma$$

where $B_i u$ is an appropriate linear combination of the functions $\partial^\alpha u$ where $|\alpha| = m + i$. The expressions $B_i u$ may then be regarded as the Neumann data for u. Notice that if L is a fourth-order constant-coefficient scalar operator, then $B_0 = -M_A$ and $B_1 = K_A$, where K_A, M_A are given by (97). Agranovich provided some brief discussion of the conditions needed to resolve the Neumann problem with this notion of inhomogeneous boundary data. Essentially the same notion of Neumann boundary data was used in [109] (a book considering only the case of constant coefficients); an explicit formula for $B_i u$ in this case may be found in [109, Proposition 4.3].

However, there are several major problems with this notion of Neumann boundary data. These difficulties arise from the fact that the different components $B_i u$ may have different degrees of smoothness. For example, in the case of the biharmonic L^p-Neumann problem of section "The Neumann Problem for the Biharmonic Equation," the term $M_\rho u = -B_0 u$ is taken in the space $L^p(\partial\Omega)$, while the term $K_\rho u = B_1 u$ is taken in the negative smoothness space $W^p_{-1}(\partial\Omega)$.

If Ω is a Lipschitz domain, then the space $W^q_1(\partial\Omega)$ of functions with one degree of smoothness on the boundary is meaningful, and so we may define $W^p_{-1}(\partial\Omega)$, $1/p + 1/q = 1$, as its dual space. However, higher degrees of smoothness on the boundary and thus more negative smoothness spaces $W^p_{-k}(\partial\Omega)$ are not meaningful, and so this notion of boundary data is difficult to formulate on Lipschitz domains. (This difficulty may in some sense be circumvented by viewing the Neumann boundary data as lying in the dual space to $WA^p_{m-1+s}(\partial\Omega)$; see [108, 109]. However, this approach has some limits; for example, there is a rich theory of boundary value problems with boundary data in Hardy or Besov spaces which do not arise as dual spaces (i.e., with $p < 1$) and which is thus unavailable in this context.)

Furthermore, observe that as we discussed in section "Boundary Value Problems with Constant Coefficients," a core result needed to approach Neumann and regularity problems is a Rellich identity-type estimate, that is, an equivalence of norms of the Neumann and regularity boundary data of a solution; in the second-order case this may be stated as

$$\|\nabla_\tau u\|_{L^2(\partial\Omega)} \approx \|\nu \cdot A\nabla u\|_{L^2(\partial\Omega)}$$

whenever $\operatorname{div} A\nabla u = 0$ in Ω, for at least some domains Ω and classes of coefficients A. In section "Open Questions and Preliminary Results," we will discuss some possible approaches and preliminary results concerning higher-order boundary value problems, in which a higher-order generalization of the Rellich identity is crucial; thus, it will be highly convenient to have notions of regularity and Neumann boundary values that can both be reasonably expected to lie in the space L^2.

Thus, it is often convenient to formulate Neumann boundary values in the following way. Observe that, if $Lu = 0$ in Ω, and $\partial\Omega$ is connected, then for all nice test functions φ, the quantity

$$\sum_{j,k=1}^{\ell} \sum_{|\alpha|=|\beta|=m} \int_\Omega \partial^\alpha \varphi_j\, a^{jk}_{\alpha\beta}\, \partial^\beta u_k$$

depends only on the values of $\nabla^{m-1}\varphi$ on $\partial\Omega$; thus, there exist functions $M^{j,\gamma}_A u$ such that

$$\sum_{j,k=1}^{\ell} \sum_{|\alpha|=|\beta|=m} \int_\Omega \partial^\alpha \varphi_j\, a^{jk}_{\alpha\beta}\, \partial^\beta u_k = \sum_{|\gamma|=m-1} \sum_{j=1}^{\ell} \int_{\partial\Omega} M^{j,\gamma}_A u\, \partial^\gamma \varphi_j\, d\sigma. \tag{126}$$

We may then consider the array of functions $\dot{M}_A u$ to be the Neumann boundary values of u. This formulation only requires dealing with a single order of smoothness, but is somewhat less intuitive as there is no explicit formula for $\dot{M}_A u(X)$ in terms of the derivatives of u evaluated at X.

Also observe that, if we adopt this notion of Neumann boundary data, it is more natural to view the Dirichlet boundary values of u as the array $\{\partial^\gamma u\big|_{\partial\Omega} : |\gamma| = m-1\}$, and not $\{\partial^\gamma u\big|_{\partial\Omega} : |\gamma| \leq m-1\}$, as was done in sections "Boundary Value Problems with Constant Coefficients" and "The Dirichlet Problem for Operators in Divergence Form." The natural notion of regularity boundary values is then $\{\nabla_\tau \partial^\gamma u\big|_{\partial\Omega} : |\gamma| = m-1\}$; again, all components conveniently may then be expected to have the same degree of smoothness.

Open Questions and Preliminary Results

The well-posedness results of section "The Dirichlet Problem for Operators in Divergence Form" cover only a few classes of elliptic differential operators and some special boundary value problems; the theory of boundary value problems for higher-order divergence-form operators currently contains many open questions.

Some efforts are underway to investigate these questions. Using the Lax–Milgram theorem, it is straightforward to establish that the Poisson problems

$$Lu = h \text{ in } \Omega, \qquad \partial^\alpha u\big|_{\partial\Omega} = 0, \qquad \|u\|_{W^p_{m,1-s-1/p}} \leq C\|h\|_{V^p_{-m,1-s-1/p}} \tag{127}$$

$$Lu = h \text{ in } \Omega, \qquad \dot{M}_A u = 0, \qquad \|u\|_{W^p_{m,1-s-1/p}} \leq C\|h\|_{V^p_{-m,1-s-1/p}}. \tag{128}$$

are well-posed for $p = 2$ and $s = 1/2$. It is possible to show that these problems are well-posed whenever $|p-2|$ and $|s-1/2|$ are small enough; recall from [109] that if L has constant coefficients then the Dirichlet problem (102) and a similar Neumann problem are well-posed for this range of p and s.

Turning to a broader range of exponents p and s, we observe that perturbative results for the Poisson problems (127) and (128) are often fairly straightforward to establish. That is, with some modifications to the relevant function spaces, it is possible to show that if (127) or (128) is well-posed in some bounded domain Ω, for some operator L_0 and for some $1 < p < \infty$, $0 < s < 1$, and certain technical assumptions are satisfied, then the same problem must also be well-posed for any operator L_1 whose coefficients are sufficiently close to those of L_0 (in the L^∞ norm).

Recall from section "The Dirichlet Problem for Operators in Divergence Form" that for any $1 < p < \infty$ and $0 < s < 1$, the Dirichlet problem

$$Lu = 0 \text{ in } \Omega, \qquad \partial^\alpha u\big|_{\partial\Omega} = f_\alpha, \qquad \|u\|_{W^p_{m,1-s-1/p}} \leq C\|\dot{f}\|_{WA^p_{m-1+s}(\partial\Omega)} \tag{129}$$

for boundary data $\dot{f}$ in the fractional smoothness space $WA^p_{m-1+s}(\partial\Omega)$ can be reduced to well-posedness of the Poisson problem (127). (Some results are also available at the endpoint $p = \infty$ and in the case $p \leq 1$; the integer smoothness endpoints $s = 0$ and $s = 1$ generally must be studied using entirely different approaches.) A similar argument shows that well-posedness of the Neumann problem

$$Lu = 0 \text{ in } \Omega, \qquad M_A^{j,\gamma} u = g_{j,\gamma}, \qquad \|u\|_{W^p_{m,1-s-1/p}} \leq C\|\dot{g}\|_{(WA^q_{m-1+s}(\partial\Omega))^*} \tag{130}$$

follows from well-posedness of the Poisson problem (128) for $1 < p \leq \infty$. (In the case of the Neumann problem results for $p \leq 1$ are somewhat more involved.) A paper [20] containing these perturbative results was recently submitted by the first author of the present paper.

A key component in the construction of solutions of [20] are appropriate layer potentials, specifically, the Newton potential and the double and single layer potentials given by

$$(\Pi^L \dot{h})_j(x) = \sum_{k=1}^{\ell} \sum_{|\alpha|=m} \int_{\mathbb{R}^n} \partial_y^\alpha F_{j,k}^L(x,y)\, h_{k,\alpha}(y)\, dy,$$

$$(\mathcal{D}_\Omega^A \dot{f})_i(x) = \mathbf{1}_{\bar{\Omega}^C}(x)\tilde{f}_i(x) - \sum_{|\alpha|=|\beta|=m} \sum_{j,k=1}^{\ell} \int_{\bar{\Omega}^C} \partial_y^\alpha F_{i,j}^L(x,y)\, a_{\alpha\beta}^{jk}(y)\, \partial^\beta \tilde{f}_k(y)\, dy,$$

$$(\mathcal{S}_\Omega^A \dot{g})_j(x) = \sum_{k=1}^{\ell} \sum_{|\gamma|=m-1} \int_{\partial\Omega} \partial_y^\gamma F_{j,k}^L(x,y)\, g_{k,\gamma}(y)\, d\sigma(y)$$

where F^L denotes the fundamental solution discussed in section "The Fundamental Solution" and where $\tilde{f}$ is any function that satisfies $\partial^\gamma \tilde{f}_k = f_{k,\gamma}$ on $\partial\Omega$. We remark that these are very natural generalizations of layer potentials in the second-order case, and also of various potential operators used in the theory of constant-coefficient higher-order differential equations; see in particular [104, 109].

In particular, to solve the Dirichlet or Neumann problems (129) or (130), it was necessary to establish the bounds on layer potentials

$$\|\mathcal{D}_\Omega^A \dot{f}\|_{W^p_{m,1-s-1/p}} \leq C\|\dot{f}\|_{WA^p_{m-1+s}(\partial\Omega)}, \qquad \|\mathcal{S}_\Omega^A \dot{g}\|_{W^p_{m,1-s-1/p}} \leq C\|\dot{g}\|_{B^{p,p}_{s-1}(\partial\Omega)}$$

In [20], these bounds are derived from the bound

$$\|\Pi^L \dot{h}\|_{W^p_{m,1-s-1/p}} \leq C\|\dot{h}\|_{W^p_{0,1-s-1/p}}.$$

This bound is the technical assumption mentioned above; we remark that it is stable under perturbation and is always valid if $p = 2$ and $s = 1/2$.

Analogy with the second-order case suggests that, in order to establish well-posedness of the L^2-Dirichlet, L^2-Neumann, and L^2-regularity problems, a good first step would be to establish the estimates

$$\|\mathcal{D}^A_\Omega \dot{f}\|_{\mathfrak{X}} \leq C\|\dot{f}\|_{L^2(\partial\Omega)}, \quad \|\mathcal{D}^A_\Omega \dot{f}\|_{\mathfrak{Y}} \leq C\|\nabla_\tau \dot{f}\|_{L^2(\partial\Omega)}, \quad \|\mathcal{S}^A_\Omega \dot{g}\|_{\mathfrak{Y}} \leq C\|\dot{g}\|_{L^2(\partial\Omega)}$$

for some spaces $\mathfrak{X}$ and $\mathfrak{Y}$. (We remark that the corresponding bounds for constant-coefficient operators are Theorem 4.7 and Proposition 5.2 in [109], and therein were used to establish well-posedness results.) In the paper, the recently submitted paper [25], Steve Hofmann together with the authors of the present paper has established the bounds

$$\int_{\mathbb{R}^n}\int_0^\infty \left|\nabla^m \partial_t \mathcal{S}^A \dot{g}(x,t)\right|^2 t\,dt\,dx \leq C\|\dot{g}\|^2_{L^2(\partial\Omega)}, \tag{131}$$

$$\int_{\mathbb{R}^n}\int_0^\infty \left|\nabla^m \partial_t \mathcal{D}^A \dot{f}(x,t)\right|^2 t\,dt\,dx \leq C\|\nabla_x \dot{f}\|^2_{L^2(\partial\Omega)} \tag{132}$$

where $\mathcal{D}^A = \mathcal{D}^A_{\mathbb{R}^n_+}$, $\mathcal{S}^A = \mathcal{S}^A_{\mathbb{R}^n_+}$, for scalar operators L, provided that the coefficients $a_{\alpha\beta}$ are pointwise bounded, elliptic in the sense of (108), and are constant in the t-direction, that is, the direction transverse to the boundary of $\mathbb{R}^n_+$. As discussed in section "The Dirichlet Problem for Operators in Composition Form," this assumption of t-independent coefficients is very common in the theory of second-order differential equations.

We hope that in a future paper we may be able to extend this result to domains of the form $\Omega = \{(x,t) : t > \varphi(x)\}$ for some Lipschitz function φ; in the second-order case, this generalization may be obtained automatically via a change of variables, but in the higher-order case this technique is only available for equations in composition form (section "The Dirichlet Problem for Operators in Composition Form") and not in divergence form.

In the case of second-order equations $-\operatorname{div} A\nabla u = 0$, where the matrix A of coefficients is real (or self-adjoint) and t-independent, a straightforward argument involving Green's theorem establishes the Rellich identity

$$\|\nabla_x u(\,\cdot\,,0)\|_{L^2(\partial\mathbb{R}^n_+)} \approx \|-\vec{e}\cdot A\nabla u\|_{L^2(\partial\mathbb{R}^n_+)}$$

where $-\vec{e}$ is the unit outward normal to $\mathbb{R}^n_+$; that is, we have an equivalence of norms between the regularity and Neumann boundary values of a solution u to $\operatorname{div} A\nabla u = 0$. Together with boundedness and certain other properties of layer potentials, this estimate leads to well-posedness of the L^2-regularity and L^2-Neumann problems; it is then a straightforward argument to derive well-posedness of the Dirichlet problem. See [8, 13, 22, 59], and others.

We have hopes that a similar argument will yield the higher-order Rellich identity

$$\|\nabla_x \nabla^{m-1} u(\,\cdot\,,0)\|_{L^2(\partial\mathbb{R}^n_+)} \approx \|\dot{M}_A u\|_{L^2(\partial\mathbb{R}^n_+)}$$

where the Neumann boundary values $\dot{M}_A u$ are as in section "Formulation of Neumann Boundary Data," for solutions u to divergence-form equations with t-independent and self-adjoint coefficients (that is, coefficients that satisfy $a_{\alpha\beta} = \overline{a_{\beta\alpha}}$), and that a similar argument will imply well-posedness of higher-order L^2 boundary value problems.

The results of section "The Kato Problem and the Riesz Transforms" may also lead to well-posedness of L^2 boundary value problems for a different class of operators, namely, operators of block type. Again, this argument would proceed by establishing a Rellich-type identity.

Let us review the theory of second-order divergence-form operators $\mathbb{L} = -\operatorname{div} \mathbb{A}\nabla$ in $\mathbb{R}^{n+1}$, where $\mathbb{A}$ is an $(n+1) \times (n+1)$, t-independent matrix in block form; that is, $\mathbb{A}_{j,n+1} = \mathbb{A}_{n+1,j} = 0$ for $1 \leq j \leq n$, and $\mathbb{A}_{n+1,n+1} = 1$. It is fairly easy to see that one can formally realize the solution to $\mathbb{L}u = 0$ in $\mathbb{R}^{n+1}_+$, $u\big|_{\mathbb{R}^n} = f$, as the Poisson semigroup $u(x,t) = e^{-t\sqrt{L}}f(x)$, $(x,t) \in \mathbb{R}^{n+1}_+$. Then the Kato estimate (109) essentially provides an analogue of the Rellich identity-type estimate for the block operator $\mathbb{L}$, that is, the L^2-equivalence between normal and tangential derivatives of the solution on the boundary

$$\|\partial_t u(\cdot\,,0)\|_{L^2(\mathbb{R}^n)} \approx \|\nabla_x u(\cdot\,,0)\|_{L^2(\mathbb{R}^n)}.$$

Boundedness of layer potentials for block matrices also follows from the Kato estimate.

Following the same line of reasoning, one can build a higher-order "block-type" operator $\mathbb{L}$, for which the Kato estimate (109) of section "The Kato Problem and the Riesz Transforms" would imply a certain comparison between normal and tangential derivatives on the boundary

$$\|\partial_t^m u(\cdot\,,0)\|_{L^2(\mathbb{R}^n)} \approx \|\nabla_x^m u(\cdot\,,0)\|_{L^2(\mathbb{R}^n)}.$$

It remains to be seen whether these bounds lead to standard well-posedness results. However, we would like to emphasize that such a result would be restricted to very special, block-type, operators.

Acknowledgements Svitlana Mayboroda is partially supported by the NSF grants DMS 1220089 (CAREER), DMS 1344235 (INSPIRE), DMR 0212302 (UMN MRSEC Seed grant), and the Alfred P. Sloan Fellowship.

References

1. D. Adams, L^p potential theory techniques and nonlinear PDE, in *Potential theory (Nagoya, 1990)* (de Gruyter, Berlin, 1992), pp. 1–15 MR 1167217 (93e:31014)
2. D. Adams, Potential and capacity before and after Wiener, in *Proceedings of the Norbert Wiener Centenary Congress, 1994 (East Lansing, MI, 1994)*. Proceedings of Symposia in Applied Mathematics, vol. 52 (American Mathematical Society, Providence, RI, 1997), pp. 63–83. MR 1440907 (98k:31003)

3. D. Adams, L. Hedberg, *Function Spaces and Potential Theory*. Grundlehren der Mathematischen Wissenschaften. Fundamental Principles of Mathematical Sciences, vol. 314 (Springer, Berlin, 1996). MR 1411441 (97j:46024)
4. V. Adolfsson, J. Pipher, The inhomogeneous Dirichlet problem for Δ^2 in Lipschitz domains. J. Funct. Anal. **159**(1), 137–190 (1998). MR 1654182 (99m:35048)
5. S. Agmon, Multiple layer potentials and the Dirichlet problem for higher order elliptic equations in the plane. I. Commun. Pure Appl. Math. **10**, 179–239 (1957). MR 0106323 (21 #5057)
6. S. Agmon, Maximum theorems for solutions of higher order elliptic equations. Bull. Am. Math. Soc. **66**, 77–80 (1960). MR 0124618 (23 #A1930)
7. M.S. Agranovich, On the theory of Dirichlet and Neumann problems for linear strongly elliptic systems with Lipschitz domains. Funktsional. Anal. i Prilozhen. **41**(4), 1–21, 96 (2007). English translation: Funct. Anal. Appl. **41**(4), 247–263 (2007). MR 2411602 (2009b:35070)
8. M.A. Alfonseca, P. Auscher, A. Axelsson, S. Hofmann, S. Kim, Analyticity of layer potentials and L^2 solvability of boundary value problems for divergence form elliptic equations with complex L^∞ coefficients. Adv. Math. **226**(5), 4533–4606 (2011). MR 2770458
9. S. Antman, *Nonlinear Problems of Elasticity*, 2nd edn. Applied Mathematical Sciences, vol. 107 (Springer, New York, 2005). MR 2132247 (2006e:74001)
10. N. Aronszajn, T. Creese, L. Lipkin, *Polyharmonic Functions*. Oxford Mathematical Monographs (The Clarendon Press, Oxford University Press, New York, 1983). Notes taken by Eberhard Gerlach, Oxford Science Publications. MR 745128 (86g:31001)
11. P. Auscher, On L^p estimates for square roots of second order elliptic operators on $\mathbb{R}^n$. Publ. Mat. **48**(1), 159–186 (2004). MR 2044643 (2005m:35065)
12. P. Auscher, On necessary and sufficient conditions for L^p-estimates of Riesz transforms associated to elliptic operators on $\mathbb{R}^n$ and related estimates. Mem. Am. Math. Soc. **186**(871), xviii+75 (2007). MR 2292385 (2007k:42025)
13. P. Auscher, A. Axelsson, Weighted maximal regularity estimates and solvability of nonsmooth elliptic systems I. Invent. Math. **184**(1), 47–115 (2011). MR 2782252
14. P. Auscher, M. Qafsaoui, Equivalence between regularity theorems and heat kernel estimates for higher order elliptic operators and systems under divergence form. J. Funct. Anal. **177**(2), 310–364 (2000). MR 1795955 (2001j:35057)
15. P. Auscher, P. Tchamitchian, Square root problem for divergence operators and related topics. Astérisque **249**, viii+172 (1998). MR 1651262 (2000c:47092)
16. P. Auscher, S. Hofmann, A. McIntosh, P. Tchamitchian, The Kato square root problem for higher order elliptic operators and systems on $\mathbb{R}^n$. J. Evol. Equ. **1**(4), 361–385 (2001). Dedicated to the memory of Tosio Kato. MR 1877264 (2003a:35046)
17. P. Auscher, A. Axelsson, S. Hofmann, Functional calculus of Dirac operators and complex perturbations of Neumann and Dirichlet problems. J. Funct. Anal. **255**(2), 374–448 (2008). MR 2419965 (2009h:35079)
18. P. Auscher, A. Axelsson, A. McIntosh, Solvability of elliptic systems with square integrable boundary data. Ark. Mat. **48**(2), 253–287 (2010). MR 2672609 (2011h:35070)
19. I. Babuška, The theory of small changes in the domain of existence in the theory of partial differential equations and its applications, in *Differential Equations and Their Applications*. Proceedings of Conference, Prague, 1962 (Czechoslovak Academic Science, Prague, Academic Press, New York, 1963), pp. 13–26. MR 0170133 (30 #373)
20. A. Barton, Gradient estimates and the fundamental solution for higher-order elliptic systems with rough coefficients. Manuscripta Math., DOI: 10.1007/s00229-016-0839-x
21. A. Barton, Perturbation of well-posedness and layer potentials for higher-order elliptic systems with rough coefficients March (2016). http://arxiv.org/abs/1604.00062v1 `arXiv:1604.00062v1 [math.AP]`.
22. A. Barton, S. Mayboroda, Layer potentials and boundary-value problems for second order elliptic operators with data in Besov spaces. Mem. Amer. Math. Soc. **243** (2016), no. 1149, iv+109

23. A. Barton, S. Mayboroda, The Dirichlet problem for higher order equations in composition form. J. Funct. Anal. **265**(1), 49–107 (2013). MR 3049881
24. A. Barton, S. Mayboroda, Boundary-value problems for higher-order elliptic equations in non-smooth domains, in *Concrete Operators, Spectral Theory, Operators in Harmonic Analysis and Approximation*. Operator Theory: Advances and Applications, vol. 236 (Birkhäuser/Springer, Basel, 2014), pp. 53–93. MR 3203053
25. A. Barton, S. Hofmann, S. Mayboroda, Square function estimates on layer potentials for higher-order elliptic equations, August (2015). http://arxiv.org/abs/1508.04988v1arXiv:1508.04988v1 [math.AP]
26. S. Blunck, P. Kunstmann, Weak type (p, p) estimates for Riesz transforms. Math. Z. **247**(1), 137–148 (2004) MR 2054523 (2005f:35071)
27. L. Caffarelli, E. Fabes, C. Kenig, Completely singular elliptic-harmonic measures. Indiana Univ. Math. J. **30**(6), 917–924 (1981). MR 632860 (83a:35033)
28. A. Calderón, E. Fabes, Y. Sagher, N.M. Rivière, June 10, 1940–January 3, 1978, Harmonic analysis in Euclidean spaces, in *Proceedings of Symposia in Pure Mathematics Williams College, Williamstown, MA, 1978*. Part 1, Proceedings of Symposia in Pure Mathematics, vol. XXXV, Part (American Mathematical Society, Providence, RI, 1979), pp. vii–xvii (1 plate). MR 545234 (83a:01041)
29. S. Campanato, Sistemi ellittici in forma divergenza. Regolarità all'interno. Quaderni. [Publications], Scuola Normale Superiore Pisa, Pisa (1980). MR 668196 (83i:35067)
30. S.-Y. Chang, Conformal invariants and partial differential equations. Bull. Am. Math. Soc. (N.S.) **42**(3), 365–393 (2005). MR 2149088 (2006b:53045)
31. S.-Y. Chang, P. Yang, Non-linear partial differential equations in conformal geometry, in *Proceedings of the International Congress of Mathematicians, Vol. I (Beijing, 2002)* (Higher Education Press, Beijing, 2002), pp. 189–207. MR 1989185 (2004d:53031)
32. P. Ciarlet, Theory of plates, in *Mathematical Elasticity*, vol. II. Studies in Mathematics and its Applications, vol. 27 (North-Holland, Amsterdam, 1997). MR 1477663 (99e:73001)
33. J. Cohen, J. Gosselin, The Dirichlet problem for the biharmonic equation in a C^1 domain in the plane. Indiana Univ. Math. J. **32**(5), 635–685 (1983). MR 711860 (85b:31004)
34. J. Cohen, J. Gosselin, Adjoint boundary value problems for the biharmonic equation on C^1 domains in the plane. Ark. Mat. **23**(2), 217–240 (1985). MR 827344 (88d:31006)
35. B. Dahlberg, On the Poisson integral for Lipschitz and C^1-domains. Stud. Math. **66**(1), 13–24 (1979). MR 562447 (81g:31007)
36. B. Dahlberg, Weighted norm inequalities for the Lusin area integral and the nontangential maximal functions for functions harmonic in a Lipschitz domain. Stud. Math. **67**(3), 297–314 (1980). MR 592391 (82f:31003)
37. B. Dahlberg, C. Kenig, L^p estimates for the three-dimensional systems of elastostatics on Lipschitz domains, in *Analysis and Partial Differential Equations*. Lecture Notes in Pure and Applied Mathematics, vol. 122 (Dekker, New York, 1990), pp. 621–634. MR 1044810 (91h:35053)
38. B. Dahlberg, D. Jerison, C. Kenig, Area integral estimates for elliptic differential operators with nonsmooth coefficients. Ark. Mat. **22**(1), 97–108 (1984). MR 735881 (85h:35021)
39. B. Dahlberg, C. Kenig, G. Verchota, The Dirichlet problem for the biharmonic equation in a Lipschitz domain. Ann. Inst. Fourier (Grenoble) **36**(3), 109–135 (1986). MR 865663 (88a:35070)
40. B. Dahlberg, C. Kenig, J. Pipher, G. Verchota, Area integral estimates for higher order elliptic equations and systems. Ann. Inst. Fourier (Grenoble) **47**(5), 1425–1461 (1997). MR 1600375 (98m:35045)
41. G. Dal Maso, U. Mosco, Wiener criteria and energy decay for relaxed Dirichlet problems. Arch. Ration. Mech. Anal. **95**(4), 345–387 (1986). MR 853783 (87m:35021)
42. M. Dalla Riva, A family of fundamental solutions of elliptic partial differential operators with real constant coefficients. Integr. Equ. Oper. Theory **76**(1), 1–23 (2013). MR 3041718
43. M. Dalla Riva, J. Morais, P. Musolino, A family of fundamental solutions of elliptic partial differential operators with quaternion constant coefficients. Math. Methods Appl. Sci. **36**(12), 1569–1582 (2013). MR 3083261

44. A. Dall'Acqua, G. Sweers, Estimates for Green function and Poisson kernels of higher-order Dirichlet boundary value problems. J. Differ. Equ. **205**(2), 466–487 (2004). MR 2092867 (2005i:35065)
45. G. Dolzmann, S. Müller, Estimates for Green's matrices of elliptic systems by L^p theory. Manuscripta Math. **88**(2), 261–273 (1995). MR 1354111 (96g:35054)
46. R. Duffin, On a question of Hadamard concerning super-biharmonic functions. J. Math. Phys. **27**, 253–258 (1949). MR 0029021 (10,534h)
47. L. Evans, R. Gariepy, Wiener's criterion for the heat equation. Arch. Ration. Mech. Anal. **78**(4), 293–314 (1982). MR 653544 (83g:35047)
48. E. Fabes, M. Jodeit Jr., J. Lewis, Double layer potentials for domains with corners and edges. Indiana Univ. Math. J. **26**(1), 95–114 (1977). MR 0432899 (55 #5879)
49. E. Fabes, M. Jodeit Jr., N. Rivière, Potential techniques for boundary value problems on C^1-domains. Acta Math. **141**(3–4), 165–186 (1978). MR 501367 (80b:31006)
50. E. Fabes, D. Jerison, C. Kenig, The Wiener test for degenerate elliptic equations. Ann. Inst. Fourier (Grenoble) **32**(3), vi, 151–182 (1982). MR 688024 (84g:35067)
51. E. Fabes, N. Garofalo, E. Lanconelli, Wiener's criterion for divergence form parabolic operators with C^1-Dini continuous coefficients. Duke Math. J. **59**(1), 191–232 (1989). MR 1016884 (90k:35115)
52. J. Frehse, An irregular complex valued solution to a scalar uniformly elliptic equation. Calc. Var. **33**(3), 263–266 (2008). MR 2429531 (2009h:35084)
53. M. Fuchs, The Green matrix for strongly elliptic systems of second order with continuous coefficients. Z. Anal. Anwendungen **5**(6), 507–531 (1986). MR 894243 (89a:35069)
54. P. Garabedian, A partial differential equation arising in conformal mapping. Pac. J. Math. **1**, 485–524 (1951). MR 0046440 (13,735a)
55. M. Grüter, K.-O. Widman, The Green function for uniformly elliptic equations. Manuscripta Math. **37**(3), 303–342 (1982). MR 657523 (83h:35033)
56. J. Hadamard, Mémoire sur le problème d'analyse relatif à l'équilibre des plaques élastiques encastrées. Institute de France, Académie des Sciences, Mémoires présentés par divers savants, vol. 33, no. 4, (1908)
57. S. Hofmann, S. Kim, The Green function estimates for strongly elliptic systems of second order. Manuscripta Math. **124**(2), 139–172 (2007). MR 2341783 (2008k:35110)
58. S. Hofmann, C. Kenig, S. Mayboroda, J. Pipher, Square function/non-tangential maximal function estimates and the Dirichlet problem for non-symmetric elliptic operators. J. Amer. Math. Soc. **28** (2015), no. 2, 483–529. MR3300700
59. S. Hofmann, C. Kenig, S. Mayboroda, J. Pipher, The regularity problem for second order elliptic operators with complex-valued bounded measurable coefficients. Math. Ann. **361** (2015), no. 3–4, 863–907. MR3319551
60. L. Hörmander, Distribution theory and Fourier analysis, in *The analysis of linear partial differential operators. I.* Classics in Mathematics (Springer, Berlin, 2003). Reprint of the second edition (1990) (Springer, Berlin). MR1065993 (91m:35001a)]. MR 1996773
61. D. Jerison, C. Kenig, The Dirichlet problem in nonsmooth domains. Ann. Math. (2) **113**(2), 367–382 (1981). MR 607897 (84j:35076)
62. D. Jerison, C. Kenig, The Neumann problem on Lipschitz domains. Bull. Am. Math. Soc. (N.S.) **4**(2), 203–207 (1981). MR 598688 (84a:35064)
63. F. John, *Plane Waves and Spherical Means Applied to Partial Differential Equations* (Interscience Publishers, New York, London, 1955). MR 0075429 (17,746d)
64. C. Kenig, *Progress on Two Problems Posed by Rivière*. Harmonic analysis and partial differential equations (Boca Raton, FL, 1988). Contemporary Mathematics, vol. 107 (American Mathematical Society, Providence, RI, 1990), pp. 101–107. MR 1066473
65. C. Kenig, Harmonic analysis techniques for second order elliptic boundary value problems, in *CBMS Regional Conference Series in Mathematics*, vol. 83 (Published for the Conference Board of the Mathematical Sciences, Washington, DC, 1994). MR 1282720 (96a:35040)
66. C. Kenig, W.-M. Ni, On the elliptic equation $Lu - k + K\exp[2u] = 0$. Ann. Scuola Norm. Sup. Pisa Cl. Sci. (4) **12**(2), 191–224 (1985). MR 829052 (87f:35065)

67. C. Kenig, J. Pipher, The Neumann problem for elliptic equations with nonsmooth coefficients. Invent. Math. **113**(3), 447–509 (1993). MR 1231834 (95b:35046)
68. C. Kenig, D. Rule, The regularity and Neumann problem for non-symmetric elliptic operators. Trans. Am. Math. Soc. **361** (1), 125–160 (2009). MR 2439401 (2009k:35050)
69. C. Kenig, H. Koch, J. Pipher, T. Toro, A new approach to absolute continuity of elliptic measure, with applications to non-symmetric equations. Adv. Math. **153**(2), 231–298 (2000). MR 1770930 (2002f:35071)
70. J. Kilty, Z. Shen, A bilinear estimate for biharmonic functions in Lipschitz domains. Math. Ann. **349**(2), 367–394 (2011). MR 2753826
71. J. Kilty, Z. Shen, The L^p regularity problem on Lipschitz domains. Trans. Am. Math. Soc. **363**(3), 1241–1264 (2011). MR 2737264 (2012a:35072)
72. V. Kozlov, V. Maz'ya, J. Rossmann, *Spectral Problems Associated with Corner Singularities of Solutions to Elliptic Equations*. Mathematical Surveys and Monographs, vol. 85 (American Mathematical Society, Providence, RI, 2001). MR 1788991 (2001i:35069)
73. Ju.P. Krasovskiĭ, Isolation of the singularity in Green's function. Izv. Akad. Nauk SSSR Ser. Mat. **31** (1967), 977–1010. MR 0223740 (36 #6788)
74. D. Labutin, Potential estimates for a class of fully nonlinear elliptic equations. Duke Math. J. **111**(1), 1–49 (2002). MR 1876440 (2002m:35053)
75. H. Lebesgue, Sur des cas d'impossibilité du problème de Dirichlet ordinaire. Société Mathématique de France C.R. des Séances de la, p .17 (1913)
76. W. Littman, G. Stampacchia, H. Weinberger, Regular points for elliptic equations with discontinuous coefficients. Ann. Scuola Norm. Sup. Pisa (3) **17**, 43–77 (1963). MR 0161019 (28 #4228)
77. G. Luo, V. Maz'ya, Wiener type regularity of a boundary point for the 3D Lamé system. Potential Anal. **32**(2), 133–151 (2010). MR 2584981 (2011a:35119)
78. J. Malý, W. Ziemer, *Fine Regularity of Solutions of Elliptic Partial Differential Equations*. Mathematical Surveys and Monographs, vol. 51 (American Mathematical Society, Providence, RI, 1997). MR 1461542 (98h:35080)
79. S. Mayboroda, The connections between Dirichlet, regularity and Neumann problems for second order elliptic operators with complex bounded measurable coefficients. Adv. Math. **225**(4), 1786–1819 (2010). MR 2680190
80. S. Mayboroda, V. Maz'ya, Boundedness of the Hessian of a biharmonic function in a convex domain. Commun. Partial Diff. Equ. **33**(7–9), 1439–1454 (2008). MR 2450165 (2010d:35064)
81. S. Mayboroda, V. Maz'ya, Boundedness of the gradient of a solution and Wiener test of order one for the biharmonic equation. Invent. Math. **175**(2), 287–334 (2009). MR 2470109 (2009m:35093)
82. S. Mayboroda, V. Maz'ya, Pointwise estimates for the polyharmonic Green function in general domains, in *Analysis, Partial Differential Equations and Applications*. Operator Theory: Advances and Applications, vol. 193 (Birkhäuser Verlag, Basel, 2009), pp. 143–158. MR 2766075 (2012c:35102)
83. S. Mayboroda, V. Maz'ya, Polyharmonic capacity and wiener test of higher order, November (2014). http://arxiv.org/abs/1411.4202v1`arXiv:1411.4202v1 [math.AP]`
84. S. Mayboroda, V. Maz'ya, Regularity of solutions to the polyharmonic equation in general domains. Invent. Math. **196**(1), 1–68 (2014). MR 3179572
85. V. Maz'ya, On the behavior near the boundary of solutions of the Dirichlet problem for the biharmonic operator. Dokl. Akad. Nauk SSSR **18**, 15–19 (1977). English translation: Soviet Math. Dokl. **18**(4), 1152–1155 (1977).
86. V. Maz'ya, Behaviour of solutions to the Dirichlet problem for the biharmonic operator at a boundary point, in *Equadiff IV*. Proceedings of Czechoslovak Conference Differential Equations and their Applications, Prague, 1977. Lecture Notes in Mathematics, vol. 703 (Springer, Berlin, 1979), pp. 250–262. MR 535346 (80e:35026)

87. V. Maz'ya, Unsolved problems connected with the Wiener criterion, in *Proceedings of Symposia in Pure Mathematics*, vol. 60. The Legacy of Norbert Wiener: A Centennial Symposium (Cambridge, MA, 1994 (American Mathematical Society, Providence, RI, 1997), pp. 199–208. MR 1460283 (98e:35040)
88. V. Maz'ya, On the Wiener type regularity of a boundary point for the polyharmonic operator. Appl. Anal. **71**(1–4), 149–165 (1999). MR 1690096 (2000b:35097)
89. V. Maz'ya, On Wiener's type regularity of a boundary point for higher order elliptic equations, in *Nonlinear analysis, function spaces and applications, vol. 6, Prague, 1998* (Academy of Sciences of the Czech Republic, Prague, 1999), pp. 119–155. MR 1777714 (2001m:35129)
90. V. Maz'ya, The Wiener test for higher order elliptic equations. Duke Math. J. **115**(3), 479–512 (2002). MR 1940410 (2003i:35065)
91. V. Maz'ya, T. Donchev, Regularity in the sense of Wiener of a boundary point for a polyharmonic operator. C. R. Acad. Bulgare Sci. **36**(2), 177–179 (1983). MR 709006 (84m:31009)
92. V. Maz'ya, S. Nazarov, Paradoxes of the passage to the limit in solutions of boundary value problems for the approximation of smooth domains by polygons. Izv. Akad. Nauk SSSR Ser. Mat. **50**(6), 1156–1177, 1343 (1986). MR 883157 (88i:35016)
93. V. Maz'ya, S. Nazarov, The apex of a cone can be irregular in Wiener's sense for a fourth-order elliptic equation. Mat. Zametki **39**(1), 24–28 (1986), 156. English translation: Math. Notes **39**(1–2), 14–16 (1986). MR 830840 (87h:35088)
94. V. Maz'ya, B. Plamenevskiĭ, Asymptotic behavior of the fundamental solutions of elliptic boundary value problems in domains with conical points, in *Boundary value problems.* Spectral theory (Russian). Problems in Mathematical Analysis, vol. 7 (Leningrad University, Leningrad, 1979), pp. 100–145, 243. MR 559106 (81e:35042)
95. V. Maz'ya, B. Plamenevskiĭ, On the maximum principle for the biharmonic equation in a domain with conical points. Izv. Vyssh. Uchebn. Zaved. Mat. (2), 52–59 (1981). English translation: Soviet Math. (Iz. VUZ) **25**(2), 61–70 (1981). MR 614817 (84b:35037)
96. V. Maz'ya, J. Rossmann, On the Agmon-Miranda maximum principle for solutions of elliptic equations in polyhedral and polygonal domains. Ann. Glob. Anal. Geom. **9**(3), 253–303 (1991). MR 1143406 (92h:35027)
97. V. Maz'ya, J. Rossmann, On the Agmon-Miranda maximum principle for solutions of strongly elliptic equations in domains of $\mathbf{R}^n$ with conical points. Ann. Glob. Anal. Geom. **10**(2), 125–150 (1992). MR 1175915 (93i:35025)
98. V. Maz'ya, S. Nazarov, B. Plamenevskiĭ, Singularities of solutions of the Dirichlet problem in the exterior of a thin cone. Mat. Sb. (N.S.) **122 (164)**(4), 435–457 (1983). English translation: Math. USSR-Sb. **50**(2), 415–437 (1985). MR 725451 (85h:35074)
99. V. Maz'ya, S. Nazarov, B. Plamenevskiĭ, *Asymptotische Theorie elliptischer Randwertaufgaben in singulär gestörten Gebieten. I.* Mathematische Lehrbücher und Monographien, II. Abteilung: Mathematische Monographien [Mathematical Textbooks and Monographs, Part II: Mathematical Monographs], vol. 82 (Akademie, Berlin, 1991). Störungen isolierter Randsingularitäten. [Perturbations of isolated boundary singularities]. MR 1101139 (92g:35059)
100. V. Maz'ya, M. Mitrea, T. Shaposhnikova, The Dirichlet problem in Lipschitz domains for higher order elliptic systems with rough coefficients. J. Anal. Math. **110** (2010), 167–239. MR 2753293 (2011m:35088)
101. V.V. Meleshko, Selected topics in the history of the two-dimensional biharmonic problem. Appl. Mech. Rev. **56**(1), 33–85 (2003)
102. C. Miranda, Formule di maggiorazione e teorema di esistenza per le funzioni biarmoniche de due variabili. Giorn. Mat. Battaglini (4) **2(78)**, 97–118 (1948). MR 0030058 (10,706f)
103. C. Miranda, Teorema del massimo modulo e teorema di esistenza e di unicità per il problema di Dirichlet relativo alle equazioni ellittiche in due variabili. Ann. Mat. Pura Appl. (4) **46**, 265–311 (1958). MR 0124615 (23 #A1927)

104. I. Mitrea, Mapping properties of layer potentials associated with higher-order elliptic operators in Lipschitz domains, in *Topics in Operator Theory. Volume 2. Systems and Mathematical Physics*. Operator Theory: Advances and Applications, vol. 203 (Birkhäuser Verlag, Basel, 2010), pp. 363–407. MR 2683248 (2011g:35087)
105. I. Mitrea, M. Mitrea, On the Dirichlet and regularity problems for the bi-Laplacian in Lipschitz domains, in *Integral Methods in Science and Engineering*, vol. 1 (Birkhäuser, Boston, MA, 2010), pp. 245–254. MR 2663136 (2011d:35143)
106. D. Mitrea, I. Mitrea, On the regularity of Green functions in Lipschitz domains. Commun. Partial Differ. Equ. **36**(2), 304–327 (2011). MR 2763343
107. I. Mitrea, M. Mitrea, M. Wright, Optimal estimates for the inhomogeneous problem for the bi-Laplacian in three-dimensional Lipschitz domains. J. Math. Sci. (N. Y.) **172**(1), 24–134 (2011). Problems in mathematical analysis. No. 51. MR 2839870 (2012h:35056)
108. I. Mitrea, M. Mitrea, Boundary value problems and integral operators for the bi-Laplacian in non-smooth domains. Atti Accad. Naz. Lincei Rend. Lincei Mat. Appl. **24**(3), 329–383 (2013). MR 3097019
109. I. Mitrea, M. Mitrea, *Multi-Layer Potentials and Boundary Problems for Higher-Order Elliptic Systems in Lipschitz Domains*. Lecture Notes in Mathematics, vol. 2063 (Springer, Heidelberg, 2013). MR 3013645
110. C. Morrey Jr., Second-order elliptic systems of differential equations, in *Contributions to the Theory of Partial Differential Equations*. Annals of Mathematics Studies, vol. 33 (Princeton University Press, Princeton, NJ, 1954), pp. 101–159. MR 0068091 (16,827e)
111. J. Moser, On Harnack's theorem for elliptic differential equations. Commun. Pure Appl. Math. **14**, 577–591 (1961). MR 0159138 (28 #2356)
112. A. Nádai, *Theory of Flow and Fracture of Solids*, vol. II (McGraw-Hill, New York, 1963)
113. J. Nečas, Les méthodes directes en théorie des équations elliptiques. Masson et Cie, Éditeurs, Paris (1967). MR 0227584 (37 #3168)
114. N. Ortner, P. Wagner, A short proof of the Malgrange-Ehrenpreis theorem, in *Functional analysis* (Trier, 1994) (de Gruyter, Berlin, 1996), pp. 343–352. MR 1420460 (97g:35021)
115. J. Pipher, G. Verchota, Area integral estimates for the biharmonic operator in Lipschitz domains. Trans. Am. Math. Soc. **327**(2), 903–917 (1991). MR 1024776 (92a:35052)
116. J. Pipher, G. Verchota, The Dirichlet problem in L^p for the biharmonic equation on Lipschitz domains. Am. J. Math. **114**(5), 923–972 (1992). MR 1183527 (94g:35069)
117. J. Pipher, G. Verchota, A maximum principle for biharmonic functions in Lipschitz and C^1 domains. Comment. Math. Helv. **68**(3), 385–414 (1993). MR 1236761 (94j:35030)
118. J. Pipher, G. Verchota, Dilation invariant estimates and the boundary Gårding inequality for higher order elliptic operators. Ann. Math. (2) **142**(1), 1–38 (1995). MR 1338674 (96g:35052)
119. J. Pipher, G. Verchota, Maximum principles for the polyharmonic equation on Lipschitz domains. Potential Anal. **4**(6), 615–636 (1995). MR 1361380 (96i:35021)
120. H. Poincaré, Sur les Equations aux Derivees Partielles de la Physique Mathematique. Am. J. Math. **12**(3), 211–294 (1890). MR 1505534
121. A. Rosén, Layer potentials beyond singular integral operators. Publ. Mat. **57**(2), 429–454 (2013). MR 3114777
122. D. Rule, Non-symmetric elliptic operators on bounded Lipschitz domains in the plane. Electron. J. Differ. Equ. **144**, 1–8 (2007). MR 2366037 (2008m:35070)
123. B.-W. Schulze, A priori estimates in uniform norms for strongly elliptic systems. Sibirsk. Mat. Ž. **16**, 384–394, 422 (1975). English translation: Siberian Math. J. 16(2), 297–305 (1975). MR 0470468 (57 #10222)
124. T. Sedrakyan, L. Glazman, A. Kamenev, Absence of Bose condensation on lattices with moat bands. **89**(20), 201112 (2014). ArXiv e-prints
125. R. Selvaggi, I. Sisto, An existence theorem for the Dirichlet problem with respect to the operator Δ^2 in certain domains of class C^1. Boll. Un. Mat. Ital. B (5) **18**(2), 473–483 (1981). MR 629418 (84f:35044)
126. Z. Shapiro, On elliptical systems of partial differential equations. C. R. (Doklady) Acad. Sci. URSS (N.S.) **46**, 133–135 (1945). MR 0012357 (7,14f)

127. H. Shapiro, M. Tegmark, An elementary proof that the biharmonic Green function of an eccentric ellipse changes sign. SIAM Rev. **36**(1), 99–101 (1994). MR 1267051 (94m:35096)
128. Z. Shen, Necessary and sufficient conditions for the solvability of the L^p Dirichlet problem on Lipschitz domains. Math. Ann. **336**(3), 697–725 (2006). MR 2249765 (2008e:35059)
129. Z. Shen, On estimates of biharmonic functions on Lipschitz and convex domains. J. Geom. Anal. **16**(4), 721–734 (2006). MR 2271951 (2008a:35062)
130. Z. Shen, The L^p Dirichlet problem for elliptic systems on Lipschitz domains. Math. Res. Lett. **13**(1), 143–159 (2006). MR 2200052 (2007f:35067)
131. Z. Shen, A relationship between the Dirichlet and regularity problems for elliptic equations. Math. Res. Lett. **14**(2), 205–213 (2007). MR 2318619 (2008c:35043)
132. Z. Shen, The L^p boundary value problems on Lipschitz domains. Adv. Math. **216**(1), 212–254 (2007). MR 2353255 (2009a:35064)
133. I. Skrypnik, *Methods for Analysis of Nonlinear Elliptic Boundary Value Problems*. Translations of Mathematical Monographs, vol. 139 (American Mathematical Society, Providence, RI, 1994). Translated from the 1990 Russian original by Dan D. Pascali. MR 1297765 (95i:35109)
134. V. Solonnikov, The Green's matrices for elliptic boundary value problems. I. Trudy Mat. Inst. Steklov. **110**, 107–145 (1970). MR 0289935 (44 #7120)
135. V. Solonnikov, The Green's matrices for elliptic boundary value problems. II. Trudy Mat. Inst. Steklov. **116**, 181–216, 237 (1971). Boundary value problems of mathematical physics, 7. MR 0364854 (51 #1108)
136. G. Sweers, A survey on boundary conditions for the biharmonic. Complex Var. Elliptic Equ. **54**(2), 79–93 (2009). MR 2499118 (2010d:35065)
137. N. Trudinger, X.-J. Wang, On the weak continuity of elliptic operators and applications to potential theory. Am. J. Math. **124**(2), 369–410 (2002). MR 1890997 (2003c:35025)
138. G. Verchota, The Dirichlet problem for the biharmonic equation in C^1 domains. Indiana Univ. Math. J. **36**(4), 867–895 (1987). MR 916748 (88m:35051)
139. G. Verchota, The Dirichlet problem for the polyharmonic equation in Lipschitz domains. Indiana Univ. Math. J. **39**(3), 671–702 (1990). MR 1078734 (91k:35073)
140. G. Verchota, Potentials for the Dirichlet problem in Lipschitz domains, in *Potential theory—ICPT 94 (Kouty, 1994)* (de Gruyter, Berlin, 1996), pp. 167–187. MR 1404706 (97f:35041)
141. G. Verchota, The biharmonic Neumann problem in Lipschitz domains. Acta Math. **194**(2), 217–279 (2005). MR 2231342 (2007d:35058)
142. G. Verchota, Boundary coerciveness and the Neumann problem for 4th order linear partial differential operators, in *Around the research of Vladimir Maz'ya. II*. International Mathematical Series (New York), vol. 12 (Springer, New York, 2010), pp. 365–378. MR 2676183 (2011h:35065)
143. N. Wiener, The Dirichlet problem. J. Math. Phys. **3**, 127–146 (1924)
144. S. Zaremba, Sur le principe du minimum. Bulletin Internationale de l'Académie des Sciences de Cracovie, Classe des Sciences, Mathématiques et Naturelles. **7**, 197–264 (1909)

Victor Shapiro and the Theory of Uniqueness for Multiple Trigonometric Series

J. Marshall Ash

Abstract In 1870, Georg Cantor proved that if a trigonometric series converges to 0 everywhere, then all its coefficients must be 0. In the twentieth century this result was extended to higher dimensional trigonometric series when the mode of convergence is taken to be spherical convergence and also when it is taken to be unrestricted rectangular convergence. We will describe the path to each result. An important part of the first path was Victor Shapiro's seminal 1957 paper, *Uniqueness of multiple trigonometric series*. This paper also was an unexpected part of the second path.

1991*Mathematics Subject Classification:* Primary 42-02; Secondary 42A63, 42B99, 42A20

Dedication

My thesis advisor, Antoni Zygmund, and his student, Albert Calderón, created a wonderful school of mathematical analysis that radiated outward from the University of Chicago in the 1950s and 1960s. Their intellectual curiosity, complemented by their generosity and friendliness, induced similar good feelings among their direct mathematical descendants. This is a large community, numbering 65 Ph.D. mathematicians. Zygmund had 40 Ph.D. students, including Calderón, Calderón had 27, and they had one student in common.

Cora Sadosky was the one student who had both Zygmund and Calderón listed for thesis advisors [13]. Cora was my fellow student at the University of Chicago. When I heard of the loss of this cheerful, enthusiastic mathematician in 2010, it was sad news indeed.

J.M. Ash (✉)
Mathematics Department, DePaul University Chicago, IL 60614, USA
e-mail: mash@math.depaul.edu
http://condor.depaul.edu/mash

M.C. Pereyra et al. (eds.), *Harmonic Analysis, Partial Differential Equations, Complex Analysis, Banach Spaces, and Operator Theory (Volume 1)*, Association for Women in Mathematics Series 4, DOI 10.1007/978-3-319-30961-3_5

Victor Shapiro was a student of Zygmund, and therefore a fellow "mathematical sibling" of Cora's and mine. His death in 2013 was followed later in that year by a conference to honor him held in Riverside, California. I gave a talk at that conference. The paper that I present here is based on that talk. I suspect that Cora, whose first question upon meeting me after the passage of some time was usually about my mathematics, would have enjoyed seeing this.

Two Theorems and a Conjecture

Let $\{d_n\}_{-\infty<n<\infty}$ be a sequence of complex numbers and let $x \in \mathbb{T}^1 = [0, 2\pi)$. Suppose a function has a representation of the form

$$\sum d_n e^{inx} = \lim_{N\to\infty} d_0 + \sum_{n=1}^{N} \left(d_{-n}e^{-inx} + d_n e^{inx}\right).$$

It is natural to combine the nth and $-n$th terms, for if a_n and b_n are real, $d_n = (a_n + ib_n)/2$ and d_{-n} is the complex conjugate of d_n, then $d_n e^{inx} + d_{-n}e^{-inx} = a_n \cos nx + b_n \sin nx$, the "natural" nth term of a real valued trigonometric series. Is this representation unique? In other words, if $\sum d_n e^{inx} = \sum d'_n e^{inx}$ for every x, does it necessarily follow that $d_n = d'_n$ for every n? Subtract and set $c_n = d_n - d'_n$ to get a cleaner formulation: Does $\sum c_n e^{inx} = 0$ imply that $c_n = 0$ for every n? Here is Georg Cantor's answer.

Theorem 2.1. *Let $\sum c_n e^{inx} = 0$ for every $x \in \mathbb{T}^1$. Then $c_n = 0$ for every n [9].*

He proved this in 1870. Notice that in the statement of his theorem, Cantor made a choice of what it means for a trigonometric series to represent the function $z(x)$ where z has domain $\mathbb{T}^1$ and range $\{0\}$, namely that it converges to that point at every point of $\mathbb{T}^1$. Many other notions of "represent" have been considered since then; many are discussed in Chap. IX of Antoni Zygmund's book Trigonometric Series [16]. We will mostly focus on this pointwise everywhere notion of representation.

The entire subject of this broad survey concerns attempts to extend this result to higher dimensions. In all dimensions we will always combine terms whose indices differ only by signs. This reduction in dimension 1 converts a two-sided numerical series $\sum_{n=-\infty}^{\infty} C_n$ to the series $\sum_{n\in\mathbb{Z}^+} T_n$, where for each $n \in \mathbb{Z}^+ = \{0, 1, 2, \ldots\}$, $T_n = \sum_{\{\nu:\, |\nu|=n\}} C_\nu$. Since the nonnegative integers have a natural ordering, Cantor's theorem's hypothesis is unambiguous. When $d \geq 2$, the corresponding reduction of $\sum_{n\in\mathbb{Z}^d} C_n$ to $\sum_{n\in(\mathbb{Z}^+)^d} T_n$ where $T_n = \sum_{\{\nu:\, |\nu_i|=n_i \text{ for } 1\leq i\leq d\}} C_\nu$ does not produce a "natural ordering" because $\left(\mathbb{Z}^+\right)^d$ does not have a natural ordering, so many conjectures arise in each dimension.

Here are three important distinct ways of adding the elements of the numerical series $\sum_{n\in(\mathbb{Z}^+)^d} T_n$.

Spherical convergence: The Nth partial sum contains all terms with indices in the intersection of the sphere of radius $\sqrt{N}$ with the positive cone $(\mathbb{Z}^+)^d$. The natural norm on $\mathbb{Z}^d$ for spherical convergence will be denoted by $|\nu|$ and is given by

$$|\nu| := \sqrt{\nu_1^2 + \cdots + \nu_d^2}.$$

The spherical sum is defined to be

$$\begin{aligned} SPH \sum_{n\in(\mathbb{Z}^+)^d} T_n &:= \lim_{N\to\infty} \sum_{\{\nu:\text{all } \nu_i\geq 0 \text{ and } \nu_1^2+\cdots+\nu_d^2\leq N\}} T_\nu \\ &= \lim_{N\to\infty} \sum_{\{\nu:\text{all } \nu_i\geq 0 \text{ and } |\nu|\leq\sqrt{N}\}} T_\nu . \end{aligned}$$

Square convergence: The Nth partial sum contains all terms with indices in the rectangular parallelepiped with opposite corners $(0,\ldots,0)$ and $(N,\ldots,N)$. The natural norm on $\mathbb{Z}^d$ for square convergence will be denoted by $\|\nu\|$ and is given by

$$\|\nu\| := \max\{|\nu_1|,\ldots,|\nu_d|\}.$$

The square sum is defined as

$$SQ \sum_{n\in(\mathbb{Z}^+)^d} T_n := \lim_{N\to\infty} \sum_{\nu_1=0}^{N} \cdots \sum_{\nu_d=0}^{N} T_\nu = \lim_{N\to\infty} \sum_{\{\nu:\text{all } \nu_i\geq 0 \text{ and } \|\nu\|\leq N\}} T_\nu .$$

Unrestricted rectangular convergence: This is no longer a one variable process. Assign to each point n of $(\mathbb{Z}^+)^d$ the rectangular partial sum S_n of all terms whose indices are in the rectangular parallelepiped with corners $(0,\ldots,0)$ and $(n_1,\ldots,n_d)$. The unrestricted rectangular limit of $\sum T_\nu$ is a number L such that for each $\epsilon > 0$, there is a number $N(\epsilon)$ so that for every n with $\min\{n_1,\ldots,n_d\} > N(\epsilon)$, $|S_n - L| < \epsilon$. When such an L exists, we call it the unrestricted rectangular sum of $\sum T_n$ and write

$$UR \sum_{n\in(\mathbb{Z}^+)^d} T_n := \lim_{\min\{N_1,\cdots,N_d\}\to\infty} \sum_{\nu_1=0}^{N_1} \cdots \sum_{\nu_d=0}^{N_d} T_\nu .$$

It is obvious that if a numerical series is unrestrictedly rectangularly convergent, then it is square convergent to the same sum; i.e.,

$$UR \sum_{n\in(\mathbb{Z}^+)^d} T_n = L \text{ implies } SQ \sum_{n\in(\mathbb{Z}^+)^d} T_n = L. \tag{1}$$

It is easy to give examples of double series of numbers $\sum_{n\in(\mathbb{Z}^+)^2} T_n$ which show that each of the five other possible connections between these three methods of convergence is, in general, false.

In dimension $d \geq 2$, with $n = (n_1, \ldots, n_d) \in \mathbb{Z}^d$, $x = (x_1, \ldots, x_d) \in \mathbb{T}^d$, and $nx = n_1x_1 + \cdots + n_dx_d$, and $T_n(x) = \sum_{\{\nu:\, |\nu_i|=n_i \text{ for } 1\leq i\leq d\}} c_\nu e^{i\nu x}$, there are three distinct and natural hypotheses for direct generalizations of Cantor's theorem. Here is the present state of knowledge.

Theorem 2.2. *Let*

$$SPH \sum_{n\in\mathbb{Z}^d} c_n e^{inx} = SPH \sum_{n\in(\mathbb{Z}^d)^+} \sum_{\{:\,|_i|=n_i \text{ for } 1\leq i\leq d\}} ce^{ix} = 0$$

for every $x \in \mathbb{T}^d$. Then $c_n = 0$ for every n[8].

Theorem 2.3. *Let*

$$UR \sum_{n\in\mathbb{Z}^d} c_n e^{inx} = UR \sum_{n\in(\mathbb{Z}^d)^+} \sum_{\{:\,|_i|=n_i \text{ for } 1\leq i\leq d\}} ce^{ix} = 0$$

for every $x \in \mathbb{T}^d$. Then $c_n = 0$ for every n[4, 15].

Conjecture 2.4. Let

$$SQ \sum_{n\in\mathbb{Z}^d} c_n e^{inx} = SQ \sum_{n\in(\mathbb{Z}^d)^+} \sum_{\{:\,|_i|=n_i \text{ for } 1\leq i\leq d\}} ce^{ix} = 0$$

for every $x \in \mathbb{T}^d$. Then $c_n = 0$ for every n.

We will later need to mention one higher dimensional extension of Theorem 2.2 which involves replacing the condition of spherical convergence, namely that $SPH \sum c_n e^{inx}$ exists, by the weaker condition of spherical Abel summability:

Theorem 2.5. *If for every $x \in \mathbb{T}^d$, $SPH \sum c_n e^{inx} r^{|n|}$ exists for all positive $r < 1$ and $\lim_{r\to 1^-} SPH \sum c_n e^{inx} r^{|n|} = 0$, and if*

$$\sum_{R-1<|n|\leq R} |c_n| = o(R) \text{ as } R \to \infty; \tag{2}$$

then $c_n = 0$ for every n[14].

History of the Two Theorems

The steps of Cantor's brilliant proof are well known. Our discussion here will be informed by drawing comparisons with them. Here are the four major steps of his proof.

(1) Establish the Cantor–Lebesgue theorem, which implies that everywhere convergence ensures that

$$\epsilon(R) = \sum_{\|n\|=R} |c_n|^2 \to 0 \text{ as } R \to \infty. \tag{3}$$

In dimension 1, $\sum_{\|n\|=R} |c_n|^2 = |c_R|^2 + |c_{-R}|^2$.

(2) Show that the Riemann function, the formal second integral, $F(x) = c_0 \frac{x^2}{2} + \sum_{n\neq 0} \frac{c_n}{(in)^2} e^{inx}$, is continuous. (Formal means that $\frac{d^2}{dx^2}\frac{x^2}{2} = 1$ and for each $n \neq 0$, $\frac{d^2}{dx^2}\frac{e^{inx}}{(in)^2} = e^{inx}$.)

(3) Establish the consistency of Riemann summability, that the Schwarz second derivative D^2 defined by

$$D^2F(x) = \lim_{h\to 0} \frac{F(x+h) - 2F(x) + F(x-h)}{h^2} \tag{4}$$

satisfies at every x

$$D^2F(x) = \lim_{h\to 0} c_0 + \sum_{n\neq 0} c_n e^{inx} \left(\frac{\sin n\frac{h}{2}}{n\frac{h}{2}} \right)^2 = 0.$$

(4) Use Schwarz's theorem, that continuous functions with identically zero Schwarz second derivative are of the form $ax + b$.

Step (2), the proof that F is continuous, is immediate from Step (1) and the Weierstrass M-test.

The Spherical Uniqueness Theorem, Theorem 2.2

The first big step towards Theorem 2.2 was taken in 1957, when Victor Shapiro proved a powerful d dimensional theorem. Shapiro worked in a more general context, also considering questions of summability. He did not prove Theorem 2.2 because his proof required an extra assumption on the coefficient size[14]. A corollary of one of Shapiro's results was a weaker version of Theorem 2.2 which required the additional hypothesis of condition (2). This condition is quite natural when

$d = 2$: since there are $O(r)$ lattice points being summed over in condition (2), in dimension 2 this assumption asserts that the c_m tend to zero "on the average" as $|m| \to \infty$. But the assumption becomes much stronger as the dimension increases; specifically, in dimension d there are $O\left(r^{d-1}\right)$ terms in the sum, so that the coefficients are required to be decaying like $o\left(r^{2-d}\right)$ on the average.

Shapiro's 1957 proof is a direct extension of Cantor's in the sense that he follows the same four steps.

(1) He controls the coefficient size by simply adding a second hypothesis, namely that condition (2) holds is true *by assumption*.
(2) His Riemann function $F(x)$ is the formal anti-Laplacian of the series $\sum_{n\in\mathbb{Z}^d} c_n e^{inx}$ appearing in the hypothesis of Theorem 2.2,

$$F(x) = c_0 \frac{\|x\|^2}{2d} - \sum_{n\neq 0} c_n \frac{e^{inx}}{\|n\|^2} = \lim_{t\to 1-} c_0 \frac{\|x\|^2}{2d} - \sum_{n\neq 0} \frac{c_n}{\|n\|^2} e^{inx-\|n\|t},$$

which he proves to be continuous. (Now formal means that $\Delta \frac{\|x\|^2}{2d} = 1$ and $\Delta \frac{e^{inx}}{\|n\|^2} = e^{inx}$ for all non-zero n. For the Laplacian $\Delta = \sum_{i=1}^{d} \left(\frac{\partial}{\partial x_i}\right)^2$ and $\|x\|^2 = \sum_{j=1}^{d} x_j^2$; and for each $n \neq 0$, $\Delta e^{inx} = \|n\|^2 e^{inx}$.)
(3) He establishes a kind of consistency of Riemann summability by showing that at every x, the generalized Laplacian of F,

$$\lim_{h\to 0+} \frac{8}{h^2} \left\{ \frac{1}{\pi h^2} \int_{\|\eta\|<h} F(x+\eta)\, d\eta - F(x) \right\}$$

is zero.
(4) He uses a well-known theorem that continuous functions with identically zero generalized Laplacian are harmonic.

By far the most delicate and difficult part of his work is Step (2), the proof that F is continuous.

Because the original series is only required to be Abel summable to 0 everywhere, it is not possible to weaken the hypothesis by replacing $o(R)$ by $O(R)$. For the proposed stronger theorem would be contradicted by the fact that the one-dimensional series

$$\delta'(x) = \sum ine^{inx} = -2\sum n \sin nx$$

has Abel limit 0 everywhere. To my mind, the most beautiful thing about Theorem 2.2 is that there is no hypothesis about coefficient size. In 1971, 14 years after Shapiro's theorem was proved, Roger Cooke proved a two-dimensional Cantor–Lebesgue theorem[12]. An immediate consequence of his result is that everywhere two-dimensional spherical convergence ensures that condition (3) must hold. But in dimension 2 (and *not* in higher dimensions), condition (3) implies condition (2). So the two-dimensional version of Theorem 2.2 was proved.

The first precursor to a higher dimensional spherical theory came in 1976, when Bernard Connes extended the Cantor–Lebesgue result of Cooke, whose proof was exceedingly two-dimensional, to all dimensions[11]. At this point, we knew that if we wanted to prove Theorem 2.2, we could use the fact that $\epsilon(R) \to 0$ without having to add a second hypothesis involving coefficient size.

But in dimension 3, there is a very large gap between condition (3) and the much stronger condition (2), and this gap becomes ever larger as the dimension increases. So it seemed likely that when dimension $d \geq 3$, Step (2) of Shapiro's proof, the proof that his Riemann function is continuous, would be inaccessible. Shapiro and I discussed this problem. We agreed that it was totally unclear if there was a proof or a counterexample ahead for the cases of $d \geq 3$, and he speculated that there might be a century of mathematical analysis development required to bring this question within reach.

Victor Shapiro had one more major contribution to make towards the solution of Theorem 2.2. Victor told me that 1 day in the middle 1990s, he and Jean Bourgain happened to be strolling across the University of California, Riverside campus and Victor mentioned this problem. Their conversation inspired Bourgain to look at the problem and solve it! He stayed entirely within the framework established by Cantor and generalized by Shapiro. Assuming only the everywhere spherical convergence to zero, and bringing together Connes result, hard analysis, harmonic measure, and some probability theory (martingales), he was able to prove that Shapiro's Riemann function was continuous. In 1996, Bourgain published his proof that the hypothesis of Theorem 2.2 implies the continuity of Shapiro's Riemann function[8]. Together, these two excellent papers, the first by Shapiro and the latter by Bourgain, published 39 years apart, provide the complete proof of Theorem 2.2.

Bourgain's paper is only 15 pages long. Although it is absolutely correct and says everything that should be said in just the right order, it is extremely terse. In fact, Gang Wang and I took 9 months to read it, but once we got it, we were able to reproduce a lot of the substantial collection of one- dimensional extensions of Cantor's theorem that can be found in Chap. IX of Antoni Zygmund's *Trigonometric Series* [5, 6, 16]. To see an expansion of Bourgain's proof of continuity, see the 22 page version in [5]; and to see his proof expanded to 42 pages while being specialized down to two dimensions, see [2].

The Unrestricted Rectangular Uniqueness Theorem, Theorem 2.3

Let $M(x)$ be a one-dimensional trigonometric series that converges to zero a.e. and let $\delta(y) = \sum e^{iny}$ be the trigonometric series associated with the unit mass at the origin. Because the partial sums of $\delta(y)$ are bounded by $\csc y$, the double trigonometric series $M(x)\,\delta(y)$ is unrestrictedly rectangularly convergent to 0 a.e.[7]. But by Zygmund's extension of Cooke's theorem, $M\delta$ cannot converge circularly on a set of positive measure, since its coefficients do not tend to 0 as $\|n\| \to \infty$ [17]. So the proof of Theorem 2.3 seems to have nothing to do with Shapiro's 1957 theorem. However, by a strange quirk of fate, it does.

Theorem 2.3 was announced in 1919. The announced proof also followed the model of Cantor's theorem. It was a very simple induction that seemed to indicate that there were no interesting things to do in this direction, so for many years nothing involving uniqueness for unrestricted rectangular convergence appeared in the literature. Grant Welland and I studied the proof around 1970 and could not follow the step that generalized Schwarz's theorem. In fact, some years later Chris Freiling and Dan Rinne showed me the counterexample function $(x+y)\,|x+y|$. This function satisfies the hypotheses (being continuous and having a certain generalized F_{xxyy} identically 0), but not the conclusion (having the expected form

$$a\,(y)\,x + b\,(y) + c\,(x)\,y + d\,(x)\,,$$

with a, b, c, d being twice differentiable), of the generalized Schwarz's theorem necessary for the 1919 paper's proof to be valid.

So Grant Welland and I tried to prove Theorem 2.3. We were able to prove a version of the Cantor–Lebesgue theorem stating that when a multiple trigonometric series converges unrestrictedly rectangularly a.e., "most" coefficients tend to zero, while all coefficients are bounded. From this control of the coefficient size, it follows that Shapiro's coefficient size condition (2) holds in dimension 2. In view of the $M\delta$ example just mentioned, one cannot expect the hypothesis of

$$UR \sum_{n \in \mathbb{Z}^d} c_n e^{inx} = 0 \text{ for all } x \in \mathbb{T}^d \tag{5}$$

to easily imply that $SPH \sum_{n \in \mathbb{Z}^d} c_n e^{inx} = 0$ for all $x \in \mathbb{T}^d$. Nevertheless, it turns out to be easy to prove that hypothesis (5) does imply spherical Abel summability to zero everywhere. (This was quite an unexpected and happy surprise for Welland and me.) Thus Shapiro's more general Theorem 2.5 does apply here, so that in 1972 the two-dimensional case of Theorem 2.3 was shown to be another consequence of Shapiro's 1957 results [7].

After a gap of about 20 years, during which there was no activity at all in the area of uniqueness for multiple trigonometric series, two completely different proofs of Theorem 2.3 for all dimensions appeared [4, 15]. The Tetunashvili proof involves a clever induction. Some ideas from [7, 10] play a role. The Ash–Freiling–Rinne proof extensively renovates the 1919 attempted proof, and uses a complicated covering argument. (See [1] to see this covering argument applied in a much simpler situation.) It is ironical that our very complicated covering proof probably would not have happened if Tetunashvili's previously published and more direct proof had come to our attention before our article had appeared. Only time will tell if the covering techniques we developed will eventually have useful applications elsewhere.

The Conjecture

There is an obvious inclusion. If a multiple numerical series converges Unrestricted Rectangularly, then it converges Square. So

$$\text{Hypothesis of UR Theorem} \Longrightarrow \text{Hypothesis of Square Conjecture.}$$

This explains how it can be that UR uniqueness can be known while Square uniqueness remains an open question. One hint of the problems here is that it is possible for a double trigonometric series to square converge everywhere to a finite valued function while having coefficients that are not $O\left(n^J\right)$ no matter how large J may be [6]. I have discussed the square uniqueness conjecture in several places [3].

The first seven references below can be found using links from http://condor.depaul.edu/mash/realvita.html.

Acknowledgements This research was partially supported by a grant from the Faculty and Development Program of the College of Liberal Arts and Sciences, DePaul University.

References

1. J.M. Ash, A new, harder proof that continuous functions with Schwarz derivative 0 are lines, in *Contemporary Aspects of Fourier Analysis*, ed. by W.O. Bray, C.V. Stanojevic, Pure and Appl. Math. vol. 157 (1994). Marcel Dekker, New York, pp. 35–46
2. J.M. Ash, *Uniqueness for Spherically Convergent Multiple Trigonometric Series*, ed. by G. Anastassiou. Handbook of Analytic-Computational Methods in Applied Mathematics (Chapman & Hall/CRC, Boca Raton, 2000), pp. 309–355
3. J.M. Ash, Uniqueness for higher dimensional trigonometric series. Cubo **4**, 97–125 (2002)
4. J.M. Ash, C. Freiling, D. Rinne, Uniqueness of rectangularly convergent trigonometric series. Ann. Math. **137**, 145–166 (1993)
5. J.M. Ash, G. Wang, A survey of uniqueness questions in multiple trigonometric series, in *A Conference in Harmonic Analysis and Nonlinear Differential Equations in Honor of Victor L. Shapiro*, vol. 208. Contemporary Mathematics (1997), pp. 35–71
6. J.M. Ash, G. Wang, One and two dimensional Cantor-Lebesgue type theorems. Trans. Am. Math. Soc. **349**, 1663–1674 (1997)
7. J.M. Ash, G.V. Welland, Convergence, uniqueness, and summability of multiple trigonometric series. Trans. Am. Math. Soc. **163** , 401–436 (1972). MR **45** # 9057
8. J. Bourgain, Spherical summation and uniqueness of multiple trigonometric series. Int. Math. Res. Not. **1996**, 93–107 (1996)
9. G. Cantor, Beweis, das eine fur jeden reellen Wert von x durch eine trigonometrische Reihe gegebene Funktion f(x) sich nur auf eine einzige Weise in dieser Form darstellen lasst, Crelles J. fuir Math. **72**, 139–142 (1870); also in Georg Olms, Gesammelte Abhandlungen, Hildesheim, 1962, pp. 80–83
10. P.J. Cohen, Topics in the theory of uniqueness of trigonometrical series, doctoral thesis, University of Chicago, Chicago, IL (1958)
11. B. Connes, Sur les coefficients des séries trigonometriques convergents sphériquement. CRA Sc Paris t. Ser. A **283**, 159–161 (1976)

12. R.L. Cooke, A Cantor-Lebesgue theorem in two dimensions. Proc. Am. Math. Soc. **30**, 547–550 (1971). MR **43** #7847
13. Mathematics Genealogy Project. http://genealogy.math.ndsu.nodak.edu/
14. V.L. Shapiro, Uniqueness of multiple trigonometric series. Ann. Math. **66**, 467–480 (1957). MR **19**, 854; 1432
15. S. Tetunashvili, On some multiple function series and the solution of the uniqueness problem for Pringsheim convergence of multiple trigonometric series. Mat. Sb. **182**, 1158–1176 (1991) (Russian), Math. USSR Sbornik **73** , 517–534 (1992) (English). MR 93b:42022
16. A. Zygmund, Trigonometric Series, 2nd edn. vol. 1 (Cambridge University Press, Cambridge, 1958)
17. A. Zygmund, A Cantor-Lebesgue theorem for double trigonometric series. Stud. Math. **43**, 173–178 (1972). MR **47** #711

A Last Conversation with Cora

Aline Bonami

Abstract I cannot contribute to this volume without speaking of Cora and what her friendship meant to me. But I know that, would she be here, she would ask: "Raconte-moi tes maths," that is, "what are you doing right now?" Because, first of all, she was a mathematician. My mathematical contribution tends to answer her question.

I probably met Cora for the first time at El Escorial in 1979. We immediately started a friendly conversation, which went over years. Each time we met again, at a conference or when she visited Paris area, we continued as if there had been no interruption. We always spoke French, which she liked. She had no accent at all, which was at first surprising: it was only from some hesitation in the choice of words that one understood that French was not her mother language. She had spent 1 or 2 years in Paris while a very young child, in the immediate after war period and she liked to remind that time. I remember her, for instance, telling me that their house caretaker had never crossed the Seine River. I had the impression she was speaking of some mythic city, where time stays still, even if I was only a few years younger and lived in Paris area since the early 1950s. During her last stays in Paris she visited regularly Marie-Hélène Schwartz, whom she knew from her early childhood. She came back happy, and somehow serene, to have spent some time with this kind figure of her past, then a very old lady—who finally survived Cora a couple of years.

The conversation of Cora was brilliant, witty, full of life, sometimes intense. We spoke of everything: science or careers, politics and history, colleagues and women in mathematics, and harmonic analysis, of course. She was curious of everything. "Raconte-moi" (tell me), she would ask. And I went to the blackboard and explained what I was doing. We passed from one subject to another, leaving maths to come back 5 min later, as one can do between friends. It is sometimes easier to confide to

A. Bonami (✉)
Fédération Denis Poisson, MAPMO CNRS-UMR 7349, Université d'Orléans,
45067 Orléans Cedex 2, France
e-mail: Aline.Bonami@univ-orleans.fr

M.C. Pereyra et al. (eds.), *Harmonic Analysis, Partial Differential Equations, Complex Analysis, Banach Spaces, and Operator Theory (Volume 1)*, Association for Women in Mathematics Series 4, DOI 10.1007/978-3-319-30961-3_6

a friend one does not meet on an everyday basis. This is how it was between us: a real friendship. At times she spoke of the exile, and the injury it causes, the loss one feels for life. I had only lived it in books. I then lived it through her words. But the next moment we joked gaily again.

I visited her at Howard University in 1981 and met Daniel for the first time. I remind the three of us laughing together after a performance of "A Midsummer Night's Dream," the staging of which we had found ridiculous. We met at MSRI in 1988 (I recall the beach, with Cora, Daniel, and Corasol), then in Oberwolfach. I remember her persuading me to leave for Paris early on Friday afternoon. I had my car and we drove back with Guy David and Stéphane Jaffard, stopping for a lively and cheerful dinner. Cora came to my home that night and succeeded immediately to charm the whole family. In the late 1990s Daniel and Cora spent nearly each year a few days in Paris and we met regularly. I took them once to a horse show, which was not that successful and became another source of jokes between us. But unfortunately illness imperceptibly changed them year after year.

I called her from time to time, not enough, certainly. She replied in her deep, warm voice. "Comme je suis contente de t'entendre !"[1] We met for the last time in 2005, in Washington, where I stopped for half a day on my way back from Buenos Aires. I called her early in the morning. We met and walked quietly through the city, chatting happily as ever. It was our last conversation.

[1]I am so glad to hear you!

Fourier Multipliers of the Homogeneous Sobolev Space $\dot{W}^{1,1}$

Aline Bonami

In memory of Cora

Abstract We prove that the restriction to an affine subspace of such a Fourier multiplier is still a Fourier multiplier, generalizing a celebrated theorem of de Leeuw for Fourier multipliers of L^p. This may be seen as a complement to the spectacular result that such Fourier multipliers are continuous, which has been recently proved by Kazaniecki and Wojciechowski.

1991*Mathematics Subject Classification:* 42B15 42B35

Introduction

Let $W^{1,p}(\mathbb{R}^d)$ be the usual Sobolev space, consisting of functions $f \in L^p(\mathbb{R}^d)$ such that f and $|\nabla f|$ are in $L^p(\mathbb{R}^d)$. Here $|\cdot|$ stands for the Euclidean norm of a vector in $\mathbb{R}^d$. The homogenous Sobolev space $\dot{W}^{1,p}(\mathbb{R}^d)$ consists of tempered distributions f such that $|\nabla f|$ belongs to $L^p(\mathbb{R}^d)$. We define its seminorm by

$$\|f\|_{\dot{W}^{1,p}} := \|\nabla f\|_p.$$

The Schwartz class $\mathcal{S}(\mathbb{R}^d)$ is dense in $\dot{W}^{1,p}$. Let $\mathcal{F}$ be the Fourier transform, defined on $L^1(\mathbb{R}^d)$ by

$$\mathcal{F}f(\xi) := \hat{f}(\xi) := \int_{\mathbb{R}^d} f(x)e^{-2i\pi x\cdot\xi}\, dx.$$

A. Bonami (✉)
Fédération Denis Poisson, MAPMO CNRS-UMR 7349, Université d'Orléans, 45067 Orléans Cedex 2, France
e-mail: Aline.Bonami@univ-orleans.fr

M.C. Pereyra et al. (eds.), *Harmonic Analysis, Partial Differential Equations, Complex Analysis, Banach Spaces, and Operator Theory (Volume 1)*, Association for Women in Mathematics Series 4, DOI 10.1007/978-3-319-30961-3_7

For $1 \leq p < \infty$, let X be either the Lebesgue space $L^p(\mathbb{R}^d)$ or the homogeneous Sobolev space $\dot{W}^{1,p}(\mathbb{R}^d)$. A function $m \in L^\infty(\mathbb{R}^d)$ is said to be a Fourier multiplier of X if there exists a bounded linear operator $T : X \mapsto X$ such that $\widehat{Tf} = m\hat{f}$ for all $f \in \mathcal{S}(\mathbb{R}^d)$. The operator T is the convolution operator by some tempered distribution S. We denote it by T_m in the sequel.

When $p > 1$, Fourier multipliers of $\dot{W}^{1,p}(\mathbb{R}^d)$ coincide with Fourier multipliers of $L^p(\mathbb{R}^d)$. So we will concentrate on the case when $p = 1$. Recall that Fourier multipliers of $L^1(\mathbb{R}^d)$ are given by Fourier transforms of bounded measures. While Fourier multipliers of $\dot{W}^{1,1}(\mathbb{R})$ coincide with those of $L^1(\mathbb{R})$, it was observed by Poornima [8] that this is not the case in higher dimensions. Indeed, there are examples of multipliers of $\dot{W}^{1,1}(\mathbb{R}^d)$ for $d > 1$ given by Fourier transforms of distributions S which are not bounded measures but whose partial derivatives $\partial_{x_j} S$ may be written as a linear combination of partial derivatives of bounded measures,

$$\partial_{x_j} S = \sum_{k=1}^{d} \partial_{x_k} \mu_{jk}.$$

Clearly $\partial_{x_j}(S * f) = \sum \mu_{jk} * \partial_{x_j} f$, from which one concludes that $\mathcal{F}S$ is a Fourier multiplier of $\dot{W}^{1,1}(\mathbb{R}^d)$. The existence of such distributions relies on a very difficult example of Ornstein [7]. Such examples have attracted a lot of interest, in particular in relation with Korn's inequality. Another proof has been given by Conti, Faraco, and Maggi in [3] and generalized in [5].

Let us mention that for $d > 1$ the space $\dot{W}^{1,1}(\mathbb{R}^d)$ possesses other unexpected properties, such as the fact that (up to a constant) it is contained in the Lorentz space $L^{\frac{d}{d-1},1}$ (see [1, 8, 10] and also the recent paper [9]).

At the same time, one may expect that there are not so many Fourier multipliers of $\dot{W}^{1,1}(\mathbb{R}^d)$. This was the object of a joint paper [2] with Poornima, where it was proved that nonconstant homogeneous functions of degree 0 are not such Fourier multipliers. Very recently Kazaniecki and Wojciechowski have proved a much stronger statement in their preprint [4], namely,

Theorem 1.1 (Kazaniecki and Wojciechowski). *Fourier multipliers of* $\dot{W}^{1,1}(\mathbb{R}^d)$ *are continuous functions on* $\mathbb{R}^d$.

Their proof uses the fact that the continuity is already known for homogeneous functions of degree 0. It also relies on a tricky construction based on Riesz products, which gives another family of possible counterexamples. Since the counterexamples of Ornstein were the basis of Bonami and Poornima [2], two kinds of difficult constructions are required to prove the continuity.

It may be of independent interest to generalize in this context a well-known property of the Fourier multipliers. The L^p version is known as the theorem of de Leeuw (see [6]). It was in particular used by C. Fefferman to prove that the characteristic function of the unit ball is not a Fourier multiplier of $L^p(\mathbb{R}^d)$ for $d \geq 3$ once it is proved for $\mathbb{R}^2$. Namely, we prove the following.

Theorem 1.2. *The restriction of a Fourier multiplier of $\dot{W}^{1,1}(\mathbb{R}^d)$ to any affine subspace of dimension k identifies with a Fourier multiplier of $\dot{W}^{1,1}(\mathbb{R}^k)$.*

The proof does not use Theorem 1.1 but implies the continuity of Fourier multipliers on all lines.

We list properties of the space of Fourier multipliers of $\dot{W}^{1,1}(\mathbb{R}^d)$ in the next section. We give some proofs, which are not so easily available in [2, 4, 8]. We prove Theorem 1.2 in the third section.

We end this introduction with a remark. In Dimension 2, the complexification of the Sobolev space $\dot{W}^{1,1}(\mathbb{R}^2)$ identifies with the space of functions f such that $\partial_z f$ and $\partial_{\bar{z}} f$ are in $L^1(\mathbb{R}^2)$. Since one passes from one derivative to the other one by the Beurling transform $\mathcal{B}$, which is given by the multiplier $\frac{\xi_1 - i\xi_2}{\xi_1 + i\xi_2}$, it identifies also with the space of functions f in $L^1(\mathbb{R}^2)$ such that their Beurling transform $\mathcal{B}f$ is in $L^1(\mathbb{R}^2)$. This is reminiscent of the definition of the Hardy space $\mathcal{H}^1$, with the Beurling transform that plays the role of the Hilbert transform. But Theorem 1.1 proves that this space has very few Fourier multipliers, contrarily to the Hardy space.

First Properties of Fourier Multipliers of $\dot{W}^{1,1}(\mathbb{R}^d)$

We call $\mathcal{S}_0(\mathbb{R}^d)$ the space of smooth functions whose Fourier transform is compactly supported in $\mathbb{R}^d \setminus \{0\}$. It is proved in [2] that they are dense in $\dot{W}^{1,1}(\mathbb{R}^d)$.

Next, it is proved in [4] (by an elementary proof) that Fourier multipliers of $\dot{W}^{1,1}(\mathbb{R}^d)$ are continuous in $\mathbb{R}^d \setminus \{0\}$ and uniformly bounded by the norm of T_m as a convolution operator, which we simply note $|||m|||$. More precisely, the norm $|||m|||$ is the smallest constant C such that, for all $f \in \mathcal{S}_0(\mathbb{R}^d)$, we have

$$\|\nabla(T_m f)\|_1 \leq C\|\nabla f\|_1 \tag{1}$$

or, equivalently, for h with values in $\mathbb{R}^d$ that is bounded and compactly supported,

$$2\pi \left| \int_{\mathbb{R}^d} m(\xi)\hat{f}(\xi)\xi \cdot \hat{h}(\xi)d\xi \right| \leq C\|\nabla f\|_1 \|h\|_\infty. \tag{2}$$

Using this characterization and the Lebesgue dominated convergence theorem, we obtain the next lemma (a weaker version may be found in [4]). The analog for Fourier multipliers of $L^p(\mathbb{R}^d)$ when $p \geq 1$ is well known.

Lemma 2.1. *If m_n is a sequence of Fourier multipliers of $\dot{W}^{1,1}(\mathbb{R}^d)$ which are bounded in norm and if m_n tends to m a.e., then m is a Fourier multiplier of $\dot{W}^{1,1}(\mathbb{R}^d)$ and*

$$|||m||| \leq \sup_n |||m_n|||. \tag{3}$$

Next, for all functions ϕ defined on $\mathbb{R}^d$ and $\lambda > 0$, we define the dilated functions $\theta_\lambda(\phi)$ by $\theta_\lambda(\phi)(x) := \phi(\lambda x)$. We immediately have that

$$\|\theta_\lambda(f)\|_{\dot{W}^{1,1}(\mathbb{R}^d)} = \lambda^{-d+1}\|f\|_{\dot{W}^{1,1}(\mathbb{R}^d)}. \tag{4}$$

Moreover, from the identity

$$T_{\theta_\lambda(m)}f = \theta_{\lambda^{-1}}\left(T_m(\theta_\lambda(f))\right), \tag{5}$$

we deduce the invariance of the space of Fourier multipliers by dilation: $\theta_\lambda(m)$ is a Fourier multiplier at the same time as m, with norm $|||\theta_\lambda(m)||| = |||m|||$.

We now state and prove for completeness the characterization of Fourier multipliers of $\dot{W}^{1,1}(\mathbb{R})$ (it is only proved for the space $W^{1,1}(\mathbb{R})$ in [8]). The same kind of proof will be used later on.

Proposition 2.2. *Fourier multipliers of $\dot{W}^{1,1}(\mathbb{R})$ are Fourier transforms of bounded measures and, if m is the Fourier transform of the measure μ, then $|||m||| \approx \|\mu\|_{M(\mathbb{R})}$.*

Here $M(\mathbb{R})$ stands for the space of bounded measures on $\mathbb{R}$.

Proof. Since the convolution by bounded measures preserves the space $\dot{W}^{1,1}(\mathbb{R}^d)$, we only have to prove the converse. Assume that m is a Fourier multiplier of $\dot{W}^{1,1}(\mathbb{R})$. Let φ be a fixed smooth function, which is supported in $(-2,+2)$ and equal to 1 in $(-1,+1)$. Because of the analog of Lemma 2.1 for Fourier transforms of bounded measures, it is sufficient to prove that $(1-\varphi(n\cdot))m$ is uniformly the Fourier transform of a bounded measure. Using invariance by dilation for Fourier transforms of bounded measures, this is equivalent to the fact that $(1-\varphi)\theta_{1/n}(m)$ is uniformly the Fourier transform of a bounded measure. Again, using invariance by dilation of the Fourier multipliers of $\dot{W}^{1,1}(\mathbb{R})$, it is sufficient to prove that there exists some constant C such that, for all Fourier multipliers m, the function $(1-\varphi)m$ is the Fourier transform of a measure whose norm is bounded by $C|||m|||$. Let us prove this inequality when m is a Fourier multiplier of $\dot{W}^{1,1}(\mathbb{R})$ and m vanishes in $(-1,+1)$. It is sufficient to prove that m is a multiplier of $L^1(\mathbb{R})$, that is,

$$2\pi\left|\int_{\mathbb{R}} m(\xi)\hat{g}(\xi)\hat{h}(\xi)d\xi\right| \leq C|||m|||\,\|g\|_1\|h\|_\infty \tag{6}$$

for all g with compactly supported and smooth Fourier transform, and h bounded with compact support. This is a consequence of (2), written with f smooth with compact support outside $(-1/2,+1/2)$, such that $\xi\hat{f}(\xi) = \hat{g}(\xi)$, with $\|f'\|_1 \leq C\|g\|_1$. We can take $\hat{f}(\xi) := (1-\varphi(2\xi))\frac{\hat{g}(\xi)}{\xi}$. Indeed the function $\frac{1-\varphi(2\xi)}{\xi}$, which is in $L^2(\mathbb{R})$ as well as its derivative, is the Fourier transform of an L^1 function. □

The Restriction Theorem

We now prove Theorem 1.2. Since the space of Fourier multipliers is not invariant through translation, let us be more explicit. Because of the invariance by rotation, it is sufficient to consider the following situation: we fix the integer k with $1 \leq k < d$. For ξ in $\mathbb{R}^d$, we denote by ξ' its projection on the subspace generated by the k first coordinates and $\xi'' = \xi - \xi'$. We then consider the affine subspaces $\xi'' = a$, for some $a \in \mathbb{R}^{d-k}$. We claim that for m a Fourier multiplier of $\dot{W}^{1,1}(\mathbb{R}^d)$, the function $\xi' \mapsto m(\xi', a)$ is a Fourier multiplier of $\dot{W}^{1,1}(\mathbb{R}^k)$. Because of the invariance by dilation, it is sufficient to prove the theorem in two particular cases, namely $a = 0$ and $a = (0, \cdots, 0, 1)$.

Traces of Fourier Multipliers on Subspaces of $\mathbb{R}^d$

In this sub-section we consider the case when $a = 0$. Recall that we know, by the theorem of Kazaniecki and Wojciechowski, that m is a continuous function, but we will only need the continuity outside 0. Without loss of generality we assume that $|||m||| = 1$. With an obvious change of notations, because of (2), we have to prove that

$$2\pi \left| \int_{\mathbb{R}^k} m(\xi', 0)\hat{f}(\xi')\xi' \cdot \hat{h}(\xi')d\xi' \right| \leq C\|\nabla_{x'} f\|_1 \|h\|_\infty.$$

Now f and h are defined on $\mathbb{R}^k$. Let φ be a fixed function on $\mathbb{R}^{d-k}$ whose Fourier transform is smooth and compactly supported, and is equal to 1 at 0. Because of continuity, the integral on the left-hand side is the limit, for λ tending to 0, of

$$\frac{2\pi}{\lambda^{d-k}} \int_{\mathbb{R}^d} m(\xi', \xi'')\hat{f}(\xi')\xi' \cdot \hat{f}(\xi')\varphi^2(\xi''/\lambda)d\xi' d\xi''.$$

We have used the fact that $\frac{1}{\lambda^{d-k}}\varphi^2(\xi''/\lambda)d\xi''$ tends weakly to the Dirac mass at 0. But it is bounded by $\|\nabla F\|_1 \|H\|_\infty$, with $F(x) = \lambda^d f(x')\mathcal{F}^{-1}(\varphi)(\lambda x'')$ and $H(x) = h(x')(\mathcal{F}^{-1}\varphi)(\lambda x'')$. The norm of H is directly bounded by $C\|h\|_\infty$. For F, the norm of its gradient in x' is also directly bounded in terms of $\|\nabla_{x'} f\|_1$. It remains to consider the gradient in x'', which is bounded by $\lambda\|f\|_1 \|\nabla_{x''}\varphi\|_1$. When we let λ tend to 0, we find the required inequality.

Remark that we deduce from the previous proof that the norm of $m(\cdot, 0)$ as a Fourier multiplier of $\dot{W}^{1,1}(\mathbb{R}^k)$ is bounded by $|||m|||$.

Traces of Fourier Multipliers on Other Affine Subspaces of $\mathbb{R}^d$

We have seen that it is sufficient to consider the case when $a = (0, \cdots, 0, 1)$. We adapt the previous proof by replacing the function $\varphi^2(\xi''/\lambda)$ by $\varphi^2((\xi'' - a)/\lambda)$, so that now $\frac{1}{\lambda^{d-k}}\varphi^2((\xi'' - a)/\lambda)d\xi''$ tends to the Dirac mass at a. The proof goes the same way, except for a new term that appears in the gradient of the new function F that we obtain when derivating $e^{-2i\pi a.x''}$. Up to a constant, it is equal to $\lambda^{d-k}f(x')e^{-2i\pi a.x''}(\mathcal{F}^{-1}\varphi)(\lambda x'')$. This gives another term in $\|f\|_1$ for the norm $\|\nabla_{x''}F\|_1$, with no small constant before. We need to get rid of this term, which we can do when $m(\cdot, a)$ is supported outside some ball $|\xi'| < \alpha$ because of the following lemma.

Lemma 3.1. *There is a constant C such that, for every f on $\mathbb{R}^d$ whose Fourier transform is smooth and compactly supported, there exists g with the same properties, such that their Fourier transforms coincide outside the ball $|\xi| < 1$ and such that*

$$\|g\|_1 + \|\nabla g\|_1 \leq C\|\nabla f\|_1.$$

Proof. The proof is classical. Let us consider a function ψ, which is smooth and compactly supported in the ball $|\xi| < 1$, and is equal to 1 when $|\xi| < 1/2$. We take g such that

$$\hat{g}(\xi) = (1 - \psi(\xi))\hat{f}(\xi) = \sum_j \frac{(1 - \psi(\xi))\xi_j}{|\xi|^2}\xi_j\hat{f}(\xi).$$

The inequality $\|\nabla g\|_1 \leq C\|\nabla f\|_1$ follows directly from the fact that the Fourier transform of φ is in $L^1(\mathbb{R}^d)$. Let us prove the bound for the norm of g itself. It is sufficient to prove that for each j the function $\nu(\xi) := \frac{(1-\psi(\xi))\xi_j}{|\xi|^2}$ is also the Fourier transform of an L^1 function. For this we write $1 - \psi(\xi) = \sum_{j\geq 1}(\psi(\xi/2^j) - \psi(\xi/2^{j-1}))$, and see ν as a sum of functions ν_j. It is easily seen that

$$|D^\alpha \nu_j(\xi)| \leq C_\alpha 2^{-(|\alpha|+1)j}.$$

So the ν_j is the Fourier transform of an L^1 function with norm bounded by $C2^{-j}$. This finishes the proof. □

To conclude that the restriction of m to the affine subspace $\xi'' = a$ is a Fourier multiplier, we write m as the sum of two Fourier multipliers. We take this time a function θ in $\mathbb{R}^d$ whose Fourier transform is smooth and supported in the ball centered at a of radius $1/2$. We assume moreover that $\theta = 1$ in the ball centered at a of radius $1/4$. Clearly θm and $(1-\theta)m$ are Fourier multipliers of norm bounded by $C|||m|||$. Next $(1-\theta)m$ vanishes at points (ξ', a) when $|\xi'| < 1/4$. So we can use for it our previous argument. It remains to deal with Fourier multipliers that are supported in the ball centered at a of radius $1/2$. We conclude from the following lemma, using the fact that the restriction to an affine subspace of the Fourier transform of a bounded measure is also the Fourier transform of a bounded measure, and in particular a Fourier multiplier of $\dot{W}^{1,1}(\mathbb{R})$.

Lemma 3.2. *There exists a constant C such that every Fourier multiplier of $\dot{W}^{1,1}(\mathbb{R})$ which is compactly supported in the shell $1/2 \leq |\xi| \leq 3/2$ is the Fourier transform of a measure μ, with $\|\mu\|_{M(\mathbb{R}^d)} \leq C|||m|||$.*

Proof. We assume that $|||m||| = 1$. It is sufficient to prove that m is a Fourier multiplier of $L^1(\mathbb{R}^d)$, that is, for $\hat{g}$ which is smooth and compactly supported and k which is bounded and compactly supported,

$$2\pi \left| \int_{\mathbb{R}^d} m(\xi)\hat{g}(\xi)\hat{k}(\xi)d\xi \right| \leq C\|g\|_1 \|k\|_\infty.$$

This is deduced from (2) once we have found f and h_j for $j = 1, \cdots, d$ such that

$$\hat{g}(\xi)\hat{k}(\xi) = \hat{f}(\xi) \sum_j \xi_j \hat{h}_j(\xi).$$

We fix a function φ, which is smooth, equal to 1 in the shell that contains the support of m and compactly supported in a slightly larger shell, say, $1/2 \leq |\xi| \leq 3/2$. We choose $\hat{f}(\xi) = \frac{\varphi(\xi)}{|\xi|^2}\hat{g}(\xi)$, $\hat{h}_j(\xi) = \varphi(\xi)\xi_j\hat{k}(\xi)$. It remains to prove that $\|\nabla f\|_1 \leq C\|g\|_1$, and also $\|h_j\|_\infty \leq \|k\|_\infty$. This is a direct consequence that all functions $\frac{\xi_j\varphi(\xi)}{|\xi|^2}$ and $\xi_j\varphi(\xi)$ are Fourier transforms of L^1 functions. The functions h_j are not compactly supported but decrease rapidly, which is sufficient to conclude. □

The restriction mapping is surjective when $k = 0$: the Fourier transform of a bounded measure on $\mathbb{R}$ extends into the Fourier transform of a bounded measure on $\mathbb{R}^d$. When $k > 0$ and $a \neq 0$ it is not surjective: a Fourier multiplier of $\dot{W}^{1,1}(\mathbb{R}^d)$ coincides locally with the Fourier transform of a bounded measure outside 0 by Lemma 3.2, while it is not the case for a Fourier multiplier of $\dot{W}^{1,1}(\mathbb{R}^k)$ in a neighborhood of 0 for $k > 0$. We do not know whether the restriction mapping is surjective for $a = 0$.

Final Remark

The proof of Theorem 1.2 may be considered as very simple compared to the one of Theorem 1.1. One would like to have a *Functional analysis* proof for this last one. Due to Theorem 1.2 and the fact that the class of Fourier multipliers of $\dot{W}^{1,1}(\mathbb{R})$ coincides with that of $L^1(\mathbb{R})$, the restriction of a Fourier multiplier of $\dot{W}^{1,1}(\mathbb{R}^d)$ to a line in $\mathbb{R}^d$ is the Fourier transform of a bounded measure on $\mathbb{R}$. As a consequence, multipliers are continuous on rays. This simplifies one step of the proof of Kazaniecki and Wojciechowski. Remark also that to prove that the limit at 0 on rays does not depend on the direction it is sufficient to do it in Dimension 2, which is another simplification. Unfortunately we could not go further in the establishment of a simple proof of Theorem 1.1.

References

1. A. Alvino, Sulla diseguaglianza di Sobolev in spazi di Lorentz. Boll. Un. Mat. Ital. A (5) **14**(1), 148–156 (1977)
2. A. Bonami, S. Poornima, Nonmultipliers of the Sobolev spaces $W^{k,1}(\mathbf{R}^n)$. J. Funct. Anal. **71**(1), 175–181 (1987)
3. S. Conti, D. Faraco, F. Maggi, A new approach to counterexamples to L^1 estimates: Korn's inequality, geometric rigidity, and regularity for gradients of separately convex functions. Arch. Ration. Mech. Anal. **175**(2), 287–300 (2005)
4. K. Kazaniecki, M. Wojciechowski, On the continuity of Fourier multipliers on the homogeneous Sobolev spaces $\dot{W}_1^1(\mathbb{R}^d)$. Ann. Inst. Fourier (Grenoble) **66**(3), 1247–1260 (2016). Preprint available at arXiv:1306.1437v4 [math.FA]
5. K. Kazaniecki, D. Stolyarov, M. Wojciechowski, Anisotropic Ornstein non inequalities (2015). Preprint available at arXiv:1505.05416 [math.CA]
6. K. de Leeuw, On L_p multipliers. Ann. Math. **81**, 364–379 (1965)
7. D. Ornstein, A non-equality for differential operators in the L^1 norm. Arch. Ration. Mech. Anal. **11**, 40–49 (1962)
8. S. Poornima, On the Sobolev spaces $W^{k,1}(\mathbf{R}^n)$, in *Harmonic Analysis (Cortona, 1982)*. Lecture Notes in Mathematics, vol. 992 (Springer, Berlin, 1983), pp. 161–173
9. A. Schikorra, D. Spector, J. Van Schaftingen, *An L^1-type estimate for Riesz potentials* (2015). Accepted for publication in Rev. Mat. Iberoam. Preprint available at arXiv:1411.2318 [math.FA]
10. L. Tartar, Imbedding theorems of Sobolev spaces into Lorentz spaces. Boll. Unione Mat. Ital. Sez. B Artic. Ric. Mat. (8) **1**(3), 479–500 (1998)

A Note on Nonhomogenous Weighted Div-Curl Lemmas

Der-Chen Chang, Galia Dafni, and Hong Yue

Dedicated to the memory of Cora Sadosky

Abstract We prove some nonhomogeneous versions of the div-curl lemma in the context of weighted spaces. Namely, assume the vector fields $\mathbf{V}, \mathbf{W}$: $\mathbb{R}^n \to \mathbb{R}^n$, along with their distributional divergence and curl, respectively, lie in L^p_μ and L^q_ν, $\frac{1}{p} + \frac{1}{q} = 1$, where μ and ν are in certain Muckenhoupt weight classes. Then the resulting scalar product $\mathbf{V} \cdot \mathbf{W}$ is in the weighted local Hardy space $h^1_\omega(\mathbb{R}^n)$, for $\omega = \mu^{\frac{1}{p}} \nu^{\frac{1}{q}}$ in $A_{1+\frac{1}{n}}$.

2000*Mathematics Subject Classification:* Primary 42B30, 46E35

Introduction and Background

The following article describes the results presented by the second author at the Special Session on Harmonic Analysis and Operator Theory (in memory of Cora Sadosky), the AMS Western Spring Sectional Meeting, Albuquerque, NM, April 2014. All three authors have been influenced by Cora, both personally (the first two

D.-C. Chang
Department of Mathematics and Statistics, Georgetown University, Washington, DC 20057, USA

Fu Jen Catholic University, Taipei 242, Taiwan
e-mail: chang@math.georgetown.edu

G. Dafni (✉)
Department of Mathematics and Statistics, Concordia University, Montreal, QC, Canada H3G-1M8
e-mail: galia.dafni@concordia.ca

H. Yue
Department of Mathematics, Georgia College and State University, Milledgeville, GA 31061, USA
e-mail: hong.yue@gcsu.edu

M.C. Pereyra et al. (eds.), *Harmonic Analysis, Partial Differential Equations, Complex Analysis, Banach Spaces, and Operator Theory (Volume 1)*, Association for Women in Mathematics Series 4, DOI 10.1007/978-3-319-30961-3_8

authors have been her coauthors—see [3, 4]) and through her work. We would like to dedicate this note to her memory, in recognition of her contributions to harmonic analysis and to mathematics in general.

Since the work of Coifman, Lions, Meyer, and Semmes [5], there have been many results by a variety of authors on div-curl lemmas in the context of Hardy spaces (the article [4] also falls into this body of work). The idea is to show that certain quantities occurring in nonlinear PDE lie in a better space than expected, in many cases the Hardy space H^1 as opposed to L^1, enabling one to prove weak convergence of these quantities and leading to improved regularity results. Here we prove analogues of some of these theorems in the case of weighted Hardy spaces on $\mathbb{R}^n$. Previous work by Tsutsui [17] has already dealt with the homogeneous case, and we present a nonhomogeneous version, although the proofs can also be applied to the homogeneous case in order to obtain Tsutsui's conclusions in a more straightforward way.

Before stating our results, we review quickly some definitions and theorems relevant to this work. We do not attempt to give an overview of the subject. To define real Hardy spaces, we will use the maximal function characterization: for $0 < p \leq \infty$, we say a tempered distribution f lies in $H^p(\mathbb{R}^n)$ if the maximal function $\mathcal{M}_\varphi(f)$ belongs to $L^p(\mathbb{R}^n)$, where φ is a fixed Schwartz function with $\int \varphi = 1$, and

$$\mathcal{M}_\varphi(f)(x) := \sup_{t>0} |f * \varphi_t(x)|, \quad \varphi_t(\cdot) = t^{-n}\varphi(t^{-1}\cdot). \tag{1}$$

The choice of φ does not affect the space H^p, only the norm, defined by

$$\|f\|_{H^p} := \|\mathcal{M}_\varphi(f)\|_{L^p}.$$

We will be dealing with the case $p = 1$, where the Hardy space H^1 is a proper subspace of L^1. One of the main results of Coifman et al. [5] can be stated as follows:

Theorem 1.1 ([5]). *If $n \geq 2$, $1 < p < \infty$, $\frac{1}{p} + \frac{1}{q} = 1$,*

$$\mathbf{V} \in L^p(\mathbb{R}^n, \mathbb{R}^n), \quad \mathbf{W} \in L^q(\mathbb{R}^n, \mathbb{R}^n)$$

with

$$\operatorname{div} \mathbf{V} = 0, \quad \operatorname{curl} \mathbf{W} = 0$$

in the sense of distributions, then

$$\mathbf{V} \cdot \mathbf{W} \in H^1(\mathbb{R}^n),$$

with

$$\|\mathbf{V} \cdot \mathbf{W}\|_{H^1} \leq C\|\mathbf{V}\|_{L^p}\|\mathbf{W}\|_{L^q}.$$

A *local*, or *nonhomogeneous*, version of real Hardy spaces, $h^p(\mathbb{R}^n)$, $0 < p \leq \infty$, was defined in [8]: a tempered distribution $f \in h^p(\mathbb{R}^n)$ if the *localized* maximal function $\mathcal{M}_\varphi^{\rm loc}(f)$ belongs to $L^p(\mathbb{R}^n)$, where φ is fixed as in the definition of $\mathcal{M}_\varphi$ in (1), and

$$\mathcal{M}_\varphi^{\rm loc}(f)(x) := \sup_{0<t<1} |f * \varphi(x)|.$$

Here the choice of the number 1 is arbitrary: any finite number will give an equivalent norm. The term *local* is misleading because there is another (truly) local real Hardy space $H^p_{\rm loc}$, consisting of distributions which when multiplied by smooth functions of compact support become elements of h^p (or elements of H^p if an extra step is included in order to guarantee the moment conditions—see Proposition 1.92 in [14] for the case $p = 1$). A nonhomogeneous version of the div-curl lemma for vector fields in $L^p_{\rm loc}$, $L^q_{\rm loc}$, resulting in a dot product in $H^1_{\rm loc}$, was given in [5], assuming some improved "integrability" of the distributional divergence and curl, respectively, and using the Hodge decomposition.

An analogue of the result of Coifman et al. [5] above for the space h^1 was given in [6], where it was shown that the dot product $\mathbf{V}\cdot\mathbf{W}$ belongs to h^1, assuming certain conditions on the distributions div $\mathbf{V}$ and $(\text{curl }\mathbf{W})_{ij}$. A special case (corresponding to the hypotheses of the div-curl lemma of Murat [12]) is as follows:

Theorem 1.2 ([6]). *For vector fields* $\mathbf{V} \in L^p(\mathbb{R}^n, \mathbb{R}^n)$, $\mathbf{W} \in L^q(\mathbb{R}^n, \mathbb{R}^n)$, *if* div $\mathbf{V}$ *is a function in* $L^p(\mathbb{R}^n)$ *and* $(\text{curl }\mathbf{W})_{ij}$ *are functions in* $L^q(\mathbb{R}^n)$, $1 \leq i,j \leq n$, *we have* $\mathbf{V}\cdot\mathbf{W} \in h^1(\mathbb{R}^n)$ *with*

$$\|\mathbf{V}\cdot\mathbf{W}\|_{h^1(\mathbb{R}^n)} \leq C\Big[\|\mathbf{V}\|_{L^p(\mathbb{R}^n)}\|\mathbf{W}\|_{L^q(\mathbb{R}^n)} + \|\text{div }\mathbf{V}\|_{L^p(\mathbb{R}^n)}\|\mathbf{W}\|_{L^q(\mathbb{R}^n)}$$
$$+\|\mathbf{V}\|_{L^p(\mathbb{R}^n)}\sum_{i,j}\|(\text{curl }\mathbf{W})_{ij}\|_{L^q(\mathbb{R}^n)}\Big].$$

Hardy spaces can be defined in much more general settings than $\mathbb{R}^n$, such as metric measure spaces or spaces of homogeneous type. We will consider the special case of doubling measures λ on $\mathbb{R}^n$ for which the Hardy–Littlewood maximal function is bounded on L^p_λ, $1 < p < \infty$, which are measures of the form $d\lambda = \omega dx$, where $\omega > 0$ satisfies the Muckenhoupt A_p condition

$$[\omega]_{A_p} := \sup_B \left(⨍_B \omega dx\right)\left(⨍_B \omega^{-1/(p-1)dx}\right)^{p-1} < \infty.$$

Here we take the supremum over all balls B in $\mathbb{R}^n$ and use $⨍_B \omega dx$ to denote the average of ω over B with respect to Lebesgue measure. When $p = 1$, the A_1 condition becomes

$$[\omega]_{A_1} := \sup_B \left(⨍_B \omega dx\right)\left(\operatorname*{ess\,inf}_{x\in B}\omega(x)\right)^{-1} < \infty,$$

while for $p = \infty$ it is

$$[\omega]_{A_\infty} := \sup_B \left(\fint_B \omega dx \right) \left[\exp \left(\fint_B \log \omega dx \right) \right]^{-1} < \infty.$$

Recall that the A_p classes are increasing in p, so that A_1 is the smallest and A_∞ is the biggest class, which can be written as the union of all the A_p classes for $p < \infty$. For some recent instructive expositions of weights and related topics see [10, 13].

The theory of weighted Hardy spaces using A_p weights was developed in [7, 16, 18]. Local weighted Hardy spaces h^p_ω were studied by Bui [2], who proved analogues of the results of Goldberg [8] on boundary functions of harmonic functions on the strip, with the restriction that the weight ω belong to A_1.

Starting with $\omega \in A_\infty$, we will use the conditions

$$\mathcal{M}_\varphi(f) \in L^p_\omega, \quad \text{resp. } \mathcal{M}^{\text{loc}}_\varphi(f) \in L^p_\omega,$$

to define H^p_ω, resp. h^p_ω. These correspond to the *radial* maximal function, while other equivalent characterizations can be shown using *grand* and *nontangential* maximal functions. For H^p_w, Tsutsui proves the following version of the (homogeneous) div-curl lemma (Theorem 1.2 in [17]):

Theorem 1.3 ([17]). *Let $p, q \in (\frac{n}{n+1}, \infty)$ be such that $\frac{1}{r} = \frac{1}{p} + \frac{1}{q} < 1 + \frac{1}{n}$. Suppose further that $\tau \in (1, p(1+1/n))$ and $\rho \in (1, q(1+1/n))$ are such that $\frac{\tau}{p} + \frac{\rho}{q} < 1 + \frac{1}{n}$. Then for any weights $\mu \in A_\tau$ and $\nu \in A_\rho$ we have*

$$\|(\mathbf{u} \cdot \nabla) v\|_{H^r_\omega} \le c \|\mathbf{u}\|_{H^p_\mu} \|\nabla v\|_{H^q_\nu}$$

when $\operatorname{div} \mathbf{u} = 0$ *and* $\omega = \mu^{r/p} \nu^{r/q}$.

Note that ω belongs to $A_{r(\frac{\tau}{p}+\frac{\rho}{q})}$ by an interpolation result for weights (see Corollary 28 in [10]). The proof of the theorem relies on using the divergence-free condition on $\mathbf{u}$ to express $\mathbf{u} \cdot \nabla v$ in terms of some bilinear forms involving Riesz transforms, and then applying a pointwise estimate of Miyachi [11] and the boundedness of the Riesz transforms on H^p_μ. Tsutsui also proves an endpoint estimate ($p = \infty$) following the ideas in [1].

Statement and Proof of Results

We now state and prove two nonhomogeneous versions of the div-curl lemma for weighted local Hardy spaces, corresponding to the irrotational and the incompressible cases, respectively. Here φ is fixed as in definition (1) above, and we may assume that its support is contained in the ball $B(0, 1)$.

Theorem 2.1. *Let* $1 < p, q < \infty$ *be conjugate exponents,* $\frac{1}{p} + \frac{1}{q} = 1$*, and let* $\tilde{p} \in (1, p)$*,* $\tilde{q} \in (\max(\frac{q}{n}, 1), 1 + \frac{q}{n})$ *such that* $\frac{\tilde{p}}{p} + \frac{\tilde{q}}{q} = 1 + \frac{1}{n}$*. Assume that* μ *and* ν *are two weights in* $A_{\tilde{p}}$ *and* $A_{\tilde{q}}$*, respectively, and that* $\omega = \mu^{\frac{1}{p}} \nu^{\frac{1}{q}}$*. If the vector fields* $\mathbf{V}$ *in* $L^p_\mu(\mathbb{R}^n, \mathbb{R}^n)$ *and* $\mathbf{W}$ *in* $L^q_\nu(\mathbb{R}^n, \mathbb{R}^n)$ *satisfy*

$$\text{div } \mathbf{V} \in L^p_\mu, \quad \text{curl } \mathbf{W} = 0$$

in the sense of distributions, then the scalar product $f = \mathbf{V} \cdot \mathbf{W}$ *satisfies* $\mathcal{M}^{\text{loc}}_\varphi(f) \in L^1_\omega$*. Moreover,*

$$\|\mathcal{M}^{\text{loc}}_\varphi(f)\|_{L^1_\omega} \leq C\Big(\|\mathbf{V}\|_{L^p_\mu} + \|\text{div } \mathbf{V}\|_{L^p_\mu}\Big)\|\mathbf{W}\|_{L^q_\nu}. \tag{2}$$

Theorem 2.2. *Let* $1 < p, q < \infty$ *be conjugate exponents,* $\frac{1}{p} + \frac{1}{q} = 1$*, and let* $\tilde{q} \in (1, q)$*,* $\tilde{p} \in (\max(\frac{p}{n}, 1), \min(p, 1 + \frac{p}{n}))$ *such that* $\frac{\tilde{p}}{p} + \frac{\tilde{q}}{q} = 1 + \frac{1}{n}$*. Assume* μ *and* ν *are two weights in* $A_{\tilde{p}}$ *and* $A_{\tilde{q}}$*, respectively, and that* $\omega = \mu^{\frac{1}{p}} \nu^{\frac{1}{q}}$*. If the vector fields* $\mathbf{V}$ *in* $L^p_\mu(\mathbb{R}^n, \mathbb{R}^n)$ *and* $\mathbf{W}$ *in* $L^q_\nu(\mathbb{R}^n, \mathbb{R}^n)$ *satisfy*

$$\text{div } \mathbf{V} = 0, \quad (\text{curl } \mathbf{W})_{ij} \in L^q_\nu(\mathbb{R}^n, \mathbb{R}^n),\ i, j \in \{1, \ldots, n\},$$

in the sense of distributions, then the scalar product $f = \mathbf{V} \cdot \mathbf{W}$ *satisfies* $\mathcal{M}^{\text{loc}}_\varphi(f) \in L^1_\omega$*. Moreover,*

$$\|\mathcal{M}^{\text{loc}}_\varphi(f)\|_{L^1_\omega} \leq C\|\mathbf{V}\|_{L^p_\mu}\Big(\|\mathbf{W}\|_{L^q_\nu} + \sum_{i,j} \|(\text{curl } \mathbf{W})_{ij}\|_{L^q_\nu}\Big). \tag{3}$$

Note that our results correspond roughly to the case $r = 1$ in Tsutsui's result, Theorem 1.3 above. In our case interpolation gives $\omega \in A_{1+\frac{1}{n}}$, but by the self-improvement property of weights, ω will lie in some class A_s with $s < 1 + \frac{1}{n}$. Moreover, even if we took $\tilde{p}$ and $\tilde{q}$ with $\frac{\tilde{p}}{p} + \frac{\tilde{q}}{q} < 1 + \frac{1}{n}$ we can always increase them (since the A_p classes are increasing) so without loss of generality we may assume equality.

Conversely, starting with a weight ω in $A_{1+\frac{1}{n}}$, the factorization of weights (see [15], V.5.3, Proposition 9 or [10], Theorem 29) allows us to write ω as $\omega_1 \omega_2^{-1/n}$ for $\omega_1, \omega_2 \in A_1$. Then $\omega_2^{-1/n} \in A_{1+\frac{1}{n}}$ by scaling (Corollary 16 in [10]) and setting $\mu^{1/p} = \omega_1^\theta \omega_2^{-(1/n)(1-\theta)}$ and $\nu^{1/q} = \omega_1^{1-\theta} \omega_2^{-(1/n)\theta}$, one can again use interpolation and scaling (Corollary 17 in [10]) for an appropriate θ to get μ in $A_{\tilde{p}}$ and ν in $A_{\tilde{q}}$.

When both div $\mathbf{V}$ and curl $\mathbf{W}$ vanish, as in Tsutsui's hypotheses, we can use the theorem with the widest range for $\tilde{p}$ and $\tilde{q}$, namely when $1 < p < \frac{n}{n-1}$ we can use Theorem 2.2 which allows for $\tilde{q} \in (1, q)$ and $\tilde{p} \in (\max(\frac{p}{n}, 1), \min(p, 1 + \frac{p}{n})) = (1, p)$ for this range of p. When $p \geq \frac{n}{n-1}$ we can use Theorem 2.1 which allows for $\tilde{p} \in (1, p)$, $\tilde{q} \in (\max(\frac{q}{n}, 1), 1 + \frac{q}{n}) = (1, 1 + \frac{q}{n})$ since $q \leq n$. Note that in Theorem 1.3, the condition $\frac{\tau}{p} + \frac{\rho}{q} < 1 + \frac{1}{n}$ restricts the ranges of τ and ρ to less than the ones specified.

A more significant difference with Theorem 1.3 is that the right-hand side involves the weighted L^p (and L^q), rather than H^p (resp., H^q), norms. For $p > 1$, these coincide when the weight is in A_p, but otherwise, depending on the weight, the Hardy space can contain singular measures and distributions.

When both div $\mathbf{V}$ and curl $\mathbf{W}$ do not vanish, we can still apply the theorems above after first applying a Hodge decomposition, as in [5], to get an analogue of Theorem 1.2.

Proof of Theorem 2.1. Let $\alpha = \frac{p}{\tilde{p}}$ and $\beta = \frac{q}{\tilde{q}}$, then $\frac{1}{\alpha} + \frac{1}{\beta} = \frac{\tilde{p}}{p} + \frac{\tilde{q}}{q} = 1 + \frac{1}{n}$. The key inequality proved in [5] (see (10) after Lemma II.1) is a bound on the maximal function $\mathcal{M}_\varphi(f)$ in terms of the Hardy–Littlewood maximal function, denoted by M:

$$\mathcal{M}_\varphi(f)(x) \leq C\,(\mathrm{M}(|\mathbf{V}|^\alpha)(x))^{1/\alpha}\left(\mathrm{M}(|\mathbf{W}|^\beta)(x)\right)^{1/\beta} \tag{4}$$

under the assumption that $1 < \alpha < p$, $1 < \beta < q$. By our hypotheses we still have $1 < \alpha < p$ but now β lies in the range $(\frac{qn}{q+n}, \min(q, n))$ which means that as q approaches 1 we may have $\beta < 1$.

Fix $x \in \mathbb{R}^n, t > 0$. From the hypothesis on $\mathbf{W}$ we write $\mathbf{W} = \nabla\pi$ and proceed as in [5]. Since we assume φ is supported in $B(0, 1)$, we have that $\varphi_t(x - \cdot)$ is supported in $B = B(x, t)$. Then $\mathbf{W} = \nabla(\pi - \pi_B)$ and we can write, denoting by g the function in L^p_μ which is the divergence of $\mathbf{V}$ in the sense of distributions,

$$\begin{aligned}
\varphi_t * f(x) &= \int_B \varphi_t(x-y)\mathbf{V}(y)\cdot\nabla(\pi - \pi_B)(y)dy \\
&= -\int_B \operatorname{div}\,[\varphi_t(x-y)\mathbf{V}(y)](\pi - \pi_B)(y)dy \\
&= -\int_B [\nabla_y\varphi_t(x-y)\cdot\mathbf{V}(y))](\pi - \pi_B)(y)dy \\
&\quad + \int_B \varphi_t(x-y)g(y)(\pi - \pi_B)(y)dy.
\end{aligned}$$

Denote by α' the conjugate exponent of α. By Hölder's inequality, and using the fact that $t \leq 1$, we have that

$$\begin{aligned}
&|\varphi_t * f(x)| \\
&\leq \|\nabla\varphi\|_\infty t^{-n-1}\int_B |\mathbf{V}||\pi - \pi_B|dy + \|\varphi\|_\infty t^{-n}\int_B |g||\pi - \pi_B|dy \\
&\leq \frac{C_\varphi}{t}\left(\fint_B |\pi - \pi_B|^{\alpha'}\right)^{1/\alpha'}\left\{\left(\fint_B |\mathbf{V}|^\alpha\right)^{1/\alpha} + \left(\fint_B |g|^\alpha\right)^{1/\alpha}\right\}.
\end{aligned} \tag{5}$$

As pointed out above, $\frac{n}{n+1} < \frac{nq}{n+q} < \beta < n$, and since $\alpha' = \frac{\beta n}{n-\beta} = \beta^*$ is the Sobolev conjugate exponent of β, we have $\beta^* \geq 1$. Thus we can apply the version of the Sobolev–Poincaré inequality found in [9], Theorem 8.7 and Remark 8.8 (with the dimension $s = n$ and $\sigma > 1$) to π:

$$\left(\fint_B |\pi(y) - \pi_B|^{\alpha'} dy \right)^{1/\alpha'} \leq Ct \left(\fint_{\sigma B} |\nabla \pi(y)|^{\beta} dy \right)^{1/\beta} . \tag{6}$$

It is important to note that we are not using here any of the weighted versions of the Sobolev–Poincaré inequality found in [13], for example. Inserting this in (5) and taking the supremum over all $t \leq 1$ give the analogue of (4):

$$\mathcal{M}_{\varphi}^{\text{loc}}(f)(x) \leq C \left(\mathrm{M}(|\mathbf{W}|^{\beta})(x) \right)^{1/\beta} \left\{ (\mathrm{M}(|\mathbf{V}|^{\alpha})(x))^{1/\alpha} + (\mathrm{M}(|g|^{\alpha})(x))^{1/\alpha} \right\}.$$

We can again use Hölder's inequality, as well as the boundedness of the Hardy–Littlewood maximal function M on $L_{\mu}^{\tilde{p}}, L_{\nu}^{\tilde{q}}$ (see [15] Chap. V, Theorem 1, noting that $\tilde{p} = \frac{p}{\alpha} > 1$ and $\tilde{q} = \frac{q}{\beta} > 1$) to conclude:

$$\begin{aligned}
\|\mathcal{M}_{\varphi}^{\text{loc}}(f)\|_{L_{\omega}^1} &= \int \mathcal{M}_{\varphi}^{\text{loc}}(f)(x) \mu^{\frac{1}{p}}(x) \nu^{\frac{1}{q}}(x) dx \\
&\leq C \left[\int \left(\mathrm{M}(|\mathbf{W}|^{\beta})(x) \right)^{q/\beta} \nu(x) dx \right]^{1/q} \times \\
&\left\{ \left[\int (\mathrm{M}(|\mathbf{V}|^{\alpha})(x))^{p/\alpha} \mu(x) dx \right]^{1/p} + \left[(\mathrm{M}(|g|^{\alpha})(x))^{p/\alpha} \mu(x) dx \right]^{1/p} \right\} \\
&\leq C \|\mathrm{M}(|\mathbf{W}|^{\beta})\|_{L_{\nu}^{\tilde{q}}}^{\tilde{q}/q} \left\{ \|\mathrm{M}(|\mathbf{V}|^{\alpha})\|_{L_{\mu}^{\tilde{p}}}^{\tilde{p}/p} + \|\mathrm{M}(|g|^{\alpha})\|_{L_{\mu}^{\tilde{p}}}^{\tilde{p}/p} \right\} \\
&\leq C \|\mathbf{W}\|_{L_{\nu}^q} \left(\|\mathbf{V}\|_{L_{\mu}^p} + \|g\|_{L_{\mu}^p} \right).
\end{aligned}$$

∎

Proof of Theorem 2.2. We will follow the proof of Theorem 4 in [6]. As shown there, given the hypothesis div $\mathbf{V} = 0$, we can find a matrix $A = (a_{ij})$ whose columns $\vec{A}_j$ satisfy

$$\text{div}\, \vec{A}_j = v_j, \tag{7}$$

so that in the sense of distributions

$$\mathbf{V} \cdot \mathbf{W} = \sum_{j=1}^{n} \text{div}\, (\vec{A}_j w_j) + \sum_{i<j} a_{ij} (\text{curl}\, \mathbf{W})_{ij}. \tag{8}$$

Denoting $B(x,t)$ by B and replacing a_{ij} by $a_{ij}-(a_{ij})_B$ in (8), we can proceed, as in the proof of Theorem 2.1, to write

$$\varphi_t * (\mathbf{V}\cdot\mathbf{W})(x) = -\sum_{i,j}\int \frac{1}{t^{n+1}}\frac{\partial\varphi}{\partial y_i}(\frac{x-y}{t})(a_{ij}(y)-(a_{ij})_B)\, w_j(y)\, dy$$

$$+\sum_{i<j}\int_B \varphi_t(x-y)(a_{ij}(y)-(a_{ij})_B)(\text{curl }\mathbf{W})_{ij}(y)$$

For α,β as in the proof of Theorem 2.1, the assumptions on q force $1<\beta<q$, so we can use Hölder's inequality, and the fact that $t\le 1$, to get

$$|\varphi_t * f(x)| \le \frac{\|\nabla\varphi\|_\infty}{t^{n+1}}\sum_{i,j}\int_B |a_{ij}-(a_{ij})_B||w_j|\, dy$$

$$+\frac{\|\varphi\|_\infty}{t^n}\sum_{i<j}\int_B |a_{ij}-(a_{ij})_B||(\text{curl }\mathbf{W})_{ij}|dy$$

$$\le \sum_{i,j}\frac{C_\varphi}{t}\left(\fint_B |a_{ij}(y)-(a_{ij})_B|^{\beta'}dy\right)^{1/\beta'}\times$$

$$\left\{\left(\fint_B |\mathbf{W}|^\beta\right)^{1/\beta}+\left(\fint_B |(\text{curl }\mathbf{W})_{ij}|^\beta\right)^{1/\beta}\right\}.$$

For each i,j, we can again apply the Sobolev–Poincaré inequality (with the ranges for α and β reversed, so that now $\frac{n}{n+1}<\alpha<n$ and $\beta'=\alpha^*\ge 1$) to get

$$\left(\fint_B |a_{ij}(y)-(a_{ij})_B)|^{\beta'}dy\right)^{1/\beta'} \le Ct\left(\fint_{\sigma B} |\nabla a_{ij}(y)|^\alpha dy\right)^{1/\alpha}. \tag{9}$$

The resulting estimate for the maximal function is

$$|\mathcal{M}_\varphi^{\text{loc}}(f)(x)|$$
$$\le C\sum_{i,j}\text{M}(|\nabla a_{ij}|^\alpha)(x)^{1/\alpha}\left(\text{M}(|w_j|^\beta)(x)^{1/\beta}+\text{M}(|(\text{curl }\mathbf{W})_{ij}|^\beta)(x)^{1/\beta}\right).$$

Since the a_{ij} were defined (see [6]) in terms of second derivatives of the solution of Poisson's equation for the components of $\mathbf{V}$, the boundedness of the Riesz transforms on L^p_μ (see [15] Chap. V, Corollary to Theorem 2), provided $\mu\in A_{\tilde{p}}\subset A_p$ (which is why we need to assume $\tilde{p}<p$), gives $\|\nabla a_{ij}\|_{L^p_\mu}\le C\|\mathbf{V}\|_{L^p_\mu}$. This and the boundedness of the maximal function allow us to conclude the proof:

$$
\begin{aligned}
\|\mathcal{M}_{\varphi}^{\mathrm{loc}}(f)\|_{L^1_\omega} &\le C\sum_{i,j}\int \left[\mathrm{M}(|\nabla a_{ij}|^{\alpha})(x)\right]^{\frac{1}{\alpha}} \times \\
&\qquad \left[\mathrm{M}(|w_j|^{\beta})(x)^{1/\beta} + \mathrm{M}(|(\mathrm{curl}\ \mathbf{W})_{ij}|^{\beta})(x)^{1/\beta}\right]\mu^{1/p}(x)\nu^{1/q}(x)dx \\
&\le C\sum_{i,j}\|\mathrm{M}(|\nabla a_{ij}|^{\alpha})\|_{L_\mu^{\tilde{p}}}^{\tilde{p}/p}\left\{\|\mathrm{M}(|\mathbf{W}|^{\beta})\|_{L_\nu^{\tilde{q}}}^{\tilde{q}/q} + \mathrm{M}(|(\mathrm{curl}\ \mathbf{W})_{ij}|^{\beta})\|_{L_\nu^{\tilde{q}}}^{\tilde{q}/q}\right\} \\
&\le C\sum_{i,j}\|\nabla a_{ij}\|_{L^p_\mu}\left\{\|\mathbf{W}\|_{L^q_\nu} + \|(\mathrm{curl}\ \mathbf{W})_{ij}\|_{L^q_\nu}\right\} \\
&\le C\|\mathbf{V}\|_{L^p_\mu}\Big(\|\mathbf{W}\|_{L^q_\nu} + \sum_{i,j}\|(\mathrm{curl}\ \mathbf{W})_{ij}\|_{L^q_\nu}\Big).
\end{aligned}
$$

■

The authors wish to thank the editors for organizing this volume in memory of Cora Sadosky, and the referee for some helpful suggestions.

Acknowledgements Der-Chen Chang is supported by the NSF grant DMS-1408839 and a McDevitt Endowment Fund at Georgetown University.

References

1. P. Auscher, E. Russ, P. Tchamitchian, Hardy-Sobolev spaces on strongly Lipschitz domains of $\mathbb{R}^n$. J. Funct. Anal. **218**, 54–109 (2005)
2. H.-Q. Bui, Weighted Hardy spaces. Math. Nachr. **103**, 45–62 (1981)
3. D.-C. Chang, C. Sadosky, Functions of bounded mean oscillation. Taiwan. J. Math. **10**, 573–601 (2006)
4. D.-C. Chang, G. Dafni, C. Sadosky, A Div-Curl lemma in BMO on a Domain, in *Harmonic Analysis, Signal Processing, and Complexity*. Progress in Mathematics, vol. 238 (Birkhauser Boston, Boston, MA, 2005), pp. 55–65
5. R. Coifman, P.-L. Lions, Y. Meyer, S. Semmes, Compensated compactness and Hardy spaces. J. Math. Pures Appl. **72**, 247–286 (1993)
6. G. Dafni, Nonhomogeneous div-curl lemmas and local Hardy spaces. Adv. Differ. Equ. **10**, 505–526 (2005)
7. J. García-Cuerva, Weighted H^p spaces, Dissertationes Mathematicae, Dissertationes Mathematicae (Rozprawy Matematyczne), vol. 162 (1979)
8. D. Goldberg, A local version of real Hardy spaces. Duke Math. J. **46**, 27–42 (1979)
9. P. Hajłasz, Sobolev spaces on metric-measure spaces, in *Heat Kernels and Analysis on Manifolds, Graphs, and Metric Spaces (Paris, 2002)*. Contemporary Mathematics, vol. 338 (American Mathematical Society, Providence, RI, 2003), pp. 173–218
10. S. Indratno, D. Maldonado, S. Silwal, A visual formalism for weights satisfying reverse inequalities. Expo. Math. **33**, 1–29 (2015)
11. A. Miyachi, Hardy space estimate for the product of singular integrals. Canad. J. Math. **52**, 381–411 (2000)
12. F. Murat, Compacité par compensation. Ann. Scuola Norm. Sup. Pisa Cl. Sci. **5**, 489–507 (1978)

13. C. Pérez, Calderón-Zygmund theory related to Poincaré-Sobolev inequalities, fractional integrals and singular integral operators, in *Function spaces, Nonlinear Analysis and Applications*. ed. by J. Lukes, P. Lubos. Lectures Notes of the Spring Lectures in Analysis (Charles University and Academy of Sciences, 1999). The publishing house Matfyzpress of the Faculty of Mathematics and Physics, Charles University, Prague, pp. 31–94
14. S. Semmes, A primer on Hardy spaces and some remarks on a theorem of Evans and Müller. Commun. Partial Differ. Equ. **19**, 277–319 (1994)
15. E.M. Stein, *Harmonic Analysis: Real-Variable Methods, Orthogonality, and Oscillatory Integrals* (Princeton University Press, Princeton, 1993)
16. J.-O. Strömberg, A. Torchinsky, *Weighted Hardy Spaces*. Lecture Notes in Mathematics, vol. 1381 (Springer, Berlin, 1989)
17. Y. Tsutsui, An application of weighted Hardy spaces to the Navier-Stokes equations. J. Funct. Anal. **266**, 1395–1420 (2014)
18. R.L. Wheeden, A boundary value characterization of weighted H^1. Enseignement Math. **22**, 121–134 (1976)

A Remark on Bilinear Square Functions

Loukas Grafakos

Abstract We provide some remarks concerning a bilinear square function formed by products of Littlewood–Paley operators over arbitrary intervals. For $1 < p_1, p_2 < \infty$ with $1/p = 1/p_1 + 1/p_2$, we show that this square function is bounded from $L^{p_1}(\mathbf{R}) \times L^{p_2}(\mathbf{R})$ to $L^p(\mathbf{R})$ when $p > 2/3$ and unbounded when $p < 2/3$.

2000*Mathematics Subject Classification*: 42Bxx

Little work is known in the area of bilinear Littlewood–Paley square functions besides the articles of Lacey [6], Diestel [3], and Bernicot [1]. In this note, we study a bilinear square function formed by products of Littlewood–Paley operators over arbitrary intervals.

Given an interval $I = [a, b)$ on $\mathbf{R}$, let Δ_I be the Littlewood–Paley operator defined by multiplication by the characteristic function of I on the Fourier transform side. The Fourier transform of an integrable function g on $\mathbf{R}$ is defined by

$$\hat{g}(\xi) = \int_{\mathbf{R}} g(x) e^{-2\pi i x \xi} \, dx$$

and its inverse Fourier transform is defined by $g^\vee(\xi) = \hat{g}(-\xi)$. In terms of these operators we have $\Delta_I(g) = \left(\hat{g}\chi_I\right)^\vee$.

The Littlewood–Paley square function associated with the function f on $\mathbf{R}$ is given by

$$S(f) = \Big(\sum_{j \in \mathbf{Z}} |\Delta_{I_j}(f)|^2 \Big)^{\frac{1}{2}}, \tag{1}$$

L. Grafakos (✉)
University of Missouri, Columbia, MO, USA
e-mail: grafakosl@missouri.edu

M.C. Pereyra et al. (eds.), *Harmonic Analysis, Partial Differential Equations, Complex Analysis, Banach Spaces, and Operator Theory (Volume 1)*, Association for Women in Mathematics Series 4, DOI 10.1007/978-3-319-30961-3_9

where $I_j = [-2^{j+1}, -2^j) \cup [2^j, 2^{j+1})$ and the classical Littlewood–Paley theorem says that

$$\big\|S(f)\big\|_{L^p(\mathbf{R})} \le C_p \|f\|_{L^p(\mathbf{R})}$$

where $1 < p < \infty$ and C_p is a constant independent of the function f in $L^p(\mathbf{R})$ (but depends on p).

In this note, we are interested in estimates for Littlewood–Paley square functions formed by products of Littlewood–Paley operators acting on two functions. To be precise, let a_j and b_j be strictly increasing sequences on the real line with the properties $\lim_{j\to\infty} a_j = \lim_{j\to\infty} b_j = \infty$ and $\lim_{j\to-\infty} a_j = \lim_{j\to-\infty} b_j = -\infty$ and consider the bilinear square function

$$S_2(f,g) = \Big(\sum_{j\in\mathbf{Z}} |\Delta_{[a_j,a_{j+1})}(f)\Delta_{[b_j,b_{j+1})}(g)|^2\Big)^{\frac{1}{2}}$$

defined for suitable functions f, g on the line. We consider the question whether S_2 satisfies the inequality

$$\big\|S_2(f,g)\big\|_{L^p(\mathbf{R})} \le C_{p_1,p_2} \|f\|_{L^{p_1}(\mathbf{R})} \|g\|_{L^{p_2}(\mathbf{R})} \tag{2}$$

for some constant C_{p_1,p_2} independent of f, g where $1 < p_1, p_2 < \infty$ and $1/p = 1/p_1 + 1/p_2$. We have the following result concerning this operator:

Theorem 1. *Let $1 < p_1, p_2 < \infty$ be given and define p by setting $1/p = 1/p_1 + 1/p_2$. Then if $p > 2/3$, there is a constant C_{p_1,p_2} such that* (2) *holds for all functions f, g on the line. Conversely, if* (2) *holds, then we must have $p \ge 2/3$.*

Proof. Introduce the maximal function

$$\mathcal{M}(f) = \sup_{-\infty<a<b<\infty} |\Delta_{[a,b)}(f)|$$

and notice that is pointwise controlled by

$$2 \sup_{a\in\mathbf{R}} |\Delta_{(-\infty,a)}(f)|$$

and thus is controlled by the following version of the Carleson operator

$$\mathcal{C}(f)(x) = \sup_{N>0} \left| \int_{-\infty}^{N} \hat{f}(\xi) e^{2\pi i x\xi}\, d\xi \right|.$$

In view of the Carleson–Hunt theorem [2, 5] we have that $\mathcal{C}$ is bounded on $L^r(\mathbf{R})$ for $1 < r < \infty$.

Consider the case where $2 \le p_1 < \infty$ and $1 < p_2 < \infty$. Then we have that

$$S_2(f,g) \le \Big(\sum_{j\in\mathbf{Z}} |\Delta_{[a_j,a_{j+1})}(f)|^2\Big)^{\frac{1}{2}} \sup_{j\in\mathbf{Z}} |\Delta_{[b_j,b_{j+1})}(g)| = S(f)\mathcal{M}(g)$$

where S is defined as in (1) with $[a_j, a_{j+1})$ in place of I_j. In view of the Rubio de Francia inequality [7] we have that S is bounded on $L^r(\mathbf{R})$ for $2 \le r < \infty$. An application of Hölder's inequality yields the inequality

$$\big\|S_2(f,g)\big\|_{L^p(\mathbf{R})} \le \|S(f)\|_{L^{p_1}(\mathbf{R})}\|\mathcal{M}(g)\|_{L^{p_2}(\mathbf{R})} \tag{3}$$

and this (2) follows from the preceding inequality combined with the boundedness of S on $L^{p_1}(\mathbf{R})$ and $\mathcal{M}$ on $L^{p_2}(\mathbf{R})$.

An analogous argument holds with the roles of p_1 and p_2 are reversed, i.e., when we have $1 < p_1 < \infty$ and $2 \le p_2 < \infty$. Thus boundedness holds for all pairs (p_1, p_2) for which either $p_1 \ge 2$ or $p_2 \ge 2$. But there exist points (p_1, p_2) with $p = (1/p_1 + 1/p_2)^{-1} > 2/3$ for which neither p_1 nor p_2 is at least 2. (For instance, $p_1 = p_2 = 7/5$). To deal with these intermediate points we use interpolation.

Given a pair of points (p_1, p_2) with $p = (1/p_1 + 1/p_2)^{-1} > 2/3$ and $1 < p_1, p_2 < 2$, we pick two pairs of points (p_1^1, p_2^1) and (p_1^2, p_2^2) with

$$p > p^1 = (1/p_1^1 + 1/p_2^1)^{-1} = p^2 = (1/p_1^2 + 1/p_2^2)^{-1} > 2/3$$

and $1 < p_2^1 < 2 < p_1^1 < \infty, < 2$ and $1 < p_2^2 < 2 < p_1^2 < \infty$. For instance, we take (p_1^1, p_2^1, p^1) near $(1, 2, 2/3)$ and (p_1^2, p_2^2, p^2) near $(2, 1, 2/3)$. Then consider the three points $W_1 = (1/p_1^1, 1/p_2^1, 1/p^1)$, $W_2 = (1/p_1^2, 1/p_2^2, 1/p^2)$, and $W_3 = (1/2, 1/2, 1)$ and notice that the point $(1/p_1, 1/p_2, 1/p)$ lies in the interior of the convex hull of W_1, W_2, and W_3. We consider the bi-sublinear operator

$$(f,g) \mapsto S_2(f,g)$$

which is bounded at the points W_1, W_2, and W_3. Using Corollary 7.2.4 in [4] we obtain that S_2 is bounded from $L^{p_1}(\mathbf{R}) \times L^{p_2}(\mathbf{R})$ to $L^p(\mathbf{R})$. This completes the proof in the remaining case.

Next, we turn to the converse assertion of the theorem. Suppose that for some $1 < p_1, p_2 < \infty$ with $1/p = 1/p_1 + 1/p_2$ estimate (2) holds for some constant C_{p_1,p_2} and all suitable functions f, g on the line. Now consider the sequences $a_j = b_j = j$ and the functions

$$f_N = g_N = \chi_{[0,N)}^{\vee}.$$

Then we have

$$f_N(x) = \chi_{[0,N]}^{\vee}(x) = \int_0^N e^{2\pi i x\xi}\, d\xi = \frac{e^{2\pi i N x} - 1}{2\pi i x}$$

and for $j = 0, 1, \dots, N-1$ we have

$$\Delta_{[j,j+1)}(f_N)(x) = \int_j^{j+1} e^{2\pi i x\xi} d\xi = e^{2\pi i xj} \int_0^1 e^{2\pi i x\xi} d\xi = \frac{e^{2\pi i xj}(e^{2\pi i x}-1)}{2\pi i x}.$$

Consequently,

$$\Big(\sum_{j=0}^{N-1} \big|\Delta_{[j,j+1)}(f_N)(x)\Delta_{[j,j+1)}(g_N)(x)\big|^2\Big)^{\frac{1}{2}} = \sqrt{N}\,\left|\frac{e^{2\pi i x}-1}{2\pi i x}\right|^2$$

and thus

$$\big\|S_2(f_N,g_N)\big\|_{L^p} \geq \sqrt{N}\left\|\frac{(e^{2\pi i x}-1)^2}{4\pi^2 x^2}\right\|_{L^p} = c\sqrt{N}$$

as long as $p > 1/2$. On the other hand we have

$$\|f_N\|_{L^{p_1}} = N^{1-\frac{1}{p_1}}\left\|\frac{e^{2\pi i x}-1}{2\pi i x}\right\|_{L^{p_1}} = c_{p_1} N^{1-\frac{1}{p_1}}$$

whenever $1 < p_1 < \infty$.

Now suppose that (2) holds. Then we must have

$$\big\|S_2(f_N,g_N)\big\|_{L^p(\mathbf{R})} \leq C_{p_1,p_2}\|f_N\|_{L^{p_1}(\mathbf{R})}\|g_N\|_{L^{p_2}(\mathbf{R})} \tag{4}$$

and this implies that

$$c\sqrt{N} \leq C_{p_1,p_2} c_{p_1} N^{1-\frac{1}{p_1}} c_{p_2} N^{1-\frac{1}{p_2}} = C_{p_1,p_2} c_{p_1} c_{p_2} N^{2-\frac{1}{p}}$$

which forces $p \geq 2/3$ by letting $N \to \infty$. □

It is unclear to us at the moment as to what happens when $p = 2/3$.

We now discuss a related larger square function. Let $1 < p_1, p_2 < \infty$ with $1/p_1 + 1/p_2 = 1/p$. It is not hard to see that the square function

$$S_{22}(f,g) = \Big(\sum_{j\in\mathbf{Z}}\sum_{k\in\mathbf{Z}} |\Delta_{[a_j,a_{j+1})}(f)\Delta_{[b_k,b_{k+1})}(g)|^2\Big)^{\frac{1}{2}}$$

is bounded from $L^{p_1}(\mathbf{R}) \times L^{p_2}(\mathbf{R})$ to $L^p(\mathbf{R})$ if and only if $p_1, p_2 \geq 2$. Indeed, one direction is a trivial consequence of Hölder's inequality; for the other direction, let

$$f_N(x) = g_N(x) = \chi_{[0,N]}^{\vee}(x) = \int_0^N e^{2\pi i x\xi}\, d\xi = \frac{e^{2\pi i N x}-1}{2\pi i x}.$$

The preceding argument shows that

$$\left\| S_{22}(f_M, g_N) \right\|_{L^p} \geq c^2 \sqrt{M} \sqrt{N}$$

and we also have

$$\|f_M\|_{L^{p_1}(\mathbf{R})} \|g_N\|_{L^{p_2}(\mathbf{R})} = c_{p_1} c_{p_2} M^{1-\frac{1}{p_1}} N^{1-\frac{1}{p_2}} .$$

Hence, letting $M \to \infty$ with N fixed or $N \to \infty$ with M fixed, we obtain that both p_1 and p_2 satisfy $p_1, p_2 \geq 2$.

I would like to end this note by expressing a few feelings about Cora Sadosky. Although, I have not had a very close personal relationship with her, I have always admired the great dedication and enthusiasm Cora has displayed in mathematics and the sincere love and support she has provided to young people who wished to pursue a research career in harmonic analysis. I warmly recall the personal interest she showed in my search for a permanent position in the USA. Cora's untimely passing away was a big loss for our harmonic analysis community and we are all proud of the strong legacy she has left behind.

Acknowledgements The author would like to acknowledge the Simons Foundation.

References

1. F. Bernicot, L^p estimates for non smooth bilinear Littlewood–Paley square functions. Math. Ann. **351**, 1–49 (2011)
2. L. Carleson, On convergence and growth of partial sums of Fourier series. Acta Math. **116**(1), 135–157 (1966)
3. G. Diestel, Some remarks on bilinear Littlewood–Paley theory. J. Math. Anal. Appl. **307**, 102–119 (2005)
4. L. Grafakos, *Modern Fourier Analysis*, 3rd edn. Graduate Text in Mathematics, vol. 250 (Springer, New York, 2015)
5. R. Hunt, *On the Convergence of Fourier Series*. 1968 Orthogonal Expansions and Their Continuous Analogues (Proceedings of the Conference, Edwardsville, Ill, 1967), pp. 235–255. Southern Illinois University Press, Carbondale Ill
6. M. Lacey, On bilinear Littlewood–Paley square functions. Publ. Mat. **40**, 387–396 (1996)
7. J.-L. Rubio de Francia, A Littlewood–Paley inequality for arbitrary intervals. Rev. Mat. Iberoam. **1**(2), 1–14 (1985)

Unique Continuation for the Elasticity System and a Counterexample for Second-Order Elliptic Systems

Carlos Kenig and Jenn-Nan Wang

Dedicated to the memory of Cora Sadosky.

Abstract In this paper we study the global unique continuation property for the elasticity system and the general second-order elliptic system in two dimensions. For the isotropic and the anisotropic systems with measurable coefficients, under certain conditions on coefficients, we show that the global unique continuation property holds. On the other hand, for the anisotropic system, if the coefficients are Lipschitz, we can prove that the global unique continuation is satisfied for a more general class of media. In addition to the positive results, we also present counterexamples to unique continuation and strong unique continuation for general second elliptic systems.

Introduction

In this work, we study the unique continuation property for the elasticity system and the general second-order elliptic system in two dimensions. We begin with the elasticity system. Let $u = (u_1, u_2)^T$ be a vector-valued function satisfying

$$\partial_j(a_{ijkl}(x)\partial_k u_l) = 0 \quad \text{in} \quad \mathbb{R}^2, \tag{1}$$

where $a_{ijkl}(x)$ is a rank four tensor satisfying the symmetry properties:

$$a_{ijkl} = a_{klij} = a_{jikl}. \tag{2}$$

C. Kenig
Department of Mathematics, University of Chicago, Chicago, IL 60637, USA
e-mail: cek@math.uchicago.edu.

J.-N. Wang (✉)
Institute of Applied Mathematical Sciences, NCTS, National Taiwan University, Taipei 106, Taiwan
e-mail: jnwang@math.ntu.edu.tw

M.C. Pereyra et al. (eds.), *Harmonic Analysis, Partial Differential Equations, Complex Analysis, Banach Spaces, and Operator Theory (Volume 1)*, Association for Women in Mathematics Series 4, DOI 10.1007/978-3-319-30961-3_10

Throughout, the Latin indices range from 1 to 2. Also, the summation convention is imposed. For isotropic media, we have that $a_{ijkl} = \lambda\delta_{ij}\delta_{kl} + \mu(\delta_{ik}\delta_{jl} + \delta_{il}\delta_{jk})$, where λ and μ are called Lamé coefficients. In this case, (1) is written as

$$\nabla \cdot (\mu(\nabla u + (\nabla u)^T)) + \nabla(\lambda \nabla \cdot u) = 0 \quad \text{in} \quad \mathbb{R}^2. \tag{3}$$

We say that u, a solution of (3), satisfies the *global unique continuation property* if whenever u vanishes in the lower half plane, it vanishes identically in $\mathbb{R}^2$. Recall that any solution u of the partial differential equation defined in an open connected set Ω is said to satisfy the unique continuation property if whenever u vanishes in a non-empty open subset of Ω, it is zero in Ω. On the other hand, u satisfies the strong unique continuation property if whenever u vanishes of infinite order at any point of Ω, it vanishes in Ω.

For the isotropic system (3) with nice Lamé coefficients, there are a lot of results on the unique continuation property and the strong unique continuation property (for dimension $n \geq 2$). We will not review the detailed development here. To motivate our study, we only mention the recent result in [10], where the strong unique continuation property was proved for $\mu \in W^{1,\infty}$ and $\lambda \in L^\infty$, which is the best known assumption on the coefficients by far. For the scalar second-order elliptic equation in nondivergence or divergence form

$$A\nabla^2 u = 0 \quad \text{in} \quad \mathbb{R}^2 \tag{4}$$

or

$$\nabla \cdot (A(x)\nabla u) = 0 \quad \text{in} \quad \mathbb{R}^2, \tag{5}$$

the strong unique continuation property is satisfied for $A \in L^\infty$ (see, for example, [1–3, 6, 13]). The proof is based on the intimate connection between (4) or (5) and quasiregular mappings. Therefore, it is a natural question to ask whether the unique continuation or the strong unique continuation holds for (3) or even for (1) when all coefficients are only measurable.

When μ of (3) is Lipschitz, it is known that (3) is weakly coupled. Hence, the usual Carleman method will lead us to the unique continuation properties. However, if μ is only measurable, (3) is strongly coupled. To the best of our knowledge, the Carleman method has never been successfully applied to strongly coupled systems. The general elasticity system (1) is always strongly coupled, regardless of the regularity of coefficients.

In this work we would like to show that solutions u of (1) satisfy the global unique continuation property under some restrictions on the measurable coefficients a_{ijkl}. Our approach to prove this result relies on the connection between (1) and the Beltrami system with matrix-valued coefficients (see (22)). When this matrix-valued coefficient is sufficiently small (which is satisfied when the coefficients do not deviate too much from a set of constant coefficients), we can follow the arguments in [9] and use the L^p-norm of the Beurling–Ahlfors transform to conclude the result. When the coefficients are only measurable, the set of constant coefficients is rather restricted, see the conditions in Theorems 0.1 and 0.3. If some coefficients of the general system (1) are Lipschitz, the global unique continuation is true for coefficients near a larger set of constant coefficients, see Theorem 0.1.

In addition to the positive results mentioned above, we also present a counterexample to unique continuation for a second-order elliptic system (in the sense of (62)) with measurable coefficients based on the example derived in [9]. The main idea in the construction of the counterexample is to convert the second-order elliptic system to a first-order elliptic system and match coefficients of the first-order system obtained from the example in [9]. Based on the example given in [8] (also see related article [12]), using the same argument, we can also construct a counterexample to strong unique continuation for second-order elliptic systems with continuous coefficients satisfying the Legendre–Hadamard condition (27) or even the strong convexity condition:

$$a_{ijkl}(x)\xi_k^l\xi_j^i \geq c|\xi|^2 \tag{6}$$

for any 2×2 matrix $\xi = (\xi_k^l)$, where c is a positive constant. Note that (6) implies the Legendre–Hadamard condition.

We would like to remark that the nontrivial solution u of the counterexample to unique continuation described above vanishes in the lower half plane. It was shown in [11] that there exists nontrivial $W^{1,2}(\mathbb{R}^2)$ solution u or Lipschitz solution u whose supports are compact solving second-order elliptic system with measurable coefficients satisfying the Legendre–Hadamard condition. This is another counterexample to unique continuation for second-order elliptic systems with measurable coefficients. We want to point out that the examples in [11] do not exist in second-order elliptic systems satisfying the strong convexity condition (6). This can be easily seen by the integration by parts. Nonetheless, the strong convexity condition (6) does not rule out the existence of the counterexample to unique continuation we constructed in this paper since this nontrivial solution does not necessarily have compact support.

The paper is organized as follows. In section "Elasticity System with Measurable Coefficients," we prove the global unique continuation property for the Lamé and general anisotropic systems when the measurable elastic coefficients are close to some constant values. In section "Anisotropic System with Regular Coefficients," we expand the set of constant values when the elastic coefficients are Lipschitz. Finally, in section "Counterexample to Unique Continuation," we construct counterexamples to unique continuation and strong unique continuation for general second-order elliptic systems with measurable coefficients and continuous coefficients, respectively.

Elasticity System with Measurable Coefficients

It is instructive to begin with the isotropic system, i.e., Lamé system (3). Assume that $\lambda, \mu \in L^\infty$ satisfy the ellipticity condition

$$\mu \geq \delta > 0, \quad \lambda + 2\mu \geq \delta, \quad \forall\, x \in \mathbb{R}^2. \tag{7}$$

The key to proving the global unique continuation property lies in arranging the coefficients nicely. We write the system (3) componentwise

$$\partial_1(2\mu\partial_1 u_1) + \partial_2(\mu(\partial_1 u_2 + \partial_2 u_1)) + \partial_1(\lambda\nabla\cdot u) = 0 \tag{8}$$

and

$$\partial_1(\mu(\partial_1 u_2 + \partial_2 u_1)) + \partial_2(2\mu\partial_2 u_2) + \partial_2(\lambda\nabla\cdot u) = 0. \tag{9}$$

Let $v = \nabla\cdot u = \partial_1 u_1 + \partial_2 u_2$ and $w = \nabla\times u = \partial_1 u_2 - \partial_2 u_1$, then (8) is written as

$$\partial_1(2\mu\partial_1 u_1 + \lambda v) + \partial_2(2\mu\partial_2 u_1 + \mu w) = 0. \tag{10}$$

Similarly, (9) is equivalent to

$$\partial_1(2\mu\partial_1 u_2 - \mu w) + \partial_2(2\mu\partial_2 u_2 + \lambda v) = 0. \tag{11}$$

By taking advantage of the relation

$$\Delta u_1 = \partial_1 v - \partial_2 w \quad \text{and} \quad \Delta u_2 = \partial_2 v + \partial_1 w,$$

we obtain from (10) that

$$\partial_1((2\mu-1)\partial_1 u_1 + (\lambda+1)v) + \partial_2((2\mu-1)\partial_2 u_1 + (\mu-1)w) = 0 \tag{12}$$

and from (11) that

$$\partial_1((2\mu-1)\partial_1 u_2 - (\mu-1)w) + \partial_2((2\mu-1)\partial_2 u_2 + (\lambda+1)v) = 0. \tag{13}$$

Therefore, there exist u_1' and u_2' such that

$$\begin{cases} \partial_2 u_1' = (2\mu-1)\partial_1 u_1 + (\lambda+1)v, \\ -\partial_1 u_1' = (2\mu-1)\partial_2 u_1 + (\mu-1)w, \end{cases} \tag{14}$$

and

$$\begin{cases} \partial_2 u_2' = (2\mu-1)\partial_1 u_2 - (\mu-1)w, \\ -\partial_1 u_2' = (2\mu-1)\partial_2 u_2 + (\lambda+1)v. \end{cases} \tag{15}$$

Setting $f_1 = u_1 + iu_1'$ and $f_2 = u_2 + iu_2'$, (14) and (15) become

$$\begin{cases} \bar{\partial} f_1 = \sigma \overline{\partial f_1} + h, \\ \bar{\partial} f_2 = \sigma \overline{\partial f_2} + ih, \end{cases} \tag{16}$$

where

$$\sigma = \frac{1-\mu}{\mu} \quad \text{and} \quad h = -\frac{\lambda+1}{2\mu} v - i\frac{\mu-1}{2\mu} w.$$

As usual, we define

$$\bar{\partial} = \frac{1}{2}(\partial_1 + i\partial_2), \quad \partial = \frac{1}{2}(\partial_1 - i\partial_2).$$

Using the obvious relations

$$\partial_1 u_1 = \frac{1}{2}(\bar{\partial} f_1 + \partial f_1 + \partial \bar{f}_1 + \overline{\partial f_1}), \quad \partial_1 u_2 = \frac{1}{2}(\bar{\partial} f_2 + \partial f_2 + \partial \bar{f}_2 + \overline{\partial f_2}),$$
$$\partial_2 u_1 = \frac{1}{2i}(\bar{\partial} f_1 - \partial f_1 - \partial \bar{f}_1 + \overline{\partial f_1}), \quad \partial_2 u_2 = \frac{1}{2i}(\bar{\partial} f_2 - \partial f_2 - \partial \bar{f}_2 + \overline{\partial f_2}),$$

we can compute

$$\begin{aligned} h &= -\frac{\lambda+1}{2\mu} v - i\frac{\mu-1}{2\mu} w \\ &= -\frac{\lambda+1}{4\mu}\Big((\bar{\partial} f_1 + \partial f_1 + \partial \bar{f}_1 + \overline{\partial f_1}) - i(\bar{\partial} f_2 - \partial f_2 - \partial \bar{f}_2 + \overline{\partial f_2})\Big) \\ &\quad - i\frac{\mu-1}{4\mu}\Big((\bar{\partial} f_2 + \partial f_2 + \partial \bar{f}_2 + \overline{\partial f_2}) + i(\bar{\partial} f_1 - \partial f_1 - \partial \bar{f}_1 + \overline{\partial f_1})\Big) \\ &= \left(\frac{\mu-\lambda-2}{4\mu}\right)\bar{\partial} f_1 + \left(\frac{-\lambda-\mu}{4\mu}\right)\partial f_1 + \left(\frac{-\lambda-\mu}{4\mu}\right)\partial \bar{f}_1 + \left(\frac{\mu-\lambda-2}{4\mu}\right)\overline{\partial f_1} \\ &\quad + i\left(\frac{\lambda-\mu+2}{4\mu}\right)\bar{\partial} f_2 + i\left(\frac{-\lambda-\mu}{4\mu}\right)\partial f_2 + i\left(\frac{-\lambda-\mu}{4\mu}\right)\partial \bar{f}_2 + i\left(\frac{\lambda-\mu+2}{4\mu}\right)\overline{\partial f_2}. \end{aligned}$$

For simplicity, let us denote

$$\alpha = \frac{\mu-\lambda-2}{4\mu}, \quad \beta = \frac{-\lambda-\mu}{4\mu},$$

then

$$h = \alpha \bar{\partial} f_1 + \beta \partial f_1 + \beta \partial \bar{f}_1 + \alpha \overline{\partial f_1} - i\alpha \bar{\partial} f_2 + i\beta \partial f_2 + i\beta \partial \bar{f}_2 - i\alpha \overline{\partial f_2}$$

and

$$ih = i\alpha\bar\partial f_1 + i\beta\partial f_1 + i\beta\partial\bar f_1 + i\alpha\overline{\partial f_1} + \alpha\bar\partial f_2 - \beta\partial f_2 - \beta\partial\bar f_2 + \alpha\overline{\partial f_2}.$$

Therefore, (16) is equivalent to

$$\begin{aligned} &\left(I_2 + \begin{pmatrix} -\alpha & i\alpha \\ -i\alpha & -\alpha \end{pmatrix}\right)\bar\partial\begin{pmatrix} f_1 \\ f_2 \end{pmatrix} + \begin{pmatrix} -\beta & -i\beta \\ -i\beta & \beta \end{pmatrix}\partial\begin{pmatrix} \bar f_1 \\ \bar f_2 \end{pmatrix} \\ &= \begin{pmatrix} \beta & i\beta \\ i\beta & -\beta \end{pmatrix}\partial\begin{pmatrix} f_1 \\ f_2 \end{pmatrix} + \begin{pmatrix} \sigma+\alpha & -i\alpha \\ i\alpha & \sigma+\alpha \end{pmatrix}\overline{\partial\begin{pmatrix} f_1 \\ f_2 \end{pmatrix}}, \end{aligned} \tag{17}$$

where I_n denotes the $n \times n$ unit matrix. Setting $\frac{\partial\bar f}{\bar\partial f} = 0$, if $\bar\partial f = 0$ and $\frac{\overline{\partial f}}{\partial f} = 0$, and if $\partial f = 0$, (17) can be written as

$$\begin{aligned} &\left(I_2 + \begin{pmatrix} -\alpha & i\alpha \\ -i\alpha & -\alpha \end{pmatrix}\right)\bar\partial\begin{pmatrix} f_1 \\ f_2 \end{pmatrix} + \begin{pmatrix} -\beta & -i\beta \\ -i\beta & \beta \end{pmatrix}\begin{pmatrix} \frac{\partial\bar f_1}{\bar\partial f_1} & 0 \\ 0 & \frac{\partial\bar f_2}{\bar\partial f_2} \end{pmatrix}\bar\partial\begin{pmatrix} f_1 \\ f_2 \end{pmatrix} \\ &= \begin{pmatrix} \beta & i\beta \\ i\beta & -\beta \end{pmatrix}\partial\begin{pmatrix} f_1 \\ f_2 \end{pmatrix} + \begin{pmatrix} \sigma+\alpha & -i\alpha \\ i\alpha & \sigma+\alpha \end{pmatrix}\begin{pmatrix} \frac{\overline{\partial f_1}}{\partial f_1} & 0 \\ 0 & \frac{\overline{\partial f_2}}{\partial f_2} \end{pmatrix}\partial\begin{pmatrix} f_1 \\ f_2 \end{pmatrix}. \end{aligned} \tag{18}$$

Finally, let U and V be two 2×2 matrices

$$U = I_2 + \begin{pmatrix} -\alpha & i\alpha \\ -i\alpha & -\alpha \end{pmatrix} + \begin{pmatrix} -\beta & -i\beta \\ -i\beta & \beta \end{pmatrix}\begin{pmatrix} \frac{\partial\bar f_1}{\bar\partial f_1} & 0 \\ 0 & \frac{\partial\bar f_2}{\bar\partial f_2} \end{pmatrix},$$

$$V = \begin{pmatrix} \beta & i\beta \\ i\beta & -\beta \end{pmatrix} + \begin{pmatrix} \sigma+\alpha & -i\alpha \\ i\alpha & \sigma+\alpha \end{pmatrix}\begin{pmatrix} \frac{\overline{\partial f_1}}{\partial f_1} & 0 \\ 0 & \frac{\overline{\partial f_2}}{\partial f_2} \end{pmatrix},$$

then (18) can be written as

$$U\bar\partial\begin{pmatrix} f_1 \\ f_2 \end{pmatrix} = V\,\partial\begin{pmatrix} f_1 \\ f_2 \end{pmatrix}. \tag{19}$$

It is clear that

$$\|U - I_2\|_{L^\infty} \le C(\|\alpha\|_{L^\infty} + \|\beta\|_{L^\infty}) \quad \text{and} \quad \|V\|_{L^\infty} \le C(\|\beta\|_{L^\infty} + \|\sigma\|_{L^\infty} + \|\alpha\|_{L^\infty}), \tag{20}$$

where C is an absolute constant. Hereafter, for any matrix-valued function $A(x)$: $\mathbb{C}^n \to \mathbb{C}^n$ with $x \in \mathbb{R}^2$, the norm $\|A\|_{L^\infty}$ is defined by

$$\|A\|_{L^\infty} = \sup_{x\in\mathbb{R}^2} \|A(x)\|,$$

where $\|\cdot\|$ is the usual matrix norm derived by treating $\mathbb{C}^n$ as an inner-product space.

Theorem 0.1. *There exists an* $\varepsilon > 0$ *such that if*

$$\|\mu - 1\|_{L^\infty} \le \varepsilon \quad \textit{and} \quad \|\lambda + 1\|_{L^\infty} \le \varepsilon, \tag{21}$$

then for any Lipschitz solution u *of* (3) *vanishing in the lower half plane, we must have* $u \equiv 0$.

Remark 0.2. It is clear that if the Lamé coefficients λ and μ satisfy (21), then the ellipticity condition (7) holds.

Proof. Since u is Lipschitz and vanishes in the lower half plane, so does the vector-valued function

$$F = \begin{pmatrix} f_1 \\ f_2 \end{pmatrix}.$$

In view of the definitions of σ, α, β, if ε of (21) is sufficiently small, then all of them are sufficiently small as well. By (20), U is invertible and V is small, consequently, $\|U^{-1}V\|_{L^\infty} \le \varepsilon' \ll 1$. Therefore, from (19) we have

$$\bar{\partial} F = \Psi \partial F \quad \text{in} \quad \mathbb{C}, \tag{22}$$

where $\Psi = U^{-1}V$ and we have identified $\mathbb{R}^2$ as the complex plane $\mathbb{C}$. Note that here $F : \mathbb{C} \to \mathbb{C}^2$. Equation (22) is a Beltrami system studied in [9]. It was proved in [9] that if $\|\Psi\|_{L^\infty}$ is sufficiently small and F vanishes in the lower half plane, then F is trivial, i.e., u is trivial. For the sake of completeness, we sketch the proof here. We refer to [9, Sect. 7] for more details. Without loss of generality, we assume that F vanishes for $\Im z \le 1$. Define $G(z) = F(\sqrt{z})$. Then G satisfies

$$\bar{\partial} G = \frac{z}{|z|} \Psi(\sqrt{z}) \partial G. \tag{23}$$

We can see that the differential of G, DG, lies in $L^p(\mathbb{C})$ for some $p > 4$ (see (7.13) in [9]). In other words, we have

$$\|\bar{\partial} G\|_{L^p} \le \varepsilon' \|\partial G\|_{L^p}. \tag{24}$$

Let S be the Beurling–Ahlfors transform, i.e., $S\bar{\partial} = \partial$. It is known from the Calderón–Zygmund theory that

$$\|S\|_{L^p \to L^p} \le a_p \tag{25}$$

for some constant a_p, depending only on p (see [5] for a more precise bound on a_p). Hence, (25) implies

$$\|\partial G\|_{L^p} \le a_p \|\bar{\partial} G\|_{L^p}. \tag{26}$$

Combining (24) and (26), we conclude that $\|\bar{\partial} G\|_{L^p} = \|\partial G\|_{L^p} = 0$ provided $a_p \varepsilon' < 1$ for some $p > 4$. The proof of theorem then follows. □

Now we turn to the general elasticity system (1). Assume that $a_{ijkl} \in L^\infty$ satisfies the ellipticity condition

$$a_{ijkl}(x)\xi_i\xi_l\rho_j\rho_k \geq \delta|\xi|^2|\rho|^2 \quad \forall\, \xi,\ \rho \in \mathbb{R}^2, \tag{27}$$

i.e., the Legendre–Hadamard condition. Due to the symmetry properties (2), for simplicity, we denote

$$\begin{cases} a_{1111} = a, a_{2222} = b, a_{1112} = a_{1211} = a_{2111} = a_{1121} = c, \\ a_{1122} = a_{2211} = d, a_{1212} = a_{2112} = a_{1221} = a_{2121} = e, \\ a_{1222} = a_{2122} = a_{2212} = a_{2221} = g. \end{cases}$$

Componentwise, (1) is written as

$$\begin{cases} \partial_1(a\partial_1u_1{+}c\partial_1u_2{+}c\partial_2u_1{+}d\partial_2u_2){+}\partial_2(c\partial_1u_1 + e\partial_1u_2 + e\partial_2u_1 + g\partial_2u_2) = 0, \\ \partial_1(c\partial_1u_1{+}e\partial_1u_2{+}e\partial_2u_1{+}g\partial_2u_2){+}\partial_2(d\partial_1u_1 + g\partial_1u_2 + g\partial_2u_1 + b\partial_2u_2) = 0. \end{cases} \tag{28}$$

For our purpose, we will express (28) as

$$\begin{cases} \partial_1((a-d-1)\partial_1u_1 + (d+1)v + c\partial_1u_2 + c\partial_2u_1) \\ \quad + \partial_2((2e-1)\partial_2u_1 + (e-1)w + c\partial_1u_1 + g\partial_2u_2) = 0, \\ \partial_1((2e-1)\partial_1u_2 - (e-1)w + c\partial_1u_1 + g\partial_2u_2) \\ \quad + \partial_2((b-d-1)\partial_2u_2 + (d+1)v + g\partial_1u_2 + g\partial_2u_1) = 0. \end{cases} \tag{29}$$

Comparing (29) with (12) and (13), it is not difficult to see that (29) can be transformed to (22) with similar smallness condition on Ψ if $|a-d-2| \ll 1, |b-d-2| \ll 1, |d+1| \ll 1, |e-1| \ll 1, |c| \ll 1, |g| \ll 1$. Therefore, we can prove that

Theorem 0.3. *There exists $\varepsilon > 0$ such that if*

$$\begin{cases} \|a-d-2\|_{L^\infty} \leq \varepsilon, \|b-d-2\|_{L^\infty} \leq \varepsilon, \|d+1\|_{L^\infty} \leq \varepsilon, \\ \|e-1\|_{L^\infty} \leq \varepsilon, \|c\|_{L^\infty} \leq \varepsilon, \|g\|_{L^\infty} \leq \varepsilon, \end{cases} \tag{30}$$

then if u is a Lipschitz function solving (1) *and vanishes in the lower half plane, then u vanishes identically.*

Remark 0.4. Under the assumptions (30), (1) is a slightly perturbed system of the Lamé system with λ, μ satisfying (21). Therefore, the ellipticity condition (27) holds.

Anisotropic System with Regular Coefficients

One may wonder if $|d+1| \ll 1$ and $|e-1| \ll 1$ in Theorem 0.3 can be replaced by $|d+k_0| \ll 1$ and $|e-k_0| \ll 1$ for $k_0 \neq 1$. For measurable coefficients, it is not possible since the requirement of $|e-k_0| \ll 1$ and $2e-k_0 \sim 1$ will force $k_0 = 1$. However, if a, b, d, e are Lipschitz, we can extend Theorem 0.3 to a larger class of system. Let k_0 be any fixed constant. Similarly to (29), we obtain that

$$\begin{cases} \partial_1((a-d-k_0)\partial_1 u_1 + (d+k_0)v + c\partial_1 u_2 + c\partial_2 u_1) \\ \quad + \partial_2((2e-k_0)\partial_2 u_1 + (e-k_0)w + c\partial_1 u_1 + g\partial_2 u_2) = 0, \\ \partial_1((2e-k_0)\partial_1 u_2 - (e-k_0)w + c\partial_1 u_1 + g\partial_2 u_2) \\ \quad + \partial_2((b-d-k_0)\partial_2 u_2 + (d+k_0)v + g\partial_1 u_2 + g\partial_2 u_1) = 0. \end{cases} \tag{31}$$

Denote $\tilde{a} = a-d-k_0, \tilde{e} = 2e-k_0, \tilde{b} = b-d-k_0$. Suppose that $\tilde{a} \neq 0$ and $\tilde{b} \neq 0$. Then (31) is equivalent to

$$\begin{cases} \partial_1(\partial_1(\tilde{a}u_1) - \partial_1\tilde{a}u_1 + (d+k_0)v + c\partial_1 u_2 + c\partial_2 u_1) \\ \quad + \partial_2(\tilde{e}\tilde{a}^{-1}(\partial_2(\tilde{a}u_1) - \partial_2\tilde{a}u_1) + (e-k_0)w + c\partial_1 u_1 + g\partial_2 u_2) = 0, \\ \partial_1(\tilde{e}\tilde{b}^{-1}(\partial_1(\tilde{b}u_2) - \partial_1\tilde{b}u_2) - (e-k_0)w + c\partial_1 u_1 + g\partial_2 u_2) \\ \quad + \partial_2(\partial_2(\tilde{b}u_2) - \partial_2\tilde{b}u_2 + (d+k_0)v + g\partial_1 u_2 + g\partial_2 u_1) = 0. \end{cases} \tag{32}$$

We set $\tilde{u}_1 = \tilde{a}u_1, \tilde{u}_2 = \tilde{b}u_2$, then (32) becomes

$$\begin{cases} \partial_1(\partial_1(\tilde{u}_1) - \partial_1\tilde{a}\tilde{a}^{-1}\tilde{u}_1 + (d+k_0)v + c\partial_1(\tilde{b}^{-1}\tilde{u}_2) + c\partial_2(\tilde{a}^{-1}\tilde{u}_1)) \\ \quad + \partial_2(\tilde{e}\tilde{a}^{-1}(\partial_2(\tilde{u}_1) - \partial_2\tilde{a}\tilde{a}^{-1}\tilde{u}_1) + (e-k_0)w + c\partial_1(\tilde{a}^{-1}\tilde{u}_1) + g\partial_2(\tilde{b}^{-1}\tilde{u}_2)) = 0, \\ \partial_1(\tilde{e}\tilde{b}^{-1}(\partial_1(\tilde{u}_2) - \partial_1\tilde{b}\tilde{b}^{-1}\tilde{u}_2) - (e-k_0)w + c\partial_1(\tilde{a}^{-1}\tilde{u}_1) + g\partial_2(\tilde{b}^{-1}\tilde{u}_2)) \\ \quad + \partial_2(\partial_2(\tilde{u}_2) - \partial_2\tilde{b}\tilde{b}^{-1}\tilde{u}_2 + (d+k_0)v + g\partial_1(\tilde{b}^{-1}\tilde{u}_2) + g\partial_2(\tilde{a}^{-1}\tilde{u}_1)) = 0. \end{cases} \tag{33}$$

We then express

$$\begin{cases} v = \partial_1 u_1 + \partial_2 u_2 = \partial_1\tilde{a}^{-1}\tilde{u}_1 + \tilde{a}^{-1}\partial_1\tilde{u}_1 + \partial_2\tilde{b}^{-1}\tilde{u}_2 + \tilde{b}^{-1}\partial_2\tilde{u}_2 \\ w = \partial_1 u_2 - \partial_2 u_1 = \partial_1\tilde{b}^{-1}\tilde{u}_2 + \tilde{b}^{-1}\partial_1\tilde{u}_2 - \partial_2\tilde{a}^{-1}\tilde{u}_1 - \tilde{a}^{-1}\partial_2\tilde{u}_1. \end{cases}$$

Theorem 0.1. *Let $k_0 > 0$. Assume that a, b, d, e are Lipschitz and c, g are measurable. Moreover, suppose that a, b, d, e are constants in $\mathbb{R}^2 \setminus K$, where K is compact set. Then there exists an $\varepsilon = \varepsilon(k_0, K) > 0$ such that if*

$$\begin{cases} \|\nabla(a-d)\|_{L^\infty} \le \varepsilon,\ \|\nabla(b-d)\|_{L^\infty} \le \varepsilon,\ \|a-d-2e\|_{L^\infty} \le \varepsilon,\ \|b-d-2e\|_{L^\infty} \le \varepsilon, \\ \|d+k_0\|_{L^\infty} \le \varepsilon,\ \|e-k_0\|_{L^\infty} \le \varepsilon,\ \|c\|_{L^\infty} \le \varepsilon,\ \|g\|_{L^\infty} \le \varepsilon, \end{cases} \tag{34}$$

then if u vanishes in the lower half plane, then u is identically zero.

Proof. We first note that when ε, depending on k_0, is sufficiently small, $\tilde{a}$ and $\tilde{b}$ are strictly positive. Let $f_1 = \tilde{u}_1 + \tilde{u}_1'$ and $f_2 = \tilde{u}_2 + i\tilde{u}_2'$, where $\tilde{u}_1'$ and $\tilde{u}_2'$ are conjugate functions of $\tilde{u}_1$ and $\tilde{u}_2$ defined as above. In view of (34), (33) is reduced to

$$\bar{\partial}F = \tilde{\Psi}\partial F + HF + \tilde{H}\bar{F}, \tag{35}$$

where $\|\tilde{\Psi}\|_{L^\infty} \le \varepsilon'$, $\|H\|_{L^\infty} \le \varepsilon'$, $\|\tilde{H}\|_{L^\infty} \le \varepsilon'$, and $H, \tilde{H}$ are supported in K. Note that $\varepsilon' \to 0$ as $\varepsilon \to 0$. As before, let $G(z) = F(\sqrt{z})$, then G satisfies

$$\bar{\partial}G = \frac{z}{|z|}\tilde{\Psi}(\sqrt{z})\partial G + H(\sqrt{z})G + \tilde{H}(\sqrt{z})\bar{G}.$$

By the Poincaré inequality, we have that

$$\|HG\|_{L^p} + \|\tilde{H}\bar{G}\|_{L^p} \le \varepsilon' C(\|\bar{\partial}G\|_{L^p} + \|\partial G\|_{L^p}) \tag{36}$$

for $p \ge 2$, where C depends on K (and p). Using (36), we have from (35) that

$$\|\bar{\partial}G\|_{L^p} \le \varepsilon''\|\partial G\|_{L^p}$$

with $\varepsilon'' \to 0$ as $\varepsilon \to 0$. Next, using the same arguments as in the proof of Theorem 0.1, the result follows. □

Remark 0.2. From the ellipticity condition (7) for isotropic media, it is readily seen that if $k_0 > 0$ and ε is sufficiently small, then the ellipticity condition (27) is satisfied.

Counterexample to Unique Continuation

In this section we will construct a counterexample to the unique continuation property, which vanishes in the lower half plane, for second-order elliptic systems with measurable coefficients. Precisely, we consider

$$\partial_j(a_{ijkl}(x)\partial_k u_l) = 0 \quad \text{in} \quad \mathbb{R}^2, \tag{37}$$

where the coefficients a_{ijkl} do not necessarily satisfy the symmetry conditions (2). For simplicity, we use the following short-hand notations:

$$11 \to 1,\ 12 \to 2,\ 21 \to 3,\ 22 \to 4,$$

i.e.,

$$a_{1111} = a_{11}, a_{1112} = a_{12}, a_{1121} = a_{13}, a_{1122} = a_{14}, \cdots \text{ etc.}$$

So, the system (37) is written as

$$\begin{cases} \partial_1(a_{11}\partial_1 u_1 + a_{12}\partial_1 u_2 + a_{13}\partial_2 u_1 + a_{14}\partial_2 u_2) \\ \qquad + \partial_2(a_{21}\partial_1 u_1 + a_{22}\partial_1 u_2 + a_{23}\partial_2 u_1 + a_{24}\partial_2 u_2) = 0, \\ \partial_1(a_{31}\partial_1 u_1 + a_{32}\partial_1 u_2 + a_{33}\partial_2 u_1 + a_{34}\partial_2 u_2) \\ \qquad + \partial_2(a_{41}\partial_1 u_1 + a_{42}\partial_1 u_2 + a_{43}\partial_2 u_1 + a_{44}\partial_2 u_2) = 0. \end{cases} \tag{38}$$

As before, we can find v_1 and v_2 such that

$$\begin{cases} \partial_2 v_1 = a_{11}\partial_1 u_1 + a_{13}\partial_2 u_1 + a_{12}\partial_1 u_2 + a_{14}\partial_2 u_2, \\ -\partial_1 v_1 = a_{21}\partial_1 u_1 + a_{23}\partial_2 u_1 + a_{22}\partial_1 u_2 + a_{24}\partial_2 u_2, \end{cases} \tag{39}$$

and

$$\begin{cases} \partial_2 v_2 = a_{32}\partial_1 u_2 + a_{34}\partial_2 u_2 + a_{31}\partial_1 u_1 + a_{33}\partial_2 u_1, \\ -\partial_1 v_2 = a_{42}\partial_1 u_2 + a_{44}\partial_2 u_2 + a_{41}\partial_1 u_1 + a_{43}\partial_2 u_1. \end{cases} \tag{40}$$

Here we will use a different reduction from the one used in section "Elasticity System with Measurable Coefficients." The method is inspired by Bojarski's work [7]. Denote

$$\alpha_1 = \frac{(a_{11} + a_{23}) + i(a_{21} - a_{13})}{2}, \quad \beta_1 = \frac{(a_{11} - a_{23}) + i(a_{21} + a_{13})}{2},$$
$$\zeta_1 = a_{12} + ia_{22}, \quad \eta_1 = a_{14} + ia_{24}.$$

Let $f_1 = u_1 + iv_1$ and $f_2 = u_2 + iv_2$, then we can compute that

$$\begin{aligned} 0 &= (1+\alpha_1)\bar{\partial} f_1 + \beta_1 \partial \overline{f_1} + \beta_1 \partial f_1 - (1-\alpha_1)\overline{\partial f_1} + \zeta_1 \partial_1 u_2 + \eta_1 \partial_2 u_2 \\ &= (1+\alpha_1)\bar{\partial} f_1 + \beta_1 \partial \overline{f_1} + \beta_1 \partial f_1 - (1-\alpha_1)\overline{\partial f_1} + \frac{\zeta_1}{2}(\bar{\partial} f_2 + \partial f_2 + \partial \overline{f_2} + \overline{\partial f_2}) \\ &\quad + \frac{\eta_1}{2i}(\bar{\partial} f_2 - \partial f_2 - \partial \overline{f_2} + \overline{\partial f_2}) \\ &= (1+\alpha_1)\bar{\partial} f_1 + \beta_1 \partial \overline{f_1} + \beta_1 \partial f_1 - (1-\alpha_1)\overline{\partial f_1} + (\frac{\zeta_1}{2} + \frac{\eta_1}{2i})\bar{\partial} f_2 + (\frac{\zeta_1}{2} - \frac{\eta_1}{2i})\partial f_2 \\ &\quad + (\frac{\zeta_1}{2} - \frac{\eta_1}{2i})\partial \overline{f_2} + (\frac{\zeta_1}{2} + \frac{\eta_1}{2i})\overline{\partial f_2}. \end{aligned} \tag{41}$$

Likewise, we denote

$$\alpha_2 = \frac{(a_{32}+a_{44})+i(a_{42}-a_{34})}{2}, \quad \beta_2 = \frac{(a_{32}-a_{44})+i(a_{42}+a_{34})}{2},$$
$$\zeta_2 = a_{31}+ia_{41}, \quad \eta_2 = a_{33}+ia_{43},$$

then we obtain

$$\begin{aligned} 0 = &(1+\alpha_2)\bar\partial f_2 + \beta_2\partial\overline{f_2} + \beta_2\partial f_2 - (1-\alpha_2)\overline{\partial f_2} + (\frac{\zeta_2}{2}+\frac{\eta_2}{2i})\bar\partial f_1 + (\frac{\zeta_2}{2}-\frac{\eta_2}{2i})\partial f_1 \\ &+ (\frac{\zeta_2}{2}-\frac{\eta_2}{2i})\partial\overline{f_1} + (\frac{\zeta_2}{2}+\frac{\eta_2}{2i})\overline{\partial f_1}. \end{aligned} \tag{42}$$

Putting (41) and (42) in matrix form gives

$$A\bar\partial F + B\partial\overline{F} + C\partial F + D\overline{\partial F} = 0 \quad \text{in} \quad \mathbb{C}, \tag{43}$$

where $F = \begin{pmatrix} f_1 \\ f_2 \end{pmatrix}$ and

$$\begin{aligned} A &= \begin{pmatrix} 1+\alpha_1 & \frac{\zeta_1}{2}-i\frac{\eta_1}{2} \\ \frac{\zeta_2}{2}-i\frac{\eta_2}{2} & 1+\alpha_2 \end{pmatrix}, \quad B = \begin{pmatrix} \beta_1 & \frac{\zeta_1}{2}+i\frac{\eta_1}{2} \\ \frac{\zeta_2}{2}+i\frac{\eta_2}{2} & \beta_2 \end{pmatrix}, \\ C &= \begin{pmatrix} \beta_1 & \frac{\zeta_1}{2}+i\frac{\eta_1}{2} \\ \frac{\zeta_2}{2}+i\frac{\eta_2}{2} & \beta_2 \end{pmatrix} (= B), \quad D = \begin{pmatrix} -1+\alpha_1 & \frac{\zeta_1}{2}-i\frac{\eta_1}{2} \\ \frac{\zeta_2}{2}-i\frac{\eta_2}{2} & -1+\alpha_2 \end{pmatrix}. \end{aligned} \tag{44}$$

Note that $D = A - 2I_2$. Conversely, it is easy to see that, given any 2×2 complex-valued matrices A, B, C, D satisfying $B = C$ and $D = A - 2I_2$ and (43) with $F = \begin{pmatrix} u_1+iv_1 \\ u_2+iv_2 \end{pmatrix}$, then, writing A, B, C, D as in (44), we can find real numbers $a_{11}, a_{12}, \cdots, a_{44}$ such that (39) and (40) hold, and hence (38) is satisfied.

It was proved in [9] that there exists a 2×2 complex-valued matrix $Q \in L^\infty(\mathbb{C})$ with

$$\|Q\|_{L^\infty(\mathbb{C})} \le \kappa < 1 \tag{45}$$

and a nontrivial Lipschitz function $\tilde F : \mathbb{C} \to \mathbb{C}^2$ vanishing in the lower half plane of $\mathbb{C}$ such that

$$\bar\partial\tilde F + Q\partial\tilde F = 0 \quad \text{in} \quad \mathbb{C}. \tag{46}$$

Adding $A\times$(46) and $B\times\overline{(46)}$, for any A, B, gives

$$A\bar\partial\tilde F + B\partial\overline{\tilde F} + AQ\partial\tilde F + B\overline{Q}\,\overline{\partial\tilde F} = 0 \quad \text{in} \quad \mathbb{C}. \tag{47}$$

Comparing (43) with (47), we hope to find A, B, C, D satisfying

$$B = C = AQ, \quad D = B\overline{Q}, \quad D = A - 2I_2. \tag{48}$$

To fulfill (48), we begin with

$$A - 2I_2 = D = AQ\overline{Q},$$

which implies

$$A(I_2 - Q\overline{Q}) = 2I_2. \tag{49}$$

In view of (45), we have that $\|Q\overline{Q}\|_{L^\infty(\mathbb{C})} \leq \kappa^2 < 1$ and hence $I_2 - Q\overline{Q}$ is invertible. In other words, (49) gives

$$A = 2(I_2 - Q\overline{Q})^{-1}.$$

Once A is determined, we can find C and, of course, B. Hence the relations in (48) hold. Finally, in view of the definitions of $\alpha_j, \beta_j, \zeta_j, \eta_j, j = 1, 2$, there exists a unique fourth rank-tensor $(a_{ijkl}(x))$ producing A, B, C, D which were determined above.

With such A, B, C, D obtained above, there exists a nontrivial solution $F : \mathbb{C} \to \mathbb{C}^2$, i.e., $F = \tilde{F}$, vanishing in the lower half plane of $\mathbb{C}$ and satisfying (43) (and hence, (47)), i.e.,

$$A\bar{\partial}F + AQ\partial\overline{F} + AQ\partial F + AQ\overline{Q}\,\overline{\partial F} = 0 \quad \text{in} \quad \mathbb{C}. \tag{50}$$

As mentioned above, (50) is equivalent to the second-order system (38) with corresponding coefficients $(a_{ijkl}(x))$. Now we would like to verify that this second-order system is elliptic. The meaning of ellipticity will be specified later. We first show that $L_0F := \bar{\partial}F + Q\partial F$ is equivalent to a first-order uniformly elliptic system. Let us denote

$$F = \begin{pmatrix} u_1 + iv_1 \\ u_2 + iv_2 \end{pmatrix},$$

then

$$2\bar{\partial}F = \begin{pmatrix} (\partial_1 u_1 - \partial_2 v_1) + i(\partial_1 v_1 + \partial_2 u_1) \\ (\partial_1 u_2 - \partial_2 v_2) + i(\partial_1 v_2 + \partial_2 u_2) \end{pmatrix}$$

and

$$2\partial F = \begin{pmatrix} (\partial_1 u_1 + \partial_2 v_1) + i(\partial_1 v_1 - \partial_2 u_1) \\ (\partial_1 u_2 + \partial_2 v_2) + i(\partial_1 v_2 - \partial_2 u_2) \end{pmatrix}.$$

Let $Q = Q_r + iQ_i$, then $2L_0F$ can be put into the following equivalent system

$$L_1 \begin{pmatrix} u_1 \\ u_2 \\ v_1 \\ v_2 \end{pmatrix} := \begin{pmatrix} I_2 + Q_r & -Q_i \\ Q_i & I_2 + Q_r \end{pmatrix} \partial_1 \begin{pmatrix} u_1 \\ u_2 \\ v_1 \\ v_2 \end{pmatrix} + \begin{pmatrix} Q_i & -I_2 + Q_r \\ I_2 - Q_r & Q_i \end{pmatrix} \partial_2 \begin{pmatrix} u_1 \\ u_2 \\ v_1 \\ v_2 \end{pmatrix}, \tag{51}$$

i.e., $2L_0F = G = \begin{pmatrix} g_1 \\ g_2 \end{pmatrix} = \begin{pmatrix} w_1 + iz_1 \\ w_2 + iz_2 \end{pmatrix}$ is equivalent to

$$L_1 \begin{pmatrix} u_1 \\ u_2 \\ v_1 \\ v_2 \end{pmatrix} = \begin{pmatrix} w_1 \\ w_2 \\ z_1 \\ z_2 \end{pmatrix}.$$

For simplicity, we denote

$$R = \begin{pmatrix} I_2 + Q_r & -Q_i \\ Q_i & I_2 + Q_r \end{pmatrix} \quad \text{and} \quad S = \begin{pmatrix} Q_i & -I_2 + Q_r \\ I_2 - Q_r & Q_i \end{pmatrix}.$$

Now we want to show that (51) is uniformly elliptic, i.e.,

$$\det(\alpha R + \beta S) \geq c(\alpha^2 + \beta^2)^2, \quad \forall\ z \in \mathbb{C},\ (\alpha, \beta) \in \mathbb{R}^2 \neq 0, \tag{52}$$

where $c = c(\kappa) > 0$. To prove (52), we first observe that

$$S = \begin{pmatrix} Q_i & -I_2 + Q_r \\ I_2 - Q_r & Q_i \end{pmatrix} = \begin{pmatrix} I_2 - Q_r & Q_i \\ -Q_i & I_2 - Q_r \end{pmatrix} J$$

where

$$J = \begin{pmatrix} 0 & -I_2 \\ I_2 & 0 \end{pmatrix}.$$

Therefore, we obtain that

$$\alpha R + \beta S = \alpha(I_4 + E) + \beta(I_4 - E)J = (\alpha I_4 + \beta J) + E(\alpha I_4 - \beta J), \tag{53}$$

where

$$E = \begin{pmatrix} Q_r & -Q_i \\ Q_i & Q_r \end{pmatrix}.$$

From (45), we have that for $E : \mathbb{R}^4 \to \mathbb{R}^4$

$$\|E\|_{L^\infty(\mathbb{R}^4)} \leq \kappa. \tag{54}$$

It is easy to see that

$$\|(\alpha I_4 + \beta J)Z\| = \|(\alpha I_4 - \beta J)Z\| = \sqrt{\alpha^2 + \beta^2}\,\|Z\|, \quad \forall\ Z \in \mathbb{R}^4. \tag{55}$$

Combining (54) and (55) gives

$$\det(\alpha R + \beta S) = \det(\alpha I_4 + \beta J)\det(I_4 + (\alpha I_4 + \beta J)^{-1}E(\alpha I_4 - \beta J)) \geq c(\alpha^2 + \beta^2)^2$$

with $c = c(\kappa)$ and (52) is proved.

Now we want to consider

$$2(\bar{\partial} F + Q\partial F + Q(\partial \overline{F} + \overline{Q}\,\overline{\partial F})) = 2L_0 F + 2Q\overline{L_0 F}. \tag{56}$$

It is easy to see that

$$2\overline{L_0 F} = \begin{pmatrix} w_1 - iz_1 \\ w_2 - iz_2 \end{pmatrix},$$

which is equivalent to

$$\hat{I} L_1 \begin{pmatrix} u_1 \\ u_2 \\ v_1 \\ v_2 \end{pmatrix} = \hat{I} \begin{pmatrix} w_1 \\ w_2 \\ z_1 \\ z_2 \end{pmatrix},$$

where

$$\hat{I} = \begin{pmatrix} I_2 & 0 \\ 0 & -I_2 \end{pmatrix}.$$

Consequently, (56) can be written as

$$(R + E\hat{I}R)\partial_1 \begin{pmatrix} u_1 \\ u_2 \\ v_1 \\ v_2 \end{pmatrix} + (S + E\hat{I}S)\partial_2 \begin{pmatrix} u_1 \\ u_2 \\ v_1 \\ v_2 \end{pmatrix}.$$

It is clear that

$$\begin{aligned} \det(\alpha(R + E\hat{I}R) + \beta(S + E\hat{I}S)) &= \det([\alpha R + \beta S] + E\hat{I}[\alpha R + \beta S]) \\ &= \det(\alpha R + \beta S) \cdot \det(I_4 + E\hat{I}) \\ &\geq c(\alpha^2 + \beta^2)^2. \end{aligned}$$

Finally, let us denote $A = A_r + iA_i$ and

$$\hat{A} = \begin{pmatrix} A_r & -A_i \\ A_i & A_r \end{pmatrix},$$

then

$$A\bar{\partial}F + AQ\partial\overline{F} + AQ\partial F + AQ\overline{Q}\,\overline{\partial F} = 0 \tag{57}$$

is equivalent to the first-order system

$$\hat{A}(R + E\hat{I}R)\partial_1 \begin{pmatrix} u_1 \\ u_2 \\ v_1 \\ v_2 \end{pmatrix} + \hat{A}(S + E\hat{I}S)\partial_2 \begin{pmatrix} u_1 \\ u_2 \\ v_1 \\ v_2 \end{pmatrix} = 0. \tag{58}$$

Since $\det\hat{A} \geq c > 0$, we immediately obtain that

$$\det(\alpha\hat{A}(R + E\hat{I}R) + \beta\hat{A}(S + E\hat{I}S)) \geq c(\alpha^2 + \beta^2)^2. \tag{59}$$

We would like to remind the reader that (57) is equivalent to (39) and (40) (and (38)).

Now we return to the system (39) and (40), i.e.,

$$\begin{cases} \partial_2 v_1 = a_{11}\partial_1 u_1 + a_{13}\partial_2 u_1 + a_{12}\partial_1 u_2 + a_{14}\partial_2 u_2, \\ -\partial_1 v_1 = a_{21}\partial_1 u_1 + a_{23}\partial_2 u_1 + a_{22}\partial_1 u_2 + a_{24}\partial_2 u_2, \end{cases}$$

and

$$\begin{cases} \partial_2 v_2 = a_{32}\partial_1 u_2 + a_{34}\partial_2 u_2 + a_{31}\partial_1 u_1 + a_{33}\partial_2 u_1, \\ -\partial_1 v_2 = a_{42}\partial_1 u_2 + a_{44}\partial_2 u_2 + a_{41}\partial_1 u_1 + a_{43}\partial_2 u_1. \end{cases}$$

We put this system as

$$\begin{pmatrix} a_{11} & a_{12} & 0 & 0 \\ a_{21} & a_{22} & 1 & 0 \\ a_{31} & a_{32} & 0 & 0 \\ a_{41} & a_{42} & 0 & 1 \end{pmatrix} \partial_1 \begin{pmatrix} u_1 \\ u_2 \\ v_1 \\ v_2 \end{pmatrix} + \begin{pmatrix} a_{13} & a_{14} & -1 & 0 \\ a_{23} & a_{24} & 0 & 0 \\ a_{33} & a_{34} & 0 & -1 \\ a_{43} & a_{44} & 0 & 0 \end{pmatrix} \partial_2 \begin{pmatrix} u_1 \\ u_2 \\ v_1 \\ v_2 \end{pmatrix} = 0, \tag{60}$$

which is equivalent to (58). From (59), we have that

$$
\begin{aligned}
&c(\alpha^2+\beta^2)^2\\
&\le \det(\alpha\hat{A}(R+E\hat{I}R)+\beta\hat{A}(S+E\hat{I}S))\\
&=\det\left(\alpha\begin{pmatrix} a_{11} & a_{12} & 0 & 0\\ a_{21} & a_{22} & 1 & 0\\ a_{31} & a_{32} & 0 & 0\\ a_{41} & a_{42} & 0 & 1\end{pmatrix}+\beta\begin{pmatrix} a_{13} & a_{14} & -1 & 0\\ a_{23} & a_{24} & 0 & 0\\ a_{33} & a_{34} & 0 & -1\\ a_{43} & a_{44} & 0 & 0\end{pmatrix}\right)\\
&=-\det\left(\alpha\begin{pmatrix} a_{11} & a_{12} & 0 & 0\\ a_{31} & a_{32} & 0 & 0\\ a_{21} & a_{22} & 1 & 0\\ a_{41} & a_{42} & 0 & 1\end{pmatrix}+\beta\begin{pmatrix} a_{13} & a_{14} & -1 & 0\\ a_{33} & a_{34} & 0 & -1\\ a_{23} & a_{24} & 0 & 0\\ a_{43} & a_{44} & 0 & 0\end{pmatrix}\right)\\
&=\det\begin{pmatrix} a_{12}\alpha+a_{14}\beta & a_{11}\alpha+a_{13}\beta & -\beta & 0\\ a_{32}\alpha+a_{34}\beta & a_{31}\alpha+a_{33}\beta & 0 & -\beta\\ a_{22}\alpha+a_{24}\beta & a_{21}\alpha+a_{23}\beta & \alpha & 0\\ a_{42}\alpha+a_{44}\beta & a_{41}\alpha+a_{43}\beta & 0 & \alpha\end{pmatrix}\\
&=\det\begin{pmatrix} a_{12}\alpha^2+a_{14}\alpha\beta+a_{22}\alpha\beta+a_{24}\beta^2 & a_{11}\alpha^2+a_{13}\alpha\beta+a_{21}\alpha\beta+a_{23}\beta^2\\ a_{32}\alpha^2+a_{34}\alpha\beta+a_{42}\alpha\beta+a_{44}\beta^2 & a_{31}\alpha^2+a_{33}\alpha\beta+a_{41}\alpha\beta+a_{43}\beta^2\end{pmatrix}.
\end{aligned}
\tag{61}
$$

It follows from (61) that for any $\xi=(\xi_1,\xi_2)\neq 0$, the 2×2 matrix $(\sum_{j,k} a_{ijkl}(z)\xi_j\xi_k)$ satisfies

$$
|\det(\sum_{j,k} a_{ijkl}(z)\xi_j\xi_k)|\ge c|\xi|^4,\quad \forall\ z\in\mathbb{C}. \tag{62}
$$

In summary, we have shown that

Theorem 0.1. *There exists a nontrivial vector-valued function $u=(u_1,u_2)^T:\mathbb{R}^2\to\mathbb{R}^2$ vanishing in the lower half plane solving a second-order uniformly elliptic system* (37), *in the sense of* (62), *with essentially bounded coefficients.*

To prove Theorem 0.1, we used a reduction different from the one given in section "Elasticity System with Measurable Coefficients." It is natural to investigate whether the reduction used here can be applied to prove positive results stated in section "Elasticity System with Measurable Coefficients." We only discuss the Lamé system. Comparing (8), (9), and (38) implies

$$
\begin{cases}
a_{11}=\lambda+2\mu,\ a_{12}=a_{13}=0,\ a_{14}=\lambda,\\
a_{21}=a_{24}=0,\ a_{22}=a_{23}=\mu,\\
a_{31}=a_{34}=0,\ a_{32}=a_{33}=\mu,\\
a_{41}=\lambda,\ a_{42}=a_{43}=0,\ a_{44}=\lambda+2\mu.
\end{cases}
$$

By the definitions, we see that

$$\begin{cases} \alpha_1 = \dfrac{\lambda+3\mu}{2},\ \beta_1 = \dfrac{\lambda+\mu}{2},\ \zeta_1 = i\mu,\ \eta_1 = \lambda, \\ \alpha_2 = \dfrac{\lambda+3\mu}{2},\ \beta_2 = -\dfrac{\lambda+\mu}{2},\ \zeta_2 = i\lambda,\ \eta_2 = \mu, \end{cases}$$

and thus,

$$A = \begin{pmatrix} 1+\frac{\lambda+3\mu}{2} & \frac{i}{2}(\mu-\lambda) \\ \frac{i}{2}(\lambda-\mu) & 1+\frac{\lambda+3\mu}{2} \end{pmatrix},\quad B = C = \begin{pmatrix} \frac{\lambda+\mu}{2} & \frac{i}{2}(\mu+\lambda) \\ \frac{i}{2}(\lambda+\mu) & -\frac{\lambda+\mu}{2} \end{pmatrix},$$

$$D = \begin{pmatrix} -1+\frac{\lambda+3\mu}{2} & \frac{i}{2}(\mu-\lambda) \\ \frac{i}{2}(\lambda-\mu) & -1+\frac{\lambda+3\mu}{2} \end{pmatrix}.$$

Now if $\mu \approx 1$ and $\lambda \approx -1$ as in Theorem 0.1, then $B \approx 0$, $C \approx 0$, but

$$A \approx \begin{pmatrix} 2 & i \\ -i & 2 \end{pmatrix},\ D = \begin{pmatrix} 0 & i \\ -i & 0 \end{pmatrix}.$$

In other words, (43) corresponding to the Lamé system cannot be put into the form

$$\bar{\partial} F = \Psi \partial F$$

with $\|\Psi\|_{L^\infty} \ll 1$.

On the other hand, if the coefficients (a_{pq}) of (38) satisfy

$$\begin{cases} \|a_{pq}-1\|_{L^\infty} \le \varepsilon \ \text{ for } \ pq = 11, 23, 32, 44, \\ \|a_{pq}\|_{L^\infty} \le \varepsilon \ \text{ for all other } pq'\text{s} \end{cases} \tag{63}$$

with a sufficiently small ε, then

$$\|A - 2I_2\|_{L^\infty} \le c\varepsilon,\ \|B\|_{L^\infty} = \|C\|_{L^\infty} \le c\varepsilon,\ \|D\|_{L^\infty} \le c\varepsilon$$

for some constant c. For this case, we can prove that the global unique continuation property holds as in the proof of Theorem 0.3. It is not hard to see that the second-order system (38) with coefficients satisfying (63) is elliptic in the sense of (62). In fact, we can even show that $(a_{pq}) = (a_{ijkl})$ satisfies the strong convexity condition (6) (and, of course, the Legendre–Hadamard condition (27)) provided ε is small. To see this, it suffices to consider $a_{11} = a_{1111} = 1$, $a_{23} = a_{1221} = 1$, $a_{32} = a_{2112} = 1$, $a_{44} = a_{2222} = 1$, and all other a_{pq}'s are zero. Then we have that for any 2×2 matrix $\xi = (\xi_k^l)$

$$a_{ijkl}\xi_k^l\xi_j^i = a_{1111}\xi_1^1\xi_1^1 + a_{1221}\xi_2^1\xi_2^1 + a_{2112}\xi_1^2\xi_1^2 + a_{2222}\xi_2^2\xi_2^2 = |\xi|^2,$$

which implies (6) for small ε. The class of second-order elliptic systems (38) satisfying (63) contains a special class of hyperelastic materials, where only the major symmetry property $a_{ijkl} = a_{klij}$ holds.

A counterexample to the strong unique continuation for (22) was constructed in [8] (see also related article [12]). The counterexample given in [8] shows that there exists a nontrivial function F vanishing at 0 to infinite order satisfying

$$\bar{\partial} F + Q \partial F = 0,$$

where $Q(x) \in \mathbb{C}^{2\times 2}$ is continuous and vanishes at 0 to infinite order as well. Based on this example, using the same framework as above, we can construct a counterexample to strong unique continuation for second-order elliptic systems with continuous coefficients in the plane. Observe that for the extreme case $Q = 0$, we have $A = 2I$ and $B = C = D = 0$. Consequently, we see that

$$a_{11} = a_{23} = a_{32} = a_{44} = 1$$

and all other a_{pq}'s are zero. Therefore, when x is near 0, Q is sufficiently small, which is exactly the case we discussed in (63). In other words, for the second-order elliptic system with coefficients satisfying (63) and the strong convexity condition (6), the global unique continuation property holds, in spite of the fact that there are examples showing that the strong unique continuation property fails. Furthermore, we want to point out that the counterexample to the strong unique continuation for (37) we constructed is a small perturbation of the Laplacian Δ near the origin. In section "Elasticity System with Measurable Coefficients" we have shown that the Lamé system with $\lambda \approx -1$ and $\mu \approx 1$ can be written as a small perturbation of the Laplacian. Therefore, this counterexample strongly suggests that the Lamé system with measurable coefficients, even when $\lambda \approx -1$ and $\mu \approx 1$, does not possess the strong unique continuation property. Moreover, this example or an earlier example constructed in [4] also suggests that the strong unique continuation property for the anisotropic elasticity system even with continuous coefficients is most likely not true.

Acknowledgements Carlos Kenig is supported in part by NSF Grant DMS-1265249. Jenn-Nan Wang is supported in part by MOST Grant 102-2918-I-002-009 and 102-2115-M-002-009-MY3.

References

1. G. Alessandrini, A simple proof of the unique continuation property for two dimensional elliptic equations in divergence form. Quaderni Matematici II, Serie 276 (Dipartimento di Scienze Matematiche, Trieste, 1992). http://www.dmi.units.it/~alessang/unique92.pdf
2. G. Alessandrini, Strong unique continuation for general elliptic equations in 2D. J. Math. Anal. Appl. **386**, 669–676 (2012)

3. G. Alessandrini, L. Escauriaza, Null-controllability of one-dimensional parabolic equations. ESAIM: Control Optim. Cal. Var. **14**, 284–293 (2008)
4. S. Alinhac, Non-unicité pour des opérateurs différentiels à caractéristiques complexes simples. Ann. Sci. Ecole Norm. Sup. **13**, 385–393 (1980)
5. R. Bañuelos, P. Janakiraman, L^p-bounds for the Beurling-Ahlfors transform. Trans. Am. Math. Soc. **360**, 3603–3612 (2008)
6. L. Bers, L. Nirenberg, On a representation theorem for linear elliptic systems with discontinuous coefficients and its applications, in Convegno Internazionale sullen Equazioni Lineari alle Derivate Parziali, Trieste, 1954 (Edizioni Cremonese, Roma, 1955), pp. 111–140
7. B.V. Bojarski, Generalized solutions of a system of differential equations of first order and of elliptic type with discontinuous coefficients. Mat. Sb. N.S. **43**(85), 451–503 (1957)
8. A. Coffman, Y. Pan, Smooth counterexamples to strong unique continuation for a Beltrami 375 system in $\mathbb{C}^2$. Comm. PDE. **37**, 2228–2244 (2012)
9. T. Iwaniec, G.C. Verchota, A.L. Vogel, The failure of rank-one connections. Arch. Ration. Mech. Anal. **163**, 125–169 (2002)
10. C.L. Lin, G. Nakamura, G. Uhlmann, J.N. Wang, Quantitative strong unique continuation for the Lame system with less regular coefficients. Methods Appl. Anal. **18**, 85–92 (2011)
11. S. Müller, V. Šverák, Convex integration for Lipschitz mappings and counterexamples to regularity. Ann. Math. **157**, 715–742 (2003)
12. J.P. Rosay, Uniqueness in rough almost complex structures and differential inequalities. Ann. Inst. Fourier (Grenoble). **60**, 2261–2273 (2010)
13. F. Schulz, On the unique continuation property of elliptic divergence form equations in the plane. Math. Z. **228**, 201–206 (1998)

Hardy Spaces of Holomorphic Functions for Domains in $\mathbb{C}^n$ with Minimal Smoothness

Loredana Lanzani and Elias M. Stein

Dedicated to the memory of Cora Sadosky

Abstract We prove various representations and density results for Hardy spaces of holomorphic functions for two classes of bounded domains in $\mathbb{C}^n$, whose boundaries satisfy minimal regularity conditions (namely the classes C^2 and $C^{1,1}$, respectively) together with naturally occurring notions of convexity.

2000*Mathematics Subject Classification: 30E20, 31A10, 32A26, 32A25, 32A50, 32A55, 42B20, 46E22, 47B34, 31B10*

Introduction

Here we discuss the interplay of the holomorphic Hardy spaces with the Cauchy integrals, and with the Cauchy–Szegő projection,[1] for a broad class of bounded domains $D \subset \mathbb{C}^n$ when $n \geq 2$. We make minimal assumptions on the domains' boundary regularity. While such interplay is fairly well understood in the context of one complex variable (that is, for $D \subset \mathbb{C}$, see [20, 33, 42, 43] and references therein) the situation in higher dimensions presents further obstacles, both conceptual and technical in nature, which require a different analysis.

[1] Also known as the *Szegő projection*.

L. Lanzani (✉)
Department of Mathematics, Syracuse University Syracuse, NY 13244-1150, USA
e-mail: llanzani@syr.edu

E.M. Stein
Department of Mathematics, Princeton University, Princeton, NJ 08544-100, USA
e-mail: stein@math.princeton.edu

M.C. Pereyra et al. (eds.), *Harmonic Analysis, Partial Differential Equations, Complex Analysis, Banach Spaces, and Operator Theory (Volume 1)*, Association for Women in Mathematics Series 4, DOI 10.1007/978-3-319-30961-3_11

At the root of all these questions are three basic issues.

First, the fact that for any bounded domain $D \subset \mathbb{C}^n$ whose boundary is of class C^2 there are two characterizations of the Hardy space $\mathbf{H}^p(D)$, see [61]:

The holomorphic functions F for which

$$\|\mathcal{N}(F)\|_{L^p(bD,d\sigma)} < \infty\,, \tag{1}$$

where $\mathcal{N}(F)$ is the so-called nontangential maximal function of F, see (4) below, and $d\sigma$ is the induced Lebesgue measure on the boundary of D. Alternatively,

$$\sup_{0<t<c} \int\limits_{w\in bD_t} |F(w)|^p\, d\sigma_t(w) < \infty. \tag{2}$$

where $\{D_t\}_t$ is an exhaustion of D by appropriate subdomains D_t. The quantities (1) and (2) give equivalent norms of the space $\mathbf{H}^p(D)$, which is also known as the *Smirnov class*, see [20]. Such F have (nontangential) boundary limits

$$f = \dot{F}\,.$$

If $\mathcal{H}^p(bD, d\sigma)$ denotes the space of such functions, then $\mathcal{H}^p(bD, d\sigma)$ is a closed subspace of $L^p(bD, d\sigma)$. Moreover the Cauchy–Szegő projection $\mathcal{S}$ can then be defined as the orthogonal projection of $L^2(bD, d\sigma)$ onto $\mathcal{H}^2(bD, d\sigma)$.

Second, if in addition the domain D is strongly pseudo-convex, we know (by [47]) that D supports an appropriate Cauchy integral $\mathbf{C}$ and a corresponding Cauchy transform $\mathcal{C}$, which is a bounded operator on $L^p(bD, d\sigma)$, $1 < p < \infty$.

Third, a combination of the above facts leads to the following consequences:

- The approximation theorem (Theorem 10): the class of functions holomorphic in a neighborhood of $\overline{D}$ is dense in $\mathbf{H}^p(D)$, $1 < p < \infty$, and correspondingly their restrictions to bD are dense in $\mathcal{H}^p(bD, d\sigma)$.
- The fact that when $f \in L^p(bD, d\sigma)$ then $f \in \mathcal{H}^p(bD, d\sigma)$ if and only if $f = \mathcal{C}(f)$ (Corollary 11). Related to this is the conclusion that the image of $L^p(bD, d\sigma)$ under $\mathcal{C}$ is exactly $\mathcal{H}^p(bD, d\sigma)$ (Proposition 8).
- The identities $\mathcal{S}\mathcal{C} = \mathcal{C}$ and $\mathcal{C}\mathcal{S} = \mathcal{S}$ which hold in $L^2(bD, d\sigma)$ (Proposition 12). These are crucial in the proof (given in [47]) of the $L^p(bD, d\sigma)$-boundedness of $\mathcal{S}$, $1 < p < \infty$.
- A further characterization of $\mathcal{H}^p(bD, d\sigma)$ as the space of those $f \in L^p(bD)$ for which $f = \mathcal{S}(f)$ (Proposition 13).

The results given in Proposition 8 and Corollary 11, Theorem 10, Propositions 9 and 12 also hold for $C^{1,1}$ domains that are strongly $\mathbb{C}$-linearly convex. For these one uses the L^p estimates of the Cauchy–Leray integral given in [46], together with a suitable modification of Lemma 14. These will appear in a future publication.

We point out that analogous statements apply to the weighted spaces $L^p(bD, \omega\, d\sigma)$ (resp., $\mathcal{H}^p(bD, \omega\, d\sigma)$) whenever ω is a continuous strictly positive density on bD: these weighted measures include the so-called *Leray–Levi measure* considered in [46, 47]. But note that the space $L^p(bD, \omega\, d\sigma)$ (resp., $\mathcal{H}^p(bD, \omega\, d\sigma)$) contains the same elements as $L^p(bD, d\sigma)$ (resp., $\mathcal{H}^p(bD, \omega\, d\sigma)$) and the two norms are equivalent, and so we will continue to denote both of these spaces simply by $L^p(bD)$ (resp., $\mathcal{H}^p(bD)$). However the distinction between $L^2(bD, d\sigma)$ and $L^2(bD, \omega\, d\sigma)$ becomes relevant when defining the Cauchy–Szegő projections because these spaces have different inner products that give different notions of orthogonality, and so we will distinguish between the two orthogonal projections $\mathcal{S}$ and $\mathcal{S}_\omega$.

The structure of this paper is as follows. In section "Preliminaries" we collect the main definitions along with a few, well-known features of the spaces $\mathcal{H}^p(bD)$ for any bounded domain D of class C^2. In section "The Role of the Cauchy Integral" we restrict the focus to the strongly pseudo-convex domains and establish various connections of $\mathcal{H}^p(bD)$ with the holomorphic Cauchy integrals that were studied in [47]; in particular, we give the proof of Proposition 8 and of Corollary 11, and use these results to establish an operator identity that directly links our Cauchy integrals to the Cauchy–Szegő projection (Proposition 12). The approximation theorem for $\mathcal{H}^p(bD)$ is also proved in section "The Role of the Cauchy Integral" (Theorem 10). In section "Further Results" we give a further characterization of $\mathcal{H}^p(bD)$ as the range of the Cauchy–Szegő projection for D, again for D strongly pseudo-convex and of class C^2 (Proposition 13). In the appendix we go over the more technical tools and results that are needed to prove Theorem 10.

Lastly, in the References we provide a number of papers, including [1–19, 21–32, 34–41, 48–60, 62, 63], that offer further insight on the vast literature on Hardy Spaces.

We remark that this paper complements and at the same time is complemented by the paper [47], in the following sense. Part I of [47] is needed to prove the results in section "The Role of the Cauchy Integral" of the present paper. On the other hand, section "The Role of the Cauchy Integral" in this paper (in particular Proposition 12) is needed in Part II of [47]. There is no circularity in our proofs, as Part I of [47] is independent from Part II in that same paper. Finally, the results of Part II in [47] (along with section "The Role of the Cauchy Integral" in this paper) lead to the proof of Proposition 13 in section "Further Results" of the present paper.

Preliminaries

Here we recall the main definitions and basic features of the theory of the holomorphic Hardy spaces for a domain $D \subset \mathbb{C}^n$. While all results are stated for the induced Lebesgue measure $d\sigma$, they are in fact valid for any measure that is equivalent to $d\sigma$ in the sense discussed in the previous section (in particular for the aforementioned Leray–Levi measure). Thus we will drop explicit reference to the measure and again write $L^p(bD)$, $\mathcal{H}^p(bD)$, and so forth.

Most proofs are deferred to [61, Sects. 1–5 and 9–10] and [20], where references to the earlier literature can also be found. To simplify the exposition we limit this review to the case $1 < p < \infty$, however we point out that what is needed for $p = \infty$ is a trivial consequence of $p < \infty$.

The only assumptions on D that we need to make at this stage are that D is bounded and it is of class C^2. In fact the class $C^{1+\epsilon}$ would suffice here; the requirement that the domain D is of class C^2 will be needed in section "The Role of the Cauchy Integral" and onwards, when we deal with pseudo-convex domains.

Small Perturbations of the Domain

Starting with our original domain D we will need to construct a family of domains $\{D_t\}$ where t is a small *real* parameter. We recall that since D is of class C^2 it admits a C^2-smooth defining function $\rho : \mathbb{C}^n \to \mathbb{R}$ such that

$$D = \{\rho < 0\}, \quad bD = \{\rho = 0\} \quad \text{and} \quad \nabla\rho \neq 0 \text{ on } bD.$$

Then for each real t which is small we consider

$$\rho_t = \rho + t \quad \text{and} \quad D_t = \{\rho_t < 0\}, \quad \text{so that} \quad bD_t = \{\rho_t = 0\}.$$

Note that with this notation we have $D_0 = D$ and furthermore $\overline{D_t} \subset D_0$ if $t > 0$, whereas $\overline{D}_0 \subset D_t$ if $t < 0$. We will also need an appropriate bijection between bD and bD_t. This is given by the exponential map of a normal vector field, which we denote $\Phi_t : bD \to bD_t$, and whose properties are detailed in the appendix.

Hardy Spaces

When $1 \leq p < \infty$ and c is a small positive constant, the class $\mathbf{H}^p(D)$ is defined as the space of functions F that are holomorphic in D and for which (2) holds with D_t as above and $t > 0$. (Note that therefore $\overline{D}_t \subset D$.) We take the norm $\|F\|_{\mathbf{H}^p(D)}$ to be the p-th root of the left-hand side of the expression (2).

We recall some basic properties of $\mathbf{H}^p(D)$. First, the space is independent of the specific choice of defining function that is being used (i.e., another defining function will give an equivalent norm). Also, one has the following fact about convergence in $\mathbf{H}^p(D)$:

If a sequence $\{F_k\} \subset \mathbf{H}^p(D)$ is uniformly bounded in the norm, and $F_k \to F$ uniformly on compact subsets of D, then $F \in \mathbf{H}^p(D)$.

A key feature of functions in $\mathbf{H}^p$ involves nontangential convergence. To describe this we fix a convenient nontangential approach region for each point $w \in bD$. With $\beta > 1$ fixed throughout, set $\Gamma(w) = \{z \in D \ : \ |z - w| < \beta\,\text{dist}(z, bD)\}$, where $\text{dist}(z, bD)$ is the usual Euclidean distance of $z \in D$ from the boundary of D. Then if $F \in \mathbf{H}^p(D)$, one knows (see [61, Theorem 10, Sect. 10, and its corollary]):

$$\lim_{\substack{z \to w \\ z \in \Gamma(w)}} F(z) = \dot{F}(w) \quad \text{exists for a.e. } w \in bD\,. \tag{3}$$

$$\mathcal{N}(F)(w) := \sup_{z \in \Gamma(w)} |F(z)| \in L^p(bD) \tag{4}$$

$$\|F\|_{\mathbf{H}^p(D)} \approx \|\dot{F}\|_{L^p(bD)} \approx \|\mathcal{N}(F)\|_{L^p(bD)} \tag{5}$$

The above allow us to define the Hardy space in terms of the boundary values $\dot{F}$ of $F \in \mathbf{H}^p(D)$. More precisely, the *Hardy space* $\mathcal{H}^p(bD)$ consists of those function $f \in L^p(bD, d\sigma)$ which arise as

$$\dot{F} = f\,, \quad \text{for some} \quad F \in \mathbf{H}^p(D).$$

If we take

$$\|f\|_{\mathcal{H}^p(bD)} := \|f\|_{L^p(bD)}$$

then (5) shows that

$$\|f\|_{\mathcal{H}^p(bD)} \approx \|F\|_{\mathbf{H}^p(D)} \tag{6}$$

and in this sense the spaces $\mathbf{H}^p(D)$ and $\mathcal{H}^p(bD)$ are identical with equivalent norms, and we will henceforth refer to either one of $\mathbf{H}^p$ and $\mathcal{H}^p$ as "the Hardy space." It follows that (1) gives an alternate characterization of $\mathbf{H}^p$.

Connected with this is the fact that whenever $F \in \mathbf{H}^p(D)$, one has

$$F(z) = \mathbf{P}_z(f)\,, \quad \text{with} \quad z \in D,$$

where $\mathbf{P}_z(f)$ is the Poisson integral of f, that is,

$$F(z) = \mathbf{P}_z(f)(z) = \int\limits_{w \in bD} \mathcal{P}_z(w) f(w)\, d\sigma(w)\,.$$

The characteristic property of the Poisson kernel $\mathcal{P}_z(w)$ is that it gives the solution of the Dirichlet problem for the Laplace operator for D with data f. Namely, if f is any continuous function on bD, and we set $u(z) = \mathbf{P}_z(f)(z)$, then u is harmonic in D, it extends continuously to $\overline{D}$ and $u(w) = f(w)$ whenever $w \in bD$. More generally, if $f \in L^p(bD)$, then $\dot{u}(w) = f(w)$ for σ-a.e. $w \in bD$ and furthermore, $\|\mathcal{N}(u)\|_{L^p(bD)} \lesssim \|f\|_{L^p(bD)}$. The fact that

$$\sup_{\substack{w \in bD \\ z \in D_t}} \mathcal{P}_z(w) \leq c_t, \quad \text{if} \quad t > 0, \tag{7}$$

implies that a Cauchy sequence $\{F_k\}_k \subset \mathbf{H}^p(D)$ has a subsequence that is uniformly convergent on any compact subset of D to a function F, which is in fact in $\mathbf{H}^p(D)$, from which the completeness of $\mathbf{H}^p(D)$ is evident. Thus (6) shows that $\mathcal{H}^p(bD)$ is a closed subspace of $L^p(bD)$. Moreover, one has the following simple characterization.

Proposition 1. *Suppose $f \in L^p(bD)$. Then $f \in \mathcal{H}^p(bD)$ if, and only if, $u(z) = \mathbf{P}_z(f)$ is holomorphic in D.*

Indeed, if $f \in \mathcal{H}^p(bD)$, then there is $F \in \mathbf{H}^p(D)$ such that $\dot{F} = f$. But one also has $\dot{\mathbf{P}}_z(f) = f$, and we conclude that $\mathbf{P}_z(f) = F(z), z \in D$, by the uniqueness of the solution of the Dirichlet problem for harmonic functions with data $f \in L^p(bD)$. Conversely, if $\mathbf{P}_z(f)$ is holomorphic in D, then by the aforementioned estimates for the solution of the Dirichlet problem with data $f \in L^p(bD)$ and by the equivalence (5), we have that $\mathbf{P}_z(f) \in \mathbf{H}^p(D)$ and furthermore, that $\dot{\mathbf{P}}_z(f) = f$, showing that $f \in \mathcal{H}^p(bD)$ with $F(z) := \mathbf{P}_z(f), z \in D$.

Proposition 1 has the following immediate consequence which, of course, is interesting only when $p_2 > p_1$.

Corollary 2. *If $f \in \mathcal{H}^{p_1}(bD) \cap L^{p_2}(bD)$, then $f \in \mathcal{H}^{p_2}(bD)$.*

Indeed, first note that $u(z) = \mathbf{P}_z(f)$ is holomorphic in D by Proposition 1 (because $f \in \mathcal{H}^{p_1}(bD)$). From this we conclude that $f \in \mathcal{H}^{p_2}(bD)$ again by Proposition 1 (because $f \in L^{p_2}(bD)$ and $\mathbf{P}_z(f)$ is holomorphic in D).

The Role of the Cauchy Integral

In this section we come to terms with the main issue that arises in the context of several complex variables, namely, the fact that there is no canonical, holomorphic Cauchy kernel for $D \subset \mathbb{C}^n$ when $n \geq 2$. For this reason we need to impose the additional restriction that our domain be strongly pseudo-convex and of class C^2, so we may apply the results of [47, Part I], where the existence (and explicit construction) of a family of holomorphic Cauchy kernels is established, along with L^p- and Hölder-regularity properties of the resulting integral operators.

Holomorphic Cauchy Integrals

The proofs of the statements in this section can be found in Part I of [47]. Here we briefly recall the main ideas and ingredients in the proofs.

- Taking $\rho : \mathbb{C}^n \to \mathbb{R}$ to be a strictly plurisubharmonic defining function of D, one begins by constructing a family of locally holomorphic kernels, denoted $\{C^1_\epsilon(w,z)\}_\epsilon$, by applying the Cauchy–Fantappiè theory, see [45], to a perturbation of the Levi polynomial of D in which the second derivatives of ρ (which are only continuous functions of $w \in \overline{D}$) are replaced by a smooth approximation τ_ϵ. One then achieves global holomorphicity by adding to each such kernel the solution of an ad-hoc $\overline{\partial}$-problem in the z-variable (for fixed w in a neighborhood of bD), whose data is defined in a (strongly pseudo-convex and smooth) neighborhood

Ω of $\overline{D}$; we denote such solution $C^2_\epsilon(w,z)$. The outcome of this procedure is a family of globally holomorphic kernels $\{C_\epsilon(w,z)\}_\epsilon$:

$$C_\epsilon(w,z) = C^1_\epsilon(w,z) + C^2_\epsilon(w,z), \quad 0 < \epsilon < \epsilon_0 \tag{8}$$

which are holomorphic in $z \in D$ whenever w is in bD. More precisely, each of the $C_\epsilon(w,z)$'s is defined in terms of a denominator $\geq^{-n}(w,z)$, where $\geq(w,z)$ is a holomorphic function of $z \in D$ (and of class C^1 in $w \in \Omega$) that satisfies the following inequalities uniformly in $0 < \epsilon < \epsilon_0$:

$$\mathrm{Re} \geq (w,z) \geq c'(-\rho(z) + |w-z|^2), \quad \text{for } z \in \overline{D},\ w \in bD\,, \tag{9}$$

and

$$\mathrm{Re} \geq (w,z) \geq c'(\rho(w) - \rho(z) + |w-z|^2) \tag{10}$$

for z and w in a neighborhood of bD. We denote the resulting integral operators by $\mathbf{C}_\epsilon$, that is,

$$\mathbf{C}_\epsilon f(z) = \int\limits_{w \in bD} f(w)\, C_\epsilon(w,z)\,, \quad z \in D.$$

From now on we are only interested in the properties of these operators for a fixed (small) value of ϵ. Thus we will drop explicit reference to ϵ and will write $\mathbf{C}$ for $\mathbf{C}_\epsilon$, $C(w,z)$ for $C_\epsilon(w,z)$, $g(w,z)$ for $\geq(w,z)$, and so forth.

The key properties of $\mathbf{C}$ are summarized in Propositions 3 and 4 below.

Proposition 3.

(1) *Whenever f is integrable, $\mathbf{C}(f)(z)$ is holomorphic for $z \in D$.*
(2) *If F is continuous in $\overline{D}$ and holomorphic in D and*

$$f = F\Big|_{bD}\,,$$

then $\mathbf{C}(f)(z) = F(z)$, $z \in D$.

- An important feature of the Levi polynomial of D is that it determines a quasi-distance function $\mathtt{d}(w,z)$, defined for $w, z \in bD$, that exhibits borderline integrability:

$$\int\limits_{w \in \mathbf{B}_r(z)} \mathtt{d}(w,z)^{-2n+\beta} d\sigma(w) \leq c_\beta\, r^\beta; \quad \int\limits_{w \in bD \setminus \mathbf{B}_r(z)} \mathtt{d}(w,z)^{-2n-\beta} d\sigma(w) \leq c_\beta\, r^{-\beta} \tag{11}$$

for $0 < r < 1$ and $\beta > 0$, where $\mathbf{B}_r(z) = \{w \in bD \mid \mathtt{d}(w,z) < r\}$.

Then one has the following extension result (proved in Part I of [47]):

Proposition 4. *If f satisfies the Hölder-type condition:*

$$|f(w) - f(z)| \leq c\, d(w,z)^{\alpha}, \quad w, z \in bD. \tag{12}$$

then $\mathbf{C}(f)$ *extends to a continuous function on* $\overline{D}$.

It follows from this proposition that one may define the Cauchy transform $\mathcal{C}$ as the restriction of $\mathbf{C}$ to the functions on bD that satisfy the Hölder-like condition (12), that is,

$$\mathcal{C}(f) = \mathbf{C}(f)\big|_{bD} \quad \text{for } f \text{ as in (12).}$$

Note that with the notation of section "Small Perturbations of the Domain" we have

$$\mathcal{C}(f) = \dot{\mathbf{C}}(f)\,.$$

The Cauchy transforms have the following regularity properties (proved in Part I of [47]).

Theorem 5. *The operator $\mathcal{C}$ initially defined for functions satisfying* (12) *extends to a bounded linear transformation on $L^p(bD, d\sigma)$, for $1 < p < \infty$.*

Proposition 6. *For any $0 < \alpha < 1$, the transform $\mathcal{C} : f \mapsto \mathcal{C}(f)$ preserves the space of Hölder-like functions satisfying condition* (12).

Stability Under Small Perturbations of the Domain

We should note that when $|t|$ is small, the approximating domains D_t that were defined in section "Small Perturbations of the Domain" are strongly pseudo-convex as a consequence of the assumption that D is strongly pseudo-convex. It turns out that the Cauchy kernel $C(w,z)$ that was introduced in the previous section for the original domain D works as well, *mutatis mutandis*, for the domains $\{D_t\}_t$. More precisely, we have:

Proposition 7. *If $|t|$ is sufficiently small, the kernel $C(w,z)$ given by* (8) *has a natural extension to $z \in D_t$ and $\zeta \in bD_t$. The corresponding Cauchy integral operator $\mathbf{C}^{(t)}$ for D_t, defined as*

$$\mathbf{C}^{(t)}(\tilde{f})(z) = \int\limits_{\zeta \in bD_t} \tilde{f}(\zeta)\, C(\zeta, z), \quad z \in D_t,$$

satisfies the following properties:

(1) *Whenever $\tilde{f}$ is integrable on bD_t, $\mathbf{C}^{(t)}(\tilde{f})(z)$ is holomorphic for $z \in D_t$.*

(2) *If $\tilde{F}$ is continuous in $\overline{D}_t$ and holomorphic in D_t and*

$$\tilde{f} = \tilde{F}\Big|_{bD_t},$$

then $\mathbf{C}^{(t)}(\tilde{f})(z) = \tilde{F}(z),\ z \in D_t.$

We point out that this proposition includes the result for the Cauchy integral of D (Proposition 3) as the case $t = 0$.

Proof. To construct the extension of the kernel we begin by noting that $g(w, z)$ still satisfies the inequality analogous to (9) for $z \in \overline{D}_t$, $\zeta \in bD_t$, when ρ is replaced by $\rho_t = \rho + t$. Namely, we have by (10) that

$$\mathrm{Re}\left(g(\zeta, z)\right) \geq c(-\rho_t(z) + |\zeta - z|^2), \quad \text{when} \quad z \in \overline{D}_t,\ \zeta \in bD_t\,.$$

(Note that using ρ_t in place of ρ does not change the definition of g.) Hence if we take

$$\mathbf{C}^{(t),1}(\tilde{f})(z) = \int\limits_{\zeta \in bD_t} \tilde{f}(\zeta)\, C^1(\zeta, z)\,, \quad z \in D_t\,,$$

where

$$C^1(\zeta, z) = \frac{G(\zeta, z) \wedge (\overline{\partial} G(\zeta, z))^{n-1}}{g(\zeta, z)^n}$$

then $\mathbf{C}^{(t),1}$ is a Cauchy–Fantappiè integral for D_t whose kernel is holomorphic for $z \in D_t$ close to $\zeta \in bD_t$. Next, as pointed out in, e.g., [45, Lemma 7] and [44, Proposition 3.2], there is a smooth, strongly pseudo-convex domain $\emptyset$ that contains $\overline{D}_t$ for $|t|$ sufficiently small, with the property that $H(\zeta, z) := -\overline{\partial}_z C^1(\zeta, z)$ is smooth when $z \in \emptyset$, and is continuous in $\zeta \in bD_t$. We may thus consider the correction operator $\mathbf{C}^2$ and its kernel $C^2(\zeta, z)$ as described in [47, Part I] and references therein, however now for $z \in \emptyset$ and $\zeta \in bD_t$, so that

$$\overline{\partial}_z C^2(\zeta, z) = -C^1(\zeta, z) \quad \text{whenever} \quad z \in \emptyset\,,\ \zeta \in bD_t\,,$$

and set

$$\mathbf{C}^{(t),2}(\tilde{f})(z) = \int\limits_{\zeta \in bD_t} \tilde{f}(\zeta)\, C^2(\zeta, z), \quad \text{for } z \in D_t\,.$$

Then if $\mathbf{C}^{(t)}(\tilde{f}) = \mathbf{C}^{(t),1}(\tilde{f}) + \mathbf{C}^{(t),2}(\tilde{f})$, we have

$$\mathbf{C}^{(t)}(\tilde{f})(z) = \int_{bD_t} \tilde{f}(\zeta)\, C(\zeta, z), \quad z \in D_t$$

where $C(\zeta, z) = C^1(\zeta, z) + C^2(\zeta, z)$ is in fact the kernel of $\mathbf{C}$ (the Cauchy integral for the domain D that was defined in section "Holomorphic Cauchy Integrals") and so it is independent of t. It should be clear from the above that (1) and (2) hold. Namely, $C(\zeta, z)$ is holomorphic in $z \in D_t$ for any $\zeta \in bD_t$; and if $\tilde{F}$ is holomorphic in D_t and continuous in $\overline{D}_t$, we have that

$$\tilde{F}(z) = \int_{w \in bD_t} \tilde{f}(\zeta)\, C(\zeta, z), \quad \text{for} \quad z \in D_t \quad \text{and } |t| \text{ small}, \tag{13}$$

where

$$\tilde{f} = F\big|_{bD_t}.$$

□

Hardy Spaces and the Cauchy Integral

The following proposition gives the first link between the Cauchy integral and $\mathbf{H}^p(D)$.

Proposition 8. *Suppose $f \in L^p(bD, d\sigma)$, $1 < p < \infty$, and let $F(z) = \mathbf{C}f(z)$, $z \in D$. Then, $F \in \mathbf{H}^p(D)$ and*

$$\|F\|_{\mathbf{H}^p(D)} \lesssim \|f\|_{L^p(bD, d\sigma)}.$$

Moreover, we have that $\mathcal{C}(f) \in \mathcal{H}^p(bD, d\sigma)$.

Proof. We let $\{g_k\}_k$ be a sequence of smooth (say, C^1) functions so that $g_k \to f$ in the $L^p(bD, d\sigma)$-norm. Let

$$F_k := \mathbf{C}(g_k).$$

Then by Proposition 4, F_k is holomorphic in D and continuous on $\overline{D}$, and

$$\dot{F}_k = F_k\big|_{bD} = \mathcal{C}(g_k),$$

by the definition of the transform $\mathcal{C}$. So $\{F_k\}_k \subset \mathbf{H}^p(D)$, and

$$\|F_k - F_j\|_{\mathbf{H}^p(D)} \lesssim \|\mathcal{C}(g_k - g_j)\|_{L^p(bD)} \lesssim \|g_k - g_j\|_{L^p(bD)},$$

with the first inequality due to (5), and the second due to the L^p-boundedness of $\mathcal{C}$ ([47, Theorem 7]). This shows that $\{F_k\}_k$ is a Cauchy sequence in $\mathbf{H}^p(D)$ and it follows from (7) and the comments thereafter that $\{F_k\}_k$ has a subsequence (which we keep denoting $\{F_k\}$) that converges uniformly on compact subsets of D and therefore in $\mathbf{H}^p(D)$ to a limit F which is thus in $\mathbf{H}^p(D)$. Since as we have seen $\|F_k\|_{\mathbf{H}^p(D)} \lesssim \|g_k\|_{L^(bD)}$, this yields the first assertion.

The fact that $\mathcal{C}(g_k) \in \mathcal{H}^p(bD, d\sigma)$ (recall that $\mathcal{C}(g_k) = \dot{F}_k$ with $F_k \in \mathbf{H}^p(D)$), and the continuity of $\mathcal{C}$ in the $L^p(bD, d\sigma)$-norm ([47, Theorem 7]), then shows that

$$\mathcal{C}(f) = \lim_{k\to\infty} \mathcal{C}(g_k) \in \mathcal{H}^p(bD, d\sigma).$$

This proves the second assertion, completing the proof of the proposition. □

The operator $\mathbf{C}^{(t)}$ that appeared in Proposition 7 will be used in the proofs of the next two results. Specifically, the situation when $t > 0$ (for which $\overline{D}_t \subset D$) arises in the proof of Proposition 9 below, whereas the case $t < 0$ (for which $\overline{D} \subset D_t$) will occur in the proof of Theorem 10 in the next section.

Proposition 9. *Suppose $F \in \mathbf{H}^p(D)$ and $f = \dot{F}$. Then,*

$$F(z) = \mathbf{C}(f)(z), \quad z \in D.$$

The assertion above is an elaboration of [47, Proposition 5], in which F was taken to be in the subspace of $\mathbf{H}^p(D)$ consisting of the functions that are holomorphic in D and continuous on $\overline{D}$.

Proof. For $t > 0$ we consider the Cauchy integral $\mathbf{C}^{(t)}$ for the region D_t that was defined in section "Small Perturbations of the Domain" and then pass to the limit as $t \to 0$.

To this end, first note that by (13), for any fixed $z \in D$, we have

$$F(z) = \int_{bD_t} F(\zeta)\, C(\zeta, z) \quad \text{for } t > 0 \quad \text{and sufficiently small}$$

because F is holomorphic in D_t and continuous on $\overline{D}_t$. We may now use the bijection $\Phi_t\colon bD \to bD_t$ that was described in section "Small Perturbations of the Domain" to express the identity above in the equivalent form

$$F(z) = \int_{bD} F(\Phi_t(w))J_t(w)\, C(\Phi_t(w), z) \tag{14}$$

via a corresponding formulation of the change of variables formula (21) (given in the appendix below). Since $J_t \to 1$ uniformly on bD, and the coefficients of $C(w, z)$ are continuous in w in a neighborhood of $\overline{D}$, then $J_t(w)C(\Phi_t(w), z)$ converges to $C(w, z)$ as $t \to 0$, uniformly in $w \in bD$; moreover, the convergence of $\Phi_t(w)$ to w

is nontangential. Thus (3) and the dominated convergence via the maximal function $\mathcal{N}(F)$ show that the integral in (14) converges to

$$\int_{bD} \dot{F}(w)\, C(w,z) \;=\; \mathbf{C}(f)(z).$$

The proposition is therefore proved. □

Density Properties of $\mathcal{H}^p(bD)$

There are two realizations of $\mathcal{H}^p(bD)$ that follow from the previous results for $\mathbf{C}$ and $\mathcal{C}$. The first is in fact a basic approximation of $\mathcal{H}^p(bD)$. We define $\mathcal{H}_\vartheta(bD)$ to consist of all functions f that arise as restrictions to bD of functions F that are holomorphic in some neighborhood of $\overline{D}$ (which need not be fixed and may, in fact, depend on F).

Theorem 10. *For each p, $1 < p < \infty$, we have that $\mathcal{H}_\vartheta(bD)$ is dense in $\mathcal{H}^p(bD)$.*

In particular, the space of functions that arise as restrictions to bD of functions that are holomorphic in D and continuous on $\overline{D}$ is dense in $\mathcal{H}^p(bD)$.

Proof of Theorem 10. We denote by $\mathcal{H}_\alpha(bD)$ the space of functions f on bD that satisfy the Hölder condition (12) and moreover, arise as

$$f = \dot{F},$$

with F holomorphic in D and continuous in $\overline{D}$. Note that $\mathcal{H}_\vartheta(bD) \subset \mathcal{H}_\alpha(bD)$.

The proof of the proposition is given in two steps: in the first step we show that $\mathcal{H}_\alpha(bD)$ is dense in $\mathcal{H}^p(bD)$. Let $f \in \mathcal{H}^p(bD)$, with $\dot{F} = f$, where $F \in \mathbf{H}^p(D)$. Note that by Proposition 9 we have

$$\mathbf{C}(f) = F.$$

Now let $\{h_n\}$ be a sequence of C^1-functions on bD (which automatically satisfy (12)) so that

$$h_n \to f$$

in the $L^p(bD)$-norm, and set $F_n = \mathbf{C}(h_n)$. As in the proof of [47, Proposition 6], we see that F_n is holomorphic in D and continuous in $\overline{D}$. Let

$$f_n = \dot{F}_n = \mathcal{C}(h_n).$$

Then by Proposition 6 we have that f_n is Hölder continuous in the sense of (12) and we conclude that

$$\{f_n\}_n \subset \mathcal{H}_\alpha(bD)\,.$$

We claim that $f_n \to f$ in $\mathcal{H}^p(bD)$. Indeed by the definitions of F_n and of F we have that

$$F_n - F = \mathbf{C}(h_n - f)\,,$$

and Proposition 8 thus grants that

$$\|F_n - F\|_{\mathbf{H}^p(D)} \lesssim \|h_n - f\|_{L^p(bD)} \to 0.$$

On the other hand by (6) we also have

$$\|F_n - F\|_{\mathbf{H}^p(D)} \approx \|\dot{F}_n - \dot{F}\|_{L^p(bD)} = \|f_n - f\|_{L^p(bD)}$$

from which the desired convergence follows.

In the second step of the proof of Theorem 10 we show that any $f \in \mathcal{H}_\alpha(bD)$ can be approximated uniformly on bD by a family $\{F_t\}_t$ of functions whose boundary values are in $\mathcal{H}_\vartheta(bD)$. To construct such a family we use the Cauchy integrals $\{\mathbf{C}^{(t)}\}_t$ for the domains D_t that were defined in section "Small Perturbations of the Domain" for *negative* t (note that then $\overline{D} \subset D_t$), and apply them to a suitable transposition of f to bD_t. More precisely, given $f \in \mathcal{H}_\alpha(bD)$ we define $\tilde{f}$ by requiring that $\tilde{f}(\Phi_t(w)) = f(w)$ if t is negative and sufficiently small. (Here $\Phi_t : bD \to bD_t$ is again as in section "Small Perturbations of the Domain.") Now define

$$F_t(z) = \mathbf{C}^{(t)}(\tilde{f})(z), \quad z \in D_t, \quad t < 0 \quad \text{and small,}$$

where we recall that

$$\mathbf{C}^{(t)}(\tilde{f})(z) := \int\limits_{w \in bD_t} \tilde{f}(\zeta)\, C(\zeta, z) \quad z \in D_t$$

is the aforementioned Cauchy integral for D_t. Then by part (1) of Proposition 7 we have that each F_t is holomorphic in D_t and so its restriction to bD belongs to $\mathcal{H}_\vartheta(bD)$.

It will suffice to show that

$$F_t(z) \to f(z) \quad \text{uniformly for } z \in bD, \quad \text{as } t \to 0.$$

To this end, note that

$$F_t(z) = f(z) + \int\limits_{\zeta \in bD_t} [\tilde{f}(\zeta) - f(z)]\, C(\zeta, z)\,, \quad z \in bD \tag{15}$$

because

$$\int_{\zeta\in bD_t} C(\zeta,z) = 1 \quad \text{for } z \in D_t$$

by conclusion (2) of Proposition 7. By an analogous formulation of the change of variables formula (21), the identity (15) can be rewritten as

$$F_t(z) = f(z) + \int_{w\in bD} \big[f(w) - f(z)\big]\, J_t(w)\, C(\Phi_t(w), z)\,, \quad z \in bD\,. \tag{16}$$

We point out that a corresponding representation for $\mathcal{C}(f)$ was given in [47, (3.2)], namely

$$\mathcal{C}(f)(z) = f(z) + \int_{w\in bD} \big[f(w) - f(z)\big]\, C(w, z)\,, \quad z \in bD\,. \tag{17}$$

We next remark that whenever $f \in \mathcal{H}_\alpha(bD)$, one has

$$\mathcal{C}(f) = f\,. \tag{18}$$

To see this, we write f as $\dot{F}$, for some $F \in \mathbf{H}^p$ (in fact for F holomorphic in D and continuous on $\overline{D}$). Then it follows by Proposition 9 that $\mathbf{C}f = F$ on D; by Proposition 4, this identity extends to $\overline{D}$, and (18) then follows by the definition of $\mathcal{C}$. Combining (18) with (17) we obtain

$$f(z) = f(z) + \int_{w\in bD} \big[f(w) - f(z)\big]\, C(w, z)\,, \quad z \in bD\,.$$

Subtracting the above from (16) we find

$$F_t(z) - f(z) = I_t(z) + II_t(z)$$

where

$$I_t(z) = \int_{w\in bD} (f(w) - f(z))\, \big[C(\Phi_t(w), z) - C(w, z)\big]\,,$$

and

$$II_t(z) = \int_{w\in bD} (f(w) - f(z))\, \big[J_t(w) - 1\big]\, C(\Phi_t(w), z),$$

To treat $I_t(z)$, we break the integral on bD into two parts: when $\mathrm{d}(w,z) \le a$ and $\mathrm{d}(w,z) \ge a$. To study the integration in w where $\mathrm{d}(w,z) \le a$, we invoke the following inequality concerning the denominators $\geq (w, z)$ that were described in section "Holomorphic Cauchy Integrals," which is a consequence of Lemma 14 in the appendix below,

$$|g(\Phi_t(w), z)| \gtrsim |g(w, z)|, \quad \text{for } w, z \in bD. \tag{19}$$

Assuming for now the truth of this inequality, we see by the Hölder-regularity of f that the integrand above is bounded by a multiple of

$$\frac{1}{\mathtt{d}(w, z)^{2n}} \mathtt{d}(w, z)^{\alpha} .$$

Thus, by (11) the integral on the set where $\mathtt{d}(w, z) \leq a$ is majorized by a multiple of

$$\int\limits_{\mathtt{d}(w,z)\leq a} \mathtt{d}(w, z)^{-2n+\alpha} \lesssim a^{\alpha} .$$

On the other hand, by the continuity of $C(w, z)$ where $\mathtt{d}(w, z) \geq a$, we see that integration over this set gives a quantity that tends to 0 uniformly as $t \to 0$. Since a can be chosen arbitrarily small, this shows that the first term $I_t(z)$ tends to 0 as $t \to 0$, uniformly in $z \in bD$. The second term $II_t(z)$ can be treated similarly to conclude that $II_t(z) \to 0$ uniformly in $z \in bD$, as well.

Combining all of the above, we conclude that

$$\sup_{z\in bD} |F_t(z) - f(z)| \to 0 \quad \text{as } t \to 0^-,$$

and the proposition is established. □

Our second characterization of $\mathcal{H}^p(bD)$ is as the range of the Cauchy transform $\mathcal{C}$.

Corollary 11. *Suppose $h \in L^p(bD, d\sigma)$. Then $h \in \mathcal{H}^p(bD, d\sigma)$ if, and only if, $h = \mathcal{C}(h)$.*

Proof. Recall that if $h \in L^p(bD, d\sigma)$ then $\mathcal{C}(h) \in \mathcal{H}^p(bD, d\sigma)$, by Proposition 8. Thus, if $h = \mathcal{C}(h)$, we have that $h \in \mathcal{H}^p(bD, d\sigma)$. Conversely if $h \in \mathcal{H}^p(bD, d\sigma)$, by the density just proved we can approximate it by a sequence $\{f_n\}$ with the property that $f_n = \dot{F}_n$ with F_n holomorphic in D and continuous in $\overline{D}$. Hence,

$$\mathcal{C}(f_n) = \mathbf{C}(\dot{F}_n)\big|_{bD} = f_n$$

where the last equality is due to the identity $\mathbf{C}(\dot{F}_n)(z) = F_n(z)$ for z in D (Proposition 9), which extends to z in $\overline{D}$ because of the continuity of F_n on $\overline{D}$. Thus

$$f_n = \mathcal{C}(f_n) \quad \text{for each } n ,$$

and the conclusion $h = \mathcal{C}(h)$ follows by the continuity of $\mathcal{C}$ in the $L^p(bD, d\sigma)$-norm. □

Comparing the Cauchy–Szegő Projection with the Cauchy Integral

Proposition 8 and Corollary 11 show that $\mathcal{C}$ is a projection: $L^p(bD, d\sigma) \to \mathcal{H}^p(bD, d\sigma)$. Thus, when $p = 2$ we may compare $\mathcal{C}$ with the Cauchy–Szegő projection $\mathcal{S}$, which is the *orthogonal* projection: $L^2(bD, d\sigma) \to \mathcal{H}^2(bD, d\sigma)$.

Proposition 12. *As operators on* $L^2(bD, d\sigma)$ *we have*

(a) $\mathcal{CS} = \mathcal{S}$
(b) $\mathcal{SC} = \mathcal{C}$

In fact, whenever $f \in L^2(bD)$, then $g = \mathcal{S}f \in \mathcal{H}^2(bD)$, by definition of $\mathcal{S}$. If we apply Corollary 11 (to $g = \mathcal{S}f$), we see that $\mathcal{C}(\mathcal{S}f) = \mathcal{C}(g) = g = \mathcal{S}f$, proving *(a)*. Next, by Proposition 8, $\mathcal{C}(f) \in \mathcal{H}^2(bD)$, for $f \in L^2(bD)$. Thus $\mathcal{SC}(f) = \mathcal{C}(f)$, which shows *(b)*, thus proving the proposition.

We point out that a corresponding version of Proposition 12 holds for the orthogonal projections: $\mathcal{S}_\omega : L^2(bD, \omega\, d\sigma) \to \mathcal{H}^2(bD, \omega\, d\sigma)$ for the densities ω discussed in the introduction (so in particular for the Leray–Levi measure).

Further Results

To conclude, we give a further characterization of $\mathcal{H}^p(bD, d\sigma)$ as the range of the Cauchy–Szegő projection $\mathcal{S}$. This uses the $L^p(bD)$-regularity of the Cauchy–Szegő projection $\mathcal{S}$, proven in [47, Theorem 16], and the approximation theorem just proved (Theorem 10).

Proposition 13. *Suppose* $f \in L^p(bD)$, $1 < p < \infty$. *Then* $\mathcal{S}(f) \in \mathcal{H}^p(bD)$. *Furthermore, we have that* $f \in \mathcal{H}^p(bD)$ *if, and only if,* $f = \mathcal{S}(f)$.

Proof. To prove the first conclusion, we consider first the case when $p \geq 2$. Then $\mathcal{S}(f) \in \mathcal{H}^2(bD) \cap L^p(bD)$ (because $L^p \subseteq L^2$ and $\mathcal{S} : L^2 \to \mathcal{H}^2(bD)$) and thus $\mathcal{S}(f) \in \mathcal{H}^p(bD)$ by Corollary 2. In the case when $p < 2$, we take $(f_n)_n \subset C^1(bD)$ with $\|f_n - f\|_p \to 0$. Then $\mathcal{S}(f_n) \in \mathcal{H}^q(bD)$ for any $q > 2$ (by the case just proved, since $f_n \in C^1(bD) \subset L^q(bD)$ and $q > 2$). But $\mathcal{H}^q(bD) \subset \mathcal{H}^p(bD)$ (because $q > 2 > p$) and so $\{\mathcal{S}(f_n)\}_n \subset \mathcal{H}^p(bD)$. Furthermore $\|\mathcal{S}(f_n - f)\|_p \lesssim \|f_n - f\|_p$ by the $L^p(bD)$-regularity of $\mathcal{S}$. We conclude that $\mathcal{S}(f) \in \mathcal{H}^p(bD)$ by the completeness of $\mathcal{H}^p(bD)$. This proves the first conclusion of the proposition. To prove the second conclusion, suppose first that $f \in \mathcal{H}^p(bD)$; then by Theorem 10 there is a sequence $\{f_n\} \subset \mathcal{H}_\vartheta(bD)$ such that $f_n \to f$ in $L^p(bD, d\sigma)$. But $\mathcal{H}_\vartheta(bD) \subset \mathcal{H}^2(bD)$ and thus $\mathcal{S}f_n = f_n$ (by the definition of $\mathcal{S}$). On the other hand, $\mathcal{S}f_n \to \mathcal{S}f$ in $L^p(bD)$ by the regularity of $\mathcal{S}$ in $L^p(bD)$. Thus $\mathcal{S}f = f$. Conversely, if $f \in L^p(bD)$ and $f = \mathcal{S}(f)$, then $f \in \mathcal{H}^p(bD)$ by the first conclusion of the proposition. □

Appendix

Consider the C^1-smooth vector field ν defined in a neighborhood of bD by

$$\nu(h)(x) = \frac{1}{|\nabla\rho(x)|^2}\sum_{j=1}^{2n}\frac{\partial\rho}{\partial x_j}(x)\,\frac{\partial h}{\partial x_j}(x) \quad x \in U, \quad h \in C^1(U).$$

Note that $\nu(\rho)(x) \equiv 1$. Let $\Phi_t(x) = \exp(t\,\nu)(x)$ be the exponential map associated with ν, defined on a small neighborhood U' of bD, for small t. One knows that

$$x \mapsto \Phi_t(x), \quad x \in U'$$

is a C^1-smooth mapping when t is small. We recall that $\Phi_t(\cdot)(x)$ arises as the solution of the time-independent differential equation

$$\frac{d}{dt}\left(\Phi_t(\cdot)(x)\right) = a\left(\Phi_t(\cdot)(x)\right) \quad \text{with} \quad a(u) = \frac{\nabla\rho(u)}{|\nabla\rho(u)|^2}, \tag{20}$$

with initial condition

$$\Phi_t(x)\Big|_{t=0} = x\,.$$

As a result,

$$\frac{d}{dt}\left(\rho(\Phi_t(x))\right) \equiv 1\,,$$

and hence $\rho(\Phi_t(x)) = \rho(x) + t = \rho_t(x)$.

The existence, uniqueness, and C^1-regularity of the solution to equations like (20) guarantee the following properties:

- $\Phi_{t_1} \circ \Phi_{t_2} = \Phi_{t_1+t_2}$ if the t_j's are small;
- $\Phi_0 =$ Identity.

Note that these properties imply that Φ_t is a C^1-bijection: $bD \to bD_t$ with inverse Φ_{-t}. Moreover $J_t(x)$ (the Jacobian determinant of Φ_t) has the following properties:

- $0 < c_1 \leq J_t(x) \leq c_2 < \infty$ uniformly in t and $x \in bD$;
- $|J_t(x) - 1| \to 0$ as $t \to 0$, uniformly in $x \in bD$;
- The following change of variable formula holds:

$$\int_{bD}\tilde{f}(\Phi_t(w))\,J_t(w)\,d\sigma(w) = \int_{bD_t}\tilde{f}(\zeta)\,d\sigma_t(\zeta) \tag{21}$$

whenever $\tilde{f}$ is integrable on bD_t.

Proof of Inequality (19)

The proof of this inequality is a consequence of the following lemma.

Lemma 14. *For $t < 0$, t small, we have*

$$|g(\Phi_t(w), z)| \approx |g(w, z)| + |t|, \quad for \quad w, z \in bD.$$

Proof. Without loss of generality we will assume that w is close to z; then, we may write $g(w, z) = \langle \partial\rho(w), w - z\rangle + Q_w(w - z)$, with $Q_w(u)$ a quadratic form in u.

Letting ν_w denote the inner unit normal vector at $w \in bD$, we claim that

$$g(w - \delta\nu_w, z) = g(w, z + \delta\nu_z) + O(\delta|w - z| + \delta^2) \tag{22}$$

when $\delta > 0$ is sufficiently small. To prove this claim, we begin by noting that

$$\begin{aligned}\langle \partial\rho(w - \delta\nu_w), w - \delta\nu_w - z\rangle &= \langle \partial\rho(w), w - \delta\nu_w - z\rangle + O(\delta|w - z| + \delta^2)\\ &= \langle \partial\rho(w), w - z - \delta\nu_z\rangle + O(\delta|w - z| + \delta^2)\,.\end{aligned}$$

One similarly has

$$Q_{w-\delta\nu_w}(w - \delta\nu_w - z) = Q_w(w - z + \delta\nu_z) + O(\delta|w - z| + \delta^2)\,,$$

which combined with the above proves the claim.

Next, we observe that the first-order Taylor expansion at $t = 0$ of the bijection $\Phi_t(w)$ (as a function of t, for fixed $w \in bD$) is

$$\Phi_t(w) = w + t\mathcal{N}_w + o(|t|) \quad \text{as } t \to 0$$

with $o(|t|)$ uniform as w ranges over bD, where

$$\mathcal{N}_w = \frac{d\Phi_t(w)}{dt}\Big|_{t=0} = \frac{\nu_w}{|\nabla\rho(w)|}\,.$$

As a result we obtain

$$g(\Phi_t(w), z) = g(w - \delta\nu_w, z) + o(\delta) \quad \text{with } \delta = -\frac{t}{|\nabla\rho(w)|} > 0 \quad \text{and small.}$$

By (22) we have

$$|g(w - \delta\nu_w, z)| = |g(w, z + \delta\nu_z)| + O(\delta|w - z| + \delta^2) + o(\delta).$$

And by Lanzani and Stein [47, Corollary 2]

$$|g(w,z+\delta v_z)| + O(\delta|w-z| + \delta^2) + o(\delta) \approx$$
$$\approx |g(w,z)| + \delta + o(\delta) + O(\delta|w-z| + \delta^2) \approx |g(w,z)| + |t|,$$

since $\delta \approx |t|$ for t, and hence δ, sufficiently small. □

Acknowledgements Loredana Lanzani is supported in part by the National Science Foundation, awards DMS-1001304 and DMS-1503612. Elias M. Stein is supported in part by the National Science Foundation, awards DMS-0901040 and DMS-1265524.

References

1. P. Ahern, R. Schneider, The boundary behavior of Henkin's kernel. Pac. J. Math. **66**, 9–14 (1976)
2. P. Ahern, R. Schneider, A smoothing property of the Henkin and Szegő projections. Duke Math. J. **47**, 135–143 (1980)
3. M. Andersson, M. Passare, R. Sigurdsson, *Complex Convexity and Analytic Functionals* (Birkhäuser, Basel, 2004)
4. D.E. Barrett, Irregularity of the Bergman projection on a smooth bounded domain. Ann. Math. **119**, 431–436 (1984)
5. D.E. Barrett, Behavior of the Bergman projection on the Diederich-Fornæworm. Acta Math. **168**, 1–10 (1992)
6. D. Barrett, L. Lanzani, The Leray transform on weighted boundary spaces for convex Reinhardt domains. J. Funct. Anal. **257**, 2780–2819 (2009)
7. D. Bekollé, A. Bonami, Inegalites a poids pour le noyau de Bergman. C. R. Acad. Sci. Paris Ser. A-B **286**(18), A775–A778 (1978)
8. S. Bell, E. Ligocka, A simplification and extension of Fefferman's theorem on biholomorphic mappings. Invent. Math. **57**(3), 283–289 (1980)
9. M. Bolt, The Möbius geometry of hypersurfaces. Mich. Math. J. **56**(3), 603–622 (2008)
10. A. Bonami, N. Lohoué, Projecteurs de Bergman et Szegő pour une classe de domaines faiblement pseudo- convexes et estimations L^p. Compos. Math. **46**(2), 159–226 (1982)
11. D. Catlin, Boundary behavior of holomorphic functions on pseudoconvex domains. J. Differ. Geom. **15**(4), 605–625 (1980)
12. P. Charpentier, Y. Dupain, Estimates for the Bergman and Szegő Projections for pseudo-convex domains of finite type with locally diagonalizable Levi forms. Publ. Mat. **50**, 413–446 (2006)
13. S.-C. Chen, M.-C. Shaw, *Partial Differential Equations in Several Complex Variables* (American Mathematical Society, Providence, 2001)
14. M. Christ, A T(b)-theorem with remarks on analytic capacity and the Cauchy integral. Colloq. Math. **60/61**(2), 601–628 (1990)
15. M. Christ, *Lectures on Singular Integral Operators* .CBMS Regional Conf. Series, vol. 77 (American Mathematical Society, Providence, 1990)
16. R. Coifman, A. McIntosh, Y. Meyer, L'intègrale de Cauchy dèfinit un opèrateur bornè sur L^2 pour les courbes lipschitziennes. Ann. Math. **116**(2), 367–387 (1982)
17. A. Cumenge, Comparaison des projecteurs de Bergman et Szegő et applications. Ark. Mat. **28**, 23–47 (1990)
18. G. David, Opérateurs intégraux singuliers sur certain courbes du plan complexe. Ann. Sci. Éc. Norm. Sup. **17**, 157–189 (1984)

19. G. David, J.L. Journé, S. Semmes, Opérateurs de Calderón-Zygmund, fonctions para-accrètives et interpolation. Rev. Mat. Iberoam. **1**(4), 1–56 (1985)
20. P.L. Duren, *Theory of H^p Spaces* (Dover, Mineola, 2000)
21. D. Ehsani, I. Lieb, L^p-estimates for the Bergman projection on strictly pseudo-convex non-smooth domains. Math. Nachr. **281**, 916–929 (2008)
22. C. Fefferman, The Bergman kernel and biholomorphic mappings of pseudo-convex domains. Invent. Math. **26**, 1–65 (1974)
23. G.B. Folland, J.J. Kohn, *The Neumann Problem for the Cauchy-Riemann Complex*. Annals of Mathematics Studies, vol. 75 (Princeton University Press, Princeton 1972)
24. N. Hanges, Explicit formulas for the Szegő kernel for some domains in $\mathbb{C}^2$. J. Funct. Anal. **88**, 153–165 (1990)
25. N. Hanges, G. Francsics, Trèves curves and the Szegő kernel. Indiana Univ. Math. J. **47**, 995–1009 (1998)
26. T. Hansson, On Hardy spaces in complex ellipsoids. Ann. Inst. Fourier (Grenoble) **49**, 1477–1501 (1999)
27. H. Hedenmalm, The dual of a Bergman space on simply connected domains. J. d' Anal. **88**, 311–335 (2002)
28. G. Henkin, Integral representations of functions holomorphic in strictly pseudo-convex domains and some applications. Mat. Sb. **78**, 611–632 (1969). Engl. Transl. Math. USSR Sb. **7**, 597–616 (1969)
29. G.M. Henkin, Integral representations of functions holomorphic in strictly pseudo-convex domains and applications to the $\overline{\partial}$-problem. Mat. Sb. **82**, 300–308 (1970). Engl. Transl. Math. USSR Sb. **11**, 273–281 (1970)
30. G.M. Henkin, J. Leiterer, *Theory of Functions on Complex Manifolds* (Birkhäuser, Basel, 1984)
31. L. Hörmander, *An Introduction to Complex Analysis in Several Variables*. North Holland Math. Library (North Holland Publishing Co., Amsterdam, 1990)
32. L. Hörmander, *Notions of Convexity* (Birkhäuser, Basel, 1994)
33. C. Kenig, Weighted $\mathbf{H}^p$ spaces on Lipschitz domains. Am. J. Math. **102**(1), 129–163 (1980)
34. C. Kenig, *Harmonic Analysis Techniques for Second Order Elliptic Boundary Value Problems*. CBMS Regional Conference Series in Mathematics, vol. 83 (American Mathematics Society, Providence RI, 1994). 0-8218-0309-3
35. N. Kerzman, E.M. Stein, The Szegö kernel in terms of the Cauchy-Fantappié kernels. Duke Math. J. **25**, 197–224 (1978)
36. K.D. Koenig, Comparing the Bergman and Szegő projections on domains with subelliptic boundary Laplacian. Math. Ann. **339**, 667–693 (2007)
37. K.D. Koenig, An analogue of the Kerzman-Stein formula for the Bergman and Szegő projections. J. Geom. Anal. **19**, 81–863 (2009)
38. K. Koenig, L. Lanzani, Bergman vs. Szegő via Conformal Mapping. Indiana Univ. Math. J. **58**(2), 969–997 (2009)
39. S. Krantz, Canonical kernels versus constructible kernels. Preprint. ArXiv: 1112.1094
40. S. Krantz, *Function Theory of Several Complex Variables*, 2nd edn. (American Mathematical Society, Providence, RI, 2001)
41. S. Krantz, M. Peloso, The Bergman kernel and projection on non-smooth worm domains. Houst. J. Math. **34**, 873–950 (2008)
42. L. Lanzani, Cauchy transform and Hardy spaces for rough planar domains, *Analysis, Geometry, Number Theory: The Mathematics of Leon Ehrenpreis* (Philadelphia, PA, 1998). Contemporary Mathematics, vol. 251 (American Mathematical Society, Providence, RI, 2000), pp. 409–428
43. L. Lanzani, E.M. Stein, Cauchy-Szegö and Bergman projections on non-smooth planar domains. J. Geom. Anal. **14**, 63–86 (2004)
44. L. Lanzani, E.M. Stein, The Bergman projection in L^p for domains with minimal smoothness. Ill. J. Math. **56**(1), 127–154 (2013)
45. L. Lanzani, E.M. Stein, Cauchy-type integrals in several complex variables. Bull. Math. Sci. **3**(2), 241–285 (2013). doi:10.1007/s13373-013-0038-y

46. L. Lanzani, E.M. Stein, The Cauchy integral in $\mathbb{C}^n$ for domains with minimal smoothness. Adv. Math. **264**, 776–830 (2014). doi:10.1016/j.aim.2014.07.016
47. L. Lanzani, E.M. Stein, The Cauchy-Szegő projection for domains in $\mathbb{C}^n$ with minimal smoothness. To appear in Duke Math. J.
48. E. Ligocka, The Hölder continuity of the Bergman projection and proper holomorphic mappings. Stud. Math. **80**, 89–107 (1984)
49. J. McNeal, Boundary behavior of the Bergman kernel function in $\mathbb{C}^2$. Duke Math. J. **58**(2), 499–512 (1989)
50. J. McNeal, Estimates on the Bergman kernel of convex domains. Adv. Math. **109**, 108–139 (1994)
51. J. McNeal, E.M. Stein, Mapping properties of the Bergman projection on convex domains of finite type. Duke Math. J. **73**(1), 177–199 (1994)
52. J. McNeal, E.M. Stein, The Szegö projection on convex domains. Math. Zeit. **224**, 519–553 (1997)
53. A. Nagel, J.-P. Rosay, E.M. Stein, S. Wainger, Estimates for the Bergman and Szegö kernels in $\mathbb{C}^2$. Ann. Math. **129**(2), 113–149 (1989)
54. D. Phong, E.M. Stein, Estimates for the Bergman and Szegö projections on strongly pseudo-convex domains. Duke Math. J. **44**(3), 695–704 (1977)
55. E. Ramirez, Ein divisionproblem und randintegraldarstellungen in der komplexen analysis. Math. Ann. **184**, 172–187 (1970)
56. M. Range, *Holomorphic Functions and Integral Representations in Several Complex Variables* (Springer, Berlin, 1986)
57. R.M. Range, An integral kernel for weakly pseudoconvex domains. Math. Ann. **356**, 793–808 (2013)
58. A.S. Rotkevich, The Aizenberg formula in nonconvex domains and some of its applications. (Russian) Zap. Nauchn. Sem. S.-Peterburg. Otdel. Mat. Inst. Steklov. (POMI) **389** (2011); Issledovaniya po Lineinym Operatoram i Teorii Funktsii. **38**, 206–231; **288**; translation in J. Math. Sci. (N.Y.) **182**(5), 699–713 (2012)
59. A.S. Rotkevich, The Cauchy-Leray-Fantappiè integral in linearly convex domains. (Russian) Zap. Nauchn. Sem. S.-Peterburg. Otdel. Mat. Inst. Steklov. (POMI) **401** (2012); Issledovaniya po Lineinym Operatoram i Teorii Funktsii **40**, 172–188; **201**; translation in J. Math. Sci. (N.Y.) **194**(6), 693–702 (2013)
60. W. Rudin, *Function Theory in the Unit Ball of* $\mathbb{C}^n$ (Springer, Berlin, 1980)
61. E.M. Stein, *Boundary Behavior of Holomorphic Functions of Several Complex Variables* (Princeton University Press, Princeton, 1972)
62. E.L. Stout, H^p-functions on strictly pseudoconvex domains. Am. J. Math. **98**(3), 821–852 (1976)
63. Y. Zeytuncu, L^p-regularity of weighted Bergman projections. Trans. AMS **365**, 2959–2976 (2013)

On the Preservation of Eccentricities of Monge–Ampère Sections

Diego Maldonado

Dedicated to the memory of Cora Sadosky

Abstract A study on the preservation of eccentricities of Monge–Ampère sections under an integral Dini-type condition on the Monge–Ampère measure is presented. The approach is based solely on $C^{2,\alpha}$-estimates for solutions to the Monge–Ampère equation. The main results are then related to the local quasi-conformal Jacobian problem and to a priori estimates for solutions to the linearized Monge–Ampère equation.

mscdot[2000]Primary 30C65, Secondary 26B25, 31B15

Introduction

The purpose of this article is to present an alternative proof of a regularity result for solutions to the Monge–Ampère equation due to Jiang and Wang in [17, pp. 611–613]. As a novelty in our proof (which sets it apart from the techniques in [15–17], for instance), no a priori C^3 or C^4-estimates are used and instead we rely on $C^{2,\alpha}$-estimates only. In addition, an effort has been made to provide full details and to quantify the role of the eccentricities of Monge–Ampère sections in each one of the a priori estimates.

The main results are the following (see Section "Preliminaries" for definitions):

Theorem 1.1. *Let $\Omega \subset \mathbb{R}^n$ be an open convex set and suppose that a strictly convex function $u \in C^2(\Omega)$ satisfies*

$$0 < \lambda \leq \det D^2 u =: f \leq \Lambda \quad in\ \Omega. \tag{1}$$

D. Maldonado (✉)
Department of Mathematics, Kansas State University, 138 Cardwell Hall, Manhattan, KS 66506, USA
e-mail: dmaldona@math.ksu.edu

M.C. Pereyra et al. (eds.), *Harmonic Analysis, Partial Differential Equations, Complex Analysis, Banach Spaces, and Operator Theory (Volume 1)*, Association for Women in Mathematics Series 4, DOI 10.1007/978-3-319-30961-3_12

Assume that the section $S_u(0,1)$ of u satisfies $S_u(0,1) \subset\subset \Omega$ and let E_0 denote its eccentricity. Then, there exists $\varepsilon_0 > 0$, depending only on λ, Λ, n, and E_0, such that the inequality

$$\sum_{k=0}^{\infty}\left(\frac{1}{|S_u(0,4^{-k})|}\int_{S_u(0,4^{-k})}|f(x)^{\frac{1}{n}}-f(0)^{\frac{1}{n}}|^n\,dx\right)^{\frac{1}{n}}<\varepsilon_0 \tag{2}$$

implies the eccentricity estimates

$$Ecc(S_u(0,4^{-k}))\le C_{13}E_0 \quad \forall k\in\mathbb{N}_0, \tag{3}$$

for a dimensional constant $C_{13} > 0$.

Theorem 1.2. *Let $\Omega \subset \mathbb{R}^n$ be an open convex set and suppose that a strictly convex function $u \in C^2(\Omega)$ satisfies (1). Assume that the section $S_u(0,1)$ of u satisfies $S_u(0,1) \subset\subset \Omega$ and let E_0 denote its eccentricity. Then, there exists $\varepsilon_1 > 0$, depending only on λ, Λ, n, and E_0, such that*

$$\sum_{k=0}^{\infty}\left(\frac{1}{|B(0,2^{-k})|}\int_{B(0,2^{-k})}|f(x)^{\frac{1}{n}}-f(0)^{\frac{1}{n}}|^n\,dx\right)^{\frac{1}{n}}<\varepsilon_1 \tag{4}$$

implies

$$B(0,\kappa_3E_0^{-n}2^{-k})\subset S_u(0,4^{-k})\le B(0,K_3E_0^{n}2^{-k}) \quad \forall k\in\mathbb{N}_0, \tag{5}$$

for some structural constants $\kappa_3, K_3 > 0$.

Corollary 1.3. *Under the hypotheses of Theorem 1.2 there is a structural constant $K_{11} > 0$ such that*

$$\|D^2u(0)\|\le K_{11}Ecc(S_u(0,1))^2. \tag{6}$$

The article is organized as follows: in Section "Preliminaries" we introduce the basic notation and the notion of eccentricity for convex sets that will be used throughout. In Section "A Priori Estimates" the main a priori estimates are proved. Built upon the classical Aleksandrov–Bakelman–Pucci and Pogorelov results, such estimates include an explicit description of the interplay between the size of the Hessian of a convex solution and the eccentricities of its sections (Lemma 3.5) and a $C^{2,\alpha}$-regularity result (Lemma 3.6). In turn, these results are proved sufficient for obtaining a comparison principle between two solutions of $\det D^2u = 1$ (Lemma 3.7). Sections "Proof of Theorem 1.1", "Proof of Theorem 1.2" and "Proof of Corollary 1.3" are devoted to the proofs of Theorem 1.1, Theorem 1.2, and Corollary 1.3, respectively. In Section "On the Local Quasi-Conformal Jacobian Problem" we relate Theorem 1.2 to the local quasi-conformal Jacobian problem and to a result by H.M. Riemann by means of local quasi-conformal mappings with convex potentials. In Section "On the Uniform Ellipticity for the Linearized Monge–Ampère Operator" we apply Corollary 1.3 to show uniform ellipticity for cofactor matrices of Hessians of solutions in terms of eccentricities of their sections.

Preliminaries

About Notation

$|A|$ as well as $\det A$ will denote the determinant of an $n \times n$ matrix A while $\|A\|$ will denote its matrix norm. Thus, either one of $\det D^2u(x)$ or $|D^2u(x)|$ will denote the determinant of the Hessian of a function u at x and $\|D^2u(x)\|$ will denote the matrix norm of $D^2u(x)$.

$B(x, r)$ will denote the Euclidean ball centered at $x \in \mathbb{R}^n$ with radius $r > 0$, $|E|$ will stand for the Lebesgue measure of a set $E \subset \mathbb{R}^n$ and we put $\omega_n := |B(0, 1)|$. $|x|$ will denote the norm of a vector $x \in \mathbb{R}^n$, and the dot product between two vectors $x, y \in \mathbb{R}^n$ will be denoted as $\langle x, y\rangle$.

Constants depending only on dimension n will be called *dimensional constants* and will be denoted as C_j's. We will also use C_n to denote a generic dimensional constant that might change from line to line. Constants depending only on λ, Λ, and n will be called *structural constants* and will be denoted as K_j's and κ_j's.

Given $\alpha \in (0, 1)$ and scalar and matrix functions w and W defined on an open set $U \subset \mathbb{R}^n$, the norms $\|w\|^*_{2,\alpha;U}$ and $\|W\|^*_{0,\alpha;U}$ are defined as

$$\|w\|^*_{2,\alpha;U} := \|w\|_{L^\infty(U)} + \sup_{x\in U} d_x|\nabla w(x)| + \sup_{x\in U} d_x^2\|D^2w(x)\| + \sup_{\substack{x,y\in U\\ x\neq y}} d_{x,y}^{2+\alpha}\frac{\|D^2w(x) - D^2w(y)\|}{|x-y|^\alpha} \tag{7}$$

and

$$\|W\|^*_{0,\alpha;U} := \|W\|_{L^\infty(U)} + \sup_{\substack{x,y\in U\\ x\neq y}} d_{x,y}^{\alpha}\frac{\|W(x) - W(y)\|}{|x-y|^\alpha}, \tag{8}$$

where $d_x := \operatorname{dist}(x, \partial U)$ and $d_{x,y} := \min\{d_x, d_y\}$.

Normalization and Eccentricity of Open, Bounded, and Convex Sets

An (open) ellipsoid $E \subset \mathbb{R}^n$ centered at $x_c \in \mathbb{R}^n$ is an open, bounded, and convex set of the form

$$E := \{x \in \mathbb{R}^n : \langle Q(x - x_c), x - x_c \rangle < 1\} \tag{9}$$

where Q is a symmetric, positive-definite $n \times n$ (real) matrix. For $\alpha > 0$ the ellipsoid αE is defined as

$$\alpha E := \{x \in \mathbb{R}^n : \langle Q(x - x_c), x - x_c \rangle < \alpha\}. \tag{10}$$

Given an open, bounded, and convex $S \subset \mathbb{R}^n$, a seminal result due to F. John establishes the existence of a unique ellipsoid E of minimum volume circumscribing S. In addition, the ellipsoid E satisfies

$$\alpha_n E \subset S \subset E$$

with $\alpha_n := 1/n$. Now, if $T : \mathbb{R}^n \to \mathbb{R}^n$ is an affine transformation satisfying $T(E) = B(0, 1)$ it follows that

$$B(0, \alpha_n) \subset T(S) \subset B(0, 1), \tag{11}$$

and T is of the form $Tx = L_T x + b$ for some symmetric, positive-definite $n \times n$ matrix L_T and some $b \in \mathbb{R}^n$. By abusing notation, we will identify T with its matrix L_T by writing $|T|$ for $|L_T|$ and T for its Jacobian L_T. Also, we will refer to the eigenvalues of L_T as the eigenvalues of T (Fig. 1).

If $S \subset \mathbb{R}^n$ is open, bounded, and convex, the *eccentricity* of S is defined as

$$Ecc(S) := |T|^{-\frac{1}{n}} \|T\| \tag{12}$$

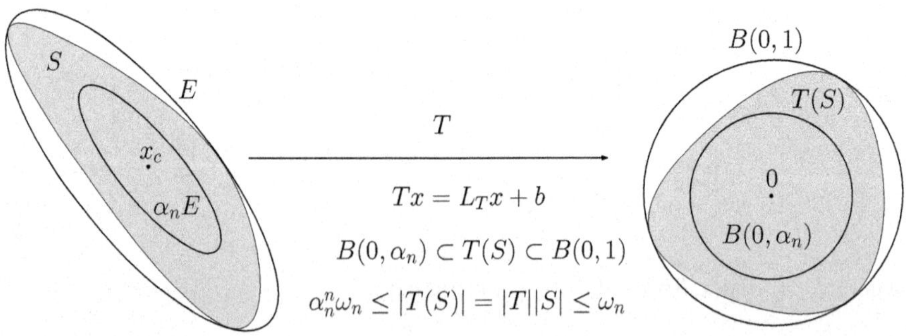

Fig. 1 Normalization of an open, bounded, and convex set $S \subset \mathbb{R}^n$ through an affine transformation T. Among all ellipsoids containing S, the ellipsoid E has minimal volume

where T normalizes S. That is, $Ecc(S)$ is comparable, with dimensional constants, to the ratio between the arithmetic and geometric means of the eigenvalues of T. Notice that $Ecc(S)$ is well-defined due to the uniqueness of T. Also, if $0 < \lambda_1 \leq \cdots \leq \lambda_n$ denote the eigenvalues of T, then $\|T\| = \lambda_n$, and if $H > 0$ satisfies $|T|^{-\frac{1}{n}}\|T\| \leq H$, then

$$\lambda_n \leq H^n \lambda_1. \tag{13}$$

If $0 < \rho_1 \leq \cdots \leq \rho_n$ denote the radii of an ellipsoid E with $\alpha_n E \subset S \subset E$ and T normalizes S, then the eigenvalues of T are $0 < \frac{1}{\rho_n} \leq \cdots \leq \frac{1}{\rho_1}$, so that $\|T\| = \frac{1}{\rho_1}$ and $\|T^{-1}\| = \rho_n = \mathrm{diam}(E)/2$ and, by (13)

$$\mathrm{diam}(S) \leq \mathrm{diam}(E) = 2\rho_n = \frac{2}{\lambda_1} \leq \frac{2Ecc(S)^n}{\lambda_n} = \frac{2Ecc(S)^n}{\|T\|}.$$

Hence,

$$\|T\| \leq \frac{2Ecc(S)^n}{\mathrm{diam}(S)}. \tag{14}$$

Monge–Ampère Sections

Given an open convex set $\Omega \subset \mathbb{R}^n$ and a convex function $u \in C^2(\Omega)$, the section of *u centered at* $x_0 \in \Omega$ and of *height* $t > 0$ is the open convex set defined by

$$S_u(x_0, t) := \{x \in \Omega : u(x) < u(x_0) + \langle \nabla u(x_0), x - x_0 \rangle + t\}. \tag{15}$$

We will use the fact that whenever $0 < \lambda \leq \det D^2 u \leq \Lambda$ is an open convex set $\Omega \subset \mathbb{R}^n$, there exist structural constants $K_1, K_2 > 0$ (that is, depending only on λ, Λ and n) such that

$$K_1 t \leq |S_u(x_0, t)|^{\frac{2}{n}} \leq K_2 t \quad \forall S_u(x_0, t) \subset\subset \Omega, \tag{16}$$

see, for instance, Corollary 3.2.4 in [12, p. 50].

Notice that if $S := S_u(x_0, t)$ is a fixed section and if E is an ellipsoid with $\alpha_n E \subset S \subset E$, from (16) we have

$$\mathrm{diam}(S)^n \geq \alpha_n^n \,\mathrm{diam}(E)^n \geq \alpha_n^n |E| \geq \alpha_n^n |S| \geq K_1^{\frac{n}{2}} t^{\frac{n}{2}},$$

so that

$$K_1 t \leq \mathrm{diam}(S_u(x_0, t))^2. \tag{17}$$

Now, given $0 < t' < t$, what can be said about the eccentricity of a section $S_u(x_0, t)$ compared to the one of $S_u(x_0, t')$? By Theorem 3.3.8 in [12, p. 57], if u satisfies $0 < \lambda \leq \det D^2 u \leq \Lambda$ (in fact, just a doubling condition suffices) there exist structural constants $\kappa_0 > 0$ and $K_0 \geq 1$ such that

$$Ecc(S_u(x_0, t')) \leq K_0 \left(\frac{t}{t'}\right)^{\kappa_0} Ecc(S_u(x_0, t)) \quad \forall t' \in (0, t). \tag{18}$$

By taking, for instance, $t' := 2^{-k}t$ for $k \in \mathbb{N}$, as we move inwards from $S_u(x_0, t)$ to $S_u(x_0, 2^{-k}t)$, eccentricities can be expected to grow exponentially in k.

The main purpose of this article is then the exploration of regularity conditions on $\det D^2 u$ that guarantee the uniform comparability

$$Ecc(S_u(x_0, t')) \lesssim Ecc(S_u(x_0, t)) \quad \forall t' \in (0, t), \tag{19}$$

where the implied constants are structural constants.

A Priori Estimates

A Version of the ABP for Open, Bounded, and Convex Sets

A version of the well-known Aleksandrov–Bakelman–Pucci maximum principle reads as follows (see Lemma 9.3 in [11, p. 222]): let $\Omega \subset \mathbb{R}^n$ be open and bounded and, for $x \in \Omega$, let $A(x)$ be a symmetric, positive-definite $n \times n$ matrix. Then, for every $g \in C^2(\Omega) \cap C(\overline{\Omega})$ we have

$$\sup_{\Omega} g \leq \sup_{\partial\Omega} g + \frac{\operatorname{diam}(\Omega)}{n\omega_n^{1/n}} \left\| \frac{\operatorname{tr}(AD^2 g)}{|A|^*} \right\|_{L^n(\Gamma^+)}, \tag{20}$$

where $|A|^* := |A|^{\frac{1}{n}}$ and Γ^+ denotes the upper contact set of g. Notice that A above need not be uniformly elliptic for (20) to hold. In the case where A is uniformly elliptic, that is, when there exist constants $0 < c_0 \leq C_0$ such that the eigenvalues of $A(x)$ belong to the interval $[c_0, C_0]$ for every $x \in \Omega$, X. Cabré [3] showed that (20) can be improved as to admit $|\Omega|^{1/n}$ instead of $\operatorname{diam}(\Omega)$, but with the L^n-norm taken in all of Ω and not just Γ^+. In this case, the relevant constants depend heavily on c_0 and C_0.

When A is just symmetric and positive-definite, Lemmas 3.1 and 3.2 below establish the validity of the ABP maximum principle (20) with $|\Omega|^{1/n}$ in place of $\operatorname{diam}(\Omega)$ provided that Ω be an open, bounded, and *convex* set. These lemmas appear as Problem 9.3 in [11, p. 255] and their short proofs are included for completeness' sake.

Lemma 3.1 (ABP for Convex Domains). *Let $\Omega \subset \mathbb{R}^n$ be open, bounded, and convex and, for every $x \in \Omega$, let $A(x)$ be a symmetric, positive-definite $n \times n$ matrix. Then, for every $g \in C^2(\Omega) \cap C(\overline{\Omega})$ we have*

$$\sup_{\Omega} g \leq \sup_{\partial\Omega} g + C_0 |\Omega|^{\frac{1}{n}} \left\| \frac{\operatorname{tr}(AD^2 g)}{|A|^*} \right\|_{L^n(\Gamma+)}, \tag{21}$$

where $C_0 := \frac{2}{n\alpha_n \omega_n^{2/n}}$, $|A|^* := |A|^{\frac{1}{n}}$, and Γ^+ denotes the upper contact set of g.

Proof. Let T be an affine transformation normalizing Ω and, for $y \in T(\Omega)$, introduce $h(y) := g(T^{-1}y)$ so that, for $x \in \Omega$ and $y = Tx$ we have

$$\begin{aligned}\operatorname{tr}(A(x)D^2 g(x)) = \operatorname{tr}(A(x)T^t D^2 h(y) T) &= \operatorname{tr}(TA(x)T^t D^2 h(y)) \\ &=: \operatorname{tr}(A_T(y) D^2 h(y)),\end{aligned}$$

where $A_T(y) := T^t A(T^{-1}y)T$ is a symmetric, positive-definite $n \times n$ matrix with $|A_T(y)| = |T|^2 |A(T^{-1}y)|$. Then, from (20) (applied to h and A_T on $T(\Omega)$, and noticing that the upper contact set of h coincides with $T(\Gamma^+)$) we have

$$\sup_{T(\Omega)} h \leq \sup_{\partial T(\Omega)} h + \frac{\operatorname{diam}(T(\Omega))}{n\omega_n^{\frac{1}{n}}} \left\| \frac{\operatorname{tr}(A_T D^2 h)}{|A_T|^*} \right\|_{L^n(T(\Gamma^+))}. \tag{22}$$

Now, since T normalizes Ω, from (11) we have $\operatorname{diam}(T(\Omega)) \leq \operatorname{diam}(B(0,1)) \leq 2$ as well as $\omega_n \alpha_n^n = |B(0,\alpha_n)| \leq |T(\Omega)| = |T||\Omega|$. Consequently,

$$\begin{aligned}\int_{T(\Gamma^+)} \left| \frac{\operatorname{tr}(A_T(y)D^2h(y))}{|A_T(y)|^*} \right|^n dy &= \frac{1}{|T|^2} \int_{T(\Gamma^+)} \left| \frac{\operatorname{tr}(A_T(y)D^2h(y))}{|A(T^{-1}y)|^*} \right|^n dy \\ = \frac{1}{|T|} \int_{\Gamma^+} \left| \frac{\operatorname{tr}(A(x)D^2g(x))}{|A(x)|^*} \right|^n dx &\leq \frac{|\Omega|}{\alpha_n^n \omega_n} \int_{\Gamma^+} \left| \frac{\operatorname{tr}(A(x)D^2g(x))}{|A(x)|^*} \right|^n dx,\end{aligned}$$

so that (21) follows from (22). □

Along the same lines, Theorem 9.1 in [11, p. 220] reads: given $\Omega \subset \mathbb{R}^n$, open and bounded; a symmetric, positive-definite $n \times n$ matrix $A(x)$ for $x \in \Omega$, and $g \in W^{2,n}_{loc}(\Omega) \cap C(\overline{\Omega})$ with $\operatorname{tr}(A(x)D^2 g(x)) \geq f(x)$ for (Lebesgue)-almost every $x \in \Omega$, we have

$$\sup_{\Omega} g \leq \sup_{\partial\Omega} g^+ + C_1^* \operatorname{diam}(\Omega) \left\| \frac{f}{|A|^*} \right\|_{L^n(\Omega)}, \tag{23}$$

for some constant $C_1^* > 0$ depending only on dimension n. Thus, assuming that Ω is open, bounded, and *convex*, we have

Lemma 3.2. *Let $\Omega \subset \mathbb{R}^n$ be open, bounded, and convex and, for every $x \in \Omega$, let $A(x)$ be a symmetric, positive-definite $n \times n$ matrix. Then, for every $g \in W^{2,n}_{loc}(\Omega) \cap C(\overline{\Omega})$ with $\operatorname{tr}(A(x)D^2g(x)) \geq f(x)$ a.e. $x \in \Omega$ for some measurable function f, we have*

$$\sup_{\Omega} g \leq \sup_{\partial\Omega} g^+ + C_1|\Omega|^{\frac{1}{n}} \left\| \frac{f}{|A|^*} \right\|_{L^n(\Omega)}, \tag{24}$$

with $C_1 := 2\alpha_n^{-1}\omega_n^{-\frac{1}{n}} C_1^*$.

Proof. The proof is similar to the one for Lemma 3.1 and, in keeping with the notation from that proof, Lemma 3.2 follows after using that for $x \in \Omega$ and $y = Tx$,

$$f(T^{-1}y) \leq \operatorname{tr}(A(x)D^2g(x)) = \operatorname{tr}(A_T(y)D^2h(y)), \quad a.e.\, y \in T(\Omega),$$

by applying (23) to h, A_T, and $f(T^{-1}\cdot)$ on $T(\Omega)$, and by changing variables from y back to x and recalling that $\operatorname{diam}(T(\Omega)) \leq 2$ and $|T|^{-1} \leq \alpha_n^{-n}\omega_n^{-1}|\Omega|$. □

A Comparison Principle

In view of Lemma 3.1 above, the next lemma stands as a version of Lemma 3.1 in [16] and Lemma 4.1 in [17], involving $|\Omega|^{1/n}$ instead of $\operatorname{diam}(\Omega)$.

Lemma 3.3. *Let $\Omega \subset \mathbb{R}^n$ be open, bounded, and convex. Let $u, v \in C^2(\Omega) \cap C(\overline{\Omega})$ be convex functions with $\det D^2u$ and $\det D^2v$ vanishing at most on set of (Lebesgue) measure zero in Ω. Then,*

$$\sup_{\Omega} |u - v| \leq \sup_{\partial\Omega} |u - v| + nC_1|\Omega|^{\frac{1}{n}} \left\| (\det D^2u)^{\frac{1}{n}} - (\det D^2v)^{\frac{1}{n}} \right\|_{L^n(\Omega)}, \tag{25}$$

where C_1 is the dimensional constant in (24).

Proof. For almost every $x \in \Omega$, let us put $f(x) := \det D^2u(x) > 0$ and $g(x) := \det D^2v(x) > 0$, and introduce the symmetric, positive-definite $n \times n$ matrices

$$A_u(x) := \frac{1}{n}|D^2u(x)|^{\frac{1}{n}} D^2u(x)^{-1} \quad \text{and} \quad A_v(x) := \frac{1}{n}|D^2v(x)|^{\frac{1}{n}} D^2u(v)^{-1}.$$

Notice that $|A_u(x)| = |A_v(x)| = n^{-n}$ and

$$\operatorname{tr}(A_uD^2u) = \det(D^2u)^{\frac{1}{n}} = f^{\frac{1}{n}}$$

$$\operatorname{tr}(A_vD^2v) = \det(D^2v)^{\frac{1}{n}} = g^{\frac{1}{n}}.$$

Now, by means of the arithmetic-geometric means inequality, we get

$$\begin{aligned} \operatorname{tr}(A_u D^2 v) &= \frac{1}{n}|D^2 u|^{\frac{1}{n}} \operatorname{tr}((D^2 u)^{-1} D^2 v) \\ &\geq |D^2 u|^{\frac{1}{n}} (|(D^2 u)^{-1}||D^2 v|)^{\frac{1}{n}} = |D^2 v|^{\frac{1}{n}} = g^{\frac{1}{n}}. \end{aligned} \tag{26}$$

Therefore, almost everywhere in Ω, we have

$$\operatorname{tr}(A_u D^2(v-u)) = \operatorname{tr}(A_u D^2 v) - \operatorname{tr}(A_u D^2 u) \geq g^{\frac{1}{n}} - f^{\frac{1}{n}}.$$

Then, by (24) applied to $v-u$, A_u, and $g^{\frac{1}{n}} - f^{\frac{1}{n}}$ on Ω,

$$\sup_{\Omega}(v-u) \leq \sup_{\partial\Omega}(v-u)^+ + nC_1|\Omega|^{\frac{1}{n}} \left\| g^{\frac{1}{n}} - f^{\frac{1}{n}} \right\|_{L^n(\Omega)}. \tag{27}$$

By interchanging the roles of u and v in (26), it similarly follows that

$$\sup_{\Omega}(u-v) \leq \sup_{\partial\Omega}(u-v)^+ + nC_1|\Omega|^{\frac{1}{n}} \left\| f^{\frac{1}{n}} - g^{\frac{1}{n}} \right\|_{L^n(\Omega)}. \tag{28}$$

Finally, (27) and (28) yield (25). □

Pogorelov's Estimates

The goal of this section is to recast the classical Pogorelov's estimates in terms of Monge–Ampère sections and their eccentricities. The following results from Sect. 4.2 of [12] will be our starting point: let $\Omega \subset \mathbb{R}^n$ be open, bounded, and convex with $B(0,\alpha_n) \subset \Omega \subset B(0,1)$ and let $u \in C^4(\Omega)$ with

$$\begin{aligned} \det D^2 u &= 1 \quad \text{in } \Omega \\ u &= 0 \quad \text{on } \partial\Omega. \end{aligned}$$

Then, there exist dimensional constants $C_3, C_4, C_5 > 0$ such that for every $\varepsilon > 0$ we have (in the sense of positive-definite matrices)

$$C(\varepsilon)^{1-n} I \leq D^2 u(x) \leq C(\varepsilon) I \quad \forall x \in \Omega_\varepsilon, \tag{29}$$

where $\Omega_\varepsilon := \{x \in \Omega : u(x) < -\varepsilon\}$ and

$$C(\varepsilon) := \frac{C_4}{\varepsilon} \exp(C_5 \varepsilon^{-2n}). \tag{30}$$

Moreover,

$$\mathrm{dist}(\Omega_\varepsilon, \partial\Omega) \geq C_3 \varepsilon^n \quad \forall \varepsilon > 0. \tag{31}$$

In the next lemma, the role of Ω in the results above will be played by normalized Monge–Ampère sections.

Lemma 3.4 (Pogorelov's Estimates on Sections). *Fix a convex domain $\Omega \subset \mathbb{R}^n$. Let $u \in C^4(\Omega)$ be convex with $\det D^2u(x) = 1$ for every $x \in \Omega$. Assume that there exists $x_0 \in \Omega$ with $\nabla u(x_0) = 0$ and given $t_0 > 0$ such that $S_u(x_0, t_0) \subset\subset \Omega$ let $T : \mathbb{R}^n \to \mathbb{R}^n$ be an affine transformation normalizing $S_u(x_0, t_0)$.*

Then, for every $\delta \in (0, 1)$ we have the estimates

$$dist(S_u(x_0, \delta t_0), \partial S_u(x_0, t_0)) \geq C_{11}(1 - \delta)^n \|T\|^{-1} \tag{32}$$

and, in the sense of positive-definite matrices,

$$F(\delta)^{1-n}|T|^{-\frac{2}{n}} T^t T \leq D^2u(x) \leq F(\delta)|T|^{-\frac{2}{n}} T^t T \quad \forall x \in S_u(x_0, \delta t_0), \tag{33}$$

where

$$F(\delta) := \frac{C_9}{(1-\delta)} \exp(C_{10}(1-\delta)^{-2n}), \tag{34}$$

and $C_9, C_{10}, C_{11} > 0$ are dimensional constants.

Proof. The contents of this lemma are illustrated in Fig. 2 and its proof follows after a change of variables based on the normalization technique; however, the behavior of the relevant constants needs to be tracked closely.

Let T be an affine transformation normalizing $S_u(x_0, t_0)$. For $y \in T(S_u(x_0, t_0))$ define

$$v(y) := |T|^{\frac{2}{n}}(u(T^{-1}y) - t_0 - u(x_0)), \tag{35}$$

so that for every $t > 0$ with $S_u(x_0, t) \subset \Omega$, setting $y_0 := Tx_0$ we have

$$S_v(y_0, t|T|^{\frac{2}{n}}) = T(S_u(x_0, t)). \tag{36}$$

Clearly, $\det D^2v = 1$ in $S_v(y_0, t_0|T|^{\frac{2}{n}})$ and, since $u(x) = t_0 + u(x_0)$ for every $x \in \partial S_u(x_0, t_0)$, we also have $v = 0$ on $\partial S_v(y_0, t_0|T|^{\frac{2}{n}})$. On the other hand, the fact that T normalizes $S_u(x_0, t_0)$ yields

$$B(0, \alpha_n) \subset T(S_u(x_0, t_0)) = S_v(y_0, t_0|T|^{\frac{2}{n}}) \subset B(0, 1)$$

and, for every $\delta \in (0, 1)$, the facts that $v(y_0) = -t_0|T|^{\frac{2}{n}}$ and $\nabla v(y_0) = 0$ give

$$\{y \in S_v(y_0, t_0|T|^{\frac{2}{n}}) : v(y) < -(1-\delta)t_0|T|^{\frac{2}{n}}\} = S_v(y_0, \delta t_0|T|^{\frac{2}{n}}). \tag{37}$$

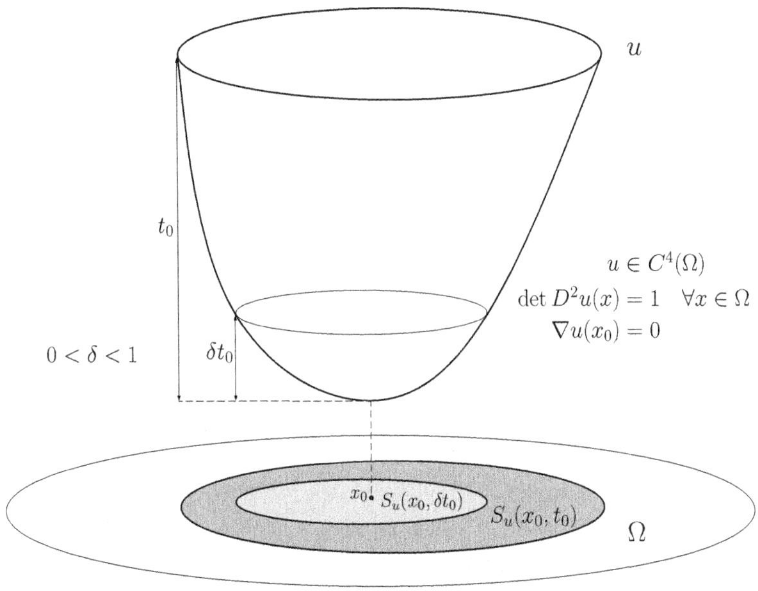

Fig. 2 Pogorelov's estimates on sections: if T normalizes $S_u(x_0, t_0)$, it follows that $D^2u(x) \simeq |T|^{-\frac{2}{n}} T^t T$ for every $x \in S_u(x_0, \delta t_0)$, with constants depending only on δ and n

By applying (29) to v on $S_v(y_0, t_0|T|^{\frac{2}{n}})$ with $\varepsilon := (1-\delta)t_0|T|^{\frac{2}{n}} > 0$ we get

$$C(\varepsilon)^{1-n} I \leq D^2 v(y) \leq C(\varepsilon) I \quad \forall y \in S_v(y_0, \delta t_0 |T|^{\frac{2}{n}}), \tag{38}$$

which, in terms of u, translates as

$$C(\varepsilon)^{1-n} I \leq |T|^{\frac{2}{n}} (T^{-1})^t D^2 u(x) T^{-1} \leq C(\varepsilon) I \quad \forall x \in S_u(x_0, \delta t_0). \tag{39}$$

Next, we relate the sizes of ε and $(1-\delta)$ in order to control $C(\varepsilon)$ in terms of δ. For this we resort to (16) which, used with v, $\lambda = \Lambda = 1$ and $S_v(y_0, t_0|T|^{\frac{2}{n}})$, yields dimensional constants $C_6, C_7 > 0$ such that

$$C_6 t_0 |T|^{\frac{2}{n}} \leq |S_v(y_0, t_0|T|^{\frac{2}{n}})|^{\frac{2}{n}} \leq C_7 t_0 |T|^{\frac{2}{n}}. \tag{40}$$

Along with the fact that, due to normalization,

$$\alpha_n^n \omega_n \leq |S_v(y_0, t_0|T|^{\frac{2}{n}})| \leq \omega_n,$$

(40) implies that

$$\alpha_n^2 \omega_n^{\frac{2}{n}} C_7^{-1} \leq t_0 |T|^{\frac{2}{n}} \leq \omega_n^{\frac{2}{n}} C_6^{-1} \tag{41}$$

and then

$$\alpha_n^2 \omega_n^{\frac{2}{n}} C_7^{-1}(1-\delta) \leq \varepsilon := (1-\delta)t_0|T|^{\frac{2}{n}} \leq \omega_n^{\frac{2}{n}} C_6^{-1}(1-\delta). \tag{42}$$

By recalling the definition of $C(\varepsilon)$ in (30) and setting $C_8 := \alpha_n^2 \omega_n^{\frac{2}{n}} C_7^{-1}$, we get the bound

$$C(\varepsilon) < \frac{C_4}{C_8(1-\delta)} \exp(C_5(C_8(1-\delta))^{-2n}) =: F(\delta)$$

and (34) follows with $C_9 := C_8^{-1}C_4$ and $C_{10} := C_5 C_8^{-2n}$. By using (31), we can write

$$\begin{aligned} C_3 C_8^n (1-\delta)^n \leq C_3 \varepsilon^n &\leq \operatorname{dist}(S_v(y_0, \delta t_0|T|^{\frac{2}{n}}), \partial S_v(y_0, t_0|T|^{\frac{2}{n}})) \\ &= \operatorname{dist}(T(S_u(x_0, \delta t_0)), \partial T(S_u(x_0, t_0))) \\ &\leq \|T\| \operatorname{dist}(S_u(x_0, \delta t_0), \partial S_u(x_0, t_0)), \end{aligned}$$

and (32) follows with $C_{11} := C_3 C_8^n$. □

Inequality (33) quantifies the interplay between the eccentricity of a section of u and the size of its Hessian. Let us explicitly state this interplay as follows:

Lemma 3.5 (The Interplay Between Hessian and Eccentricity). *Fix a convex domain $\Omega \subset \mathbb{R}^n$. Let $u \in C^4(\Omega)$ be convex with $\det D^2 u(x) = 1$ for every $x \in \Omega$. Assume that there exists $x_0 \in \Omega$ with $\nabla u(x_0) = 0$ and given $t_0 > 0$ such that $S_u(x_0, t_0) \subset\subset \Omega$ let T be an affine transformation normalizing $S_u(x_0, t_0)$.*

For every $\delta \in (0,1)$ and $z \in S_u(x_0, \delta t_0)$ we have

$$Ecc(S_u(x_0, t_0))^2 \leq F(\delta)^{n-1} \|D^2 u(z)\| \tag{43}$$

and

$$F(\delta)^{1-n} \|D^2 u(z)\| \leq Ecc(S_u(x_0, t_0))^2 \tag{44}$$

where $F(\delta)$ is as in (34).

Proof. Since Pogorelov's estimate (33) gives $|T|^{-\frac{2}{n}} T^t T \leq F(\delta)^{n-1} D^2 u(z)$ in the sense of positive-definite matrices, (43) follows immediately. □

The next lemma is based on Theorem 4.2.1 in [12, p. 67] and it makes explicit the role of eccentricities in interior $C^{2,\alpha}$-estimates for solutions of $\det D^2 u = 1$.

Lemma 3.6 (An Interior $C^{2,\alpha}$-estimate). *Fix a convex domain $\Omega \subset \mathbb{R}^n$. Let $u \in C^4(\Omega)$ be convex with $\det D^2 u(x) = 1$ for every $x \in \Omega$. Assume that there exists $x_0 \in \Omega$ with $\nabla u(x_0) = 0$ and given $t_0 > 0$ such that $S_u(x_0, t_0) \subset\subset \Omega$ and T be an affine transformation normalizing $S_u(x_0, t_0)$.*

For every $\delta \in (0,1)$ there exist constants $\alpha \in (0,1)$ and $H(\delta,n) > 0$, depending only on δ and n, such that for every $x, y \in S_u(x_0, \delta t_0)$ with $x \neq y$, we have

$$d_{x,y}^{\alpha} \frac{\|D^2u(x) - D^2u(y)\|}{|x-y|^{\alpha}} \leq H(\delta,n) Ecc(S_u(x_0,t_0))^{n\alpha}. \tag{45}$$

Consequently, (45) *and* (44) *imply that*

$$\left\|D^2u\right\|^*_{0,\alpha;(S_u(x_0,\delta t_0))} \leq F(\delta)^{n-1} Ecc(S_u(x_0,t_0))^2 + H(\delta,n) Ecc(S_u(x_0,t_0))^{n\alpha}. \tag{46}$$

Proof. In keeping with the notation from the proof of Lemma 3.4, let v be defined as in (35). Thus, v satisfies

$$\begin{aligned} \det D^2 v &= 1 \quad \text{in } S_v(y_0, t_0|T|^{\frac{2}{n}}) \\ v &= 0 \quad \text{on } \partial S_v(y_0, t_0|T|^{\frac{2}{n}}), \end{aligned}$$

with $B(0,\alpha_n) \subset S_v(y_0, t_0|T|^{\frac{2}{n}}) \subset B(0,1)$ and $y_0 := Tx_0$. Given $\delta \in (0,1)$ fix δ' with $0 < \delta < \delta' < 1$. Then the inequalities (38) imply that D^2v is uniformly elliptic in $S_v(y_0, \delta' t_0|T|^{\frac{2}{n}})$ with eigenvalues bounded between $F(\delta')^{1-n}$ and $F(\delta')$ (here $F(\delta')$ is as in (34)). Consequently, the nonlinear equation $G(D^2v) := \log(\det D^2v) = 0$ is uniformly elliptic with

$$\frac{\partial G}{v_{ij}} = v^{ij} \quad \forall i,j = 1,\ldots,n,$$

where the v^{ij}'s denote the entries of the matrix $(D^2v)^{-1}$ whose eigenvalues lie between $F(\delta')^{-1}$ and $F(\delta')^{n-1}$ in $S_v(y_0, \delta' t_0|T|^{\frac{2}{n}})$. Also, since G is concave, by L.C. Evan's Hölder estimate (see inequality (17.41) in [11, p. 456]) there are constants $C(\delta',n) > 0$ and $\alpha(\delta',n) \in (0,1)$, depending only on n and $F(\delta')$, such that for every Euclidean ball $B_{R_0} \subset S_v(y_0, \delta' t_0|T|^{\frac{2}{n}})$ and every $0 < R \leq R_0$, it holds true that

$$\operatorname{osc}_{B_R} D^2 v \leq C(\delta',n) \left(\frac{R}{R_0}\right)^{\alpha(\delta',n)} \operatorname{osc}_{B_{R_0}} D^2 v. \tag{47}$$

By Corollary 3.3.6(i) in [12, p. 55] we have

$$\operatorname{dist}(S_u(x_0,\delta t_0), \partial S_u(x_0,\delta' t_0)) \geq C_{14}(1-\delta/\delta')^n \delta' \|T\|^{-1}, \tag{48}$$

for some dimensional constant $C_{14} > 0$. Define $R_0' := C_{14}(1-\delta/\delta')^n\delta'\|T\|^{-1}/2$ and given $x_1, x_2 \in S_u(x_0, \delta t_0)$ with $R' := |x_1 - x_2| \leq R_0'$,(48) guarantees that $B(x_1, R') \subset B(x_1, R_0') \subset S_v(y_0, \delta' t_0|T|^{\frac{2}{n}})$. Let us also define

$$R := |Tx_1 - Tx_2| \leq \|T\||x_1 - x_2| = \|T\|R' \leq \|T\|R_0' =: R_0,$$

so that

$$\|D^2u(x_1) - D^2u(x_2)\| \le |T|^{-2/n}\|T\|^2\|D^2v(Tx_1) - D^2v(Tx_2)\|$$

$$\le \operatorname{osc}_{B(Tx_1,R)} D^2v \le C(\delta', n)\left(\frac{R}{R_0}\right)^{\alpha(\delta',n)} \operatorname{osc}_{B(Tx_1,R_0)} D^2v$$

$$\le 2F(\delta')C(\delta', n)\left(\frac{R}{R_0}\right)^{\alpha(\delta',n)} \le 2F(\delta')C(\delta', n)\left(\frac{|x_1 - x_2|}{R_0'}\right)^{\alpha(\delta',n)}.$$

On the other hand, we have

$$d_z := \operatorname{dist}(z, \partial S_u(x_0, t_0)) \le 2\|T\|^{-1} \quad \forall z \in S_u(x_0, \delta t_0),$$

because for every $z' \in \partial S_u(x_0, t_0)$ we can write

$$|z - z'| = |T^{-1}(Tz) - T^{-1}(Tz')| \le \|T^{-1}\||Tz - Tz'| \le 2\|T^{-1}\|$$

where we have used that $T(S_u(x_0, t_0)) \subset B(0, 1)$. Therefore, for every $x_1, x_2 \in S_u(x_0, \delta t_0)$,

$$d_{x_1,x_2}^{\alpha(\delta',n)}\|D^2u(x_1) - D^2u(x_2)\| \le (2\|T^{-1}\|)^{\alpha(\delta',n)}2F(\delta')C(\delta', n)\left(\frac{|x_1 - x_2|}{R_0'}\right)^{\alpha(\delta',n)}$$

$$= (2\|T^{-1}\|)^{\alpha(\delta',n)}2F(\delta')C(\delta', n)\left(\frac{2|x_1 - x_2|}{C_{14}(1 - \delta/\delta')^n\delta'\|T\|^{-1}}\right)^{\alpha(\delta',n)}$$

$$=: H(\delta, \delta', n)(\|T\|\|T^{-1}\|)^{\alpha(\delta',n)}|x_1 - x_2|^{\alpha(\delta',n)}$$

$$\le H(\delta, \delta', n)Ecc(S_u(x_0, t_0))^{n\alpha(\delta',n)}|x_1 - x_2|^{\alpha(\delta',n)},$$

after fixing, for instance, $\delta' := (1 + \delta)/2$, (45) follows. □

A Comparison Between Two Solutions of $\det D^2u = 1$

Lemma 3.7. *Fix a convex domain $\Omega \subset \mathbb{R}^n$. Let $u, v \in C^4(\Omega)$ be convex solutions of $\det D^2u(x) = \det D^2v(x) = 1$ for every $x \in \Omega$. Assume that there exists $x_0 \in \Omega$ with $\nabla u(x_0) = 0$ and given $t_0 > 0$ such that $S_u(x_0, t_0) \subset\subset \Omega$ let $T : \mathbb{R}^n \to \mathbb{R}^n$ be an affine transformation normalizing $S_u(x_0, t_0)$.*

Suppose that for some $\alpha \in (0, 1)$ and $N > 0$ we have

$$\|D^2u\|^*_{0,\alpha;(S_u(x_0,t_0))} + \|D^2v\|^*_{0,\alpha;(S_u(x_0,t_0))} \le N. \tag{49}$$

Then there exists a constant $M^* > 0$, depending only on N, α, and n, such that

$$\|u - v\|^*_{2,\alpha;S_u(x_0,t_0)} \le M^* \|u - v\|_{L^\infty(S_u(x_0,t_0))}. \tag{50}$$

In particular, given $\delta \in (0, 1)$, there exists a constant M, depending only on N, α, and n (in fact, M is just a dimensional multiple of M^*) such that

$$\left\|D^2u - D^2v\right\|_{L^\infty(S_u(x_0,\delta t_0))} \le M \frac{Ecc(S_u(x_0, t_0))^{2n}}{(1-\delta)^{2n} t_0} \|u - v\|_{L^\infty(S_u(x_0,t_0))}. \tag{51}$$

Proof. For $t \in [0, 1]$, set

$$A_t(x) := D^2u(x)(1 - t) + tD^2v(x) \quad \forall x \in \Omega, \tag{52}$$

so that A_t is a symmetric, positive-definite $n \times n$ matrix and, by Minkowski's inequality,

$$|A_t(x)|^{1/n} \ge (1 - t)|D^2u(x)|^{1/n} + t|D^2v(x)|^{1/n} = 1 \quad \forall x \in \Omega. \tag{53}$$

On the other hand,

$$\begin{aligned} 0 = |D^2u| - |D^2v| = \int_0^1 \frac{d|A_t|}{dt}\, dt &= \int_0^1 |A_t| \operatorname{tr}(A_t^{-1}D^2(u - v))\, dt \\ &=: \operatorname{tr}(AD^2(u - v)), \end{aligned}$$

where $A(x) := \int_0^1 |A_t(x)|A_t(x)^{-1}\, dt$. The point will be to show that $A(x)$ is uniformly elliptic in $S_u(x_0, t_0)$ and that the norm $\|A\|_{0,\alpha;S_u(x_0,t_0)}$ can be controlled in terms of N and n and then use Schauder's estimates on $u - v$.

For each fixed $t \in [0, 1]$, notice that (49) implies

$$\|A_t(x)\| \le (1 - t)\|D^2u(x)\| + t\|D^2v(x)\| \le N \quad \forall x \in S_u(x_0, t_0) \tag{54}$$

which then gives

$$\|A\|_{L^\infty(S_u(x_0,t_0))} \le N. \tag{55}$$

Next, for a fixed $t \in [0, 1]$, let $0 < a_1(x) \le \cdots \le a_n(x)$ denote the eigenvalues of $A_t(x)$. From (54) we have $a_n(x) \le N$ and, from (53),

$$1 \le |A_t(x)| = \prod_{j=1}^n a_j(x) \le a_1(x)N^{n-1} \quad \forall x \in S_u(x_0, t_0)$$

which then yields $N^{1-n} \le a_1(x) \le a_n(x) \le N$ and

$$N^{-1} \le \frac{1}{a_n(x)} \le \frac{1}{a_1(x)} \le N^{n-1} \quad \forall x \in S_u(x_0, t_0). \tag{56}$$

Consequently, the eigenvalues of $|A_t(x)|A_t(x)^{-1}$ can be bounded as follows:

$$N^{-1} \le \frac{|A_t(x)|}{a_n(x)} \le \frac{|A_t(x)|}{a_1(x)} \le N^{2n-1} \quad \forall x \in S_u(x_0, t_0). \tag{57}$$

Notice that the definition of A_t in (52) and (49) implies

$$\|A_t\|^*_{0,\alpha;S_u(x_0,t_0)} \le N \quad \forall t \in [0,1] \tag{58}$$

and, since the norm $\|\cdot\|_{0,\alpha,S_u(x_0,t_0)}$ is sub-multiplicative and $|A_t|$ is a homogeneous polynomial of degree n in the coefficients of A_t, we get

$$\||A_t|\|^*_{0,\alpha,S_u(x_0,t_0)} \le C_n N^n \quad \forall t \in [0,1]. \tag{59}$$

From (53) it follows that $|A_t|^{-1} \le 1$, so that

$$\begin{aligned} ||A_t(x)|^{-1} - |A_t(y)|^{-1}| &= |A_t(x)|^{-1}|A_t(y)|^{-1}||A_t(x)| - |A_t(y)|| \\ &\le ||A_t(x)| - |A_t(y)|| \quad \forall x, y \in S_u(x_0, t_0), \end{aligned}$$

and then

$$\left\| |A_t|^{-1} \right\|^*_{0,\alpha;(S_u(x_0,t_0)} \le C_n N^n \quad \forall t \in [0,1]. \tag{60}$$

Now, using the fact that $A_t^{-1} = |A_t|^{-1}\mathrm{adj}(A_t)$, and $\mathrm{adj}(A_t)$ is a polynomial of degree $n-1$ in the entries of A_t, it follows that

$$\left\| A_t^{-1} \right\|^*_{0,\alpha;S_u(x_0,t_0)} \le C_n N^{2n-1} \quad \forall t \in [0,1]. \tag{61}$$

Consequently, we have obtained that

$$\left\| |A_t| A_t^{-1} \right\|^*_{0,\alpha;S_u(x_0,t_0)} \le C_n N^{3n-1} \quad \forall t \in [0,1] \tag{62}$$

so that the matrix $A(x)$ satisfies

$$\|A\|^*_{0,\alpha;S_u(x_0,t_0)} \le C_n N^{3n-1}. \tag{63}$$

In addition, (57) implies that the ratio between the largest and smallest eigenvalue of $A(x)$ is bounded by N^{2n}. Next, Schauder's estimates (see, for instance, Theorem 6.2 in [11, p. 90]) yield

$$\|u - v\|^*_{2,\alpha;(S_u(x_0,t_0))} \le M^* \|u - v\|_{L^\infty(S_u(x_0,t_0))} \tag{64}$$

where $M^* > 0$ depends only on the ratio between the largest and smallest eigenvalue of $A(x)$ (in our case bounded by N^2) and the C^α-norm of A [bounded as in (63)] as well as on α and n.

Finally, given $\delta \in (0, 1)$, (64) implies that

$$\sup_{x \in S_u(x_0, \delta t_0)} d_x^2 \|D^2 u(x) - D^2 v(x)\| \leq M^* \, \|u - v\|_{L^\infty(S_u(x_0, t_0))} \tag{65}$$

and by (32) we get

$$\text{dist}(S_u(x_0, \delta t_0), \partial S_u(x_0, t_0))^{-2} \leq \frac{\|T\|^2}{C_{11}^2 (1-\delta)^{2n}},$$

in turn, by (14),

$$\|T\|^2 \leq \frac{4Ecc(S_u(x_0, t_0))^{2n}}{\text{diam}(S_u(x_0, t_0))^2} \leq \frac{4Ecc(S_u(x_0, t_0))^{2n}}{C_{12} t_0},$$

where for the last inequality we used (17) with the function u, which satisfies $\det D^2 u = 1$, hence its corresponding K_1, K_2 as in (16) are just dimensional constants. Thus, (50) follows with $M := 4M^*(C_{11}^2 C_{12})^{-1}$. □

Proof of Theorem 1.1

By subtracting a hyperplane we can assume that $u(0) = 0$ and $\nabla u(0) = 0$ (which implies $u \geq 0$). Suppose first that $f(0) = 1$ and for $k \in \mathbb{N}_0$ define u_k as the convex solution to

$$\begin{aligned} \det D^2 u_k &= 1 && \text{in } S_u(0, 4^{-k}) \\ u_k &= 4^{-k} && \text{on } \partial S_u(0, 4^{-k}). \end{aligned}$$

Since u satisfies

$$\begin{aligned} \det D^2 u &= f && \text{in } S_u(0, 4^{-k}) \\ u &= 4^{-k} && \text{on } \partial S_u(0, 4^{-k}), \end{aligned}$$

the comparison principle (25) gives

$$\begin{aligned} \|u - u_k\|_{L^\infty(S_u(0,4^{-k}))} &\leq nC_1 |S_u(0, 4^{-k})|^{\frac{1}{n}} \left\| f^{\frac{1}{n}} - 1 \right\|_{L^n(S_u(0,4^{-k}))} \\ &= nC_1 |S_u(0, 4^{-k})|^{\frac{2}{n}} \left(\frac{1}{|S_u(0, 4^{-k})|} \int_{S_u(0,4^{-k})} |f(x)^{\frac{1}{n}} - 1|^n \, dx \right)^{\frac{1}{n}} \end{aligned}$$

and, by defining

$$\beta(k) := nC_1K_2\left(\frac{1}{|S_u(0,4^{-k})|}\int_{S_u(0,4^{-k})}|f(x)^{\frac{1}{n}}-1|^n\,dx\right)^{\frac{1}{n}}\quad\forall k\in\mathbb{N}_0, \tag{66}$$

with K_2 as in (16), we obtain the bound

$$\|u-u_k\|_{L^\infty(S_u(0,4^{-k}))}\le 4^{-k}\beta(k)\quad\forall k\in\mathbb{N}_0. \tag{67}$$

Next, we relate the sections of u_k and u.

Lemma 4.1. *For $k\in\mathbb{N}_0$ let x_k denote the minimum of u_k in $S_u(0,4^{-k})$ and set $t_k := 4^{-k}-u_k(x_k)$. Then, we have that*

$$S_u(0,4^{-k}) = S_{u_k}(x_k,t_k)\quad\forall k\in\mathbb{N}_0 \tag{68}$$

and

$$0\in S_{u_k}(x_k,4^{-k+\frac{1}{2}}\beta(k))\quad\forall k\in\mathbb{N}_0. \tag{69}$$

Moreover, given $\varepsilon>0$ and $k\in\mathbb{N}_0$, we have

$$4^{-k+\frac{1}{2}}\beta(k)\le\varepsilon t_k, \tag{70}$$

and then

$$0\in S_{u_k}(x_k,\varepsilon t_k), \tag{71}$$

provided that $\beta(k)<\varepsilon/(2+\varepsilon)$. Also,

$$S_{u_k}(x_k,4^{-(k+1)})\subset\subset S_u(0,4^{-k})\subset\subset S_{u_k}(x_k,\tfrac{11}{8}4^{-k}), \tag{72}$$

provided that $\beta(k)\le 3/8$. In addition, given $0<\varepsilon<\gamma$ we have that

$$\varepsilon 4^{-k}\le\gamma t_k, \tag{73}$$

whenever $\beta(k)\le(\gamma-\varepsilon)/\gamma$, and that

$$\varepsilon t_k\le\gamma 4^{-k}, \tag{74}$$

whenever $\beta(k)\le(\gamma-\varepsilon)/\varepsilon$.

Proof. The proof of (68) follows from the facts that u and u_k coincide on $\partial S_u(0,4^{-k})$ and that $\nabla u_k(x_k)=0$. In particular, (68) implies that $0\in S_{u_k}(x_k,t_k)$ for every $k\in\mathbb{N}_0$. In order to prove (69) we need to show that

$$u_k(0)\le u_k(x_k)+4^{-k+\frac{1}{2}}\beta(k)\quad\forall k\in\mathbb{N}_0, \tag{75}$$

which follows from the inequalities

$$\begin{aligned} u_k(0) - u_k(x_k) &= u_k(0) - u(0) + u(x_k) - u_k(x_k) - u(x_k) \\ &\le u_k(0) - u(0) + u(x_k) - u_k(x_k) \le 4^{-k}\beta(k) + 4^{-k}\beta(k), \end{aligned}$$

where we have used that $u(0) = 0$, $u \ge 0$, and (67).

Notice that (70) means

$$\varepsilon u_k(x_k) + 4^{-k+\frac{1}{2}}\beta(k) \le \varepsilon 4^{-k}.$$

Since x_k is the minimum of u_k over $S_{u_k}(x_k, t_k)$ and $0 \in S_{u_k}(x_k, t_k)$, by (67) and the fact that $u(0) = 0$, we have

$$u_k(x_k) \le u_k(0) = u_k(0) - u(0) \le 4^{-k}\beta(k).$$

Hence,

$$\varepsilon u_k(x_k) + 4^{-k+\frac{1}{2}}\beta(k) \le \varepsilon 4^{-k}\beta(k) + 4^{-k+\frac{1}{2}}\beta(k) \le \frac{\varepsilon}{\varepsilon+2}4^{-k}(\varepsilon+2) = \varepsilon 4^{-k},$$

which proves (70) [and then (71)]. In order to prove the first inclusion in (72), given $y \in S_{u_k}(x_k, 4^{-(k+1)})$, which means $u_k(y) < u_k(x_k) + 4^{-(k+1)}$, we need to show that $u(y) < 4^{-k}$. For this we write

$$\begin{aligned} u(y) &= u(y) - u_k(y) + u_k(y) - u_k(x_k) + u_k(x_k) \\ &< u(y) - u_k(y) + 4^{-(k+1)} + u_k(0) \\ &= u(y) - u_k(y) + 4^{-(k+1)} + u_k(0) - u(0) \\ &\le 4^{-k}\beta(k) + 4^{-(k+1)} + 4^{-k}\beta(k) \le 4^{-k}, \end{aligned}$$

where we used that $u_k(x_k) \le u_k(0)$, $u(0) = 0$, (67), and $\beta(k) \le 3/8$. By virtue of (68) the second inclusion in (72) amounts to showing that

$$t_k := 4^{-k} - u_k(x_k) < \tfrac{11}{8}4^{-k}. \tag{76}$$

But, since $u \ge 0$ and $\beta(k) \le 3/8$, (67) yields

$$\begin{aligned} 4^{-k} - u_k(x_k) &= 4^{-k} - u_k(x_k) + u(x_k) - u(x_k) \\ &\le 4^{-k} - u_k(x_k) + u(x_k) \le 4^{-k}(1 + \beta(k)) < \tfrac{11}{8}4^{-k}, \end{aligned}$$

and (76) follows. In order to prove (73), which means

$$\gamma u_k(x_k) \le (\gamma - \varepsilon)4^{-k},$$

we just use (67) and the fact that $u_k(x_k) \le u_k(0)$. Indeed,

$$\gamma u_k(x_k) \le \gamma u_k(0) = \gamma(u_k(0) - u(0)) \le \gamma 4^{-k}\beta(k) \le (\gamma - \varepsilon)4^{-k}.$$

The proof of (74) follows from the fact that $u \ge 0$ and (67)

$$t_k := 4^{-k} - u_k(x_k) \le 4^{-k} - u_k(x_k) + u(x_k) \le 4^{-k} + 4^{-k}\beta(k) \le \frac{\gamma}{\varepsilon}4^{-k}.$$

□

Now we relate the section $S_{u_k}(x_k, t_k)$ to contractions of $S_{u_{k+1}}(x_{k+1}, t_{k+1})$.

Lemma 4.2. *Given* $1/4 < \gamma \le 1$ *and* $k \in \mathbb{N}_0$, *then*

$$S_{u_{k+1}}(x_{k+1}, t_{k+1}) \subset S_{u_k}(x_k, \gamma t_k), \tag{77}$$

provided that

$$\beta(k), \beta(k+1) \le \frac{4\gamma - 1}{4(1-\gamma) + 5}. \tag{78}$$

Proof. For $y \in S_{u_{k+1}}(x_{k+1}, t_{k+1})$, which means

$$u_{k+1}(y) < u_{k+1}(x_{k+1}) + t_{k+1}, \tag{79}$$

and we need to show that

$$u_k(y) < u_k(x_k) + \gamma t_k. \tag{80}$$

We start by using (79) and (67) to write

$$\begin{aligned} u_k(y) &= u_k(y) - u_{k+1}(y) + u_{k+1}(y) < u_k(y) - u_{k+1}(y) + u_{k+1}(x_{k+1}) + t_{k+1} \\ &= u_k(x_k) - u_k(x_k) + u_k(y) - u_{k+1}(y) + u_{k+1}(x_{k+1}) + t_{k+1}, \end{aligned}$$

and (80) will follow after proving that

$$-u_k(x_k) + u_k(y) - u_{k+1}(y) + u_{k+1}(x_{k+1}) + t_{k+1} \le \gamma t_k. \tag{81}$$

By using that $t_k = 4^{-k} - u_k(x_k)$ and $t_{k+1} = 4^{-(k+1)} - u_{k+1}(x_{k+1})$, (81) is equivalent to

$$(\gamma - 1)u_k(x_k) + u_k(y) - u_{k+1}(y) \le (\gamma - 1/4)4^{-k}. \tag{82}$$

Now, using that $u \ge 0$, we get

$$\begin{aligned} &(\gamma - 1)u_k(x_k) + u_k(y) - u_{k+1}(y) \\ &\le (\gamma - 1)(u_k(x_k) - u(x_k)) + u_k(y) - u(y) + u(y) - u_{k+1}(y) \\ &\le (1-\gamma)4^{-k}\beta(k) + 4^{-k}\beta(k) + 4^{-(k+1)}\beta(k+1). \end{aligned}$$

Putting $\hat{\beta}_k := \max\{\beta(k), \beta(k+1)\}$ and by means of (78) we then obtain

$$(\gamma - 1)u_k(x_k) + u_k(y) - u_{k+1}(y) \leq 4^{-(k+1)}[4(1-\gamma)\hat{\beta}_k + 5\hat{\beta}_k] \leq 4^{-(k+1)}(4\gamma - 1),$$

which proves (82). Finally, notice that, from the definition of $\beta(k)$ in (66) and (16) we have

$$\beta(k+1) \leq 2(K_1/K_2)^{\frac{1}{2}}\beta(k) \quad \forall k \in \mathbb{N}_0, \tag{83}$$

so that $\hat{\beta}_k \leq 2(K_1/K_2)^{\frac{1}{2}}\beta(k)$ for every $k \in \mathbb{N}_0$.

Let us introduce

$$S_\beta := \sum_{k=0}^{\infty} \beta(k) \tag{84}$$

and notice that, by assuming the condition

$$S_\beta \leq \frac{1}{6}\left(\frac{K_2}{K_1}\right)^{1/2}, \tag{85}$$

we have $\beta(k+1) < 1/3$ and $\beta(k) < 1/6$ for every $k \in \mathbb{N}_0$ as well as the following properties:

$$t_k^{-1}4^{-k} \leq 2 \quad \forall k \in \mathbb{N}_0 \tag{86}$$

after (73) with $\gamma = 2$ and $\varepsilon = 1$,

$$S_{u_{k+1}}(x_{k+1}, t_{k+1}) \subset S_{u_k}(x_k, 3t_k/4) \quad \forall k \in \mathbb{N}_0 \tag{87}$$

after (77) with $\gamma = 3/4$, and

$$0 \in S_{u_k}(x_k, t_k/2) \quad \forall k \in \mathbb{N}_0 \tag{88}$$

by (71) with $\varepsilon = 1/2$. Now, since the convex function u_0 satisfies

$$\begin{aligned} \det D^2 u_0 &= 1 \quad \text{in } S_u(0,1) \\ u_0 &= 1 \quad \text{on } \partial S_u(0,1), \end{aligned}$$

by setting

$$D_0 := \|D^2 u_0(0)\| \tag{89}$$

Pogorelov's estimate (43) applied to u_0 in $S_{u_0}(x_0, t_0)$ with $\delta = 3/4$ and $2D_0$ (instead of just D_0) gives

$$Ecc(S_u(x_0, t_0))^2 \le 2D_0 F(3/4)^{n-1}. \tag{90}$$

On the other hand, due to (68) and the fact that $S_{u_k}(x_k, t_k) = S_u(0, 4^{-k})$ for every $k \in \mathbb{N}_0$, (18) yields

$$Ecc(S_{u_{k+1}}(x_{k+1}, t_{k+1}))^2 \le K_8\, Ecc(S_{u_k}(x_k, t_k))^2 \quad \forall k \in \mathbb{N}_0, \tag{91}$$

for some structural constant $K_8 \ge 1$. Thus, from (90) and (91) applied with $k = 0$, we get the following estimate for the eccentricity of $S_{u_1}(x_1, t_1)$:

$$Ecc(S_{u_1}(x_1, t_1))^2 \le K_8\, Ecc(S_{u_0}(x_0, t_0))^2 \le 2D_0 K_8 F(3/4)^{n-1}.$$

By the interior $C^{2,\alpha}$-estimate (46) applied to u_0 and u_1, we have

$$\|D^2 u_0\|^*_{0,\alpha;S_{u_0}(x_0, 3t_0/4)} + \|D^2 u_1\|^*_{0,\alpha;S_{u_1}(x_1, 3t_1/4)} \le N_0, \tag{92}$$

for a constant $N_0 > 0$ depending only on $(2D_0 K_8 F(3/4)^{n-1})^{1/2}$ (which bounds the eccentricities of both $S_u(x_0, t_0)$ and $S_{u_1}(x_1, t_1)$) and n. Now, by (67),

$$\|u_1 - u_0\|_{L^\infty(S_u(0, 4^{-1}))} \le \beta(0) + 4^{-1}\beta(1),$$

and then Lemma 3.7 applied to u_1, u_0 on $S_{u_1}(x_1, 3t_1/4)$ (taking, for instance, that $\Omega := S_{u_1}(x_1, t_1)$ in Lemma 3.7) gives

$$\|D^2 u_0 - D^2 u_1\|_{L^\infty(S_{u_1}(x_1, t_1/2))} \le M_0 t_1^{-1}(\beta(0) + 4^{-1}\beta(1)) \tag{93}$$

where M_0 depends only on n and $Ecc(S_{u_1}(x_1, 3t_1/4))$ (and the latter is bounded by $K_8(2D_0 F(3/4)^{n-1})^{1/2}$). In particular,

$$\|D^2 u_0(0) - D^2 u_1(0)\| \le M_0 t_1^{-1}(\beta(0) + 4^{-1}\beta(1)). \tag{94}$$

Consequently, by the definition of D_0 in (89), by assuming that

$$S_\beta \le \frac{D_0}{10M_0} \tag{95}$$

we get

$$\begin{aligned}\|D^2 u_1(0)\| &\le M_0 t_1^{-1}(\beta(0) + 4^{-1}\beta(1)) + \|D^2 u_0(0)\| \\ &\le M_0(8\beta(0) + 2\beta(1)) + D_0 \le 10M_0 S_\beta + D_0 \le 2D_0.\end{aligned}$$

That is, we obtain the same bound (namely, $2D_0$) for both $\|D^2u_0(0)\|$ and $\|D^2u_1(0)\|$, which [through Pogorelov's (43)] gives the same bound for the eccentricities of $S_u(x_0,t_0)$ and $S_{u_1}(x_1,t_1)$, and we are ready to iterate the reasoning above. That is,

$$Ecc(S_u(x_1,t_1))^2 \le 2D_0F(3/4)^{n-1}, \tag{96}$$

and using (91) with $k=1$,

$$Ecc(S_{u_2}(x_2,t_2))^2 \le K_8\, Ecc(S_{u_1}(x_1,t_1))^2 \le 2D_0K_8F(3/4)^{n-1},$$

and for the same N_0 as in (92)

$$\|u_2\|_{C^{2,\alpha}(S_{u_2}(x_2,3t_2/4)} + \|u_1\|_{C^{2,\alpha}(S_{u_1}(x_1,3t_1/4))} \le N_0,$$

so that, by (67),

$$\begin{aligned}\|u_1-u_2\|_{L^\infty(S_u(0,4^{-2}))} &\le \|u_1-u\|_{L^\infty(S_u(0,4^{-1}))} + \|u_2-u_0\|_{L^\infty(S_u(0,4^{-2}))}\\ &\le 4^{-1}\beta(1)+4^{-2}\beta(2),\end{aligned}$$

and with the same M_0 as in (93),

$$\left\|D^2u_2-D^2u_1\right\|_{C^{2,\alpha}(S_{u_2}(x_2,t_2/2))} \le M_0t_2^{-1}(4^{-1}\beta(1)+4^{-2}\beta(2)).$$

In particular, by using (86)

$$\|D^2u_2(0)-D^2u_1(0)\| \le M_0t_2^{-1}(4^{-1}\beta(1)+4^{-2}\beta(2)) \le M_0(8\beta(1)+2\beta(2)),$$

which yields

$$\begin{aligned}\|D^2u_2(0)\| &\le \|D^2u_1(0)-D^2u_2(0)\| + \|D^2u_0(0)-D^2u_1(0)\| + \|D^2u_0(0)\|\\ &\le M_0(8\beta(1)+2\beta(2)+8\beta(0)+2\beta(1))+D_0\\ &\le 10M_0S_\beta + D_0 \le 2D_0.\end{aligned}$$

At the kth step, by (86) we obtain

$$\begin{aligned}\|D^2u_k(0)-D^2u_0(0)\| &\le \sum_{j=1}^{k}\|D^2u_j(0)-D^2u_{j-1}(0)\|\\ &\le M_0\sum_{j=1}^{k} t_j^{-1}(4^{-(j-1)}\beta(j-1)+4^{-j}\beta(j))\\ &\le M_0\sum_{j=1}^{k}(8\beta(j-1)+2\beta(j)) \le 10M_0\sum_{j=0}^{k}\beta(j),\end{aligned} \tag{97}$$

and then

$$\|D^2u_k(0)\| \le 10M_0S_\beta + D_0 \le 2D_0 \quad \forall k \in \mathbb{N}_0,$$

which yields the following uniform bound for the eccentricities of $S_u(0, 4^{-k})$

$$Ecc(S_u(0, 4^{-k}))^2 \le 2D_0F(3/4)^{n-1} \quad \forall k \in \mathbb{N}_0. \tag{98}$$

By Pogorelov's (44) we have that $D_0 := \|D^2u_0(0)\| \le F(3/4)^{n-1}Ecc(S_u(0,1))^2$, so that (98) yields

$$Ecc(S_u(0, 4^{-k}))^2 \le 2F(3/4)^{2(n-1)}Ecc(S_u(0,1))^2 \quad \forall k \in \mathbb{N}_0$$

and (3) follows with $C_{13} := 2F(3/4)^{2(n-1)}$.

In the case when $f(0) \neq 1$ we apply the result to $v := u/f(0)^{1/n}$ and notice that

$$S_u(0, \Lambda^{-1/n}4^{-k}) \subset S_v(0, 4^{-k}) = S_u(0, 4^{-k}f(0)^{-1/n}) \subset S_u(0, \lambda^{-1/n}4^{-k}) \tag{99}$$

and that

$$|(f(x)/f(0))^{1/n} - 1| = \frac{1}{f(0)^{1/n}}|f(x)^{1/n} - f(0)^{1/n}| \le \frac{1}{\lambda^{1/n}}|f(x)^{1/n} - f(0)^{1/n}|$$

and, by recalling the definition of $\beta(k)$ in (66), the definition of S_β in (84) and the conditions (85), (95) imposed on it, we define the ε_0 in (2) as

$$\varepsilon_0 := \frac{\lambda^{1/n}}{nC_1K_2}\min\left\{\frac{D_0}{10M_0}, \frac{1}{6}\left(\frac{K_2}{K_1}\right)^{1/2}\right\} \tag{100}$$

which concludes the proof of Theorem 1.1. □

Proof of Theorem 1.2

For $k \in \mathbb{N}_0$ let us write $E_k := Ecc(S_u(0, 4^{-k}))$. From (18) we have

$$E_1^2 \le K_8E_0^2 \tag{101}$$

and, in order to start the iteration process and obtain a similar estimate for E_2, it is enough to ask for $\beta(0) < 1/6$ and $\beta(1) < 1/3$, with $\beta(k)$ as in (66).

Now, if Ell_k is an ellipsoid centered at c_k with radii $0 < \rho_{k,1} \le \cdots \le \rho_{k,n}$ such that $\alpha_nEll_k \subset S_u(0, 4^{-k}) \subset Ell_k$, by (13) and (98) we have

$$\omega_n\alpha_n^n\rho_{k,1}^n \le |\alpha_nEll_k| \le |S_u(0, 4^{-k})| \le |Ell_k| \le \omega_n\rho_{k,n}^n \le \omega_nE_k^{n^2}\rho_{k,1}^n \tag{102}$$

which, combined with (16), yields

$$\frac{K_1^{1/2}}{\omega_n^{1/n}}2^{-k} \leq \rho_{k,n} \leq \frac{K_2^{1/2}E_k^n}{\omega_n^{1/n}\alpha_n}2^{-k} \quad \forall k \in \mathbb{N}_0. \tag{103}$$

Thus, putting $G_k := \frac{2K_2^{1/2}E_k^n}{\omega_n^{1/n}\alpha_n}$, we get $2\rho_{k,n} \leq G_k 2^{-k}$. On the other hand, we also have

$$B(c_k, \alpha_n \rho_{k,1}) \subset \alpha_n Ell_k \subset S_u(0, 4^{-k}) \subset Ell_k \subset B(c_k, \rho_{k,n}) \tag{104}$$

and, since $0 \in S_u(0, 4^{-k})$, we get $0 \in B(c_k, \rho_{k,n})$ and then $B(c_k, \rho_{k,n}) \subset B(0, 2\rho_{k,n})$. Therefore,

$$S_u(0, 4^{-k}) \subset B(0, 2\rho_{k,n})$$

and, due to (102)and (103),

$$\begin{aligned} |S_u(0,4^{-k})| &\geq \omega_n \alpha_n^n \rho_{k,1}^n \geq \omega_n \alpha_n^n \rho_{k,n}^n E_k^{-n^2} \geq \alpha_n^n K_1^{n/2} E_k^{-n^2} 2^{-kn} \\ &= \frac{\alpha_n^n K_1^{n/2}}{\omega_n E_k^{2n^2}} |B(0, E_k^n 2^{-k})| = \alpha_n^{2n} 2^{-n} \left(\frac{K_1}{K_2}\right)^{\frac{n}{2}} |B(0, G_k 2^{-k})| \end{aligned}$$

so that

$$\begin{aligned} &\left(\frac{1}{|S_u(0,4^{-k})|}\int_{S_u(0,4^{-k})} |f(x)^{\frac{1}{n}} - f(0)^{\frac{1}{n}}|^n \, dx\right)^{\frac{1}{n}} \\ &\leq 2\alpha_n^{-2} E_k^n \left(\frac{K_2}{K_1}\right)^{\frac{1}{2}} \left(\frac{1}{|B(0,G_k 2^{-k})|}\int_{B(0,G_k 2^{-k})} |f(x)^{\frac{1}{n}} - f(0)^{\frac{1}{n}}|^n \, dx\right)^{\frac{1}{n}} \end{aligned}$$

and we can take

$$\varepsilon_1 := \frac{\alpha_n^2}{2E_k^n}\left(\frac{K_1}{K_2}\right)^{\frac{1}{2}} \varepsilon_0. \tag{105}$$

Proof of Corollary 1.3

For $k \in \mathbb{N}_0$ let us put $\tau_k := 4^{-k}$, $S_k := S_u(0, 4^{-k})$, and let T_k be an affine transformation normalizing S_k. Then, by estimates on the averages of $\|D^2 u\|$ over its sections (see, for instance, [8, Lemma 3.2] or [19, p. 97]) there exists a structural constant $K_9 > 0$ such that

$$\frac{1}{|S_k|}\int_{S_k} \|D^2 u(x)\| \, dx \leq K_9 \tau_k \|T_k\|^2 \quad \forall k \in \mathbb{N}_0. \tag{106}$$

Since T_k normalizes S_k, by (16) we have

$$\omega_n \geq |S_k||T_k| \geq K_1^{\frac{n}{2}} \tau_k^{\frac{n}{2}} |T_k| \quad \forall k \in \mathbb{N}_0,$$

and, from the definition of eccentricity, it follows that

$$\frac{1}{|S_k|} \int_{S_k} \|D^2 u(x)\| \, dx \leq K_9 K_1^{-1} \omega_n^{\frac{2}{n}} |T_k|^{-\frac{2}{n}} \|T_k\|^2 =: K_{10} Ecc(S_u(0, 4^{-k}))^2,$$

for every $k \in \mathbb{N}_0$, where $K_{10} := K_9 K_1^{-1} \omega_n^{\frac{2}{n}}$. Therefore, (3) yields

$$\frac{1}{|S_k|} \int_{S_k} \|D^2 u(x)\| \, dx \leq K_{10} C_{13}^2 Ecc(S_u(0, 1))^2 \quad \forall k \in \mathbb{N}_0. \tag{107}$$

Finally, the fact that hypothesis (1) guarantees a structure of local quasi-metric space with the sections of u acting as "quasi-metric balls" (see, for instance, [1, 9, 10]), the fact that Lebesgue measure is doubling with respect to the sections of every convex function (see [6, Lemma 5.2]), and Lebesgue's differentiation theorem applied to (107) yields (6) with $K_{11} := K_{10} C_{13}^2$. □

On the Local Quasi-Conformal Jacobian Problem

Fix $0 < \lambda \leq \Lambda < \infty$ and let $f : \Omega \to \mathbb{R}$ be a continuous function with

$$0 < \lambda \leq f(x) \leq \Lambda \quad \forall x \in \Omega \tag{108}$$

and let $\omega_{f,\Omega}$ denote the modulus of continuity of f over Ω, that is,

$$\omega_{f,\Omega}(t) := \sup\{|f(x) - f(y)| : |x - y| < t, \ x, y \in \Omega\}.$$

Burago and Kleiner [2] and McMullen [20] have constructed examples of uniformly continuous functions $f : [0, 1]^2 \to \mathbb{R}^2$ satisfying (108) such that *there is no bi-Lipschitz mapping* $F : [0, 1]^2 \to \mathbb{R}^2$ verifying the equality

$$\det DF(x) = f(x) \tag{109}$$

a.e. $x \in [0, 1]^2$. Therefore, the condition $\omega_{f,\Omega}(t) \to 0$ as $t \to 0^+$ is not enough to locally realize f as the Jacobian of a bi-Lipschitz (or quasi-conformal) map and some rate of decay for $\omega_{f,\Omega}$ should be prescribed towards that end.

In this section we apply Theorem 1.2 to the solvability of the local quasi-conformal Jacobian problem with *convex potentials*. That is, in this case the sought mapping F can be written as $F = \nabla u$, where $u : B(0, 1) \to \mathbb{R}$ is a strictly convex function.

Theorem 7.1. *Suppose that* $f : B(0,4) \to \mathbb{R}$, $n \geq 2$, *satisfies*

$$0 < \lambda \leq f(x) \leq \Lambda, \quad a.e.\ x \in B(0,4), \tag{110}$$

with $f^{\frac{1}{n}}$ *satisfying the following uniform* L^n*-Dini condition:*

$$D_0 := \sup_{y \in B(0,1)} \int_0^{1/2} \left\| f^{\frac{1}{n}}(\cdot) - f(y)^{\frac{1}{n}} \right\|_{L^n(B(y,r))} \frac{dr}{r} < \infty. \tag{111}$$

Then, there exists a quasi-conformal mapping $F : B(0,1) \to \mathbb{R}^n$ *such that the equality*

$$\det DF(x) = f(x), \quad a.e.\ x \in B(0,1), \tag{112}$$

holds true. Moreover, there is a strictly convex, differentiable function $u : B(0,1) \to \mathbb{R}$ *such that* $F = \nabla u$ *in* $B(0,1)$.

Remark 7.2. Notice that if $F = \nabla u$, then (112) becomes the Monge–Ampère equation $\det D^2 u = f$. Theorem 7.1 is sharp in the sense that counterexamples in [22] show that if $\det D^2 u(x) = f(x)$, a.e. $x \in B(0,1)$, then condition (110) alone does not imply, in general, that ∇u is locally quasi-conformal (see comments after Corollary 3.8 in [18]). Also, an example in [7, Theorem 1.2] shows that the condition (111) alone does not guarantee the local quasi-conformality of ∇u.

Let u be a convex solution, in the *Aleksandrov sense*, of the boundary value problem

$$\begin{aligned} \det D^2 u &= f \quad \text{in } B(0,2) \\ u &= 1 \quad \text{on } \partial B(0,2). \end{aligned} \tag{113}$$

Recall that u being a solution to (113) in the Aleksandrov sense means that for every Borel set $E \subset\subset B(0,1)$ we have

$$|\partial_u(E)| = \int_E f(x)\, dx, \tag{114}$$

where ∂_u is the normal mapping of u and $|\cdot|$ stands for Lebesgue measure (see Chap. 1 in [12]). If u is differentiable, then $\partial_u(E) = \nabla u(E)$. In our case u will in fact be differentiable, by virtue of Caffarelli's $C^{1,\alpha}$-regularity theorem (see [5, Theorem 2], [10, Corollary 13], or [12, Sect. 5.4]). Here $\alpha \in (0,1)$ depends only on n and Λ/λ. Also, by [4, Corollary 4], u is strictly convex in $B(0,1)$.

The main task is then proving the quasi-conformality (in $B(0,1)$) of ∇u for solutions u to (114). Once quasi-conformality is established, the change of variables $x \mapsto \nabla u(x)$ (see Theorem 33.3 in [21]) in (114) implies that given any Euclidean ball $B \subset B(0,1)$, we have

$$\int_B f(x)\,dx = |\nabla u(B)| = \int_{\nabla u(B)} 1\,dx = \int_B \det D^2u(x)\,dx, \tag{115}$$

and Lebesgue's differentiation theorem yields $\det D^2u(x) = f(x)$, a.e. $x \in B(0,1)$, as required in Theorem 7.1.

Following Definition 2.1 in [18], we say that u has *has round sections* in Ω if there is a constant $\tau \in (0,1)$ (which might depend on Ω) such that for all $x_0 \in \Omega$ and $t > 0$ with $S_u(x_0,t) \subset\subset \Omega$ there exists $R > 0$ verifying

$$B(x_0,\tau R) \subset S_u(x_0,t) \subset B(x_0,R). \tag{116}$$

The local quasi-conformality of ∇u will be a consequence of Theorem 3.1 in [18], where Kovalev and the author proved the equivalence between ∇u being quasi-conformal and u having round sections. This equivalence is quantitative, in the sense that the constants involved depend only on each other and not on u, see also Remark 3.4 in [18]. In turn, the roundedness of the sections of u follows by taking any $x_0 \in B(0,1/2)$ and applying Theorem 1.2 to the function

$$u_{x_0}(x) := u(x+x_0) \quad \forall x \in B(0,1/2),$$

and considering that, due to [12, Theorem 3.3.4], given $x, x_0 \in B(0,1/2)$ the eccentricities of $S_u(x,1/2)$ and $S_u(x_0,1/2)$ are uniformly comparable (with structural constants) to that of $S_u(x_m, 1-u(x_m))$, where x_m is the minimum of u on $B(0,2)$. By (5), the roundedness of the sections of u follows with τ depending only on structural constants and $Ecc(S_u(x_m,1-u(x_m)))$. □

On the Uniform Ellipticity for the Linearized Monge–Ampère Operator

We close this article with a remark on how to use the L^n-Dini condition in (111) to turn the matrix

$$A_u(x) := \det D^2u(x)\,D^2u(x)^{-1} \quad \forall x \in \Omega,$$

into a uniformly elliptic matrix on every section $S := S_u(x_0,t) \subset S_u(x_0,2t) \subset\subset \Omega$, with ellipticity constants controlled by structural constants and $Ecc(S)$.

Suppose that $u \in C^2(\Omega)$ is a strictly convex function with

$$0 < \lambda \leq \det D^2 u(x) =: f(x) \leq \Lambda \quad \forall x \in \Omega. \tag{117}$$

Given $S := S_u(x_0, t) \subset S_u(x_0, 2t) \subset\subset \Omega$, again by [12, Theorem 3.3.4], for every $x_1 \in S_u(x_0, t/2)$ we have that

$$Ecc(S_u(x_1, t/2)) \simeq Ecc(S_u(x_0, t))$$

with comparability constants depending only on λ, Λ, and n. Now, for each fixed $x_1 \in S_u(x_0, t/2)$, an application of Theorem 1.2 to the function

$$u_{x_1}(x) := u(x + x_1) \quad \forall x \in S_u(x_0, t/2)$$

yields (as in Corollary 1.3)

$$\|D^2 u(x_1)\| \lesssim Ecc(S_u(x_0, t))^2 \quad \forall x_1 \in S_u(x_0, t/2),$$

where the implied constants are structural constants. By means of (117), this implies that the eigenvalues of $D^2 u(x_1)$ lie between $\lambda Ecc(S_u(x_0, t))^{2(1-n)}$ and $Ecc(S_u(x_0, t))^2$ for every $x_1 \in S_u(x_0, t/2)$. Thus making $\mathcal{A}_u$ uniformly elliptic on $S_u(x_0, t/2)$.

The Monge–Ampère operator linearized at a function $u \in C^2(\Omega)$, denoted by $\mathcal{L}_u$, acts as follows:

$$\mathcal{L}_u(w) := \text{trace}(\mathcal{A}_u D^2 w). \tag{118}$$

Regularity properties of solutions to $\mathcal{L}_u(w) = h$ continue to receive attention and, for instance, in [13, 14], Gutiérrez and Nguyen established interior estimates for first and second derivatives of solutions to $\mathcal{L}_u(w) = h$. As an example of such regularity results, the main result in [14] (Theorem 1.1) establishes that for every $p > 1$ and $q > \max\{p, n\}$, solutions to $\mathcal{L}_u(w) = h$ satisfy

$$\left\|D^2 w\right\|_{L^p(\Omega')} \leq C(\|w\|_{L^\infty(\Omega)} + \|h\|_{L^q(\Omega)}), \tag{119}$$

where $C > 0$ depends only on p, q, λ, Λ, n, $\text{dist}(\Omega', \partial\Omega)$, and the modulus of continuity of $f := \det D^2 u$. Furthermore, Theorem 1.1 in [14] is stated with $\Omega = S_u(x_m, t_m)$ for some $x_m \in \mathbb{R}^n$ and $t_m > 0$ (that is, Ω in Theorem 1.1 in [14] is a section of u).

By our comments above, we see that if $f^{\frac{1}{n}}$ satisfies the L^n-Dini continuity condition (111), then $\mathcal{L}_u$ turns into a uniformly elliptic operator on every section $S \subset \Omega'$ with ellipticity constants depending on structural constants and $Ecc(S)$. Also, by [12, Theorem 3.3.4] every chain of sections $S_1, S_2, \ldots S_N$ of the same height and with $S_j \cap S_{j+1} \neq \emptyset, j = 1, \ldots, N-1$, will have comparable eccentricities (with comparability constants depending on structural constants as well as on N).

In particular, under the L^n-Dini continuity condition (111) on $f^{\frac{1}{n}}$, by the comments above on the uniform ellipticity of $\mathcal{A}_u$ and the continuity estimates for D^2u in [17, Theorem 1 and Lemma 4.1], (119) can be improved to

$$\left\|D^2 w\right\|_{L^p(\Omega')} \le C(\|w\|_{L^p(\Omega)} + \|h\|_{L^p(\Omega)}), \tag{120}$$

for every $1 < p < \infty$, see, for instance, Theorem 9.11 from [11, p. 235].

Acknowledgements The author would like to thank Stefania Marcantognini, Cristina Pereyra, Alex Stokolos, and Wilfredo Urbina for the opportunity to take part in this volume to honor Cora Sadosky. Thanks also go to the anonymous referees for useful comments that improved the presentation. Author supported by the National Science Foundation under grant DMS 1361754.

References

1. H. Aimar, L. Forzani, R. Toledano, Balls and quasi-metrics: a space of homogeneous type modeling the real analysis related to the Monge-Ampère equation. J. Fourier Anal. Appl. **4**, 377–381 (1998)
2. D. Burago, B. Kleiner, Separated nets in Euclidean space and Jacobians of bi-Lipschitz maps. Geom. Funct. Anal. **8**, 273–282 (1998)
3. X. Cabré, On the Aleksandrov-Bakelman-Pucci estimate and the reversed Hölder inequality for solutions of elliptic and parabolic equations. Commun. Pure Appl. Math. **48**, 539–570 (1995)
4. L.A. Caffarelli, A localization property of viscosity solutions to the Monge-Ampère equation and their strict convexity. Ann. Math. (2) **131**(1), 129–134 (1990)
5. L.A. Caffarelli, Some regularity properties of solutions of Monge-Ampère equation. Commun. Pure Appl. Math. **44**, 965–969 (1991)
6. L.A. Caffarelli, C.E. Gutiérrez, Properties of the solutions of the linearized Monge-Ampère equation. Am. J. Math. **119**(2), 423–465 (1997)
7. P. Daskalopoulos, O. Savin, On Monge-Ampère equations with homogeneous right hand side. Commun. Pure Appl. Math. **62**(5), 639–676 (2008)
8. G. De Philippis, A. Figalli, $W^{2,1}$ regularity for solutions of the Monge-Ampère equation. Invent. Math. **192**(1), 55–69 (2013)
9. L. Forzani, D. Maldonado, On geometric characterizations for Monge-Ampère doubling measures. J. Math. Anal. Appl. **275**(2), 721–732 (2002)
10. L. Forzani, D. Maldonado, Properties of the solutions to the Monge-Ampère equation. Nonlinear Anal. **57**(5–6), 815–829 (2004)
11. D. Gilbarg, N. Trudinger, *Elliptic Partial Differential Equations of Second Order* (Springer, Berlin, 2001)
12. C.E. Gutiérrez, *The Monge-Ampère Equation* (Birkhäuser, Basel, 2001)
13. C.E. Gutiérrez, T. Nguyen, Interior gradient estimates for solutions to the linearized Monge-Ampère equation. Adv. Math. **228**, 2034–2070 (2011)
14. C.E. Gutiérrez, T. Nguyen, Interior second derivative estimates for solutions to the linearized Monge-Ampère equation. Trans. Am. Math. Soc. **367**(7), 4537–4568 (2015)
15. Q. Huang, On the mean oscillation of the Hessian of solutions to the Monge-Ampère equation. Adv. Math. **207**, 599–616 (2006)
16. Q. Huang, Sharp regularity results on second derivatives of solutions to the Monge-Ampère equation with VMO type data. Commun. Pure Appl. Math. **53**, 677–705 (2009)
17. H. Jian, X.-J. Wang, Continuity estimates for the Monge-Ampère equation. SIAM J. Math. Anal. **39**(2), 608–626 (2007)

18. L.V. Kovalev, D. Maldonado, Mappings with convex potentials and the quasiconformal Jacobian problem. Ill. J. Math. **49**(4), 1039–1060 (2005)
19. D. Maldonado, On the $W^{2,1+\varepsilon}$-estimates for the Monge-Ampère equation and related real analysis. Calc. Var. **50**, 94–114 (2014)
20. C.T. McMullen, Lipschitz maps and nets in Euclidean space. Geom. Funct. Anal. **8**, 304–314 (1998)
21. J. Väisälä, *Lectures on n-Dimensional Quasiconformal Mappings*. Lecture Notes in Math., vol. 229 (Springer, Berlin, 1971)
22. X.-J. Wang, Some counterexamples to the regularity of Monge-Ampère equations. Proc. Am. Math. Soc. **123**(3), 841–845 (1995)

BMO: Oscillations, Self-Improvement, Gagliardo Coordinate Spaces, and Reverse Hardy Inequalities

Mario Milman

Para Corita

Abstract A new approach to classical self improving results for *BMO* functions is presented. "Coordinate Gagliardo spaces" are introduced and a generalized version of the John-Nirenberg Lemma is proved. Applications are provided.

Introduction and Background

Interpolation theory provides a framework, as well as an arsenal of tools, that can help in our understanding of the properties of function spaces and the operators acting on them. Conversely, the interaction of the abstract theory of interpolation with concrete function spaces can lead to new general methods and results. In this note we consider some aspects of the interaction between interpolation theory and *BMO*, focusing on the self improving properties of *BMO* functions.

To fix the notation, in this section we shall consider functions defined on a fixed cube, $Q_0 \subset R^n$. A prototypical example of the self-improvement exhibited by *BMO* functions is the statement that a function in *BMO* automatically belongs to all L^p spaces, $p < \infty$,

$$BMO \subset \bigcap_{p \geq 1} L^p. \tag{1}$$

In fact, *BMO* is contained in the Orlicz space e^L. This is one of the themes underlying the John-Nirenberg Lemma [46]. One way to obtain this refinement is to make explicit the rates of decay of the family of embeddings implied by (1).

M. Milman (✉)
Instituto Argentino de Matematica, Buenos Aires, Argentina
e-mail: mario.milman@gmail.com https://sites.google.com/site/mariomilman

M.C. Pereyra et al. (eds.), *Harmonic Analysis, Partial Differential Equations, Complex Analysis, Banach Spaces, and Operator Theory (Volume 1)*, Association for Women in Mathematics Series 4, DOI 10.1007/978-3-319-30961-3_13

We consider in detail an inequality apparently first shown in [19],

$$\|f\|_{L^q} \leq C_n q \, \|f\|_{L^p}^{p/q} \, \|f\|_{BMO}^{1-p/q}, \, 1 \leq p < q < \infty. \tag{2}$$

With (2) at hand we can, for example, extrapolate by the Δ-method of [44], and the exponential integrability of BMO functions follows (cf. (33) below)

$$\|f\|_{e^L} \sim \sup_{q>1} \frac{\|f\|_{L^q}}{q} \leq c_n \, \|f\|_{BMO} \, . \tag{3}$$

More generally, for compatible Banach spaces, interpolation inequalities of the form

$$\|f\|_X \leq c(\theta) \, \|f\|_{X_1}^{1-\theta} \, \|f\|_{X_2}^{\theta}, \; \theta \in (0,1), \tag{4}$$

where $c(\theta)$ are constants that depend only on θ, play an important role in analysis. What is needed to extract information at the end points (e.g., by "extrapolation" [44]) is to have good estimates of the rate of decay $c(\theta)$, as θ tends to 0 or to 1. We give a brief summary of inequalities of the form (4) for the classic methods of interpolation in section "Interpolation Theory: Some Basic Inequalities" below. For example, a typical interpolation inequality of the form (4) for the Lions–Peetre real interpolation spaces can be formulated as follows. Given a compatible pair[1] of Banach spaces $\vec{X} = (X_1, X_2)$, the "K-functional" (cf. [9], [78])is defined for $f \in \Sigma(\vec{X}) = X_1 + X_2, t > 0$, by

$$K(t,f;\vec{X}) := K(t,f;X_1,X_2) = \inf_{f=f_1+f_2, f_i \in X_i} \{\|f_1\|_{X_1} + t \, \|f_2\|_{X_2}\}. \tag{5}$$

The real interpolation spaces $\vec{X}_{\theta,q}$ can be defined through the use of the K-functional. Let $\theta \in (0,1), 0 < q \leq \infty$, then we let

$$\vec{X}_{\theta,q} = \{f \in \Sigma(\vec{X}) : \|f\|_{\vec{X}_{\theta,q}} < \infty\}, \tag{6}$$

where[2]

$$\|f\|_{\vec{X}_{\theta,q}} = \left\{ \int_0^\infty \left[t^{-\theta} K(t,f;\vec{X}) \right]^q \frac{dt}{t} \right\}^{1/q}.$$

We have (cf. Lemma 2 below) that, for $f \in X_1 \cap X_2, 0 < \theta < 1, 1 \leq q \leq \infty$,

$$[(1-\theta)\theta q]^{1/q} \, \|f\|_{(X_1,X_2)_{\theta,q}} \leq \|f\|_{X_1}^{1-\theta} \, \|f\|_{X_2}^{\theta} \, . \tag{7}$$

[1] We refer to section "Interpolation Theory: Some Basic Inequalities" for more details.

[2] With the usual modification when $q = \infty$.

We combine (7), the known real interpolation theory of *BMO* (cf. [10, Theorem 6.1], [9] and the references therein), sharp reverse Hardy inequalities (cf. [80] and [69]), and the re-scaling of inequalities via the reiteration method (cf. [12, 39]) to give a new *interpolation* proof of (2) in Lemma 3 below.

Let us now recall how the study of *BMO* led to new theoretical developments in interpolation theory.[3]

A natural follow-up question to (3) was to obtain the best possible integrability condition satisfied by *BMO* functions. The answer was found by Bennett–DeVore–Sharpley [11]. They showed the inequality[4]

$$\|f\|_{L(\infty,\infty)} := \sup_t \{f^{**}(t) - f^*(t)\} \leq c_n \|f\|_{BMO} . \tag{8}$$

The refinement here is that the (nonlinear) function space $L(\infty,\infty)$, defined by the condition

$$\|f\|_{L(\infty,\infty)} < \infty,$$

is strictly contained[5] in e^L.

In their celebrated work, Bennett–DeVore–Sharpley [11] proposed the following connection between real interpolation, weak interpolation, and *BMO* (cf. [9, p. 384]). The K-functional for the pair (L^1, L^∞) is given by (cf. [9] and section "Extrapolation of Inequalities: Burkholder–Gundy–Herz Meet Calderón–Maz'ya and Cwikel et al." below)

$$K(t,f;L^1,L^\infty) = \int_0^t f^*(s)ds.$$

Therefore, $\frac{dK(t,f;L^1,L^\infty)}{dt} = K'(t,f;L^1,L^\infty) = f^*(t)$; consequently, we can compute the "norm" of weak $L^1 := L(1,\infty)$, as follows:

$$\begin{aligned}\|f\|_{L(1,\infty)} &= \sup_{t>0} tf^*(t) \\ &= \sup_{t>0} tK'(t,f;L^1,L^\infty).\end{aligned}$$

[3]Paradoxically, except for section "Bilinear Interpolation", in this paper we do not discuss interpolation theorems per se. For interpolation theorems involving *BMO* type of spaces there is a large literature. For articles that are related to the developments in this note I refer, for example, to [11, 36, 43, 55, 76, 84].

[4]Where f^* denotes the non-increasing rearrangement of f and $f^{**}(t) = \frac{1}{t}\int_0^t f^*(s)ds$.

[5]The smallest rearrangement invariant space that contains *BMO* is e^L as was shown by Pustylnik [79].

Then, in analogy with the definition of weak L^1, Bennett–DeVore–Sharpley proceeded to define $L(\infty,\infty)$ using the functional

$$\|f\|_{L(\infty,\infty)} := \sup_{t>0} tK'(t,f;L^\infty,L^1). \tag{9}$$

Note that in (9) the order of the spaces is reversed in the computation of the K-functional. These two different K-functionals are connected by the equation

$$K(t,f;L^\infty,L^1) = tK(\frac{1}{t},f;L^1,L^\infty). \tag{10}$$

Inserting (10) in (9) we readily see that

$$\begin{aligned}\|f\|_{L(\infty,\infty)} &= \sup_{t>0}\{tK(\frac{1}{t},f;L^1,L^\infty) - K'(\frac{1}{t},f;L^1,L^\infty)\}\\ &= \sup_{t>0}\{\frac{K(t,f;L^1,L^\infty)}{t} - K'(t,f;L^1,L^\infty)\}\\ &= \sup_{t}\{f^{**}(t) - f^{*}(t)\}.\end{aligned}$$

The oscillation operator, $f \to f^{**}(t) - f^{*}(t)$, turns out to play an important role in other fundamental inequalities in analysis. A recent remarkable application of the oscillation operator provides the sharp form of the Hardy–Littlewood–Sobolev–O'Neil inequality up the borderline end point $p = n$. Indeed, if we let (cf. [15])

$$\|f\|_{L(p,q)} = \begin{cases} \{\int_0^\infty (f^*(t)t^{1/p})^q \frac{dt}{t}\}^{1/q} & 1 \le p < \infty, 1 \le q \le \infty \\ \|f\|_{L(\infty,q)} & 1 \le q \le \infty, \end{cases} \tag{11}$$

where[6]

$$\|f\|_{L(\infty,q)} := \left\{\int_0^\infty (f^{**}(t) - f^{*}(t))^q \frac{dt}{t}\right\}^{1/q}, \tag{12}$$

then it was shown in [7] that

$$\|f\|_{L(\bar{p},q)} \le c_n \|\nabla f\|_{L(p,q)}, 1 \le p \le n, \frac{1}{\bar{p}} = \frac{1}{p} - \frac{1}{n}, 1 \le q \le \infty, f \in C_0^\infty(R^n). \tag{13}$$

[6] Apparently the $L(\infty,q)$ spaces for $q < \infty$ were first introduced and their usefulness shown in [7]. Note that with the usual definition $L(\infty,\infty)$ would be L^∞, and $L(\infty,q) = \{0\}$, for $q < \infty$. The key point here is that the use of the oscillation operator introduces cancellations that make the spaces defined in this fashion nontrivial (cf. section "The Rearrangement Invariant Hull of *BMO* and Gagliardo Coordinate Spaces", Example 1).

The Sobolev inequality (13) is best possible, and for $p = q = n$ it improves on the endpoint result of Brezis–Wainger–Hanson–Maz'ya[7] (cf. [13, 35, 67]) much as the Bennett–DeVore–Sharpley inequality (8) improves upon (3). The improvement over *best possible results* is feasible because, once again, the spaces that correspond to $p = n$,

$$L(\infty, q) = \{f : \|f\|_{L(\infty,q)} < \infty\},$$

are not necessarily linear![8]

Moreover, the Sobolev inequality (13) persists up to higher order derivatives,[9] as was shown in [75],

$$\|f\|_{L(\bar{p},q)} \leq c_n \left\|\nabla^k f\right\|_{L(p,q)}, 1 \leq p \leq \frac{n}{k}, \frac{1}{\bar{p}} = \frac{1}{p} - \frac{k}{n}, \; 1 \leq q \leq \infty, \; f \in C_0^{\infty}(R^n).$$

In particular, when $p = \frac{n}{k}$ and $q = \infty$, we have the *BMO* type result[10]

$$\|f\|_{L(\infty,\infty)} \leq c \left\|\nabla^k f\right\|_{L(\frac{n}{k},\infty)}, f \in C_0^{\infty}(R^n).$$

Using the space $L(\infty, \infty)$ one can improve (2) as follows (cf. [53]):

$$\|f\|_{L^q} \leq C_n q \, \|f\|_{L^p}^{p/q} \, \|f\|_{L(\infty,\infty)}^{1-p/q}, \; 1 \leq p < q < \infty. \tag{14}$$

In my work with Jawerth[11] (cf. [43]) we give a somewhat different interpretation of the $L(\infty, q)$ spaces using Gagliardo diagrams (cf. [12, 27]); this point of view turns out to be useful to explain other applications of the oscillation operator $f^{**}(t) - f^*(t)$ (cf. [29, 63] and section "Recent Uses of the Oscillation Operator and $L(\infty, q)$ Spaces in Analysis"). The idea behind the approach in [43] is that of an "optimal decomposition," which also makes it possible to incorporate the $L(\infty, q)$ spaces into the abstract theory of real interpolation, as we shall show below.

Let $t > 0$, and let $f \in \Sigma(\vec{X}) = X_1 + X_2$. Out of all the competing decompositions for the computation of $K(t, f; \vec{X})$, an optimal decomposition

---

[7] Which in turn improves upon the classical exponential integrability result by Trudinger [87].

[8] Let X be a rearrangement invariant space, Pustylnik [79] has given necessary and sufficient conditions for spaces of functions defined by conditions of the form

$$\|(f^{**} - f^*) t^{-\gamma}\|_X < \infty$$

to be linear and normable.

[9] The improvement is also valid for Besov space inequalities as well (cf. [59]).

[10] The spaces $L(\infty, q)$ allow to interpolate between $L^{\infty} = L(\infty, 1)$ and $L(\infty, \infty) \subset e^L$.

[11] The earlier work of Herz [36] and Holmstedt [38], that precedes [11], should be also mentioned here.

$$f = D_1(t)f + D_2(t)f, \text{ with } D_i(t)f \in X_i, i = 1,2,$$

satisfies

$$K(t,f;\vec{X}) = \|D_1(t)f\|_{X_1} + t\,\|D_2(t)f\|_{X_2}, \tag{15}$$

(resp. a nearly optimal decomposition obtains if in (15) we replace $=$ by $\approx$). For an optimal decomposition of f we have[12] (cf. [38, 43])

$$\|D_1(t)f\|_{X_1} = K(t,f;\vec{X}) - t\frac{d}{dt}K(t,f;\vec{X}); \tag{18}$$

$$\|D_2(t)f\|_{X_2} = \frac{d}{dt}K(t,f;\vec{X}). \tag{19}$$

In particular, for the pair (L^1, L^∞), we have (cf. section "Extrapolation of Inequalities: Burkholder–Gundy–Herz Meet Calderón–Maz'ya and Cwikel et al."),

$$\begin{aligned}\|D_1(t)f\|_{L^1} &= K(t,f;L^1,L^\infty) - t\frac{d}{dt}K(t,f;L^1,L^\infty)\\ &= tf^{**}(t) - tf^*(t),\end{aligned}$$

and

$$\|D_2(t)f\|_{L^\infty} = \frac{d}{dt}K(t,f;L^1,L^\infty) = f^*(t).$$

Thus, reinterpreting optimal decompositions using Gagliardo diagrams (cf. [38, 43]), one is led to consider spaces, which could be referred to as "Gagliardo coordinate spaces." These spaces coincide with the Lions–Peetre real interpolation spaces for the usual range of the parameters (cf. [38]), but they also make sense at the end points (cf. [43]), and in this fashion they can be used to complete the Lions–Peetre scale much as the generalized $L(p,q)$ spaces defined by (11), complete the classical scale of Lorentz spaces. The "Gagliardo coordinate spaces" $\vec{X}^{(i)}_{\theta,q}, i = 1,2$, are formally obtained replacing $K(t,f;\vec{X})$ in the definition of the $\vec{X}_{\theta,q}$ norm of f (cf. (6)

---

[12]Interpreting $(\|D_1(t)f\|_{X_1}, \|D_2(t)f\|_{X_2})$ as coordinates on the boundary of a Gagliardo diagram (cf. [12]) it follows readily that, for all $\varepsilon > 0, t > 0$, we can find nearly optimal decompositions $x = x_\varepsilon(t) + y_\varepsilon(t)$, such that

$$(1-\varepsilon)[K(t,f;\vec{X}) - t\frac{d}{dt}K(t,f;\vec{X})] \le \|x_\varepsilon(t)\|_{X_1} \le (1+\varepsilon)[K(t,f;\vec{X}) - t\frac{d}{dt}K(t,f;\vec{X})] \tag{16}$$

$$(1-\varepsilon)\frac{d}{dt}K(t,f;\vec{X}) \le \|y_\varepsilon(t)\|_{X_2} \le (1+\varepsilon)\frac{d}{dt}K(t,f;\vec{X}). \tag{17}$$

above), by $\|D_1(t)f\|_{X_1}$ (resp. $\|D_2(t)f\|_{X_2}$) (cf. [38]). In particular, we note that the $\vec{X}^{(2)}_{\theta,q}$ spaces correspond to the k-spaces studied by Bennett [8].

This point of view led us to formulate and prove the following general version of (7) (cf. Theorem 1 below):

$$\|f\|_{\vec{X}^{(2)}_{\theta,q(\theta)}} \leq cq\,\|f\|_{X_1}^{1-\theta}\,\|f\|^{\theta}_{\vec{X}^{(1)}_{1,\infty}}\,,\theta \in (0,1), 1-\theta = \frac{1}{q(\theta)}, \tag{20}$$

which can be easily seen to imply an abstract extrapolation theorem connected with the John-Nirenberg inequality.

Underlying these developments is the following computation of the K-functional for the pair (L^1, BMO) given in [10] (cf. section "Self Improving Properties of BMO and Interpolation" below):

$$K(t,f,L^1(R^n),BMO(R^n)) \approx tf^{\#*}(t), \tag{21}$$

where $f^{\#}$ is the sharp maximal function of Fefferman–Stein [32] (cf. (23) below). In this calculation $BMO(R^n)$ is provided with the seminorm[13] (cf. section "The John-Nirenberg Lemma and Rearrangements")

$$|f|_{BMO} = \left\|f^{\#}_{R^n}\right\|_{L^\infty}.$$

In [19], the authors show that (2) can be used to give a strikingly "easy" proof of an inequality first proved in [52] using paraproducts,

$$\|fg\|_{L^p} \leq c(\|f\|_{L^p}\,\|g\|_{BMO} + \|g\|_{L^p}\,\|f\|_{BMO}), 1 < p < \infty. \tag{22}$$

The argument to prove (22) given in [19] has a general character and with suitable modifications (in particular, using the *reiteration theorem* of interpolation theory) the idea[14] can be combined with (20) to yield a new endpoint result for bilinear interpolation for generalized product and convolution operators of O'Neil type acting on interpolation scales (see section "Bilinear Interpolation" below). Further, in section "Recent Uses of the Oscillation Operator and $L(\infty,q)$ Spaces in Analysis"

[13] $BMO(R^n)$ can be normed by $|f|_{BMO}$ if we identify functions that differ by a constant.

[14] We cannot resist but to offer here our slight twist to the argument

$$\begin{aligned}
\|fg\|_{L^p} &\leq \|f\|_{L^{2p}}\,\|g\|_{L^{2p}} \\
&= \|f\|_{[L^p,BMO]_{1/2,2p}}\,\|g\|_{[L^p,BMO]_{1/2,2p}} \\
&\preceq \|f\|_{L^p}^{1/2}\,\|f\|_{BMO}^{1/2}\,\|g\|_{L^p}^{1/2}\,\|g\|_{BMO}^{1/2} \\
&\preceq \|f\|_{L^p}\,\|g\|_{BMO} + \|g\|_{L^p}\,\|f\|_{BMO}.
\end{aligned}$$

we collect applications of the methods discussed in the paper, and also offer some suggestions[15] for further research. In particular, in section "On Some Inequalities for Classical Operators by Bennett–DeVore–Sharpley and Bagby and Kurtz" we give a new approach to well-known results by Bagby–Kurtz [5], Kurtz [56], on the linear (in p) rate of growth of L^p estimates for certain singular integrals; in section "Good-Lambda Inequalities" we discuss the connection between the classical good-lambda inequalities (cf. [17, 21]) and the strong good-lambda inequalities of Bagby–Kurtz (cf. [56]) with inequalities for the oscillation operator $f^{**} - f^{*}$; in section "Extrapolation of Inequalities: Burkholder–Gundy–Herz Meet Calderón–Maz'ya and Cwikel et al." we show how oscillation inequalities for Sobolev functions are connected with the Gagliardo coordinate spaces and the property of commutation of the gradient with optimal (L^1, L^∞) decompositions (cf. [29]), we also discuss briefly the characterization of the isoperimetric inequality in terms of rearrangement inequalities for Sobolev functions, in a very general context.

The intended audience for this note are, on the one hand, classical analysts that may be curious on what abstract interpolation constructions could bring to the table, and on the other hand, functional analysts, specializing in interpolation theory, that may want to see applications of the abstract theories. To balance these objectives I have tried to give a presentation full of details in what respects to interpolation theory, and provide full references to the background material needed for the applications to classical analysis. In this respect, I have compiled a large set of references but the reader should be warned that this paper is not intended to be a survey, and that the list intends only to document the material that is mentioned in the text and simply reflects my own research interests, point of view, and limitations. In fact, many important topics dear to me had to be left out, including *Garsia inequalities* (cf. [33]).

I close the note with some personal reminiscences of my friendship with Cora Sadosky.

The John-Nirenberg Lemma and Rearrangements

In this section we recall a few basic definitions and results associated with the self improving properties of *BMO* functions.[16] In particular, we discuss the John-Nirenberg inequality (cf. [46]). In what follows we always let Q denote a cube with sides parallel to the coordinate axes.

Let Q_0 be a fixed cube in R^n. For $x \in Q_0$, let

[15] However, keep in mind the epigraph of [68], originally due Douglas Adams, The Restaurant at the End of the Universe, Tor Books, 1988 :"For seven and a half million years, Deep Thought computed and calculated, and in the end announced that the answer was in fact Forty-two—and so another, even bigger, computer had to be built to find out what the actual question was."

[16] For more background information, we refer to [9] and [83].

$$f^{\#}_{Q_0}(x) = \sup_{x \ni Q, Q \subset Q_0} \frac{1}{|Q|} \int_Q |f - f_Q|\, dx, \text{ where } f_Q = \frac{1}{|Q|} \int_Q f dx. \tag{23}$$

The space of functions of bounded mean oscillation, $BMO(Q_0)$, consists of all the functions $f \in L^1(Q_0)$ such that $f^{\#}_{Q_0} \in L^\infty(Q_0)$. Generally, we use the seminorm

$$|f|_{BMO(Q_0)} = \left\| f^{\#}_{Q_0} \right\|_{L^\infty}. \tag{24}$$

The space $BMO(Q_0)$ becomes a Banach space if we identify functions that differ by a constant. Sometimes it is preferable for us to use

$$\|f\|_{BMO(Q_0)} = |f|_{BMO(Q_0)} + \|f\|_{L^1(Q_0)}.$$

The classical John-Nirenberg Lemma is reformulated in [9, Corollary 7.7, p. 381] as follows[17]: Given a fixed cube $Q_0 \subset R^n$, there exists a constant $c > 0$, such that for all $f \in BMO(Q_0)$, and for all subcubes $Q \subset Q_0$,

$$[(f - f_Q)\, \chi_Q]^*(t) \leq c\, |f|_{BMO(Q_0)} \log^+\left(\frac{6\,|Q|}{t}\right), t > 0. \tag{25}$$

In particular, BMO has the following self improving property (cf. [9, Corollary 7.8, p. 381]). Let $1 \leq p < \infty$, and let[18]

$$f^{\#}_{Q_0,p}(x) = \sup_{x \ni Q, Q \subset Q_0} \left\{ \frac{1}{|Q|} \int_Q |f - f_Q|^p\, dx \right\}^{1/p},$$

and

$$\|f\|_{BMO^p(Q_0)} = \left\| f^{\#}_{Q_0,p} \right\|_{L^\infty} + \|f\|_{L^p(Q_0)}.$$

Then, with constants independent of f,

$$\|f\|_{BMO^p(Q_0)} \approx \|f\|_{BMO(Q_0)}.$$

It follows that, for all $p < \infty$,

$$BMO(Q_0) \subset L^p(Q_0).$$

[17] For a recent new approach to the John-Nirenberg Lemma we refer to [31].

[18] Note that $f^{\#}_{Q_0,1} = f^{\#}_{Q_0}$.

Actually, from [44] we have

$$\left\| (f - f_Q)\, \chi_Q \right\|_{\Delta(\frac{L^p(Q)}{p})} \approx \sup_t \frac{[(f - f_Q)\, \chi_Q]^*(t)}{\log^+(\frac{6|Q|}{t})} \approx \|f - f_Q\|_{e^{L(Q)}},$$

which combined with (25) gives

$$\|f - f_Q\|_{e^{L(Q)}} \leq c\, |f \chi_Q|_{BMO(Q)},$$

and therefore (cf. [9])

$$BMO(Q_0) \subset e^{L(Q_0)}.$$

In other words, the functions in $BMO(Q_0)$ are exponentially integrable.

The previous results admit suitable generalizations to R^n and more general measure spaces.

Interpolation Theory: Some Basic Inequalities

In this section we review basic definitions, and discuss inequalities of the form (4) that are associated with the classical methods of interpolation.

The starting objects of interpolation theory are pairs $\vec{X} = (X_1, X_2)$ of Banach spaces that are "compatible," in the sense that both spaces are continuously embedded in a common Hausdorff topological vector space V.[19] In real interpolation we consider two basic functionals, the K-functional, already introduced in (5), associated with the construction of the sum space $\Sigma(\vec{X}) = X_1 + X_2$, and its counterpart, the J-functional, defined on the intersection space $\Delta(\vec{X}) = X_1 \cap X_2$, by

$$J(t,f;\vec{X}) := J(t,f;X_1,X_2) = \max\left\{\|f\|_{X_1}, t\,\|f\|_{X_2}\right\}, t > 0. \tag{26}$$

The K-functional is used to construct the interpolation spaces $(X_1, X_2)_{\theta,q}$ (cf. (6) above). Likewise, associated with the J-functional we have the $(X_1, X_2)_{\theta,q;J}$, spaces.

[19] We shall then call $\vec{X} = (X_0, X_1)$ a "Banach pair". In general, the space V plays an auxiliary role, since once we know that $\vec{X}$ is a Banach pair we can use $\Sigma(\vec{X})$ as the ambient space. In particular, the functional $K(t,f;\vec{X})$ is in principle only defined on $\Sigma(\vec{X})$. On the other hand, the functional $f \to \frac{d}{dt}K(t,f;\vec{X})$, can make sense for a larger class of elements than $\Sigma(\vec{X})$. This occurs for significant examples: For example, on the interval [0, 1],

$$L(1,\infty) = \{f : \sup_t t f^*(t) < \infty\} \nsubseteq L^1 + L^\infty = L^1.$$

Let $\theta \in (0,1), 1 \leq q \leq \infty$; and let $U_{\theta,q}$ be the class of functions $u : (0,\infty) \to \Delta(\vec{X})$, such that[20] $\|u\|_{U_{\theta,q}} = \left\{\int_0^\infty (t^{-\theta} J(t, u(t); X_1, X_2))^q \frac{dt}{t}\right\}^{1/q} < \infty$. The space $(X_1, X_2)_{\theta,q;J}$ consists of elements $f \in X_1 + X_2$, such that there exists $u \in U_{\theta,q}$, with

$$f = \int_0^\infty u(s) \frac{ds}{s} \text{ (in } X_1 + X_2),$$

provided with the norm

$$\|f\|_{(X_1,X_2)_{\theta,q;J}} = \inf_{f=\int_0^\infty u(s)\frac{ds}{s}} \{\|u\|_{U_{\theta,q}}\}.$$

A basic result in this context is that these two constructions give the same spaces (*the equivalence theorem*) (cf. [12])

$$(X_1, X_2)_{\theta,q} = (X_1, X_2)_{\theta,q;J},$$

where the constants of norm equivalence depend only on θ and q.

In practice the J-method is harder to compute, but nevertheless plays an important theoretical role. In particular, the following interpolation property holds for the J-method. If X is a Banach space intermediate between X_1 and X_2, in the sense that $\Delta(\vec{X}) \subset X \subset \Sigma(\vec{X})$, then an inequality of the form

$$\|f\|_X \leq \|f\|_{X_1}^{1-\theta} \|f\|_{X_2}^{\theta}, \text{ for some fixed } \theta \in (0,1), \text{ and for all } f \in \Delta(\vec{X}), \tag{27}$$

is equivalent to

$$\|f\|_X \leq \|f\|_{(X_1,X_2)_{\theta,1;J}}, \text{ for all } f \in (X_1, X_2)_{\theta,1;J}. \tag{28}$$

One way to see this equivalence is to observe that (cf. [9])

Lemma 1. *Let $\theta \in (0,1)$. Then, for all $f \in X_1 \cap X_1$,*

$$\|f\|_{X_1}^{1-\theta} \|f\|_{X_2}^{\theta} = \inf_{t>0}\{t^{-\theta} J(t, f; X_1, X_2)\}. \tag{29}$$

The preceding discussion shows that, in particular,

$$\|f\|_{(X_1,X_2)_{\theta,1,J}} \leq \|f\|_{X_1}^{1-\theta} \|f\|_{X_2}^{\theta}, 0 < \theta < 1, \text{ for } f \in X_1 \cap X_2.$$

[20] With the usual modification when $q = \infty$.

More generally (cf. [44, 70]) we have[21]: for all $f \in X_1 \cap X_2$,

$$\|f\|_{(X_1,X_2)_{\theta,q,J}} \leq \left((1-\theta)\theta q'\right)^{-1/q'} \|f\|_{X_1}^{1-\theta} \|f\|_{X_2}^{\theta}, 0 < \theta < 1, \ 1 \leq q \leq \infty, \tag{30}$$

where $\frac{1}{q} + \frac{1}{q'} = 1$.

Likewise, for the complex method of interpolation of Calderón, $[.,.]_\theta$, we have (cf. [14]),

$$\|f\|_{[X_1,.X_2]_\theta} \leq \|f\|_{X_1}^{1-\theta} \|f\|_{X_2}^{\theta}, 0 < \theta < 1, \ \text{for} f \in X_1 \cap X_2.$$

For the "K" method we also have the following result implicit[22] in [44], which we prove for sake of completeness.

Lemma 2.

$$[(1-\theta)\theta q]^{1/q} \|f\|_{(X_1,X_2)_{\theta,q}} \leq \|f\|_{X_0}^{1-\theta} \|f\|_{X_1}^{\theta}, f \in X_1 \cap X_2, 0 < \theta < 1. \tag{31}$$

Proof. Let $f \in X_1 \cap X_2$. Using decompositions of the form $f = f + 0$, or $f = 0 + f$, we readily see that

$$K(f,t;X_1,X_2) \leq \min\{\|f\|_{X_1}, t\|f\|_{X_2}\}.$$

Therefore,

$$\begin{aligned}
\|f\|_{(X_1,X_2)_{\theta,q}} &= \left(\int_0^\infty [t^{-\theta}K(f,t;X_1,X_2)]^q \frac{dt}{t}\right)^{1/q} \\
&\leq \left(\int_0^{\|f\|_{X_1}/\|f\|_{X_2}} [t^{-\theta+1}\|f\|_{X_2}]^q \frac{dt}{t} + \int_{\|f\|_{X_1}/\|f\|_{X_2}}^\infty [t^{-\theta}\|f\|_{X_1}]^q \frac{dt}{t}\right)^{1/q} \\
&= \left(\|f\|_{X_2}^q \left(\frac{\|f\|_{X_1}}{\|f\|_{X_2}}\right)^{q(1-\theta)} \frac{1}{(1-\theta)q} + \|f\|_{X_1}^q \left(\frac{\|f\|_{X_1}}{\|f\|_{X_2}}\right)^{-\theta q} \frac{1}{\theta q}\right)^{1/q} \\
&= \|f\|_{X_1}^{1-\theta}\|f\|_{X_2}^{\theta} \left(\frac{1}{(1-\theta)q} + \frac{1}{\theta q}\right)^{1/q} \\
&= [(1-\theta)\theta q]^{-1/q} \|f\|_{X_1}^{1-\theta}\|f\|_{X_2}^{\theta}.
\end{aligned}$$

□

For more results related to this section we refer to [44, 70] and [49].

[21] Here and in what follows we use the convention $\infty^0 = 1$.

[22] See also [26].

Self Improving Properties of *BMO* and Interpolation

The purpose of this section is to provide a new proof of (2) using interpolation tools.

One the first results obtained concerning interpolation properties of *BMO* is the following (cf. [9, 32] and the references therein)

$$\left[L^1, BMO\right]_\theta = L^q, \text{ with } \frac{1}{1-\theta} = q. \tag{32}$$

In particular, it follows that

$$L^1 \cap BMO \subset L^q.$$

Therefore, if we work on a cube Q_0, we have

$$L^1(Q_0) \cap BMO(Q_0) = BMO(Q_0) \subset L^q(Q_0).$$

In other words, the following self-improvement holds:

$$f \in BMO(Q_0) \Rightarrow f \in \bigcap_{q \geq 1} L^q(Q_0).$$

While it is not true that $f \in BMO(Q_0) \Rightarrow f \in L^\infty(Q_0)$, we can quantify precisely the deterioration of the L^q norms of a function in $BMO(Q_0)$ to be able to conclude by extrapolation that

$$f \in BMO(Q_0) \Rightarrow f \in e^{L(Q_0)}. \tag{33}$$

Let us go over the details. First, consider the following inequality attributed to Chen–Zhu [19]:

Lemma 3. *Let $f \in BMO(Q_0)$, and let $1 \leq p < \infty$. Then, there exists an absolute constant that depends only on n and p, such that for all $q > p$,*

$$\|f\|_{L^q} \leq C_n q \, \|f\|_{L^p}^{p/q} \, \|f\|_{BMO}^{1-p/q}. \tag{34}$$

The point of the result, of course, is the precise dependency of the constants in terms of q. Before going to the proof let us show how (33) follows from (34).

Proof (of (33)). From (34), applied to the case $p = 1$, we find that, for all $q > 1$,

$$\begin{aligned} \|f\|_{L^q} &\leq C_n q \, \|f\|_{L^1}^{1/q} \, \|f\|_{BMO}^{1-1/q} \\ &\leq C_n q \, \|f\|_{BMO}. \end{aligned}$$

Hence, using, for example, the "Δ" method of extrapolation[23] of [44], we get

$$\|f\|_{\Delta(\frac{L^q}{q})} = \sup_{q>1} \frac{\|f\|_q}{q} \approx \|f\|_{e^L} \leq c_n \|f\|_{BMO} \cdot$$

□

We now give a proof of Lemma 3 using interpolation.

Proof. It will be convenient for us to work on R^n (the same results hold for cubes: See more details in Remark 1 below). We start by considering the case $p = 1$, the general case will follow by a re-scaling argument, which we provide below.

The first step is to make explicit the way we obtain the real interpolation spaces between L^1 and BMO. It is well known (cf. [9, 10], and the references therein) that

$$(L^1, BMO)_{1-1/q,q} = L^q, q > 1. \tag{35}$$

Here the equality of the norms of the indicated spaces is within constants of equivalence that depend only on q and n. In particular, we have

$$\|f\|_{L^q} \leq c(q,n) \|f\|_{(L^1,BMO)_{1-1/q,q}} \cdot$$

The program now is to give a precise estimate of $c(q,n)$ in terms of q, and then apply Lemma 2. We shall work with BMO provided by the seminorm $|\cdot|_{BMO}$ (cf. (24) above).

The following result was proved in [10, Theorem 6.1]:

$$K(t,f,L^1(R^n),BMO(R^n)) \approx tf^{\#*}(t), \tag{36}$$

with absolute constants of equivalence, and where $f^{\#}$ denotes the sharp maximal operator[24] (cf. [9, 32, 83])

$$f^{\#}(x) := f^{\#}_{R^n}(x) = \sup_{x \ni Q} \frac{1}{|Q|} \int_Q |f(y) - f_Q|\, dy, \text{ and } f_Q = \frac{1}{|Q|} \int_Q f(y) dy.$$

Let $f \in L^1 \cap BMO$, $q > 1$, and define θ by the equation $\frac{1}{1-\theta} = q$. Combining (36) and (31), we have, with absolute constants that do not depend on q, θ, or f,

$$\begin{aligned}[(1-\theta)\theta q]^{1/q} \left\{ \int_0^\infty [t^{-1+1/q} tf^{\#*}(t)]^q \frac{dt}{t} \right\}^{1/q} &\approx [(1-\theta)\theta q]^{1/q} \|f\|_{(L^1(R^n),BMO(R^n))_{1-1/q,q}} \\ &\preceq \|f\|_{L^1}^{1/q} \|f\|_{BMO}^{1-1/q}.\end{aligned}$$

[23] For more recent developments in extrapolation theory cf. [4].

[24] For computations of related K-functionals and further references cf. [2, 42].

Thus,

$$\left\{\int_0^\infty f^{\#*}(t)^q dt\right\}^{1/q} \leq c\left(\frac{q}{q-1}\right)^{1/q} \|f\|_{L^1}^{1/q} \|f\|_{BMO}^{1-1/q}. \tag{37}$$

Now, we recall that by [11], as complemented in [85, (3.8), p. 228], we have

$$f^{**}(t) - f^*(t) \leq cf^{\#*}(t), t > 0. \tag{38}$$

Combining (38) with (37), yields

$$\left\{\int_0^\infty [f^{**}(t) - f^*(t)]^q dt\right\}^{1/q} \leq c\left(\frac{q}{q-1}\right)^{1/q} \|f\|_{L^1}^{1/q} \|f\|_{BMO}^{1-1/q}. \tag{39}$$

Observe that, since $[tf^{**}(t)]' = (\int_0^t f^*(s)ds)' = f^*(t)$, we have

$$[-f^{**}(t)]' = \frac{f^{**}(t) - f^*(t)}{t}.$$

Moreover, since $f \in L^1$, then $f^{**}(\infty) = 0$, and it follows from the fundamental theorem of calculus that we can write

$$f^{**}(t) = \int_t^\infty f^{**}(s) - f^*(s) \frac{ds}{s}.$$

Consequently, by Hardy's inequality,

$$\left\{\int_0^\infty f^{**}(t)^q dt\right\}^{1/q} \leq q \left\{\int_0^\infty [f^{**}(t) - f^*(t)]^q dt\right\}^{1/q}.$$

Inserting this information in (39) we arrive at

$$\left\{\int_0^\infty f^{**}(t)^q dt\right\}^{1/q} \leq cq\left(\frac{q}{q-1}\right)^{1/q} \|f\|_{L^1}^{1/q} \|f\|_{BMO}^{1-1/q}. \tag{40}$$

We now estimate the left-hand side of (40) from below. By the sharp reverse Hardy inequality for decreasing functions (cf. [80], [69, Lemma 2.1], see also [88]) we can write

$$\begin{aligned} \|f\|_q &= \left\{\int_0^\infty f^*(t)^q dt\right\}^{1/q} \\ &\leq \left(\frac{q-1}{q}\right)^{1/q} \left\{\int_0^\infty f^{**}(t)^q dt\right\}^{1/q}. \end{aligned} \tag{41}$$

Combining the last inequality with (40) we obtain

$$\begin{aligned}\|f\|_q &= \left\{\int_0^\infty f^*(t)^q dt\right\}^{1/q}\\ &\leq (\frac{q-1}{q})^{1/q}\left\{\int_0^\infty f^{**}(t)^q dt\right\}^{1/q}\\ &\leq (\frac{q-1}{q})^{1/q}(\frac{q}{q-1})^{1/q} cq\left\{\int_0^\infty [f^{**}(t)-f^*(t)]^q dt\right\}^{1/q}\\ &\leq cq\|f\|_{L^1}^{1/q}\|f\|_{BMO}^{1-1/q},\end{aligned}$$

as we wanted to show.

Let us now consider the case $p > 1$. Let $q > p$. By Holmstedt's reiteration theorem (cf. [12, 38]) we have

$$(L^p, BMO)_{1-p/q,q} = ((L^1, BMO)_{1-1/p,p}, BMO)_{1-p/q,q},$$

and, moreover, with absolute constants that depend only on p,

$$K(t,f;L^p,BMO) \approx \left\{\int_0^{t^p} (s^{\frac{1}{p}-1} s f^{\#*}(s))^p \frac{ds}{s}\right\}^{1/p} = \left\{\int_0^{t^p} f^{\#*}(s)^p ds\right\}^{1/p}.$$

By Lemma 2 it follows that, with constants independent of q, f, we have

$$\begin{aligned}\left\{\int_0^\infty [t^{-(1-p/q)}\left\{\int_0^{t^p} f^{\#*}(s)^p ds\right\}^{1/p}]^q \frac{dt}{t}\right\}^{1/q} &\approx \|f\|_{(L^p(R^n),BMO(R^n))_{1-p/q,q}}\\ &\preceq p^{-1/q}[q-p]^{-1/q} q^{1/q}\|f\|_{L^p}^{p/q}\|f\|_{BMO}^{1-p/q}.\end{aligned}$$

Now,

$$\begin{aligned}&\left\{\int_0^\infty [t^{-(1-p/q)q}\left\{\int_0^{t^p} f^{\#*}(s)^p ds\right\}^{q/p} \frac{dt}{t}\right\}^{1/q}\\ &= \left\{\int_0^\infty [t^{-(1-p/q)q} t^q\left\{\frac{1}{t^p}\int_0^{t^p} f^{\#*}(s)^p ds\right\}^{q/p} \frac{dt}{t}\right\}^{1/q}\\ &= \left\{\int_0^\infty t^p\left\{\frac{1}{t^p}\int_0^{t^p} f^{\#*}(s)^p ds\right\}^{q/p} \frac{dt}{t}\right\}^{1/q}\end{aligned}$$

$$= \left(\frac{1}{p}\right)^{1/q} \left\{ \int_0^\infty u \left\{ \frac{1}{u} \int_0^u f^{\#*}(s)^p ds \right\}^{q/p} \frac{du}{u} \right\}^{1/q}$$

$$= \left(\frac{1}{p}\right)^{1/q} \left[\left\{ \int_0^\infty \left\{ \frac{1}{u} \int_0^u f^{\#*}(s)^p ds \right\}^{q/p} du \right\}^{p/q} \right]^{1/p}$$

$$\geq \left(\frac{1}{p}\right)^{1/q} \left[\frac{\frac{q}{p}}{q/p - 1} \right]^{1/q} \left\{ \int_0^\infty f^{\#*}(u)^q du \right\}^{1/q},$$

where in the last step we have used the reverse sharp Hardy inequality (cf. [69, Lemma 2.1]). Consequently,

$$\left\{ \int_0^\infty f^{\#*}(u)^q du \right\}^{1/q} \preceq p^{1/q} \left[\frac{q/p - 1}{q/p} \right]^{1/q} p^{-1/q} [q-p]^{-1/q} q^{1/q} \|f\|_{L^p}^{p/q} \|f\|_{BMO}^{1-p/q}$$
$$\sim \|f\|_{L^p}^{p/q} \|f\|_{BMO}^{1-p/q}.$$

Hence, by the analysis we already did for the case $p = 1$, we see that

$$\left\{ \int_0^\infty f^*(t)^q dt \right\}^{1/q} \preceq \left(\frac{q-1}{q}\right)^{1/q} q \|f\|_{L^p}^{p/q} \|f\|_{BMO}^{1-p/q}$$
$$\preceq q \|f\|_{L^p}^{p/q} \|f\|_{BMO}^{1-p/q},$$

as we wished to show. □

Remark 1. If we work on a cube Q_0, the replacement of (38) is (cf. [9])

$$f^{**}(t) - f^*(t) \leq c f^{\#*}(t), 0 < t < |Q_0|/3.$$

In this situation, we have $BMO(Q_0) \subset L^1(Q_0)$, and we readily see that $\left\{ \int_0^{|Q_0|/3} [t^{-1+1/q} t f^{\#*}(t)]^q \frac{dt}{t} \right\}^{1/q}$ is an equivalent seminorm for $(L^1(Q_0), BMO(Q_0))_{1-1/q,q}$. The rest of the proof now follows mutatis mutandis.

Remark 2. For related Hardy inequalities for one-dimensional oscillation operators of the form $f_{\#}(t) = \frac{1}{t} \int_0^t f(s) ds - f(t)$, cf. [76] and [54].

The Rearrangement Invariant Hull of *BMO* and Gagliardo Coordinate Spaces

In this section we introduce the "Gagliardo coordinate spaces" (cf. [38, 43, 60]) and we use them to extend (2) to the setting of real interpolation.

Let $\theta \in [0,1], q \in (0,\infty]$. Following the discussion given in the Introduction, we define the "Gagliardo coordinate spaces" as follows[25]:

$$(X_1,X_2)^{(1)}_{\theta,q} = \left\{ f \in X_1 + X_2 : \|f\|_{(X_1,X_2)^{(1)}_{\theta,q}} < \infty \right\},$$

where

$$\|f\|_{(X_1,X_2)^{(1)}_{\theta,q}} = \left\{ \int_0^\infty \left(t^{1-\theta} \left[\frac{K(t,f;X_1,X_2)}{t} - K'(t,f;X_1,X_2) \right] \right)^q \frac{dt}{t} \right\}^{1/q},$$

and

$$(X_1,X_2)^{(2)}_{\theta,q} = \left\{ f \in X_1 + X_2 : \|f\|_{(X_1,X_2)^{(2)}_{\theta,q}} < \infty \right\},$$

$$\|f\|_{(X_1,X_2)^{(2)}_{\theta,q}} = \left\{ \int_0^\infty (t^{-\theta} tK'(t,f;X_1,X_2))^q \frac{dt}{t} \right\}^{1/q},$$

and we compare them to the classical Lions–Peetre spaces $(X_1,X_2)_{\theta,q}$.

The Gagliardo coordinate spaces in principle are not linear, and the corresponding functionals, $\|f\|_{(X_1,X_2)^{(i)}_{\theta,q}}, i = 1,2$, are not norms. However, it turns out that, when $\theta \in (0,1), q \in (0,\infty]$, we have, with *norm* equivalence (cf. [38, 43]),

$$(X_1,X_2)^{(1)}_{\theta,q} = (X_1,X_2)^{(2)}_{\theta,q} = (X_1,X_2)_{\theta,q}. \tag{42}$$

More precisely, the "norm" equivalence depends only on θ and q. On the other hand, at the endpoints, $\theta = 0$ or $\theta = 1$, the resulting spaces can be very different.

Example 1. Let $(X_1,X_2) = (L^1,L^\infty)$. Then, if $\theta = 1, q = \infty$, we have

$$\|f\|_{(X_1,X_2)^{(1)}_{1,\infty}} = \|f\|_{L(\infty,\infty)}, \tag{43}$$

while

$$\|f\|_{(X_1,X_2)_{1,\infty}} = \|f\|_{(X_1,X_2)^{(2)}_{1,\infty}} = \|f\|_{L^\infty}.$$

For $\theta = 1, q < \infty$,

$$\|f\|_{(X_1,X_2)^{(1)}_{1,q}} = \|f\|_{L(\infty,q)}.$$

[25] Think in terms of a Gagliardo diagram, see, for example, [12, p. 39], [43, 46].

On the other hand,

$$\|f\|_{(X_1,X_2)^{(2)}_{1,q}} = \left\{\int_0^\infty f^*(t)^q \frac{dt}{t}\right\}^{1/q} \leq \left\{\int_0^\infty f^{**}(t)^q \frac{dt}{t}\right\}^{1/q} = \|f\|_{(X_1,X_2)_{1,q}},$$

and

$$\|f\|_{(X_1,X_2)^{(2)}_{1,q}} < \infty \Leftrightarrow f = 0.$$

For $\theta = 0, q = \infty$,

$$\|f\|_{(X_1,X_2)_{0,\infty}} = \sup_t tf^{**}(t) = \|f\|_{L^1},$$

while

$$\|f\|_{(X_1,X_2)^{(2)}_{0,\infty}} = \sup_t tf^*(t) = \|f\|_{L(1,\infty)}.$$

Moreover,

$$\begin{aligned}\|f\|_{(X_1,X_2)^{(1)}_{0,\infty}} &= \sup_t t(f^{**}(t) - f^*(t)) \\ &= \sup_t \int_{f^*(t)}^\infty \lambda_f(s)ds.\end{aligned}$$

Therefore, if $f^*(\infty) = 0$, then

$$\begin{aligned}\|f\|_{(X_1,X_2)^{(1)}_{0,\infty}} &= \int_0^\infty \lambda_f(s)ds \\ &= \|f\|_{L^1}.\end{aligned}$$

Also, if $f^{**}(\infty) = 0$,

$$\begin{aligned}\|f\|_{(X_1,X_2)^{(1)}_{1,1}} &= \int_0^\infty \left[f^{**}(t) - f^*(t)\right] \frac{dt}{t} \\ &= \lim_{r\to 0} \int_r^\infty \left[f^{**}(t) - f^*(t)\right] \frac{dt}{t} \\ &= \lim_{r\to 0} \left(f^{**}(r) - f^{**}(\infty)\right) \text{ (since } \frac{d}{dr}(-f^{**}(t)) = \frac{f^{**}(t) - f^*(t)}{t}) \\ &= \|f\|_{L^\infty}.\end{aligned}$$

Theorem 1. *Let $\theta \in [0,1)$, and let $1 \leq q < \infty$. Then, there exists an absolute constant c such that*

$$\|f\|_{\vec{X}^{(2)}_{\theta,q}} \leq cq\left(1+[(1-\theta)q]^{1/q}\right)^{1-\theta}[(1-\theta)q]^{-\theta}\,\|f\|_{X_1}^{1-\theta}\,\|f\|^{\theta}_{\vec{X}^{(1)}_{1,\infty}}\,. \tag{44}$$

In particular, if $(1-\theta)q = 1$,

$$\|f\|_{\vec{X}^{(2)}_{\theta,q}} \leq cq\,\|f\|_{X_1}^{1-\theta}\,\|f\|^{\theta}_{\vec{X}^{(1)}_{1,\infty}}\,. \tag{45}$$

Before going to the proof let us argue why such a result could be termed a generalized John-Nirenberg inequality. Indeed, let Q be a cube in R^n, and consider the pair $\vec{X} = (L^1(Q), L^\infty(Q))$. As we have seen [cf. (43)]

$$\|f\|_{\vec{X}^{(1)}_{1,\infty}} = \|f\|_{L(\infty,\infty)}\,.$$

By definition, if $f \in L(\infty,\infty)(Q)$, then $f \in L^1(Q)$. We now show that, moreover, $\|f\|_{L^1} \leq |Q|\,\|f\|_{L(\infty,\infty)}$. Indeed, for all $t > 0$ we have (cf. [1, Theorem 2.1 (ii)])

$$\begin{aligned}\int_{\{|f|>t\}} |f(x)|\,dx &\leq \int_{f^*(\lambda_f(t))}^{\infty} \lambda_f(r)dr + t\lambda_f(t)\\ &= \lambda_f(t)[f^{**}(\lambda_f(t)) - f^*(\lambda_f(t))] + t\lambda_f(t)\\ &\leq \lambda_f(t)\left(\|f\|_{L(\infty,\infty)} + t\right)\\ &\leq |Q|\left(\|f\|_{L(\infty,\infty)} + t\right),\end{aligned} \tag{46}$$

where in the second step we have used the formula (do a graph!)

$$s(f^{**}(s) - f^*(s)) = \int_{f^*(s)}^{\infty} \lambda_f(u)du.$$

Let $t \to 0$ in (46) then, by Fatou's Lemma, we see that

$$\|f\|_{L^1} \leq |Q|\,\|f\|_{L(\infty,\infty)}\,. \tag{47}$$

Now, let $\theta \in (0,1)$, and $\frac{1}{q} = 1-\theta$. Then,

$$\begin{aligned}\|f\|_{\vec{X}^{(2)}_{\theta,q}} &= \left\{\int_0^\infty (t^{-\theta} tK'(t,f;L^1,L^\infty))^q \frac{dt}{t}\right\}^{1/q}\\ &= \left\{\int_0^\infty (t^{-(1-1/q)} tf^*(t))^q \frac{dt}{t}\right\}^{1/q}\\ &= \|f\|_{L^q}\,.\end{aligned}$$

Therefore, by (45) and (47),

$$\begin{aligned}\|f\|_{L^q} &\leq cq\, \|f\|_{L^1}^{1/q}\, \|f\|_{L(\infty,\infty)}^{1/q'} \\ &\leq cq\, |Q|\, \|f\|_{L(\infty,\infty)} \\ &\leq cq\, \|f\|_{L(\infty,\infty)}\,.\end{aligned}$$

Consequently,

$$\|f\|_{e^L} \approx \|f\|_{\Delta(\frac{L^q}{q})} = \sup_q \frac{\|f\|_{L^q}}{q} \leq c\, \|f\|_{L(\infty,\infty)}\,.$$

Proof (of Theorem (1)). Let us write

$$\begin{aligned}\|f\|_{\vec{X}^{(2)}_{\theta,q}} &= \left(\int_0^\infty (u^{1-\theta}\frac{d}{du}K(u,f;\vec{X}))^q \frac{du}{u}\right)^{1/q} \\ &= \left(\int_0^t (u^{1-\theta}\frac{d}{du}K(u,f;\vec{X}))^q \frac{du}{u}\right)^{1/q} + \left(\int_t^\infty (u^{1-\theta}\frac{d}{du}K(u,f;\vec{X}))^q \frac{du}{u}\right)^{1/q} \\ &= (I) + (II).\end{aligned}$$

We estimate these two terms as follows:

$$\begin{aligned}(I) &= \left(\int_0^t u^{(1-\theta)q}(\frac{d}{du}K(u,f;\vec{X}))^q \frac{du}{u}\right)^{1/q} \\ &\leq \left(\int_0^t u^{(1-\theta)q}(\frac{K(u,f;\vec{X})}{u})^q \frac{du}{u}\right)^{1/q} \quad (\text{since } \frac{d}{du}K(u,f;\vec{X}) \leq \frac{K(u,f;\vec{X})}{u}).\end{aligned}$$

On the other hand, since $\left(\frac{K(u,f;\vec{X})}{u}\right)' = \frac{K'(u,f;\vec{X})u - K(u,f;\vec{X})}{u^2}$, we have that, for $0 < u < t$,

$$\begin{aligned}\frac{K(u,f;\vec{X})}{u} &= \frac{K(t,f;\vec{X})}{t} + \left(-\left.\frac{K(\cdot,f;\vec{X})}{\cdot}\right|_u^t\right) \\ &= \frac{K(t,f;\vec{X})}{t} + \int_u^t \left(\frac{K(r,f;\vec{X})}{r} - K'(r,f;\vec{X})\right)\frac{dr}{r} \\ &\leq \frac{K(t,f;\vec{X})}{t} + \left(\log\frac{t}{u}\right)\sup_{r\leq t}\left(\frac{K(r,f;\vec{X})}{r} - K'(r,f;\vec{X})\right) \\ &\leq \frac{K(t,f;\vec{X})}{t} + \log\frac{t}{u}\, \|f\|_{\vec{X}^{(1)}_{1,\infty}}\,.\end{aligned}$$

Therefore, by the triangle inequality,

$$(I) \leq \left(\int_0^t (u^{(1-\theta)q}\{\frac{K(t,f;\vec{X})}{t} + \log\frac{t}{u}\,\|f\|_{\vec{X}^{(1)}_{1,\infty}}\}^q\frac{du}{u}\right)^{1/q}$$
$$\leq \frac{K(t,f;\vec{X})}{t}\left(\int_0^t u^{(1-\theta)q}\frac{du}{u}\right)^{1/q} + \|f\|_{\vec{X}^{(1)}_{1,\infty}}\left(\int_0^t u^{(1-\theta)q}\left(\log\frac{t}{u}\right)^q\frac{du}{u}\right)^{1/q}$$
$$= \frac{K(t,f;\vec{X})}{t}\frac{t^{(1-\theta)}}{[(1-\theta)q]^{1/q}} + \|f\|_{\vec{X}^{(1)}_{1,\infty}}\left(\int_0^t u^{(1-\theta)q}\left(\log\frac{t}{u}\right)^q\frac{du}{u}\right)^{1/q}$$
$$= (a) + (b).$$

For the term (a) we have

$$(a) \leq \frac{t^{-\theta}}{[(1-\theta)q]^{1/q}}\lim_{t\to\infty} K(t,f;\vec{X}) \text{ (since } K(\cdot,f;\vec{X}) \text{ increases)}$$
$$\leq \frac{t^{-\theta}}{[(1-\theta)q]^{1/q}}\|f\|_{X_1}.$$

To deal with (b) we use the asymptotics of the gamma function as follows: let $s = \log\frac{t}{u}$, then $u = te^{-s}$, $du = -te^{-s}ds$, $\frac{du}{u} = -ds$, $u^{(1-\theta)q} = t^{(1-\theta)q}e^{-s(1-\theta)q}$, and we have

$$(b) = \|f\|_{\vec{X}^{(1)}_{1,\infty}}\left(\int_0^t u^{(1-\theta)q}s^q\frac{du}{u}\right)^{1/q}$$
$$= \|f\|_{\vec{X}^{(1)}_{1,\infty}}\, t^{1-\theta}\left(\int_0^\infty e^{-s(1-\theta)q}s^q ds\right)^{1/q}$$
$$= \|f\|_{\vec{X}^{(1)}_{1,\infty}}\, t^{1-\theta}\left(\int_0^\infty e^{-\tau}\frac{\tau^q}{[(1-\theta)q]^q}\frac{d\tau}{[(1-\theta)q]}\right)^{1/q}, \text{ (let } \tau = s(1-\theta)q)$$
$$= \frac{\|f\|_{\vec{X}^{(1)}_{1,\infty}}}{[(1-\theta)q]}\frac{1}{[(1-\theta)q]^{1/q}}t^{1-\theta}\left(\Gamma(q+1)\right)^{1/q}$$
$$\leq \frac{\|f\|_{\vec{X}^{(1)}_{1,\infty}}}{[(1-\theta)q]}\frac{1}{[(1-\theta)q]^{1/q}}t^{1-\theta}q.$$

Combining inequalities for (a) and (b) we have

$$(I) \leq \frac{t^{-\theta}}{[(1-\theta)q]^{1/q}}\|f\|_{X_1} + \frac{\|f\|_{\vec{X}^{(1)}_{1,\infty}}}{[(1-\theta)q]}\frac{1}{[(1-\theta)q]^{1/q}}t^{1-\theta}q.$$

We now estimate (II) :

$$
\begin{aligned}
(II) &= \left(\int_t^\infty u^{(1-\theta)q} u^{-1} \left(\frac{d}{du}K(u,f;\vec{X})\right)^{q-1} \left(u\frac{d}{du}K(u,f;\vec{X})\right)\frac{du}{u}\right)^{1/q} \\
&\leq \left\{\sup_{u\geq t}(u^{\frac{(1-\theta)q-1}{q}}\left(\frac{d}{du}K(u,f;\vec{X})\right)^{\frac{q-1}{q}})\right\}\left\{\int_t^\infty \frac{d}{du}K(u,f;\vec{X})du\right\}^{1/q} \\
&= (c)(d).
\end{aligned}
$$

The factors on the right-hand side can be estimated as follows:

$$
\begin{aligned}
(d) &= \left(\lim_{u\mapsto\infty} K(u,f;\vec{X}) - K(t,f;\vec{X})\right)^{1/q} \\
&\leq \left(\lim_{u\mapsto\infty} K(u,f;\vec{X})\right)^{1/q} \\
&= \|f\|_{X_0}^{1/q}.
\end{aligned}
$$

Also, since $K(\cdot,f;\vec{X})$ is concave, $\frac{d}{du}K(u,f;\vec{X}) \leq \frac{K(u,f;\vec{X})}{u}$, consequently,

$$
\begin{aligned}
(c) &\leq \|f\|_{X_1}^{1-1/q} \sup_{u\geq t}\{u^{\frac{(1-\theta)q-1}{q}-\frac{q-1}{q}}\} \\
&\leq \|f\|_{X_1}^{1-1/q}\{\sup_{u\geq t} u^{-\theta}\} \\
&= \|f\|_{X_1}^{1-1/q} t^{-\theta}.
\end{aligned}
$$

Thus,

$$
\begin{aligned}
(II) &\leq \|f\|_{X_1}^{1/q} \|f\|_{X_1}^{1-1/q} t^{-\theta} \\
&= \|f\|_{X_1} t^{-\theta}.
\end{aligned}
$$

Combining the estimates for (I) and (II) yields

$$
\begin{aligned}
\|f\|_{\vec{X}^{(2)}_{\theta,q}} &\leq \left(\frac{1+[(1-\theta)q]^{1/q}}{[(1-\theta)q]^{1/q}}\right) t^{-\theta}\|f\|_{X_1} + \frac{1}{[(1-\theta)q]}\frac{1}{[(1-\theta)q]^{1/q}} q t^{1-\theta}\|f\|_{\vec{X}^{(1)}_{1,\infty}} \\
&\leq cq\frac{1}{[(1-\theta)q]^{1/q}}\left\{\left(1+[(1-\theta)q]^{1/q}\right)t^{-\theta}\|f\|_{X_1} + \frac{1}{[(1-\theta)q]}t^{1-\theta}\|f\|_{\vec{X}^{(1)}_{1,\infty}}\right\}.
\end{aligned}
\tag{48}
$$

We balance the terms on the right-hand side by choosing t such that

$$\left(1+[(1-\theta)q]^{1/q}\right)t^{-\theta}\left\|f\right\|_{X_1}=\frac{1}{[(1-\theta)q]}t^{1-\theta}\left\|f\right\|_{\vec{X}^{(1)}_{1,\infty}},$$

whence,

$$t=\left(1+[(1-\theta)q]^{1/q}\right)[(1-\theta)q]\frac{\left\|f\right\|_{X_1}}{\left\|f\right\|_{\vec{X}^{(1)}_{1,\infty}}}.$$

Inserting this value of t in (48) we find

$$\begin{aligned}\left\|f\right\|_{\vec{X}^{(2)}_{\theta,q}}&\leq cq\frac{1}{[(1-\theta)q]^{1/q}}\left\{\left(1+[(1-\theta)q]^{1/q}\right)^{-\theta}[(1-\theta)q]^{-\theta}\left\|f\right\|^{1-\theta}_{X_1}\left\|f\right\|^{\theta}_{\vec{X}^{(1)}_{1,\infty}}\right\}\\&\leq cq\left(\frac{1}{[(1-\theta)q]^{1/q}}\right)\left(1+[(1-\theta)q]^{1/q}\right)^{-\theta}[(1-\theta)q]^{-\theta}\left\|f\right\|^{1-\theta}_{X_1}\left\|f\right\|^{\theta}_{\vec{X}^{(1)}_{1,\infty}},\end{aligned}$$

as we wished to show □

Remark 3. As an easy application of (45), we note that, if $X_2 \subset X_1$, we can write

$$\left\|f\right\|_{\Delta\left((1-\theta)\vec{X}^{(2)}_{\theta,\frac{1}{1-\theta}}\right)}=\sup_{\theta}(1-\theta)\left\|f\right\|_{\vec{X}^{(2)}_{\theta,\frac{1}{1-\theta}}}\leq c\left\|f\right\|_{\vec{X}^{(1)}_{1,\infty}}.$$

We believe that similar arguments would lead to computations with the Δ_p method of extrapolation (cf. [44, 48]) but this lies outside the scope of this paper so we leave the issue for another occasion.

Recent Uses of the Oscillation Operator and $L(\infty, q)$ Spaces in Analysis

We present several different applications connected with the material developed in this note. The material is nothing but a sample of results. The results presented are either new or they provide a new treatment to known results.[26] This section differs from previous ones in that we proceed formally and, whenever possible, we refer the reader to the literature for background material and complete details. Further development of materials in this section will appear elsewhere, e.g., in [29, 72, 73].

[26] See also [82], Lesson #3.

On Some Inequalities for Classical Operators by Bennett–DeVore–Sharpley and Bagby and Kurtz

In this section we show how the methods developed in this paper can be applied to give a new approach to results on singular integrals and maximal operators that appeared first in [5, 11], and [56] (cf. also the references therein).

Let T and U be operators acting in a sufficiently large class of testing functions, say the space S of Schwartz testing functions on R^n. Furthermore, suppose that there exists $C > 0$, such that for all $f \in S$, the following pointwise inequality holds:

$$(Tf)^{\#}(x) \leq CUf(x).$$

Then, taking rearrangements we have

$$(Tf)^{\#*}(t) \leq C(Uf)^*(t), t > 0.$$

Therefore,

$$\left\{\int_0^\infty (Tf)^{\#*}(t)^p dt\right\}^{1/p} \leq C\left\{\int_0^\infty (Uf)^*(t)^p dt\right\}^{1/p}.$$

Now, by (38) above, and the analysis that follows it, we see that

$$\begin{aligned}\left\{\int_0^\infty (Tf)^*(t)^p dt\right\}^{1/p} &\leq \left(\frac{p-1}{p}\right)^{1/p} pC\left\{\int_0^\infty (Uf)^*(t)^p dt\right\}^{1/p} \\ &\leq Cp\left\{\int_0^\infty (Uf)^*(t)^p dt\right\}^{1/p}.\end{aligned}$$

In other words,

$$\|Tf\|_p \leq cp\,\|Uf\|_p, \tag{49}$$

and we recover the main result of [56].

We should also point out that the method of proof can be also implemented to deal with the corresponding more general inequalities for doubling measures on R^n (cf. [21, 56], and the references therein).

Remark 4. It may be appropriate to mention that once one knows (49) then one could use the extrapolation theory of [44] to show (cf. [56, Lemma 5]) that there exist absolute constants $C, \gamma > 0$, such that,

$$(Tf)^*(t) \leq C\int_{\gamma t}^\infty (Uf)^*(s)\frac{ds}{s}, \text{ for all } f \in S. \tag{50}$$

Good-Lambda Inequalities

These inequalities apparently originate in the celebrated work of Burkholder–Gundy [17] (cf. also [16]) on extrapolation of martingale inequalities. They have been used since then to great effect in probability, and also in classical harmonic analysis, probably beginning with [18] and Coifman–Fefferman [21]. Inequalities on the oscillation operator $f^{**} - f^{*}$ are closely connected with good-lambda inequalities. This connection was pointed out long ago by Neveu [77], Herz [36], Bagby and Kurtz (cf. [5, 56]), among others. In this section we formalize some of their ideas.

To fix matters, let μ be a measure on R^n, and let T and H be operators acting on a sufficiently rich class of functions. A prototypical good-lambda inequality has the following form: for all $\lambda > 0, \varepsilon > 0$, there exists $c(\varepsilon) > 0$, with $c(\varepsilon) \to 0$, as $\varepsilon \to 0$, such that

$$\mu\{|Tf| > 2\lambda, |Hf| \leq \varepsilon\lambda\} \leq c(\varepsilon)\mu\{|Tf| > \lambda\}. \tag{51}$$

The idea here is that if the behavior of H is known on r.i. spaces, say on L^p spaces, then we can also control the behavior of T. Indeed, the distribution function of Tf can be controlled by the following elementary argument:

$$\begin{aligned}\mu\{|Tf| > 2\lambda\} &\leq \mu\{|Tf| > 2\lambda, |Hf| \leq \varepsilon\lambda\} + \mu\{|Tf| > 2\lambda, |Hf| > \varepsilon\lambda\}\\ &\leq c(\varepsilon)\mu\{|Tf| > \lambda\} + \mu\{|Hf| > \varepsilon\lambda\}.\end{aligned}$$

Then, since

$$\|f\|_p^p = p\int_0^\infty \lambda^{p-1}\mu\{|f| > \lambda\}d\lambda,$$

we readily see that we can estimate the norm of $\|Tf\|_p^p$ in terms of the norm of $\|Hf\|_p^p$ by means of making ε sufficiently small in order to be able to collect the two $\|Tf\|_p^p$ terms on the left-hand side of the inequality.

In [56], the author shows the following stronger good-lambda inequality for $f^{\#}$: There exists $B > 0$, such that for all $\varepsilon > 0, \lambda > 0$, and all locally integrable f, we have

$$\mu\{|f| > Bf^{\#} + \lambda\} \leq \varepsilon\mu\{|f| > \lambda\}.$$

This inequality is used to show the following oscillation inequality (cf. [56, p. 270]):

$$f^{*}(t) - f^{*}(2t) \leq Cf^{\#*}(\frac{t}{2}), t > 0,$$

where C is an absolute constant.

More generally, the argument in [56] can be formalized as follows:

Theorem 2. *Suppose that T and H are operators acting on the Schwartz class S, such that, moreover, for all $\varepsilon > 0$, there exists $B > 0$, such that for all $\lambda > 0$,*

$$\mu\{|Tf| > B\,|Hf| + \lambda\} \leq \varepsilon\mu\{|Tf| > \lambda\}. \tag{52}$$

Then, there exists a constant $C > 0$ such that for all $t > 0$, and for all $f \in S$,

$$(Tf)^*\,(t) - (Tf)^*\,(2t) \leq C\,(Hf)^*\,(\frac{t}{2}). \tag{53}$$

Proof. Let $\varepsilon = \frac{1}{4}$, and fix $B := B(\frac{1}{4})$ such that (52) holds for all $\lambda > 0$. Let $f \in S$, and select $\lambda = (Tf)^*\,(2t)$. Then,

$$\mu\{|Tf| > B\,|Hf| + (Tf)^*\,(2t)\} \leq \frac{1}{4}\mu\{|Tf| > (Tf)^*\,(2t)\} \leq \frac{t}{2}.$$

By definition we have,

$$\mu\{|Hf| > (Hf)^*\,(\frac{t}{2})\} \leq \frac{t}{2}.$$

Consider the set $A = \{|Tf| > B(Hf)^*(\frac{t}{2}) + (Tf)^*\,(2t)\}$. Then, it is easy to see, by contradiction, that

$$A \subset \{|Tf| > B\,|Hf| + (Tf)^*\,(2t)\}\bigcup\{|Hf| > (Hf)^*\,(t/2)\}.$$

Consequently,

$$\mu(A) \leq \frac{t}{2} + \frac{t}{2}.$$

Now, since

$$(Tf)^*\,(t) = \inf\{s : \mu\{|Tf| > s\} \leq t\},$$

it follows that

$$(Tf)^*\,(t) \leq B(Hf)^*(\frac{t}{2}) + (Tf)^*\,(2t),$$

as we wished to show. □

Remark 5. It is easy to compare the oscillation operators $(Tf)^*\,(t) - (Tf)^*\,(2t)$ and $(Tf)^{**}\,(t) - (Tf)^*\,(t)$. For example, it is shown in [7, Theorem 4.1, p. 1223] that

$$(Tf)^*\,(\frac{t}{2}) - (Tf)^*\,(t) \leq 2\left((Tf)^{**}\,(t) - (Tf)^*\,(t)\right),$$

and

$$\left((Tf)^{**}(t) - (Tf)^{*}(t)\right) \leq \frac{1}{t}\int_0^t \left((Tf)^{*}(\frac{s}{2}) - (Tf)^{*}(s)\right) ds$$
$$+ (Tf)^{*}(\frac{t}{2}) - (Tf)^{*}(t). \tag{54}$$

Combining Theorem 2 and the previous remark we have the following:

Theorem 3. *Suppose that T and H satisfy the strong good-lambda inequality (52). Then,*

$$\left((Tf)^{**}(t) - (Tf)^{*}(t)\right) \leq 2B(Hf)^{**}(\frac{t}{4}).$$

Proof. The desired result follows combining (53) with (54). □

Remark 6. It is easy to convince oneself that the good-lambda inequalities of the form (52) are, in fact, stronger than the usual good-lambda inequalities, e.g., of the form (51) (cf. [56]).

Remark 7. Clearly there are many nice results lurking in the background of this section. For example, a topic that comes to mind is to explore the use of good-lambda inequalities in the interpolation theory of operator ideals and its applications (cf. [20, 66], and the references therein). On the classical analysis side it would be of interest to explore the connections of the interpolation methods with the maximal inequalities due to Muckenhoupt–Wheeden and Hedberg–Wolff (cf. [3, 40], and the references therein).

Extrapolation of Inequalities: Burkholder–Gundy–Herz Meet Calderón–Maz'ya and Cwikel et al.

The leitmotif of [17] is the extrapolation of inequalities for the classical operators acting on martingales (e.g., martingale transforms, maximal operators, square functions). There are two main ingredients to the method. First, the authors, modeling on the classical operators acting on martingales, single out properties that the operators under consideration will be required to satisfy. Then they usually assume that an L^p or weak type L^p estimate holds, and from this information they deduce a full family of L^p or even Orlicz inequalities. The main technical step of the extrapolation procedure consists of using the assumptions we have just described in order to prove suitable good-lambda inequalities. The method is thus different from the usual interpolation theory, which works for **all** operators that satisfy a **pair** of given estimates.

In [29] we have shown how to formulate some of the assumptions of [17] in terms of optimal decompositions to compute K-functionals. Then, assuming that the operators to be extrapolated act on interpolation spaces, one can extract oscillation inequalities using interpolation theory. In particular, the developments in [29] allow the extrapolation of operators that do not necessarily act on martingales, but also on function spaces, e.g., gradients, square functions, Littlewood–Paley functions, etc. The basic technique involved to achieve the extrapolation is to use the assumptions to prove an oscillation rearrangement inequality.[27]

Unfortunately, [29] is still unpublished, although some of the results have been discussed elsewhere (cf. [63]) or will appear soon (cf. [72]). In keeping with the theme of this note, in this section I want to present some more details on how one can extrapolate Sobolev inequalities and encode the information using the oscillation operator $f^{**} - f^{*}$.

Let us take as a starting point the weak type Gagliardo–Nirenberg–Sobolev inequality in R^n (cf. [58]):

$$\|f\|_{L(n',\infty)} = \|f\|_{(L^1(R^n),L^\infty(R^n))_{1/n,\infty}} \leq c_n \|\nabla f\|_{L^1}, f \in Lip_0(R^n). \tag{55}$$

Let $f \in Lip_0(R^n)$, and assume without loss that f is positive. Let $t > 0$, then an optimal decomposition for the computation of

$$K(t,f) := K(t,f;L^1(R^n),L^\infty(R^n)) = \int_0^t f^*(s)ds,$$

is given by

$$f = f_{f^*(t)} + (f - f_{f^*(t)}), \tag{56}$$

where

$$f_{f^*(t)}(x) = \begin{cases} f(x) - f^*(t) & \text{if } f^*(t) < f(x) \\ 0 & \text{if } f(x) \leq f^*(t) \end{cases}. \tag{57}$$

By direct computation we have

$$\begin{aligned} K(t,f) &\leq \|f_{f^*(t)}\|_{L^1} + t\,\|f - f_{f^*(t)}\|_{L^\infty} \\ &= \left(\int_0^t f^*(s)ds - tf^*(t)\right) + tf^*(t) \\ &= \int_0^t f^*(s)ds. \end{aligned}$$

On the other hand, if $f = f_0 + f_1$, with $f_0 \in L^1, f_1 \in L^\infty$, then

$$\begin{aligned} \int_0^t f^*(s)ds &\leq \int_0^t f_0^*(s)ds + \int_0^t f_1^*(s)ds \\ &\leq \|f_0\|_{L^1} + t\,\|f_1\|_{L^\infty}. \end{aligned}$$

[27] Herz [37] also developed a different technique to extrapolate oscillation rearrangement inequalities for martingale operators.

Therefore,

$$K(t,f) = \left\| f_{f^*(t)} \right\|_{L^1} + t \left\| f - f_{f^*(t)} \right\|_{L^\infty}.$$

Also note that, confirming (18) and (19) above, by direct computation we have

$$\begin{aligned} \left\| f_{f^*(t)} \right\|_{L^1} &= \int_0^t f^*(s)ds - tf^*(t) \\ &= t\left(f^{**}(t) - f^*(t)\right), \\ \left\| f - f_{f^*(t)} \right\|_{L^\infty} &= f^*(t). \end{aligned}$$

The commutation of the gradient with truncations[28] implies

$$\left\| \nabla f_{f^*(t)} \right\|_{L^1} \leq \int_{\{f > f^*(t)\}} |\nabla f| \, dx.$$

Therefore

$$\left\| \nabla f_{f^*(t)} \right\|_{L^1} \leq \int_0^t |\nabla f|^* (s)ds.$$

We apply the inequality (55) to $f_{f^*(t)}$. We find

$$\begin{aligned} \left\| f_{f^*(t)} \right\|_{(L^1(R^n),L^\infty(R^n))_{1/n,\infty}} &\leq c_n \left\| \nabla f_{f^*(t)} \right\|_{L^1} \\ &\leq c_n \int_0^t |\nabla f|^* (s)ds. \end{aligned}$$

We estimate the left-hand side

$$\begin{aligned} \left\| f_{f^*(t)} \right\|_{(L^1(R^n),L^\infty(R^n))_{1/n,\infty}} &= \sup_{s>0} s^{-1/n} K(s, f_{f^*(t)}) \\ &= \sup_{s>0} s^{-1/n} K(s, f - (f - f_{f^*(t)})) \\ &\geq t^{-1/n} K(t, f - (f - f_{f^*(t)})) \\ &\geq t^{-1/n} \{K(t,f) - K(t, f - f_{f^*(t)})\}, \end{aligned}$$

where the last inequality follows by the triangle inequality, since $K(t,\cdot)$ is a norm. Now,

$$K(t,f) = tf^{**}(t),$$

---

[28](cf. [6, 34, 65, 67]).

and

$$K(t,f - f_{f^*(t)}) \leq t \left\| f_{f^*(t)} - f \right\|_{L^\infty} \\ = tf^*(t).$$

Thus,

$$\left\| f_{f^*(t)} \right\|_{(L^1(R^n),L^\infty(R^n))_{1/n,\infty}} \geq t^{-1/n} \left(tf^{**}(t) - tf^*(t) \right).$$

Combining estimates we find

$$\left(tf^{**}(t) - tf^*(t) \right) t^{-1/n} \leq c_n \int_0^t |\nabla f|^* (s) ds,$$

which can be written as

$$f^{**}(t) - f^*(t) \leq c_n t^{1/n} |\nabla f|^{**}(t). \tag{58}$$

This inequality had been essentially obtained by Kolyada [50] and is equivalent (cf. [65]) to earlier inequalities by Talenti [86] but its role in the study of limiting Sobolev inequalities, and the introduction of the $L(\infty, q)$ spaces, was only pointed out in [7]. In [65] it was shown that (58) is equivalent to the isoperimetric inequality. More generally, Martin–Milman extended (58) to Gaussian measures [62], and later[29] (cf. [64]) to metric measure spaces, where the inequality takes the following form:

$$f^{**}(t) - f^*(t) \leq \frac{t}{I(t)} |\nabla f|^{**}(t), \tag{59}$$

where I is the isoperimetric profile associated with the underlying geometry. In fact they show the equivalence of (59) with the corresponding isoperimetric inequality (for a recent survey cf. [64]).

One can prove the Gaussian Sobolev version of (59) using the same extrapolation procedure as above, but taking as a starting point Ledoux's inequality [57] as a replacement of the Gagliardo–Nirenberg inequality (cf. [63]). Further recent extensions of the Martin–Milman inequality on Gaussian measure can be found in [89].

Let us now show another Sobolev rearrangement inequality for oscillations, apparently first recorded in [65]. We now take as our starting inequality for the extrapolation procedure the sharp form of the Gagliardo–Nirenberg inequality, which can be formulated as

$$\|f\|_{L(n',1)} = \|f\|_{(L^1(R^n),L^\infty(R^n))_{1/n,1}} \leq c_n \|\nabla f\|_{L^1}, f \in Lip_0(R^n).$$

[29] See also [47] for a maximal function approach to oscillation inequalities for the gradient.

We apply the inequality to $f_{f^*(t)}$. The right-hand side we have already estimated,

$$\left\|f_{f^*(t)}\right\|_{(L^1(R^n),L^\infty(R^n))_{1/n,1}} \leq c_n \int_0^t |\nabla f|^*(s)ds.$$

Now,

$$\begin{aligned}
\left\|f_{f^*(t)}\right\|_{(L^1(R^n),L^\infty(R^n))_{1/n,1}} &= \int_0^\infty s^{-1/n} K(s, f_{f^*(t)}) \frac{ds}{s} \\
&\geq \int_0^t s^{-1/n} K(s, f - (f - f_{f^*(t)})) \frac{ds}{s} \\
&\geq \int_0^t s^{-1/n} \{K(s,f) - K(s, f - f_{f^*(t)})\} \frac{ds}{s} \\
&= \int_0^t s^{-1/n} \{sf^{**}(s) - K(s, f - f_{f^*(t)})\} \frac{ds}{s}.
\end{aligned}$$

Note that since f^* decreases, for $s \leq t$, we have

$$\begin{aligned}
K(s, f - f_{f^*(t)}) &\leq s \left\|f - f_{f^*(t)}\right\|_{L^\infty} \\
&= sf^*(t) \\
&\leq sf^*(s).
\end{aligned}$$

Therefore,

$$\begin{aligned}
\|f\|_{(L^1(R^n),L^\infty(R^n))_{1/n,1}} &\geq \int_0^t s^{-1/n} \{sf^{**}(s) - sf^*(s)\} \frac{ds}{s} \\
&= \int_0^t s^{1-1/n} \{f^{**}(s) - f^*(s)\} \frac{ds}{s}.
\end{aligned}$$

Consequently,

$$\int_0^t s^{1-1/n} \{f^{**}(s) - f^*(s)\} \frac{ds}{s} \leq c_n \int_0^t |\nabla f|^*(s)ds,$$

an inequality first shown in [65].

Bilinear Interpolation

In this section we show an extension of (22) to a class of bilinear operators that have a product or convolution like structure. These operators were first introduced by O'Neil (cf. [12, Exercise 5, p. 76]).

Let $\vec{A}, \vec{B}, \vec{C}$, be Banach pairs, and let Π be a bilinear bounded operator such that

$$\Pi : \begin{cases} A_0 \times B_0 \to C_0 \\ A_0 \times B_1 \to C_1 \\ A_1 \times B_0 \to C_1 \end{cases}.$$

For example, the choice $A_0 = B_0 = C_0 = L^\infty$, and $A_1 = B_1 = C_1 = L^1$, corresponds to a regular product operator $\Pi_1(f,g) = fg$, while the choice $A_0 = B_0 = C_1 = L^1$, and $A_1 = B_1 = C_0 = L^\infty$, corresponds to a convolution operator $\Pi_2(f,g) = f * g$ on R^n, say. The main boundedness result is given by (cf. [12, Exercise 5, p. 76]):

$$\|\Pi(f,g)\|_{\vec{C}_{\theta,r}} \leq \|f\|_{\vec{A}_{\theta_1,q_1 r}} \|g\|_{\vec{B}_{\theta_2,q_2 r}}, \tag{60}$$

where $\theta, \theta_i \in (0,1), \theta = \theta_1 + \theta_2, q_i, r \in [1,\infty], i = 1,2$, and $\frac{1}{r} \leq \frac{1}{q_1 r} + \frac{1}{q_2 r}$.

Example 2. To illustrate the ideas, make it easy to compare results, and to avoid the tax of lengthy computations with indices, we shall only model an extension of (22) for Π_1. Let us thus take $\vec{A} = \vec{B} = \vec{C} = (A_0, A_1)$, and let $\vec{X} = (X_0, X_1) = (A_1, A_0)$. Further, in order to be able to use the quadratic form argument outlined in the Introduction, we choose $q_1 = q_2 = 2$, $r = p$, $\theta_1 = \theta_2 = \frac{\theta}{2}$. Then, from (60) we get

$$\|\Pi_1(f,g)\|_{\vec{X}_{1-\theta,p}} \preceq \|f\|_{\vec{X}_{1-\frac{\theta}{2},2p}} \|g\|_{\vec{X}_{1-\frac{\theta}{2},2p}}. \tag{61}$$

Since $(1-\frac{1}{2})(1-\theta) + \frac{1}{2}1 = 1 - \frac{\theta}{2}$, we can use the reiteration theorem to write

$$\vec{X}_{1-\frac{\theta}{2},2p} = (\vec{X}_{1-\theta,p}, X_1)_{\frac{1}{2},2p},$$

and find that

$$\|\Pi_1(f,g)\|_{\vec{X}_{1-\theta,p}} \preceq \|f\|_{(\vec{X}_{1-\theta,p},X_1)_{\frac{1}{2},2p}} \|g\|_{(\vec{X}_{1-\theta,p},X_1)_{\frac{1}{2},2p}}. \tag{62}$$

By the equivalence theorem[30] (cf. (42) above),

$$(\vec{X}_{1-\theta,p}, X_1)_{\frac{1}{2},2p} = (\vec{X}_{1-\theta,p}, X_1)^{(2)}_{\frac{1}{2},2p}.$$

[30] In this example we are not interested on the precise dependence of the constants of equivalence.

Therefore,

$$\begin{aligned}\|\Pi_1(f,g)\|_{\vec{X}_{1-\theta,p}} &\preceq \|f\|_{(\vec{X}_{1-\theta,p},X_1)^{(2)}_{\frac{1}{2},2p}} \|g\|_{(\vec{X}_{1-\theta,p},X_1)^{(2)}_{\frac{1}{2},2p}} \\ &\preceq \|f\|^{1/2}_{\vec{X}_{1-\theta,p}} \|f\|^{1/2}_{(\vec{X}_{1-\theta,p},X_1)^{(1)}_{1,\infty}} \|g\|^{1/2}_{\vec{X}_{1-\theta,p}} \|g\|^{1/2}_{(\vec{X}_{1-\theta,p},X_1)^{(1)}_{1,\infty}},\end{aligned}$$

where in the last step we have used (45) applied to the pair $(\vec{X}_{1-\theta,p}, X_1)$. Consequently, by the quadratic formula, we finally obtain

$$\|\Pi_1(f,g)\|_{\vec{X}_{1-\theta,p}} \preceq \|f\|_{\vec{X}_{1-\theta,p}} \|g\|_{(\vec{X}_{1-\theta,p},X_1)^{(1)}_{1,\infty}} + \|f\|_{(\vec{X}_{1-\theta,p},X_1)^{(1)}_{1,\infty}} \|g\|_{\vec{X}_{1-\theta,p}}.$$

Remark 8. The method above uses one of the more powerful tools of the real method: the re-scaling of inequalities. However, in this case there is substantial difficulty for the implementation of the result since techniques for the actual computation of the space $\left(\vec{X}_{1-\theta,p}, X_1\right)^{(1)}_{1,\infty}$ are not well developed at present time.[31] Therefore we avoid the use of (62) and instead prove directly that if[32] $\theta = \frac{1}{p}$, we have

$$\|f\|_{\vec{X}_{1-\frac{\theta}{2},2p}} \preceq \|f\|^{1/2}_{\vec{X}_{1-\theta,p}} \|f\|^{1/2}_{\vec{X}^{(1)}_{1,\infty}}. \tag{63}$$

This given, applying (63) to both terms on the right-hand side of (61) and then using the quadratic form argument we find

$$\|\Pi_1(f,g)\|_{\vec{X}_{1-\theta,p}} \preceq \|f\|_{\vec{X}_{1-\theta,p}} \|g\|_{\vec{X}^{(1)}_{1,\infty}} + \|g\|_{\vec{X}_{1-\theta,p}} \|f\|_{\vec{X}^{(1)}_{1,\infty}}. \tag{64}$$

In the particular case of product operators and L^p spaces, $1 < p < \infty$, (64) reads

$$\|fg\|_{L^p} \preceq \|f\|_{L^p} \|g\|_{L(\infty,\infty)} + \|g\|_{L^p} \|f\|_{L(\infty,\infty)},$$

which should be compared with (22) recalling that

$$\|f\|_{L(\infty,\infty)} \leq c\,\|f\|_{BMO}.$$

[31] It is an interesting open problem to modify Holmstedt's method to be able to keep track, in a nearly optimal way, both coordinates in the Gagliardo diagram, when doing reiteration. For more on the computation of Gagliardo coordinate spaces see the forthcoming [74].

[32] To simplify the computations we model the L^p case here. Another simplification is that in this argument we don't need to be fuzzy about constants.

Proof (of (63)). To simplify the notation we let $K(t,f;\vec{X}) = K(t)$. Then,

$$\begin{aligned}
\|f\|_{\vec{X}_{1-\frac{\theta}{2},2p}} &= \left\{\int_0^\infty \left(K(s)s^{-(1-\frac{\theta}{2})}\right)^{2p}\frac{ds}{s}\right\}^{1/2p} \\
&\leq \left\{\int_0^t \left(K(s)s^{-(1-\frac{\theta}{2})}\right)^{2p}\frac{ds}{s}\right\}^{1/2p} + \left\{\int_t^\infty \left(K(s)s^{-(1-\frac{\theta}{2})}\right)^{2p}\frac{ds}{s}\right\}^{1/2p} \\
&= (1) + (2).
\end{aligned}$$

We proceed to estimate each of these two terms starting with (2) :

$$\begin{aligned}
(2) &= \left\{\int_t^\infty \left(K(s)s^{-(1-\theta)}\right)^p s^{(1-\theta)p}s^{-(1-\frac{\theta}{2})p}\left(K(s)s^{-(1-\frac{\theta}{2})}\right)^p\frac{ds}{s}\right\}^{1/2p} \\
&\leq \left\{\sup_{s\geq t} s^{(1-\theta)}s^{-(1-\frac{\theta}{2})}\left(K(s)s^{-(1-\frac{\theta}{2})}\right)\right\}^{1/2}\left\{\int_t^\infty \left(K(s)s^{-(1-\theta)}\right)^p\frac{ds}{s}\right\}^{1/2p} \\
&\leq \left\{\sup_{s\geq t}\frac{K(s)}{s}\right\}^{1/2}\|f\|_{\vec{X}_{1-\theta,p}}^{1/2} \\
&= \left\{\frac{K(t)}{t}\right\}^{1/2}\|f\|_{\vec{X}_{1-\theta,p}}^{1/2} \\
&= t^{-1/2}t^{(1-\theta)/2}\left\{K(t)t^{-(1-\theta)}\right\}^{1/2}\|f\|_{\vec{X}_{1-\theta,p}}^{1/2} \\
&\leq t^{-1/p2}\|f\|_{\vec{X}_{1-\theta,p}}.
\end{aligned}$$

Moreover,

$$\begin{aligned}
(1) &= \left\{\int_0^t \left(\frac{K(s)}{s}s^{-(1-\frac{\theta}{2})+1}\right)^{2p}\frac{ds}{s}\right\}^{1/2p} \\
&\leq \left\{\int_0^t \left(\left[\frac{K(s)}{s}-\frac{K(t)}{t}\right]s^{-(1-\frac{\theta}{2})+1}\right)^{2p}\frac{ds}{s}\right\}^{1/2p} + \left\{\int_0^t \left(\frac{K(t)}{t}s^{-(1-\frac{\theta}{2})+1}\right)^{2p}\frac{ds}{s}\right\}^{1/2p} \\
&= (a) + (b).
\end{aligned}$$

The term (b) is readily estimated:

$$\begin{aligned}
(b) &= \frac{K(t)}{t}\left\{\int_0^t s^{\frac{\theta}{2}2p}\frac{ds}{s}\right\}^{1/2p} \\
&\sim \frac{K(t)}{t}t^{1/2p}
\end{aligned}$$

$$= K(t)t^{-(1-\theta)}t^{(1-\theta)}t^{1/2p-1}$$

$$\preceq \|f\|_{\vec{X}_{1-\theta,p}} t^{-1/2p}.$$

Next we use the familiar estimate

$$\left[\frac{K(s)}{s} - \frac{K(t)}{t}\right] = \int_s^t \left[\frac{K(s)}{s} - K'(s)\right]\frac{ds}{s}$$

$$\leq \|f\|_{\vec{X}_{1,\infty}} \log\frac{t}{s},$$

to see that

$$(a) \leq \|f\|_{\vec{X}_{1,\infty}} \left\{\int_0^t s\left(\log\frac{t}{s}\right)^{2p}\frac{ds}{s}\right\}^{1/2p}$$

$$\preceq \|f\|_{\vec{X}^{(1)}_{1,\infty}} t^{1/2p}.$$

Collecting estimates, we have

$$\|f\|_{\vec{X}_{1-\frac{\theta}{2},2p}} \preceq t^{-1/p2} \|f\|_{\vec{X}_{1-\theta,p}} + \|f\|_{\vec{X}^{(1)}_{1,\infty}} t^{1/2p}.$$

Balancing the two terms on the right-hand side we find

$$\|f\|_{\vec{X}_{1-\frac{\theta}{2},2p}} \preceq \|f\|^{1/2}_{\vec{X}_{1-\theta,p}} \|f\|^{1/2}_{\vec{X}^{(1)}_{1,\infty}},$$

as we wished to show. □

Remark 9. It would be of interest to implement a similar model analysis for $\Pi_2(f, g) = f * g$. We claim that it is easy, however, we must leave the task to the interested reader.

Remark 10. The results of this section are obviously connected with Leibniz rules in function spaces. This brings to mind the celebrated commutator theorem of Coifman–Rochberg–Weiss [22] which states that if T is a Calderón–Zygmund operator (cf. [83]), and b is a BMO function, then $[T, b]f = bTf - T(bf)$, defines a bounded linear operator[33]

$$[T, b] : L^p(R^n) \to L^p(R^n), 1 < p < \infty.$$

The result has been extended in many different directions. In particular, an abstract theory of interpolation of commutators has evolved from it (cf. [28, 81], for recent

[33] In fact, the boundedness of $[T, b]$ for all CZ operators implies that $b \in BMO$ (cf. [41])

surveys). In this theory a crucial role is played by a class of operators Ω such that, for bounded operators T on a given interpolation scale, the commutator $[T, \Omega]$ is also bounded. The operators Ω often satisfy some of the functional equations associated with derivation operators. At present time we know very little about how this connection comes about or how to exploit it in concrete applications (cf. [30, 51]). More in keeping with the topic of this paper, and in view of many possible interesting applications, it would be also of interest to study oscillation inequalities for commutators $[T, \Omega]$ in the context of interpolation theory. In this connection we should mention that in [61], the authors formulate a generalized version of the Coifman–Rochberg–Weiss commutator theorem, valid in the context of real interpolation, where in a suitable fashion the function space W, introduced in [76], plays the role of BMO :

$$W = \{f \in L^1_{loc}(0, \infty) : \sup_t \left| \frac{1}{t} \int_0^t f(s)ds - f(t) \right| < \infty\}.$$

Obviously, $f \in L(\infty, \infty)$ iff $f^* \in W$.

A Brief Personal Note on Cora Sadosky

I still remember well the day (circa March 1977) that I met Mischa Cotlar and Cora Sadosky at Mischa's apartment in Caracas (cf. [71]). I was coming from Sydney, en route to take my new job as a Visiting Professor at Universidad de Los Andes, in Merida. And while, of course, I had heard a lot about them, I did not know them personally. Mischa and I had exchanged some correspondence (no e-mail those days!). I had planned to come to spend an afternoon with the Cotlars, taking advantage of a 24-h stopover while en route to Merida, a colonial city in the Andes region of Venezuela. As it turns out, my flight for the next day had been cancelled, and Mischa and Yani invited me to stay over at their place.

When I came to the Cotlar's apartment, Corita and Mischa were working in the dining room. Mischa gave me a brief explanation of what they were doing mathematically. In fact, he dismissed the whole enterprise.[34] I would learn much later that what they were doing then, would turn out to be very innovative and influential research.[35]

Corita, who also knew about my impending arrival, greeted me with something equivalent to *Oh, so you really do exist*![36] Being a *porteño* myself, I quickly

[34] I would learn quickly that Mischa's modesty was legendary.

[35] From this period I can mention [23–25].

[36] Existence here to be taken in a non mathematical sense. Of course at point in time I did NOT *exist* mathematically!

found Corita's style quite congenial and the conversation took off. We formed a friendship that lasted for as long as she lived.

As I later learned, most people called her Cora, but the Cotlars, and a few other old friends, that knew her from childhood, called her Corita.[37] Having been introduced to her at the Cotlar's home, I proceeded to call her Corita too... and that was the way it would always be.[38]

By early 1979 I moved for one semester to Maracaibo and afterwards to Brasilia. Mischa was very helpful connecting me first with Jorge Lebowitz (Maracaibo), and then with Djairo Figuereido (Brasilia). In the mean time, Corita herself had moved to the USA, where we met again, at an AMS Special Session in Harmonic Analysis.

At the time I had a visiting position at UI at Chicago, and was trying to find a tenure track job. She took an interest in my situation, and gave me very useful suggestions on the job hunting process. She would remain very helpful throughout my career. In particular, when she learned,[39] on her own, about my application for a membership at the Institute for Advanced Study in 1984, she supported my case.... I did not know this until she called me to let me know that my application had been accepted!

Corita and I met again many times over the course of the years. There were conferences on Interpolation Theory[40] in Lund and Haifa, on Harmonic Analysis in Washington D.C., Madrid and Boca Raton, there were Chicago meets to celebrate various birthdays of Alberto Calderón, a special session on Harmonic Analysis in Montreal, etc. Vanda and I went to have dinner with Corita and her family, when we all coincided during a visit to Buenos Aires in 1985. She was instrumental in my participation at Mischa's 80 birthday Conference in Caracas, 1994. We wrote papers for books that each of us edited. At some point in time she and her family came to visit us in Florida.

A few months before she passed I was helping her via e-mail, with the texting[41] of a paper of hers about Mischa Cotlar.

Probably the best way I have to describe Corita is to say that she was a force of nature. She was a brilliant mathematician, with an intense but charming personality. Having had to endure herself exile, discrimination, and very difficult working conditions, she was very sensitive to the plight of others. I am sure many of the testimonies in this book will describe how much she helped to provide opportunities for younger mathematicians to develop their careers.

---

[37] In Spanish *Corita* means little Cora.

[38] Many many years later she told me that by then everyone called her Cora, except Mischa and myself and that she would prefer for me to call her Cora. I said, of course, Corita!

[39] She had a membership at IAS herself that year.

[40] Harmonic analysts trained in the 1960s had a special place in their hearts for Interpolation theory. It was after all a theory to which the great masters of the Chicago school (e.g., Calderón, Stein, Zygmund) had made fundamental contributions.

[41] I mean using TeX!

Some things are hard to change. As much as I could not train myself to call her *Cora*, in our relationship she was always "big sister." I will always be grateful to her, for all her help and her friendship.

I know that the space *BMO* had a very special place in her mathematical interests and, indeed, *BMO* spaces appear in many considerations throughout her works. For this very reason, and whatever the merits of my small contribution, I have chosen to dedicate this note on *BMO* inequalities to her memory.

Acknowledgements The author was partially supported by a grant from the Simons Foundation (#207929 to Mario Milman). I am grateful to Sergey Astashkin and Michael Cwikel for useful comments that helped improve the quality of the paper. Needless to say that I remain fully responsible for the remaining shortcomings.

References

1. D. Aalto, Weak L^{∞} and *BMO* in metric spaces. Boll. Unione Mat. Ital. **5**, 369–385 (2012)
2. J. Alvarez, M. Milman, Spaces of Carleson measures: duality and interpolation. Ark. Mat. **25**, 155–174 (1987)
3. N. Arcozzi, R. Rochberg, E. Sawyer, B.D. Wick, Potential theory on trees, graphs and Ahlfors-regular metric spaces. Potential Anal. **41**, 317–366 (2014)
4. S. Astashkin, K. Lykov, *Extrapolation Description of Rearrangement Invariant Spaces and Related Problems*. Banach and Function Spaces III (Yokohama Publishers, Yokohama, 2011), pp. 1–52
5. R.J. Bagby, D.S. Kurtz, A rearranged good-λ inequality. Trans. Am. Math. Soc. **293**, 71–81 (1986)
6. D. Bakry, T. Coulhon, M. Ledoux, L. Saloff-Coste, Sobolev inequalities in disguise. Indiana Univ. Math. J. **44**, 1033–1074 (1995)
7. J. Bastero, M. Milman, F.J. Ruiz Blasco, A note on $L(\infty, q)$ spaces and Sobolev embeddings. Indiana Univ. Math. J. **52**, 1215–1230 (2003)
8. C. Bennett, Banach function spaces and interpolation methods I. The abstract theory. J. Funct. Anal. **17**, 409–440 (1974)
9. C. Bennett, R. Sharpley, *Interpolation of Operators*. Pure and Applied Mathematics, vol. 129 (Academic Press, Boston, 1988)
10. C. Bennett, R. Sharpley, Weak-type inequalities for H^p and BMO, in *Harmonic Analysis in Euclidean Spaces (Proc. Sympos. Pure Math., Williams Coll., Williamstown, Mass., 1978)*. Proceedings of Symposia in Pure Mathematics, Part 1, vol. XXXV (American Mathematical Society, Providence, 1979), pp. 201–229
11. C. Bennett, R.A. DeVore, R. Sharpley, Weak-L $^{\infty}$ and BMO. Ann. Math **113**, 601–611 (1981)
12. J. Bergh, J. Löfstrom, *Interpolation Spaces* (Springer, New York, 1976)
13. H. Brezis, S. Wainger, A note on limiting cases of Sobolev embeddings and convolution inequalities. Commun. Partial Differ. Equ. **5**, 773–789 (1980)
14. A.P. Calderón, Intermediate spaces and interpolation, the complex method. Stud. Math. **24**, 113–190 (1964)
15. A.P. Calderón, Spaces between L^1 and L^{∞} and the theorem of Marcinkiewicz. Stud. Math. **26**, 273–299 (1966).
16. D.L. Burkholder, Distribution function inequalities for martingales. Ann. Probab. **1**, 19–42 (1973)
17. D.L. Burkholder, R.F. Gundy, Extrapolation and interpolation of quasi-linear operators on martingales. Acta Math. **124**, 249–304 (1970)

18. D.L. Burkholder, R.F. Gundy, Distribution function inequalities for the area integrals. Stud. Math. **44**, 527–544 (1972)
19. J. Chen, X. Zhu, A note on BMO and its application. J. Math. Anal. Appl. **303**, 696–698 (2005)
20. F. Cobos, M. Milman, On a limit class of approximation spaces. Numer. Funct. Anal. Optim. **11**, 11–31 (1990)
21. R.R. Coifman, C. Fefferman, Weighted norm inequalities for maximal functions and singular integrals. Stud. Math. **51**, 241–250 (1974)
22. R.R. Coifman, R. Rochberg, G. Weiss, Factorization theorems for Hardy spaces in several variables. Ann. Math. **103**, 611–635 (1976)
23. M. Cotlar, C. Sadosky, Transformée de Hilbert, Théorème de Bochner et le Problème des Moments I. C. R. Acad. Sci. Paris Ser. A-B **285** (1977)
24. M. Cotlar, C. Sadosky, Transformée de Hilbert, Théorème de Bochner et le Problème des Moments II. C. R. Acad. Sci. Paris Ser. A-B **285** (1977)
25. M. Cotlar, C. Sadosky, On the Helson-Szegö theorem and a related class of modified Toeplitz kernels, in *Harmonic Analysis in Euclidean Spaces*. Proceedings of Symposia in Pure Mathematics, vol. 35 (American Mathematical Society, Providence, 1979), pp. 383–407
26. M. Cwikel, S. Janson, Real and complex interpolation methods for finite and infinite families of Banach spaces. Adv. Math. **66**, 234–290 (1987)
27. M. Cwikel, B. Jawerth, M. Milman, On the fundamental lemma of interpolation theory. J. Approx. Theory **60**, 70–82 (1990)
28. M. Cwikel, N. Kalton, M. Milman, R. Rochberg, A unified theory of commutator estimates for a class of interpolation methods. Adv. Math. **169**, 241–312 (2002)
29. M. Cwikel, B. Jawerth, M. Milman, *A note on extrapolation of inequalities*. Preprint (2010)
30. M. Cwikel, M. Milman, R. Rochberg, *An introduction to Nigel Kalton's work on differentials of complex interpolation processes for Köthe spaces*. Arxiv: http://arxiv.org/pdf/1404.2893.pdf (2014)
31. M. Cwikel, Y. Sagher, P. Shvartsman, *A Geometrical/Combinatorical Question with Implications for the John-Nirenberg Inequality for BMO Functions*. Banach Center Publications, vol. 95 (Polish Academy of Sciences, Warszawa, 2011), pp. 45–53
32. C. Fefferman, E.M. Stein, H^p spaces of several variables. Acta Math. **129**, 137–193 (1972)
33. A.M. Garsia, E. Rodemich, Monotonicity of certain functionals under rearrangements. Ann. Inst. Fourier (Grenoble) **24**, 67–116 (1974)
34. P. Hajlasz, Sobolev inequalities, truncation method, and John domains, in *Papers on Analysis, Rep. Univ. Jyväskylä Dep. Math. Stat.*, vol. 83 (Univ. Jyväskylä, Jyväskylä, 2001), pp. 109–126
35. K. Hansson, Imbedding theorems of Sobolev type in potential theory. Math. Scand. **45**, 77–102 (1979)
36. C. Herz, *A Best Possible Marcinkiewicz Theorem with Applications to Martingales* (Mc Gill University, Montréal, 1974)
37. C. Herz, An interpolation principle for martingale inequalities. J. Funct. Anal. **22**, 1–7 (1976)
38. T. Holmstedt, Equivalence of two methods of interpolation. Math. Scand. **18**, 45–52 (1966)
39. T. Holmstedt, Interpolation of quasi-normed spaces. Math. Scand. **26**, 177–199 (1970)
40. P. Honzík, B.J. Jaye, On the good-λ inequalities for nonlinear potentials. Proc. Am. Math. Soc. **140**, 4167–4180 (2012)
41. S. Janson, Mean oscillation and commutators of singular integral operators. Ark. Math. **16**, 263–270 (1978)
42. B. Jawerth, The K-functional for H^1 and BMO. Proc. Am. Math. Soc. **92**, 67–71 (1984)
43. B. Jawerth, M. Milman, Interpolation of weak type spaces. Math. Z. **201**, 509–519 (1989)
44. B. Jawerth, M. Milman, Extrapolation theory with applications. Mem. Am. Math. Soc. **89**(440) (1991)
45. B. Jawerth, M. Milman, *Function Spaces, Inequalities and Interpolation*. Lectures on Optimization, Image Processing and Interpolation Theory (Spring School on Analysis, Paseky, 2007)
46. F. John, L. Nirenberg, On functions of bounded mean oscillation. Commun. Pure Appl. Math. **14**, 415–426 (1961)

47. J. Kalis, M. Milman, Symmetrization and sharp Sobolev inequalities in metric spaces. Rev. Mat. Complutense **22**, 499–515 (2009)
48. G.E. Karadzhov, M. Milman, Extrapolation theory: new results and applications. J. Approx. Theory **133**, 38–99 (2005)
49. G.E. Karadzhov, M. Milman, J. Xiao, Limits of higher-order Besov spaces and sharp reiteration theorems. J. Funct. Anal. **221**, 323–339 (2005)
50. V.I. Kolyada, Rearrangements of functions and embedding theorems. Russ. Math. Surv. **44**, 73–117 (1989)
51. H. König, V. Milman, Characterizing the derivative and the entropy function by the Leibniz rule, with an appendix by D. Faifman. J. Funct. Anal. **261**, 1325–1344 (2011)
52. H. Kozono, Y. Taniuchi, Bilinear estimates in BMO and the Navier-Stokes equations. Math. Z. **235**, 173–194 (2000)
53. H. Kozono, H. Wadade, Remarks on Gagliardo-Nirenberg type inequality with critical Sobolev space and BMO. Math. Z. **259**, 935–950 (2008)
54. N. Krugljak, L. Maligranda, L.E. Persson, An elementary approach to the fractional Hardy inequality. Proc. Am. Math. Soc. **128**, 727–734 (2000)
55. N. Kruglyak, Y. Sagher, P. Shvartsman, Weak-type operators and the strong fundamental lemma of real interpolation theory. Stud. Math. **170**, 173–201 (2005)
56. D.S. Kurtz, Operator estimates using the sharp function. Pac. J. Math. **139**, 267–277 (1989)
57. M. Ledoux, Isopérimétrie et inégalitées de Sobolev logarithmiques gaussiennes. C. R. Acad. Sci. Paris **306**, 79–92 (1988)
58. G. Leoni, *A First Course in Sobolev Spaces*. Graduate Studies in Mathematics, vol. 105 (American Mathematical Society, New York, 2009)
59. J. Martín, M. Milman, Symmetrization inequalities and Sobolev embeddings. Proc. Am. Math. Soc. **134**, 2335–2347 (2006)
60. J. Martin, M. Milman, A note on Sobolev inequalities and limits of Lorentz spaces, in *Interpolation Theory and Applications*, ed. by L. De Carli et al. Contemporary Mathematics, vol. 445 (American Mathematical Society, Providence, 2007), pp. 237–245
61. J. Martin, M. Milman, An abstract Coifman-Rochberg-Weiss commutator theorem, in *Classical Analysis and Applications: A Conference in Honor of Daniel Waterman*, ed. by L. De Carli, K. Kazarian, M. Milman (World Scientific, Singapore, 2008), pp. 141–160
62. J. Martin, M. Milman, Isoperimetry and symmetrization for logarithmic Sobolev inequalities. J. Funct. Anal. **256**, 149–178 (2009)
63. J. Martin, M. Milman, Sobolev inequalities, rearrangements, isoperimetry and interpolation spaces, in *Concentration, Functional Inequalities and Isoperimetry*, ed. by C. Houdre et al. Contemporary Mathematics, vol. 545 (American Mathematical Society, Providence, 2011), pp. 167–193
64. J. Martin, M. Milman, Towards a unified theory of Sobolev inequalities, in *Special Functions, Partial Differential Equations, and Harmonic Analysis*, ed. by A.M. Stokolos et al. Springer Proceedings in Mathematics & Statistics, vol. 108 (Springer, Cham, 2014), pp. 163–201
65. J. Martin, M. Milman, E. Pustylnik, Sobolev inequalities: symmetrization and self improvement via truncation. J. Funct. Anal. **252**, 677–695 (2007)
66. M. Mastylo, M. Milman, Interpolation of real method spaces via some ideals of operators. Stud. Math. **136**, 17–35 (1999)
67. V. G. Maz'ya, *Sobolev Spaces with Applications to Elliptic Partial Differential Equations*. Grundlehren der Mathematischen Wissenschaften [Fundamental Principles of Mathematical Sciences], vol. 342, Second, revised and augmented edition (Springer, Heidelberg, 2011)
68. R. Metzler, J. Klafter, The restaurant at the end of the random walk: recent developments in the description of anomalous transport by fractional dynamics. J. Phys. A **37**, R161–R208 (2004)
69. M. Milman, A note on reversed Hardy inequalities and Gehring's lemma. Commun. Pure Appl. Math. **50**, 311–315 (1997)
70. M. Milman, Notes on limits of Sobolev spaces and the continuity of interpolation scales. Trans. Am. Math. Soc. **357**, 3425–3442 (2005)

71. M. Milman, Remembering Mischa, in *An Afternoon in Honor of Mischa Cotlar* (University of New Mexico, Albuquerque, 2007), http://www.math.unm.edu/conferences/10thAnalysis/testimonies.html
72. M. Milman, Self improving inequalities via penalty methods, in *High Dimensional Probability VII*, Corsica (2014)
73. M. Milman, *Extrapolation: three lectures in Barcelona.* Lecture Notes based on a 2013 course given at Univ. Autonoma Barcelona, in preparation
74. M. Milman, *BMO: Marcinkiewicz spaces, the E functional and Reiteration*, in preparation
75. M. Milman, E. Pustylnik, On sharp higher order Sobolev embeddings. Commun. Contemp. Math. **6**, 495–511 (2004)
76. M. Milman, Y. Sagher, An interpolation theorem. Ark. Mat. **22**, 33–38 (1984)
77. J. Neveu, *Martingales a Temps Discret* (Dunod, Paris, 1972)
78. V.I. Ovchinnikov, The method of orbits in interpolation theory. Math. Rep. **1**, 349–516 (1984)
79. E. Pustylnik, On a rearrangement-invariant function set that appears in optimal Sobolev embeddings. J. Math. Anal. Appl. **344**, 788–798 (2008)
80. P.F. Renaud, A reversed Hardy inequality. Bull. Aust. Math. Soc. **34**, 225–232 (1986)
81. R. Rochberg, Uses of commutator theorems in analysis, in *Interpolation Theory and Applications*. Contemporary Mathematics, vol. 445 (American Mathematical Society, Providence, 2007), pp. 277–295
82. G.C. Rota, *Ten lessons I wish I had been taught.* Notices Amer. Math. Soc. **44**, 22–25 (1997) http://www.ams.org/notices/199701/comm-rota.pdf
83. C. Sadosky, *Interpolation of Operators and Singular Integrals. An Introduction to Harmonic Analysis*. Monographs and Textbooks in Pure and Applied Mathematics, vol. 53 (Marcel Dekker, New York, 1979)
84. Y. Sagher, An application of the approximation functional in interpolation theory, in *Conference on Harmonic Analysis in Honor of Antoni Zygmund, Vol. I, II (Chicago, Ill., 1981)*. Wadsworth Mathematics Series (Wadsworth, Belmont, 1983), pp. 802–809
85. Y. Sagher, P. Shvartsman, Rearrangement-function inequalities and interpolationtheory. J. Approx. Theory **119**, 214–251 (2002)
86. G. Talenti, Elliptic equations and rearrangements. Ann. Scuola Norm. Sup. Pisa Cl. Sci. **4**, 697–718 (1976)
87. N.S. Trudinger, On imbeddings into Orlicz spaces and some applications. J. Math. Mech. **17**, 473–483 (1967)
88. J. Xiao, A reverse BMO-Hardy inequality. Real Anal. Exch. **25**, 673–678 (1999)
89. J. Xiao, Gaussian BV capacity. Adv. Calc. Var. **9**, 187–200 (2016)

Besov Spaces, Symbolic Calculus, and Boundedness of Bilinear Pseudodifferential Operators

Jodi Herbert and Virginia Naibo

Dedicated to the memory of Cora Sadosky

Abstract Mapping properties of bilinear pseudodifferential operators with symbols of limited smoothness in terms of Besov norms are proved in the context of Lebesgue spaces. Techniques used include the development of a symbolic calculus for certain classes of symbols considered.

2010*Mathematics Subject Classification*: Primary 35S05, 47G30; Secondary 42B15, 42B20

Introduction and Main Results

In this article we present results concerning boundedness properties in the setting of Lebesgue spaces of bilinear pseudodifferential operators associated to symbols of limited smoothness measured in terms of Besov norms. Closely connected to such symbols are the bilinear Hörmander classes $BS^m_{0,0}$, $m \in \mathbb{R}$, whose elements are infinitely differentiable complex-valued functions $\sigma = \sigma(x, \xi, \eta)$, $x, \xi, \eta \in \mathbb{R}^n$, that satisfy

$$\sup_{\substack{|\alpha|\leq N \\ |\beta|,|\gamma|\leq M}} \sup_{x,\xi,\eta\in\mathbb{R}^n} \langle \xi, \eta\rangle^{-m} |\partial_x^\alpha \partial_\xi^\beta \partial_\eta^\gamma \sigma(x, \xi, \eta)| < \infty \tag{1}$$

J. Herbert
Maranatha Baptist University, 745 W Main St., Watertown, WI 53094, USA
e-mail: jodi.herbert@mbu.edu

V. Naibo (✉)
Department of Mathematics, Kansas State University, 138 Cardwell Hall, Manhattan, KS 66506, USA
e-mail: vnaibo@math.ksu.edu

M.C. Pereyra et al. (eds.), *Harmonic Analysis, Partial Differential Equations, Complex Analysis, Banach Spaces, and Operator Theory (Volume 1)*, Association for Women in Mathematics Series 4, DOI 10.1007/978-3-319-30961-3_14

for all $N, M \in \mathbb{N}_0$ and where $\langle \xi, \eta \rangle := 1 + |\xi| + |\eta|$. Given $1 \leq p_1, p_2, p \leq \infty$ verifying Hölder's condition $\frac{1}{p_1} + \frac{1}{p_2} = \frac{1}{p}$, it was proved in [3] (see also [10] for a related result) that T_σ is bounded from $L^{p_1}(\mathbb{R}^n) \times L^{p_2}(\mathbb{R}^n)$ into $L^p(\mathbb{R}^n)$ for all $\sigma \in BS^m_{0,0}$ if $m < m(p_1, p_2)$, where

$$m(p_1, p_2) := -n \max\{\tfrac{1}{2}, \tfrac{1}{p_1}, \tfrac{1}{p_2}, 1 - \tfrac{1}{p}\}.$$

It was then shown in [11] that such mapping property also holds if $m = m(p_1, p_2)$ and $1 < p_1, p_2, p < \infty$; in addition, the condition $m \leq m(p_1, p_2)$ is necessary for every operator with symbol in $BS^m_{0,0}$ to be bounded from $L^{p_1}(\mathbb{R}^n) \times L^{p_2}(\mathbb{R}^n)$ into $L^p(\mathbb{R}^n)$.

The results from [3] were improved in [8], where boundedness from $L^{p_1}(\mathbb{R}^n) \times L^{p_2}(\mathbb{R}^n)$ into $L^p(\mathbb{R}^n)$, $p \geq 2$, was shown to hold for operators with symbols in certain Besov spaces of product type $B^{s,m}_{\infty,1}(\mathbb{R}^{3n})$ (see definitions in section "Preliminaries") of which $BS^m_{0,0}$ are proper subspaces. The index s is a vector that measures regularity of the symbols in each of the variables; a by-product of the results in [8] is an explicit bound of the smallest numbers N and M for which condition (1) is sufficient for boundedness of the associated operator.

We next present the results in this article, which continue the work originated in [7, 8]. See section "Preliminaries" for notation.

Theorem 1.1. *Consider* $1 \leq p_1, p_2 \leq \infty$ *and* $p = 1$ *such that* $\frac{1}{p_1} + \frac{1}{p_2} = \frac{1}{p}$ *or,* $2 \leq p_1, p_2 \leq \infty$ *and* p *satisfying* $\frac{1}{p_1} + \frac{1}{p_2} = \frac{1}{p}$. *If* $m < m(p_1, p_2)$, $s(p)$ *is as in Definition 2.1 and* s *is a vector of the same dimension as* $s(p)$, *the following statements hold true:*

(a) If $0 < q \leq 1$ *and* $s \geq s(p)$ *component-wise, then*

$$\|T_\sigma(f, g)\|_{L^p} \lesssim \|\sigma\|_{B^{s,m}_{\infty,q}} \|f\|_{L^{p_1}} \|g\|_{L^{p_2}},$$

for all $f, g \in \mathcal{S}(\mathbb{R}^n)$ *and all* x*-independent symbols* σ *in* $B^{s,m}_{\infty,q}(\mathbb{R}^{3n})$.

(b) If $1 < q \leq \infty$ *and* $s > s(p)$ *component-wise, then*

$$\|T_\sigma(f, g)\|_{L^p} \lesssim \|\sigma\|_{B^{s,m}_{\infty,q}} \|f\|_{L^{p_1}} \|g\|_{L^{p_2}},$$

for all $f, g \in \mathcal{S}(\mathbb{R}^n)$ *and all* x*-independent symbols* σ *in* $B^{s,m}_{\infty,q}(\mathbb{R}^{3n})$.

If $2 \leq p_1, p_2, p \leq \infty$ *with* $\frac{1}{p_1} + \frac{1}{p_2} = \frac{1}{p}$, *the above statements hold for any* $\sigma \in B^{s,m}_{\infty,q}(\mathbb{R}^{3n})$.

Theorem 1.2. *Consider* $1 \leq p_1, p_2, p \leq \infty$ *such that* $\frac{1}{p_1} + \frac{1}{p_2} = \frac{1}{p}$ *and* $m < m(p_1, p_2)$.

(a) If $0 < q \leq 1$, $s = (s_1, s_2, \frac{n}{2}) \in \mathbb{R}^3$ *or* $s = (s_1, \ldots, s_{2n}, \frac{1}{2}, \ldots, \frac{1}{2}) \in \mathbb{R}^{3n}$ *and* $\sigma \in B^{s,m}_{\infty,q}(\mathbb{R}^{3n})$ *is such that*

$$\hat{\sigma}(y, a, b) = \int_{\mathbb{R}^{3n}} \sigma(x, \xi, \eta)\, e^{-2\pi i (y\cdot x + a\cdot\xi + b\cdot\eta)}\, dx\, d\xi\, d\eta, \quad y, a, b \in \mathbb{R}^n$$

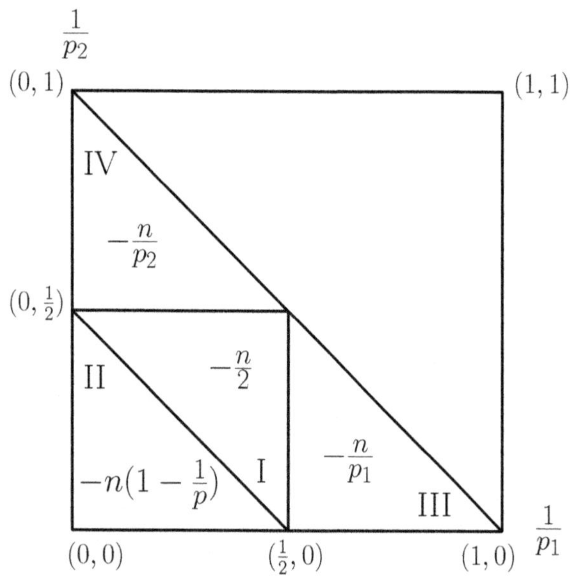

Fig. 1 Visualization of $m(p_1, p_2)$, $\frac{1}{p_1} + \frac{1}{p_2} = \frac{1}{p}$

has compact support in y *and* a *uniformly in* b, *then* T_σ *is bounded from* $L^{p_1}(\mathbb{R}^n) \times L^{p_2}(\mathbb{R}^n)$ *into* $L^p(\mathbb{R}^n)$, *where* $2 \le p_1, p_2, p \le \infty$ *or,* $1 \le p, p_1 \le 2$ *and* $2 \le p_2 \le \infty$. *The same conclusion is true for* $q > 1$ *and* $s = (s_1, s_2, s_3)$ *with* $s_3 > \frac{n}{2}$ *or* $s = (s_1, \ldots, s_{3n})$ *with* $s_j > \frac{1}{2}$ *for* $j = 2n + 1 \ldots, 3n$.

(*b*) *If* $0 < q \le 1$, $s = (s_1, \frac{n}{2}, s_3) \in \mathbb{R}^3$ *or* $s = (s_1, \ldots, s_n, \frac{1}{2}, \ldots, \frac{1}{2}, s_{2n+1}, \ldots s_{3n}) \in \mathbb{R}^{3n}$ *and* $\sigma \in B^{s,m}_{\infty,q}(\mathbb{R}^{3n})$ *is such that*

$$\hat{\sigma}(y, a, b) = \int_{\mathbb{R}^{3n}} \sigma(x, \xi, \eta)\, e^{-2\pi i(y\cdot x + a\cdot\xi + b\cdot\eta)}\, dx\, d\xi\, d\eta, \quad y, a, b \in \mathbb{R}^n$$

has compact support in y *and* b *uniformly in* a, *then* T_σ *is bounded from* $L^{p_1}(\mathbb{R}^n) \times L^{p_2}(\mathbb{R}^n)$ *into* $L^p(\mathbb{R}^n)$, *where* $2 \le p_1, p_2, p \le \infty$ *or,* $1 \le p, p_2 \le 2$ *and* $2 \le p_1 \le \infty$. *The same conclusion is true for* $q > 1$ *and* $s = (s_1, s_2, s_3)$ *with* $s_2 > \frac{n}{2}$ *or* $s = (s_1, \ldots, s_{3n})$ *with* $s_j > \frac{1}{2}$ *for* $j = n + 1 \ldots, 2n$.

Figure 1 shows the value of $m(p_1, p_2)$ according to the region in the square $[0, 1] \times [0, 1]$ to which the point $(\frac{1}{p_1}, \frac{1}{p_2})$ belongs. Theorem 1.1 states, in particular, that operators with symbols in $B^{s(p),m}_{\infty,1}(\mathbb{R}^{3n})$ are bounded from $L^{p_1}(\mathbb{R}^n) \times L^{p_2}(\mathbb{R}^n)$ into $L^p(\mathbb{R}^n)$ if $m < m(p_1, p_2)$ and $(\frac{1}{p_1}, \frac{1}{p_2})$ belongs to region II, and that operators with x-independent symbols in $B^{s(p),m}_{\infty,1}(\mathbb{R}^{3n})$ are bounded from $L^{p_1}(\mathbb{R}^n) \times L^{p_2}(\mathbb{R}^n)$ into $L^p(\mathbb{R}^n)$ if $m < m(p_1, p_2)$ and $(\frac{1}{p_1}, \frac{1}{p_2})$ belongs to region I or to the line segment joining $(1, 0)$ and $(0, 1)$. The mapping properties of Theorem 1.2 correspond to regions II and III for item (a) and to regions II and IV for item (b).

Corollary 1.3. *Consider* $1 \leq p_1, p_2 \leq \infty$ *and* $p = 1$ *such that* $\frac{1}{p_1} + \frac{1}{p_2} = \frac{1}{p}$ *or,* $2 \leq p_1, p_2 \leq \infty$ *and* p *satisfying* $\frac{1}{p_1} + \frac{1}{p_2} = \frac{1}{p}$*. If* $m < m(p_1, p_2)$*, then* T_σ *is bounded from* $L^{p_1}(\mathbb{R}^n) \times L^{p_2}(\mathbb{R}^n)$ *into* $L^p(\mathbb{R}^n)$ *for any* σ *satisfying one of the following conditions:*

(a) If $2 \leq p \leq \infty$,

$$\sup_{\substack{|\alpha| \leq [\frac{n}{p}]+1 \\ |\beta|,|\gamma| \leq [\frac{n}{2}]+1}} \sup_{x,\xi,\eta \in \mathbb{R}^n} \langle \xi, \eta \rangle^{-m} |\partial_x^\alpha \partial_\xi^\beta \partial_\eta^\gamma \sigma(x, \xi, \eta)| < \infty \tag{2}$$

or

$$\sup_{\alpha,\beta,\gamma \in \{0,1\}^n} \sup_{x,\xi,\eta \in \mathbb{R}^n} \langle \xi, \eta \rangle^{-m} |\partial_x^\alpha \partial_\xi^\beta \partial_\eta^\gamma \sigma(x, \xi, \eta)| < \infty. \tag{3}$$

(b) If $1 \leq p < 2$,

$$\sup_{|\beta|,|\gamma| \leq [\frac{n}{p}]+1} \sup_{\xi,\eta \in \mathbb{R}^n} \langle \xi, \eta \rangle^{-m} |\partial_\xi^\beta \partial_\eta^\gamma \sigma(\xi, \eta)| < \infty \tag{4}$$

or

$$\sup_{\beta,\gamma \in \{0,1\}^n} \sup_{\xi,\eta \in \mathbb{R}^n} \langle \xi, \eta \rangle^{-m} |\partial_\xi^\beta \partial_\eta^\gamma \sigma(\xi, \eta)| < \infty, \tag{5}$$

with $\{0, 1\}^n$ *replaced by* $\{0, 1, 2\}^n$ *if* $p = 1$.

Corollary 1.4. *Consider* $1 \leq p_1, p_2, p \leq \infty$ *such that* $\frac{1}{p_1} + \frac{1}{p_2} = \frac{1}{p}$ *and* $m < m(p_1, p_2)$.

(a) If $\sigma = \sigma(x, \xi, \eta)$, $x, \xi, \eta \in \mathbb{R}^n$, *is such that*

$$\hat{\sigma}(y, a, b) = \int_{\mathbb{R}^{3n}} \sigma(x, \xi, \eta)\, e^{-2\pi i (y \cdot x + a \cdot \xi + b \cdot \eta)}\, dx\, d\xi\, d\eta, \quad y, a, b \in \mathbb{R}^n$$

has compact support in y *and* a *uniformly in* b*, and*

$$\sup_{\substack{|\alpha| \leq 1, |\beta| \leq 1 \\ |\gamma| \leq [\frac{n}{2}]+1}} \sup_{x,\xi,\eta \in \mathbb{R}^n} \langle \xi, \eta \rangle^{-m} |\partial_x^\alpha \partial_\xi^\beta \partial_\eta^\gamma \sigma(x, \xi, \eta)| < \infty \tag{6}$$

or

$$\sup_{\alpha,\beta,\gamma \in \{0,1\}^n} \sup_{x,\xi,\eta \in \mathbb{R}^n} \langle \xi, \eta \rangle^{-m} |\partial_x^\alpha \partial_\xi^\beta \partial_\eta^\gamma \sigma(x, \xi, \eta)| < \infty, \tag{7}$$

then T_σ *is bounded from* $L^{p_1}(\mathbb{R}^n) \times L^{p_2}(\mathbb{R}^n)$ *into* $L^p(\mathbb{R}^n)$*, where* $2 \leq p_1, p_2, p \leq \infty$ *or,* $1 \leq p, p_1 \leq 2$ *and* $2 \leq p_2 \leq \infty$.

(b) If $\sigma = \sigma(x, \xi, \eta)$, $x, \xi, \eta \in \mathbb{R}^n$, *is such that*

$$\hat{\sigma}(y, a, b) = \int_{\mathbb{R}^{3n}} \sigma(x, \xi, \eta)\, e^{-2\pi i(y\cdot x + a\cdot\xi + b\cdot\eta)}\, dx\, d\xi\, d\eta, \quad y, a, b \in \mathbb{R}^n$$

has compact support in y *and* b *uniformly in* a, *and*

$$\sup_{\substack{|\alpha|\leq 1, |\gamma|\leq 1\\ |\beta|\leq[\frac{n}{2}]+1}} \sup_{x,\xi,\eta\in\mathbb{R}^n} \langle \xi, \eta\rangle^{-m} |\partial_x^\alpha \partial_\xi^\beta \partial_\eta^\gamma \sigma(x, \xi, \eta)| < \infty \tag{8}$$

or

$$\sup_{\alpha,\beta,\gamma\in\{0,1\}^n} \sup_{x,\xi,\eta\in\mathbb{R}^n} \langle \xi, \eta\rangle^{-m} |\partial_x^\alpha \partial_\xi^\beta \partial_\eta^\gamma \sigma(x, \xi, \eta)| < \infty, \tag{9}$$

then T_σ *is bounded from* $L^{p_1}(\mathbb{R}^n) \times L^{p_2}(\mathbb{R}^n)$ *into* $L^p(\mathbb{R}^n)$, *where* $2 \leq p_1, p_2, p \leq \infty$ *or,* $1 \leq p, p_2 \leq 2$ *and* $2 \leq p_1 \leq \infty$.

The proofs of Theorem 1.1 for the case $1 \leq p < 2$ and of Theorem 1.2 are based on results regarding a symbolic calculus for Besov spaces, which we state next. The notations σ^{*1} and σ^{*2} stand for the symbols of the first and second transposes of the bilinear operator T_σ, respectively.

Theorem 1.5. *(a) Let* $m \in \mathbb{R}$, $0 < q \leq 1$, $s \in \mathbb{R}_+^3$ *or* $s \in \mathbb{R}_+^{3n}$, *and suppose that* σ *is an x-independent symbol. Then*

$$\sigma \in B_{\infty,q}^{s,m}(\mathbb{R}^{3n}) \Rightarrow \sigma^{*j} \in B_{\infty,q}^{s^{*j},m}(\mathbb{R}^{3n}), \quad j = 1, 2,$$

where s^{*j} *is as in Definition 2.2. Moreover*

$$\left\|\sigma^{*j}\right\|_{B_{\infty,q}^{s^{*j},m}} \lesssim \|\sigma\|_{B_{\infty,q}^{s,m}}, \quad j = 1, 2, \tag{10}$$

with the implicit constant independent of σ.

(b) Let $m \in \mathbb{R}$, $0 < q \leq 1$, $s \in \mathbb{R}_+^3$ *or* $s \in \mathbb{R}_+^{3n}$, *and* $\sigma \in B_{\infty,q}^{s,m}(\mathbb{R}^{3n})$.

(i) If $\hat{\sigma}(y, a, b)$, $y, a, b \in \mathbb{R}^n$, *has compact support in* y *and* a *uniformly in* b, *then* $\sigma^{*1} \in B_{\infty,q}^{s,m}(\mathbb{R}^{3n})$.

(ii) If $\hat{\sigma}(y, a, b)$, $y, a, b \in \mathbb{R}^n$, *has compact support in* y *and* b *uniformly in* a, *then* $\sigma^{*2} \in B_{\infty,q}^{s,m}(\mathbb{R}^{3n})$.

The statements of Theorem 1.1 and Corollary 1.3 for the case $1 \leq p < 2$ correspond to operators associated to x-independent symbols, i.e., bilinear multipliers. Additional results concerning minimal smoothness conditions for the symbols of related bilinear multipliers that are sufficient for boundedness can be found in [6, 12, 13] and the references therein. In particular, a consequence of the results in [13] is that bilinear multipliers with symbols $\sigma = \sigma(\xi, \eta)$, $\xi, \eta \in \mathbb{R}^n$, satisfying

$$\sup_{|\beta+\gamma|\leq[-m(p_1,p_2)]+1} \sup_{\xi,\eta\in\mathbb{R}^n} \langle \xi, \eta\rangle^{-m(p_1,p_2)} |\partial_\xi^\beta \partial_\eta^\gamma \sigma(\xi, \eta)| < \infty$$

are bounded from $L^{p_1}(\mathbb{R}^n) \times L^{p_2}(\mathbb{R}^n)$ into $L^p(\mathbb{R}^n)$. Restricting our attention to the line segment corresponding to $p = 1$ or to region I in Fig. 1, we see that the above condition is weaker than (4) but does not compare to (5). Finally, we mention that the class $BS^m_{0,0}$ is part of the family of bilinear Hörmander classes $BS^m_{\rho,\delta}$ where $0 \leq \delta \leq \rho \leq 1$; for results regarding the Hörmander classes $BS^m_{\rho,\delta}$ and boundedness properties of the associated bilinear pseudodifferential operators consult [1–3, 8, 10, 11, 14–16].

The organization of the manuscript is as follows. In section "Preliminaries" we introduce some preliminaries and definitions. Theorem 1.5 is proved in section "Symbolic Calculus: Proof of Theorem 1.5". The proofs of Theorems 1.1 and 1.2 and Corollaries 1.3 and 1.4, along with an improved version of Theorem 1.1, are presented in section "Proof of Theorems 1.1 and 1.2 and Their Corollaries". In section "Complex Interpolation of Besov Spaces of Product Type" we show an interpolation result for Besov spaces of product type that is useful in the proof of Theorem 1.1. Section "Boundedness from $L^\infty \times L^\infty$ into L^∞ for $B^{s(\infty),m}_{\infty,1}(\mathbb{R}^{3n})$, $m < m(\infty,\infty) = -n$" is devoted to the proof of the $L^\infty(\mathbb{R}^n) \times L^\infty(\mathbb{R}^n) \to L^\infty(\mathbb{R}^n)$ mapping property for operators associated to symbols in $B^{s(\infty),m}_{\infty,1}(\mathbb{R}^n)$ with $m < m(\infty,\infty)$.

Preliminaries

The Fourier transform of $f \in \mathcal{S}'(\mathbb{R}^n)$ will be denoted by $\hat{f}$; in particular,

$$\hat{f}(\xi) := \int_{\mathbb{R}^n} f(x) e^{-2\pi i \xi \cdot x}\, dx \quad \text{if } f \in \mathcal{S}(\mathbb{R}^n),$$

where $\mathcal{S}(\mathbb{R}^n)$ denotes the Schwartz class in $\mathbb{R}^n$.

The notation $\lesssim$ means $\leq C$, where C is a constant that may only depend on some of the parameters used and not on the functions or symbols involved.

Given a complex-valued function $\sigma = \sigma(x,\xi,\eta)$, $x,\xi,\eta \in \mathbb{R}^n$, the bilinear pseudodifferential operator associated to the symbol σ, denoted by T_σ, is defined as

$$T_\sigma(f,g)(x) := \int_{\mathbb{R}^{2n}} \sigma(x,\xi,\eta)\hat{f}(\xi)\hat{g}(\eta) e^{2\pi i x\cdot(\xi+\eta)}\, d\xi d\eta \quad \forall x \in \mathbb{R}^n,\ f,g \in \mathcal{S}(\mathbb{R}^n).$$

The conditions imposed on σ will be such that the above integral is absolutely convergent for every $f,g \in \mathcal{S}(\mathbb{R}^n)$.

In the definition of Besov spaces of product type and throughout the proofs, we will consider Littlewood–Paley partitions of unity in $\mathbb{R}^N$ (for different dimensions N), $\{w_j\}_{j\in\mathbb{N}_0}$, satisfying the following conditions:

$$
\begin{aligned}
&w_0 \in \mathcal{S}(\mathbb{R}^N), \quad \operatorname{supp}(w_0) \subset \{\xi \in \mathbb{R}^N : |\xi| \le 2\},\\
&w \in \mathcal{S}(\mathbb{R}^N), \quad \operatorname{supp}(w) \subset \{\xi \in \mathbb{R}^N : \tfrac{1}{2} \le |\xi| \le 2\}, \qquad (11)\\
&w_j(\xi) := w(2^{-j}\xi) \text{ for } \xi \in \mathbb{R}^n \text{ and } j \in \mathbb{N}, \quad \sum_{j=0}^{\infty} w_j(\xi) = 1 \quad \forall \xi \in \mathbb{R}^N.
\end{aligned}
$$

Given such a Littlewood–Paley partition of unity we set

$$
\begin{aligned}
&\tilde{w}(\xi) := \sum_{l=-2}^{2} w(2^{-l}\xi) \text{ for } \xi \in \mathbb{R}^N,\\
&\tilde{w}_j(\xi) := \tilde{w}(2^{-j}\xi) = \sum_{l=j-2}^{j+2} w_l(\xi) \text{ for } j \ge 2 \text{ and } \xi \in \mathbb{R}^N, \qquad (12)\\
&\tilde{w}_j(\xi) := \sum_{l=0}^{j+2} w_l(\xi) \text{ for } j = 0, 1 \text{ and } \xi \in \mathbb{R}^N.
\end{aligned}
$$

Note that $\tilde{w}_j(\xi) \equiv 1$ for $2^{j-2} \le |\xi| \le 2^{j+2}$ and $j \ge 2$, and that $\tilde{w}_j(\xi) \equiv 1$ for $|\xi| \le 2^{j+2}$ and $j = 0, 1$; therefore $w_j\tilde{w}_j = w_j$ for all $j \in \mathbb{N}_0$.

Besov Spaces of Product Type

For $m \in \mathbb{R}$, $0 < r, q \le \infty$, and $s \in \mathbb{R}^3$ or $s \in \mathbb{R}^{3n}$, the Besov spaces $B^{s,m}_{r,q}(\mathbb{R}^{3n})$ are defined as follows:

- Given $s = (s_1, s_2, s_3) \in \mathbb{R}^3$ and $\{w_j\}_{j\in\mathbb{N}_0}$ satisfying (11) with $N = n$, $B^{s,m}_{r,q}(\mathbb{R}^{3n})$ denotes the space of complex-valued functions $\sigma(x, \xi, \eta)$, $x, \xi, \eta \in \mathbb{R}^n$, such that

$$
\|\sigma\|_{B^{s,m}_{r,q}} := \left(\sum_{k\in\mathbb{N}_0^3} \left(2^{s\cdot k} \left\| \langle \xi, \eta\rangle^{-m} \mathcal{F}^{-1}(w_k\hat{\sigma}) \right\|_{L^r}\right)^q \right)^{\frac{1}{q}} < \infty,
$$

  where for $k = (k_1, k_2, k_3)$, $w_k(y, a, b) := w_{k_1}(y)w_{k_2}(a)w_{k_3}(b)$, the inverse Fourier transform and Fourier transform as well as the L^r norm are taken in $\mathbb{R}^{3n}$, and with the corresponding modification for $q = \infty$.
- Given $s = (s_1, \ldots, s_{3n}) \in \mathbb{R}^{3n}$ and $\{w_j\}_{j\in\mathbb{N}_0}$ satisfying (11) with $N = 1$, $B^{s,m}_{r,q}(\mathbb{R}^{3n})$ denotes the space of complex-valued functions $\sigma(x, \xi, \eta)$, $x, \xi, \eta \in \mathbb{R}^n$, such that

$$
\|\sigma\|_{B^{s,m}_{r,q}} := \left(\sum_{k\in\mathbb{N}_0^{3n}} \left(2^{s\cdot k} \left\| \langle \xi, \eta\rangle^{-m} \mathcal{F}^{-1}(w_k\hat{\sigma}) \right\|_{L^r}\right)^q \right)^{\frac{1}{q}} < \infty,
$$

where for $k = (k_1, \dots, k_{3n})$, $y = (y_1, \dots, y_n)$, $a = (a_1, \dots, a_n)$ and $b = (b_1, \dots, b_n)$, $w_k(y, a, b) := w_{k_1}(y_1) \cdots w_{k_n}(y_n) w_{k_{n+1}}(a_1) \cdots w_{k_{2n}}(a_n) w_{k_{2n+1}}(b_1) \cdots w_{k_{3n}}(b_n)$, the inverse Fourier transform and Fourier transform as well as the L^r norm are taken in $\mathbb{R}^{3n}$, and with the corresponding modification for $q = \infty$.

It can be proved that, for all s, m, r, q as in the definitions above, the space $B^{s,m}_{r,q}(\mathbb{R}^{3n})$ is independent of the choice of $\{w_j\}_{j \in \mathbb{N}_0}$ satisfying (11) and is contained in $\mathcal{S}'(\mathbb{R}^{3n})$, that it is a quasi-Banach space (Banach space if $1 \le r, q \le \infty$), that it contains $\mathcal{S}(\mathbb{R}^{3n})$, and that $\mathcal{S}(\mathbb{R}^{3n})$ is dense if $0 < r, q < \infty$. We refer the reader to [17], where a variety of Besov spaces of product type are defined and many of their properties are presented; see also Proposition 2.1 below.

The Classes $C^s_m(\mathbb{R}^{3n})$

We next define classes of symbols that are closely connected with both $B^{s,m}_{r,q}(\mathbb{R}^{3n})$ and the Hörmander classes $BS^m_{0,0}$. Given $s \in \mathbb{N}_0^3$ or $s \in \mathbb{N}_0^{3n}$ and $m \in \mathbb{R}$, a complex-valued functions $\sigma(x, \xi, \eta)$, $x, \xi, \eta \in \mathbb{R}^n$, belongs to $C^s_m(\mathbb{R}^{3n})$ if it satisfies the following conditions:

- If $s = (s_1, s_2, s_3) \in \mathbb{N}_0^3$:

$$\partial_x^\alpha \partial_\xi^\beta \partial_\eta^\gamma \sigma \in C(\mathbb{R}^{3n}) \text{ for } \alpha, \beta, \gamma \in \mathbb{N}_0^n,\ |\alpha| \le s_1,\ |\beta| \le s_2,\ |\gamma| \le s_3, \text{ and}$$

$$\|\sigma\|_{C^s_m} := \sup_{\substack{|\alpha| \le s_1 \\ |\beta| \le s_2, |\gamma| \le s_3}} \sup_{x,\xi,\eta \in \mathbb{R}^n} \langle \xi, \eta \rangle^{-m} |\partial_x^\alpha \partial_\xi^\beta \partial_\eta^\gamma \sigma(x, \xi, \eta)| < \infty.$$

- If $s = (s_1, \dots, s_{3n}) \in \mathbb{N}_0^{3n}$:

$$\partial_x^{(\alpha_1,\dots,\alpha_n)} \partial_\xi^{(\alpha_{n+1},\dots,\alpha_{2n})} \partial_\eta^{(\alpha_{2n+1},\dots,\alpha_{3n})} \sigma \in C(\mathbb{R}^{3n}) \text{ for } \alpha_j \in \mathbb{N}_0, \alpha_j \le s_j, j = 1, \dots, 3n, \text{ and}$$

$$\|\sigma\|_{C^s_m} := \sup_{\substack{\alpha_j \le s_j \\ j=1,\dots,3n}} \sup_{x,\xi,\eta \in \mathbb{R}^n} \langle \xi, \eta \rangle^{-m} |\partial_x^{(\alpha_1,\dots,\alpha_n)} \partial_\xi^{(\alpha_{n+1},\dots,\alpha_{2n})} \partial_\eta^{(\alpha_{2n+1},\dots,\alpha_{3n})} \sigma(x, \xi, \eta)| < \infty.$$

We note that condition (2) corresponds to the norm in $C^s_m(\mathbb{R}^{3n})$ with $s = ([\frac{n}{p}] + 1, [\frac{n}{2}] + 1, [\frac{n}{2}] + 1)$ and that condition (3) corresponds to the norm in $C^s_m(\mathbb{R}^{3n})$ with $s = (1, 1, \dots, 1) \in \mathbb{R}^{3n}$. Analogous comments apply to the rest of the conditions appearing in Corollaries 1.3 and 1.4.

Connections Between the Spaces

The following chain of continuous proper inclusions shows the relations between the spaces introduced in this section and the bilinear Hörmander classes:

$$BS^{m}_{0,0} \subsetneqq C^{[s]+1}_{m}(\mathbb{R}^{3n}) \subsetneqq B^{s,m}_{\infty,1}(\mathbb{R}^{3n}) \subsetneqq C^{[s]}_{m}(\mathbb{R}^{3n}), \tag{13}$$

where s has positive components, $[s]$ denotes the vector of the same dimension as s and components given by the integer parts of the components of s, and adding 1 to a vector means adding 1 to each component of the vector. The first inclusion in (13) is straightforward and the rest of the inclusions are a consequence of the following proposition, which will be useful in the proof of some of our results (see [17, Theorems 1.3.2, 1.3.5, and 1.3.9 and Corollary 1.3.1] for a proof).

Proposition 2.1. *(a) Let $0 < r \le \infty$, s and $\tilde{s}$ be vectors of real numbers of the same dimension (dimension 3 or $3n$), and $m, \tilde{m} \in \mathbb{R}$. Then the following continuous inclusions hold:*

- *(i) $B^{s,m}_{r,q}(\mathbb{R}^{3n}) \subset B^{s,\tilde{m}}_{r,\tilde{q}}(\mathbb{R}^{3n})$, if $0 < q \le \tilde{q} \le \infty$ and $m \le \tilde{m}$;*
- *(ii) $B^{s,m}_{r,q}(\mathbb{R}^{3n}) \subset B^{\tilde{s},m}_{r,\tilde{q}}(\mathbb{R}^{3n})$, if $0 < q, \tilde{q} \le \infty$ and $\tilde{s} < s$ component-wise;*
- *(iii) $C^{s}_{m}(\mathbb{R}^{3n}) \subsetneqq B^{\tilde{s},m}_{\infty,q}(\mathbb{R}^{3n})$, if $0 < q \le \infty$, $0 < \tilde{s} < s$ component-wise and s has components in $\mathbb{N}$;*
- *(iv) $B^{s,m}_{\infty,1}(\mathbb{R}^{3n}) \subset C^{s}_{m}(\mathbb{R}^{3n})$, if s has components in $\mathbb{N}_0$.*

(b) If $1 \le r, \tilde{r} \le \infty$, $0 < q, \tilde{q} \le \infty$, $s, \tilde{s}$ are vectors of the same dimension (dimension 3 or $3n$) with positive components, and $m \in \mathbb{R}$, then $B^{s,m}_{r,q}(\mathbb{R}^{3n}) = B^{\tilde{s},m}_{\tilde{r},\tilde{q}}(\mathbb{R}^{3n})$ if and only if $r = \tilde{r}$, $q = \tilde{q}$ and $s = \tilde{s}$.

(c) Let $1 \le r \le \infty$, $0 < q \le \infty$, $m \in \mathbb{R}$, $s = (s_1, \dots, s_{3n}) \in \mathbb{R}^{3n}$ with $s_k > 0$ for $k = 1, \dots, 3n$, and $\tilde{s} = (s_1 + \dots + s_n, s_{n+1} + \dots + s_{2n}, s_{2n+1} + \dots + s_{3n})$. Then the following continuous inclusion holds:

$$B^{\tilde{s},m}_{r,q}(\mathbb{R}^{3n}) \subsetneqq B^{s,m}_{r,q}(\mathbb{R}^{3n}).$$

We end this section with the following definitions:

Definition 2.1. Given $1 \le p \le \infty$, $s(p)$ will denote the three-dimensional vector given by $(\min(\frac{n}{2}, \frac{n}{p}), \max(\frac{n}{2}, \frac{n}{p}), \max(\frac{n}{2}, \frac{n}{p}))$ or the $3n$-dimensional vector $(s_1, \dots, s_{3n})$ where $s_1 = \dots = s_n = \min(\frac{1}{2}, \frac{1}{p})$ and $s_{n+1} = \dots = s_{3n} = \max(\frac{1}{2}, \frac{1}{p})$. This is

$$s(p) = (\tfrac{n}{2}, \tfrac{n}{p}, \tfrac{n}{p}) \text{ or } s(p) = (\underbrace{\tfrac{1}{2}, \dots, \tfrac{1}{2}}_{n}, \underbrace{\tfrac{1}{p}, \dots, \tfrac{1}{p}}_{2n}), \quad 1 \le p < 2,$$

$$s(p) = (\tfrac{n}{p}, \tfrac{n}{2}, \tfrac{n}{2}) \text{ or } s(p) = (\underbrace{\tfrac{1}{p}, \dots, \tfrac{1}{p}}_{n}, \underbrace{\tfrac{1}{2}, \dots, \tfrac{1}{2}}_{2n}), \quad 2 \le p \le \infty.$$

It will be clear from the context which of these definitions of $s(p)$ is being used in each case.

Definition 2.2. If $s = (s_1, s_2, s_3) \in \mathbb{R}^3_+$, then $s^{*1} := (s_1, s_2 - s_3, s_3)$ and $s^{*2} := (s_1, s_2, s_3 - s_2)$. If $s = (s_1, \ldots, s_{3n}) \in \mathbb{R}^{3n}_+$, then $s^{*1} := (s_1, \ldots, s_n, s_{n+1} - s_{2n+1}, \ldots, s_{2n} - s_{3n}, s_{2n+1}, \ldots, s_{3n})$ and $s^{*2} := (s_1, \ldots, s_n, s_{n+1}, \ldots, s_{2n}, s_{2n+1} - s_{n+1}, \ldots, s_{3n} - s_{2n})$.

Symbolic Calculus: Proof of Theorem 1.5

We start with some preliminaries in relation to the symbols of the transposes of T_σ for $\sigma \in B^{s,m}_{\infty,1}(\mathbb{R}^{3n})$, where $m \in \mathbb{R}$ and s is a vector in $\mathbb{R}^3_+$ or $\mathbb{R}^{3n}_+$. For such σ let $\Sigma \in \mathcal{S}'(\mathbb{R}^{3n})$ be given by

$$\Sigma := \mathcal{F}^{-1}(\hat{\sigma}(y, -a, b-a)e^{2\pi i a\cdot y}),$$

where $\mathcal{F}^{-1}$ and $\hat{}$ are taken in $\mathbb{R}^{3n}$ and the tempered distribution $\hat{\sigma}(y, -a, b-a)e^{2\pi i a\cdot y}$ is understood in the usual sense, this is

$$\langle \hat{\sigma}(y, -a, b-a)e^{2\pi i a\cdot y}, F\rangle = \langle \hat{\sigma}, F(y, -a, b-a)e^{-2\pi i a\cdot y}\rangle, \quad F \in \mathcal{S}(\mathbb{R}^{3n}),$$

where $\langle T, \varphi\rangle$ denotes the action of a tempered distribution T on a Schwartz function φ. Assume that $\Sigma \in B^{\tilde{s},m}_{\infty,1}(\mathbb{R}^{3n})$ for some $\tilde{s}$ with positive components; let us see that $\sigma^{*1} = \Sigma$. Indeed, define

$$\begin{aligned} V(f, g, h)(y, a, b) &:= \int_{\mathbb{R}^{3n}} e^{2\pi i(y\cdot x + a\cdot\xi + b\cdot\eta)} \hat{f}(\xi)\hat{g}(\eta)e^{2\pi i x\cdot(\xi+\eta)}\bar{h}(x)\, dx d\xi d\eta \\ &= \int_{\mathbb{R}^n} e^{2\pi i y\cdot x}\bar{h}(x) f(x+a) g(x+b)\, dx, \end{aligned}$$

which belongs to $\mathcal{S}(\mathbb{R}^{3n})$ for any $f, g, h \in \mathcal{S}(\mathbb{R}^n)$. Set

$$\tilde{V}(f, g, \bar{h})(y, a, b) := e^{-2\pi i a\cdot y} V(f, g, \bar{h})(y, -a, b-a) = V(h, g, \bar{f})(y, a, b)$$

and note that since $\sigma \in B^{s,m}_{\infty,1}(\mathbb{R}^{3n}) \subset C^{[s]}_m(\mathbb{R}^{3n})$ the action of σ as a tempered distribution in $\mathbb{R}^{3n}$ is given through absolutely convergent integrals in $\mathbb{R}^{3n}$. We then have

$$\begin{aligned} \int_{\mathbb{R}^n} T_\sigma(h, g)(x) f(x)\, dx &= \int_{\mathbb{R}^{3n}} \sigma(x, \xi, \eta)\hat{h}(\xi)\hat{g}(\eta)e^{2\pi i x\cdot(\xi+\eta)} f(x)\, dx d\xi d\eta \\ &= \langle \hat{\sigma}, V(h, g, \bar{f})\rangle = \langle \hat{\sigma}, \tilde{V}(f, g, \bar{h})\rangle = \langle \hat{\Sigma}, V(f, g, \bar{h})\rangle \\ &= \int_{\mathbb{R}^{3n}} \Sigma(x, \xi, \eta)\hat{f}(\xi)\hat{g}(\eta)e^{2\pi i x\cdot(\xi+\eta)} h(x)\, dx d\xi d\eta \\ &= \int_{\mathbb{R}^n} T_\Sigma(f, g)(x) h(x)\, dx, \end{aligned}$$

which shows that $\sigma^{*1} = \Sigma$. In the second to the last equality we have used the fact that $\Sigma \in B^{\bar{s},m}_{\infty,1}(\mathbb{R}^{3n}) \subset C^{[\bar{s}]}_m(\mathbb{R}^{3n})$ and therefore the action of Σ over the function in $\mathcal{S}(\mathbb{R}^{3n})$ given by $\hat{f}(\xi)\hat{g}(\eta)e^{2\pi ix\cdot(\xi+\eta)}h(x)$ is through integration.

An analogous explanation leads to $\sigma^{*2} = \mathcal{F}^{-1}(\hat{\sigma}(y, a-b, -b)e^{2\pi ib\cdot y})$. We also observe that in the particular case when σ is x-independent, then $\sigma^{*1}(\xi,\eta) = \sigma(-\xi-\eta,\eta)$ and $\sigma^{*2}(\xi,\eta) = \sigma(\xi,-\eta-\xi)$.

We next make a remark about the norm of x-independent symbols belonging to the Besov spaces $B^{s,m}_{\infty,q}(\mathbb{R}^{3n})$. Let $\{w_j\}_{j\in\mathbb{N}_0}$ be as in (11) with $N = n$, $s = (s_1, s_2, s_3) \in \mathbb{R}^3$, $m \in \mathbb{R}$ and $0 < q \leq \infty$. If $\sigma = \sigma(\xi,\eta)$, $\xi,\eta \in \mathbb{R}^n$, is an x-independent symbol in $B^{s,m}_{\infty,q}(\mathbb{R}^{3n})$ it easily follows that

$$\|\sigma\|_{B^{s,m}_{\infty,q}} \sim \left(\sum_{k\in\mathbb{N}_0^2} \left(2^{\bar{s}\cdot k}\left\|\langle\xi,\eta\rangle^{-m}\mathcal{F}^{-1}(w_k\hat{\sigma})\right\|_{L^\infty}\right)^q\right)^{\frac{1}{q}}, \tag{14}$$

where $\bar{s} = (s_2, s_3)$, $w_k(a,b) = w_{k_1}(a)w_{k_2}(b)$ for $k = (k_1,k_2) \in \mathbb{N}_0^2$, $\mathcal{F}^{-1}$ and $\hat{}$ denote inverse Fourier transform and Fourier transform in $\mathbb{R}^{2n}$, respectively, and the L^∞ norm is taken in $\mathbb{R}^{2n}$. This is, an x-independent symbols $\sigma = \sigma(\xi,\eta)$, $\xi,\eta \in \mathbb{R}^n$, belongs to $B^{s,m}_{\infty,1}(\mathbb{R}^{3n})$ if and only if $\sigma \in B^{\bar{s},m}_{\infty,1}(\mathbb{R}^{2n})$, where $B^{\bar{s},m}_{\infty,1}(\mathbb{R}^{2n})$ is defined as the class of symbols $\tau = \tau(\xi,\eta)$, $\xi,\eta \in \mathbb{R}^n$, such that the right-hand side of (14), with τ replacing σ, is finite. An analogous remark corresponds to the case $s \in \mathbb{R}^{3n}$, with $\bar{s}$ being the vector in $\mathbb{R}^{2n}$ whose components are the last $2n$ components of s.

Lemma 3.1 below will be useful in the proof of Theorem 1.5. We state it in $\mathbb{R}^{2n}$ and $\mathbb{R}^{3n}$ because it is convenient for our settings, but more general versions also hold (compare with [17] or [18, pp. 25–28]). If h is a function defined in $\mathbb{R}^N$ and $d = (d_1,\ldots,d_N) \in \mathbb{R}^N$, denote $S_d(h)(y_1,\ldots,y_N) := h(d_1y_1,\ldots,d_Ny_N)$ and $S_{d^{-1}}(h)(y_1,\ldots,y_N) := h(d_1^{-1}y_1,\ldots,d_N^{-1}y_N)$. In particular, if $h(\xi,\eta) = \langle\xi,\eta\rangle$ for $\xi,\eta \in \mathbb{R}^n$ and $d \in \mathbb{R}^{2n}$, we write $\langle\xi,\eta\rangle_{d^{-1}}$ instead of $S_{d^{-1}}(h)$ and we have $\langle\xi,\eta\rangle_{d^{-1}} := \langle(d_1^{-1}\xi_1,\ldots,d_n^{-1}\xi_n),(d_{n+1}^{-1}\eta_1,\ldots,d_{2n}^{-1}\eta_n)\rangle$. In Lemma 3.1, integration is in $\mathbb{R}^{2n}$ in part (a) and in $\mathbb{R}^{3n}$ in part (b); additionally, the inverse Fourier transform and the Fourier transform are taken in $\mathbb{R}^{2n}$ in part (a) and in $\mathbb{R}^{3n}$ in part (b).

Lemma 3.1. *Let $1 \leq r \leq \infty$ and $t \in \mathbb{R}$.*

(a) For every continuous function $g(\xi,\eta)$ defined for $\xi,\eta \in \mathbb{R}^n$ such that $\|\langle\xi,\eta\rangle^t g\|_{L^r} < \infty$ and every $M \in \mathcal{S}(\mathbb{R}^{2n})$,

$$\left\|\langle\xi,\eta\rangle^t_{d^{-1}}\mathcal{F}^{-1}M\hat{g}\right\|_{L^r} \leq \left\|\langle\xi,\eta\rangle^{|t|}_{d^{-1}}\mathcal{F}^{-1}M\right\|_{L^1}\left\|\langle\xi,\eta\rangle^t_{d^{-1}}g\right\|_{L^r} \tag{15}$$

for any $d \in \mathbb{R}^{2n}$. In particular, if $d = (d_1,\ldots,d_{2n})$ and $d_i \geq 1$ for $i = 1,\ldots,2n$,

$$\left\|\langle\xi,\eta\rangle^t\mathcal{F}^{-1}M\hat{g}\right\|_{L^r} \leq \left\|\langle\xi,\eta\rangle^{|t|}\mathcal{F}^{-1}S_d(M)\right\|_{L^1}\left\|\langle\xi,\eta\rangle^t g\right\|_{L^r}. \tag{16}$$

(b) *For every continuous function* $g(x,\xi,\eta)$ *defined for* $x,\xi,\eta \in \mathbb{R}^n$ *such that* $\|\langle\xi,\eta\rangle^t g\|_{L^r} < \infty$ *and every* $M \in \mathcal{S}(\mathbb{R}^{3n})$,

$$\left\| \langle \xi,\eta\rangle^t_{\bar{d}^{-1}} \mathcal{F}^{-1} M\hat{g}\right\|_{L^r} \leq \left\| \langle \xi,\eta\rangle^{|t|}_{\bar{d}^{-1}} \mathcal{F}^{-1} M\right\|_{L^1} \left\| \langle \xi,\eta\rangle^t_{\bar{d}^{-1}} g\right\|_{L^r} \tag{17}$$

for any $d \in \mathbb{R}^{3n}$ *and where* $\bar{d}$ *is the vector in* $\mathbb{R}^{2n}$ *whose components are the last* $2n$ *components of* d*. In particular, if* $d = (d_1,\ldots,d_{3n})$ *and* $d_i \geq 1$ *for* $i = n+1,\ldots,3n$,

$$\left\| \langle \xi,\eta\rangle^t \mathcal{F}^{-1} M\hat{g}\right\|_{L^r} \leq \left\| \langle \xi,\eta\rangle^{|t|} \mathcal{F}^{-1} S_d(M)\right\|_{L^1} \left\| \langle \xi,\eta\rangle^t g\right\|_{L^r}. \tag{18}$$

Proof of Lemma 3.1. We prove item (a), with part (b) following analogously. We have $\langle u+y, v+z\rangle^t \lesssim \langle u,v\rangle^{|t|}\langle y,z\rangle^t$ for all $u,v,y,z \in \mathbb{R}^n$. Then

$$|\langle\xi,\eta\rangle^t_{d^{-1}}\mathcal{F}^{-1}(M\hat{g})(\xi,\eta)| \lesssim \int_{\mathbb{R}^{2n}} \langle a,b\rangle^{|t|}_{d^{-1}} |\check{M}(a,b)| \langle \xi-a,\eta-b\rangle^t_{d^{-1}} |g(\xi-a,\eta-b)|\, da\, db,$$

from where (15) follows by Minkowski's integral inequality. For (16), apply (15) with M replaced by $S_d(M)$ and g replaced by $S_{d^{-1}}(g)$ and note that $\langle\xi,\eta\rangle^{|t|}_{d^{-1}} \leq \langle\xi,\eta\rangle^{|t|}$ since $d_i \geq 1$ for $i = 1,\ldots,2n$. □

Proof of Part (a) of Theorem 1.5. We will prove the result for σ^{*1}, with the result for σ^{*2} following in an analogous way. We will also assume $q = 1$; the case $0 < q < 1$ can be deduced in the same way. Fix $m \in \mathbb{R}$ and let σ be an x-independent symbol in $B^{s,m}_{\infty,1}(\mathbb{R}^{3n})$. We have $\sigma^{*1}(\xi,\eta) = \sigma(-\xi-\eta,\eta)$ for $\xi,\eta \in \mathbb{R}^n$.

Consider first the case when $s = (s_1,s_2,s_3) \in \mathbb{R}^3_+$, then $s^{*1} = (s_1, s_2-s_3, s_3)$. As already observed, since σ is x-independent, s_1 does not play any role here. Let w and w_0 be radial functions that satisfy (11) for $N = n$ and let $\{w_j\}_{j\in\mathbb{N}_0}$ be the corresponding Littlewood–Paley partition of unity. In view of (14) we have to prove that

$$\sum_{k\in\mathbb{N}_0^2} 2^{(s_2-s_3,s_3)\cdot k} \left\| \langle\xi,\eta\rangle^{-m}\mathcal{F}^{-1}(w_k\widehat{\sigma^{*1}})\right\|_{L^\infty} \lesssim \sum_{k\in\mathbb{N}_0^2} 2^{(s_2,s_3)\cdot k} \left\| \langle\xi,\eta\rangle^{-m}\mathcal{F}^{-1}(w_k\hat{\sigma})\right\|_{L^\infty} \tag{19}$$

where $w_k(a,b) = w_{k_1}(a)w_{k_2}(b)$ for $k = (k_1,k_2) \in \mathbb{N}_0^2$, $\mathcal{F}^{-1}$ and $\hat{}$ denote inverse Fourier transform and Fourier transform in $\mathbb{R}^{2n}$, respectively, and the L^∞ norm is taken in $\mathbb{R}^{2n}$.

Given $k = (k_1,k_2) \in \mathbb{N}_0^2$ and noting that $\widehat{\sigma^{*1}}(a,b) = \hat{\sigma}(-a, b-a)$, a change of variables gives

$$\mathcal{F}^{-1}(w_k\widehat{\sigma^{*1}})(\xi,\eta) = \mathcal{F}^{-1}(w_{k_1}(a)w_{k_2}(b-a)\hat{\sigma}(a,b))(-\eta-\xi,\eta).$$

Since $\langle \xi, \eta \rangle \sim \langle \xi + \eta, \eta \rangle$, it then follows that

$$\left\| \langle \xi, \eta \rangle^{-m} \mathcal{F}^{-1}(\widehat{w_k \sigma^{*1}}) \right\|_{L^\infty} \sim \left\| \langle \xi, \eta \rangle^{-m} \mathcal{F}^{-1}(w_{k_1}(a) w_{k_2}(b-a)\hat{\sigma}(a,b)) \right\|_{L^\infty}. \tag{20}$$

We will split the summation in $(k_1, k_2) \in \mathbb{N}_0^2$ according to the following regions:

$$R = \mathbb{N}^2, \quad R_1 = \{(k_1, 0) : k_1 \geq 3\}, \quad R_2 = \{(0, k_2) : k_2 \geq 3\},$$
$$R_3 = \{(0,0), (0,1), (0,2), (1,0), (2,0)\}.$$

Let $\{\tilde{w}_j\}_{j \in \mathbb{N}_0}$ be as in (12). Recall that $\tilde{w}_j(a) \equiv 1$ for $2^{j-2} \leq |a| \leq 2^{j+2}$ and $j \geq 2$, and that $\tilde{w}_j(a) \equiv 1$ for $|a| \leq 2^{j+2}$ and $j = 0, 1$; in particular, $w_j \tilde{w}_j = w_j$ for all $j \in \mathbb{N}_0$. For $(k_1, k_2) \in \mathbb{N}_0^2$ and $a, b \in \mathbb{R}^n$ set $h_{(k_1,k_2)}(a,b) := w_{k_1}(a)\tilde{w}_{k_2}(b)$.

Summation in Region *R* Consider the following subregions:

$$R_A = \{(k_1, k_2) \in \mathbb{N}^2 : k_1 - k_2 > 2\}, \quad R_B = \{(k_1, k_2) \in \mathbb{N}^2 : k_1 - k_2 < -2\},$$
$$R_C = \{(k_1, k_2) \in \mathbb{N}^2 : -2 \leq k_1 - k_2 \leq 2\}.$$

We first estimate the summation in region R_A. If $(k_1, k_2) \in R_A$,

$$\text{supp}(w_{k_1}(a) w_{k_2}(b`-a)) \subset \{(a,b) : 2^{k_1-1} \leq |a| \leq 2^{k_1+1} \text{ and } \tfrac{1}{2}\, 2^{k_1-1} \leq |b| \leq \tfrac{9}{8}\, 2^{k_1+1}\}$$

and therefore $w_{k_1}(a) w_{k_2}(b-a) = \tilde{w}_{k_1}(a) w_{k_2}(b-a) w_{k_1}(a) \tilde{w}_{k_1}(b)$. Then (20) implies

$$\left\| \langle \xi, \eta \rangle^{-m} \mathcal{F}^{-1}(\widehat{w_k \sigma^{*1}}) \right\|_{L^\infty} \sim \left\| \langle \xi, \eta \rangle^{-m} \mathcal{F}^{-1}[\tilde{w}_{k_1}(a) w_{k_2}(b-a) \mathcal{F}(\mathcal{F}^{-1}(h_{(k_1,k_1)}\hat{\sigma}))(a,b)] \right\|_{L^\infty}.$$

By (16) in Lemma 3.1 it follows that,

$$\begin{aligned}
&\left\| \langle \xi, \eta \rangle^{-m} \mathcal{F}^{-1}[\tilde{w}_{k_1}(a) w_{k_2}(b-a) \mathcal{F}(\mathcal{F}^{-1}(h_{(k_1,k_1)}\hat{\sigma}))(a,b)] \right\|_{L^\infty} \\
&\quad \lesssim \left\| \langle \xi, \eta \rangle^{|m|} \mathcal{F}^{-1}[\tilde{w}_{k_1}(2^{k_2}a) w_{k_2}(2^{k_2}(b-a))] \right\|_{L^1} \left\| \langle \xi, \eta \rangle^{-m} \mathcal{F}^{-1}(h_{(k_1,k_1)}\hat{\sigma}) \right\|_{L^\infty}.
\end{aligned}$$

An elementary computation shows that for $1 \leq k_2 \leq k_1 - 3$,

$$\left\| \langle \xi, \eta \rangle^{|m|} \mathcal{F}^{-1}[\tilde{w}_{k_1}(2^{k_2}a) w_{k_2}(2^{k_2}(b-a))] \right\|_{L^1} \lesssim 1,$$

therefore

$$\begin{aligned}
&\sum_{(k_1,k_2) \in R_A} 2^{(s_2 - s_3, s_3)\cdot(k_1,k_2)} \left\| \langle \xi, \eta \rangle^{-m} \mathcal{F}^{-1}(\widehat{w_k \sigma^{*1}}) \right\|_{L^\infty} \\
&\qquad \lesssim \sum_{k_1=4}^{\infty} \sum_{k_2=1}^{k_1-3} 2^{(s_2 - s_3, s_3)\cdot(k_1,k_2)} \left\| \langle \xi, \eta \rangle^{-m} \mathcal{F}^{-1}(h_{(k_1,k_1)}\hat{\sigma}) \right\|_{L^\infty}.
\end{aligned}$$

We have then obtained

$$\sum_{k\in R_A} 2^{(s_2-s_3,s_3)\cdot k} \left\| \langle \xi,\eta\rangle^{-m}\mathcal{F}^{-1}(w_k\widehat{\sigma^{*1}})\right\|_{L^\infty}$$

$$\lesssim \sum_{k_1=4}^{\infty} 2^{(s_2-s_3,s_3)\cdot(k_1,k_1)} \left\| \langle \xi,\eta\rangle^{-m}\mathcal{F}^{-1}(h_{(k_1,k_1)}\hat{\sigma})\right\|_{L^\infty}. \quad (21)$$

We now look at the summation in the region R_B. Note that if $(k_1,k_2)\in R_B$, then

$$\mathrm{supp}(w_{k_1}(a)w_{k_2}(b-a)) \subset \{(a,b) : 2^{k_1-1} \le |a| \le 2^{k_1+1} \text{ and } \tfrac{1}{2}\, 2^{k_2-1} \le |b| \le \tfrac{9}{8}\, 2^{k_2+1}\}.$$

For $1\le k_1\le k_2-3$ we have $w_{k_1}(a)w_{k_2}(b-a) = w_{k_2}(b-a)\tilde{w}_{k_2}(b)w_{k_1}(a)\tilde{w}_{k_2}(b)$, and (20) implies

$$\left\| \langle \xi,\eta\rangle^{-m}\mathcal{F}^{-1}(w_k\widehat{\sigma^{*1}})\right\|_{L^\infty} \sim \left\| \langle \xi,\eta\rangle^{-m}\mathcal{F}^{-1}[w_{k_2}(b-a)\tilde{w}_{k_2}(b)\mathcal{F}(\mathcal{F}^{-1}(h_{(k_1,k_2)}\hat{\sigma}))(a,b)]\right\|_{L^\infty}.$$

By (16) in Lemma 3.1 and noting that

$$\left\| \langle \xi,\eta\rangle^{|m|}\mathcal{F}^{-1}(w_{k_2}(2^{k_2}(b-a))\tilde{w}_{k_2}(2^{k_2}b))\right\|_{L^1} \lesssim 1$$

we obtain

$$\sum_{k\in R_B} 2^{(s_2-s_3,s_3)\cdot k} \left\| \langle \xi,\eta\rangle^{-m}\mathcal{F}^{-1}(w_k\widehat{\sigma^{*1}})\right\|_{L^\infty}$$

$$\lesssim \sum_{k_2=4}^{\infty}\sum_{k_1=1}^{k_2-3} 2^{(s_2-s_3,s_3)\cdot(k_1,k_2)} \left\| \langle \xi,\eta\rangle^{-m}\mathcal{F}^{-1}(h_{(k_1,k_2)}\hat{\sigma})\right\|_{L^\infty}. \quad (22)$$

For $(k_1,k_2)\in R_C$ it follows that

$$\mathrm{supp}(w_{k_1}(a)w_{k_2}(b-a)) \subset \{(a,b) : 2^{k_1-1} \le |a| \le 2^{k_1+1} \text{ and } |b| \le 10\cdot 2^{k_1}\}.$$

Set $\chi_{k_1} := \sum_{j=0}^{k_1+4} w_j$ for $k_1\in\mathbb{N}$; since $\chi_{k_1}(b)=1$ for all b in the set $\{b : |b|\le 10\cdot 2^{k_1}\}$ then

$$w_{k_1}(a)w_{k_2}(b-a) = \sum_{j=0}^{k_1+4} w_{k_2}(b-a)\chi_{k_1}(b)w_{k_1}(a)w_j(b).$$

From this and (20) it follows that for $k_1 \in \mathbb{N}$,

$$\sum_{k_2=\max(0,k_1-2)}^{k_1+2} 2^{(s_2-s_3,s_3)\cdot(k_1,k_2)} \left\| \langle \xi, \eta \rangle^{-m} \mathcal{F}^{-1}(\widehat{w_{(k_1,k_2)}\sigma^{*1}}) \right\|_{L^\infty}$$

$$\lesssim \sum_{j=0}^{k_1+4} \sum_{k_2=\max(0,k_1-2)}^{k_1+2} 2^{(s_2-s_3,s_3)\cdot(k_1,k_2)} \| \langle \xi, \eta \rangle^{-m} \mathcal{F}^{-1}[w_{k_2}(b-a)\chi_{k_1}(b)\mathcal{F}(\mathcal{F}^{-1}(w_{(k_1,j)}\hat{\sigma}))(a,b)] \|_{L^\infty}.$$

By (16) in Lemma 3.1 and since

$$\sum_{k_2=\max(0,k_1-2)}^{k_1+2} 2^{(s_2-s_3,s_3)\cdot(k_1,k_2)} \left\| \langle \xi, \eta \rangle^{|m|} \mathcal{F}^{-1}(w_{k_2}(2^{k_1}(b-a))\chi_{k_1}(2^{k_1}b)) \right\|_{L^1} \lesssim 2^{(s_2-s_3,s_3)\cdot(k_1,k_1)},$$

it follows that

$$\sum_{k \in R_C} 2^{(s_2-s_3,s_3)\cdot k} \left\| \langle \xi, \eta \rangle^{-m} \mathcal{F}^{-1}(\widehat{w_k\sigma^{*1}}) \right\|_{L^\infty} \lesssim \sum_{k_1=1}^{\infty} \sum_{j=0}^{k_1+4} 2^{(s_2,0)\cdot(k_1,j)} \| \langle \xi, \eta \rangle^{-m} \mathcal{F}^{-1}(w_{(k_1,j)}\hat{\sigma}) \|_{L^\infty}. \tag{23}$$

Summation in Region R_1 In view of (20), we have to estimate

$$\sum_{k_1=3}^{\infty} 2^{(s_2-s_3)k_1} \| \langle \xi, \eta \rangle^{-m} \mathcal{F}^{-1}(w_{k_1}(a)w_0(b-a)\hat{\sigma}(a,b)) \|_{L^\infty}.$$

For $k_1 \geq 3$ it holds that

$$\mathrm{supp}(w_{k_1}(a)w_0(b-a)) \subset \{(a,b) : 2^{k_1-1} \leq |a| \leq 2^{k_1+1} \text{ and } 2^{k_1-2} \leq |b| \leq 2^{k_1+2}\}$$

and therefore

$$w_{k_1}(a)w_0(b-a) = w_0(b-a)\tilde{w}_{k_1}(b)w_{k_1}(a)\tilde{w}_{k_1}(b).$$

It easily follows that

$$\left\| \langle \xi, \eta \rangle^{|m|} \mathcal{F}^{-1}[w_0(b-a)\tilde{w}_{k_1}(b)] \right\|_{L^1} \lesssim 1,$$

and reasoning as above we obtain

$$\sum_{k \in R_1} 2^{(s_2-s_3,s_3)\cdot k} \left\| \langle \xi, \eta \rangle^{-m} \mathcal{F}^{-1}(\widehat{w_k\sigma^{*1}}) \right\|_{L^\infty} \lesssim \sum_{k_1=3}^{\infty} 2^{(s_2-s_3)k_1} \| \langle \xi, \eta \rangle^{-m} \mathcal{F}^{-1}(h_{(k_1,k_1)}\hat{\sigma}) \|_{L^\infty}. \tag{24}$$

Summation in Region R_2 In this case we have to estimate, again by (20),

$$\sum_{k_2=3}^{\infty} 2^{s_3 k_2} \left\| \langle \xi, \eta \rangle^{-m} \mathcal{F}^{-1}(w_0(a) w_{k_2}(b-a) \hat{\sigma}(a,b)) \right\|_{L^\infty}.$$

For $k_2 \geq 3$ it holds that

$$\text{supp}(w_0(a) w_{k_2}(b-a)) \subset \{(a,b) : |a| \leq 2 \text{ and } 2^{k_2-2 \leq} |b| \leq 2^{k_2+2}\}$$

and therefore

$$w_0(a) w_{k_2}(b-a) = w_{k_2}(b-a) \tilde{w}_{k_2}(b) w_0(a) \tilde{w}_{k_2}(b).$$

Since

$$\left\| \langle \xi, \eta \rangle^{|m|} \mathcal{F}^{-1}[w_{k_2}(2^{k_2}(b-a)) \tilde{w}_{k_2}(2^{k_2} b)] \right\|_{L^1} \lesssim 1,$$

by (16) in Lemma 3.1, we get

$$\sum_{k \in R_2} 2^{(s_2 - s_3, s_3) \cdot k} \left\| \langle \xi, \eta \rangle^{-m} \mathcal{F}^{-1}(\widehat{w_k \sigma^{*1}}) \right\|_{L^\infty} \lesssim \sum_{k_2=3}^{\infty} 2^{s_3 k_2} \left\| \langle \xi, \eta \rangle^{-m} \mathcal{F}^{-1}(h_{(0,k_2)} \hat{\sigma}) \right\|_{L^\infty}. \tag{25}$$

Summation in Region R_3 We first observe that for $(k_1, k_2) \in R_3$

$$\text{supp}(w_{k_1}(a) w_{k_2}(b-a)) \subset \{(a,b) : |a| \leq 8 \text{ and } |b| \leq 16\}.$$

Therefore, for (k_1, k_2) in region R_3 it follows that

$$\sum_{k \in R_3} 2^{(s_2 - s_3, s_3) \cdot k} \left\| \langle \xi, \eta \rangle^{-m} \mathcal{F}^{-1}(\widehat{w_k \sigma^{*1}}) \right\|_{L^\infty}$$

$$\lesssim \sum_{k_1=0}^{3} \sum_{k_2=0}^{4} 2^{(s_2 - s_3, s_3) \cdot (k_1, k_2)} \left\| \langle \xi, \eta \rangle^{-m} \mathcal{F}^{-1}(w_{(k_1,k_2)} \hat{\sigma}) \right\|_{L^\infty}. \tag{26}$$

Inequalities (21)–(26) then lead to the desired estimate (19).

We now briefly describe the proof corresponding to σ^{*1} when $s = (s_1, \ldots, s_{3n}) \in \mathbb{R}_+^{3n}$ in which case $s^{*1} := (s_1, \ldots, s_n, s_{n+1} - s_{2n+1}, \ldots, s_{2n} - s_{3n}, s_{2n+1}, \ldots, s_{3n})$. Let w and w_0 be radial functions that satisfy (11) for $N = 1$ and $\{w_j\}_{j \in \mathbb{N}_0}$ be the corresponding Littlewood–Paley partition of unity. It must be proved that

$$\sum_{k \in \mathbb{N}_0^{2n}} 2^{\overline{s^{*1}} \cdot k} \left\| \langle \xi, \eta \rangle^{-m} \mathcal{F}^{-1}(\widehat{w_k \sigma^{*1}}) \right\|_{L^\infty} \lesssim \sum_{k \in \mathbb{N}_0^{2n}} 2^{\bar{s} \cdot k} \left\| \langle \xi, \eta \rangle^{-m} \mathcal{F}^{-1}(w_k \hat{\sigma}) \right\|_{L^\infty},$$

where for $k = (k_1, \dots, k_{2n})$, $a = (a_1, \dots, a_n)$, $b = (b_1, \dots, b_n)$ we have $w_k(a, b) = w_{k_1}(a_1) \cdots w_{k_n}(a_n) w_{k_{n+1}}(b_1) \cdots w_{k_{2n}}(b_n)$, $\bar{s}$ and $\overline{s^{*1}}$ are the vectors in $\mathbb{R}^{2n}$ made of the last $2n$ components of s and s^{*1}, respectively, $\mathcal{F}^{-1}$ and $\hat{}$ denote inverse Fourier transform and Fourier transform in $\mathbb{R}^{2n}$, respectively, and the L^∞ norm is taken in $\mathbb{R}^{2n}$. For such k, we have

$$\left\| \langle \xi, \eta \rangle^{-m} \mathcal{F}^{-1}(w_k \widehat{\sigma^{*1}}) \right\|_{L^\infty} \sim \left\| \langle \xi, \eta \rangle^{-m} \mathcal{F}^{-1}(w_{K_1}(a) w_{K_2}(b-a) \hat{\sigma}(a, b)) \right\|_{L^\infty},$$

where $w_{K_1}(a) = w_{k_1}(a_1) \cdots w_{k_n}(a_n)$ and $w_{K_2}(x) = w_{k_{n+1}}(b_1) \cdots w_{k_{2n}}(b_n)$. The process is now similar to the case previously treated but much heavier in notation and the result follows by splitting the summation in $(k_1, \dots, k_{2n}) \in \mathbb{N}_0^{2n}$ based on the regions R, R_1, R_2, and R_3 for each pair (k_j, k_{n+j}), $j = 1, \dots, n$. □

Proof of Part (b) of Theorem 1.5. We prove item (i), with item (ii) following in an analogous way. Without loss of generality, we assume $q = 1$.

Let $m \in \mathbb{R}$, $s = (s_1, s_2, s_3) \in \mathbb{R}^3_+$, $\{w_j\}_{j \in \mathbb{N}_0}$ as in (11) for $N = n$, $\{\tilde{w}_j\}_{j \in \mathbb{N}_0}$ as in (12) for $N = n$, $w_k(x, \xi, \eta) = w_{k_1}(x) w_{k_2}(\xi) w_{k_3}(\eta)$ and $h_k(x, \xi, \eta) := w_{k_1}(x) w_{k_2}(\xi) \tilde{w}_{k_3}(\eta)$ for $k = (k_1, k_2, k_3) \in \mathbb{N}_0$, and R_A, R_B, R_C, R_1, R_2, R_3 as in the proof of part (a) of Theorem 1.5. Consider $\sigma \in B^{s,m}_{\infty,1}(\mathbb{R}^{3n})$; recall that $\widehat{\sigma^{*1}}(y, a, b) = \hat{\sigma}(y, -a, b-a) e^{2\pi i a \cdot y}$ and define $\Gamma \in \mathcal{S}'(\mathbb{R}^{3n})$ by $\hat{\Gamma}(y, a, b) = \hat{\sigma}(y, a, b) e^{-2\pi i a \cdot y}$. It then follows that

$$\left\| \langle \xi, \eta \rangle^{-m} \mathcal{F}^{-1}(w_k \widehat{\Sigma}) \right\|_{L^\infty} \sim \left\| \langle \xi, \eta \rangle^{-m} \mathcal{F}^{-1}(w_{k_1}(y) w_{k_2}(a) w_{k_3}(b-a) \hat{\Gamma}(y, a, b)) \right\|_{L^\infty},$$

where $k = (k_1, k_2, k_3) \in \mathbb{N}_0^3$. Similar calculations to those in the proof of part (a) of Theorem 1.5 (using item (b) of Lemma 3.1) give

$$\sum_{k_1=0}^{\infty} \sum_{(k_2,k_3) \in R_A} 2^{(s_1,s_2,s_3)\cdot(k_1,k_2,k_3)} \left\| \langle \xi, \eta \rangle^{-m} \mathcal{F}^{-1}(w_{(k_1,k_2,k_3)} \widehat{\sigma^{*1}}) \right\|_{L^\infty}$$
$$\lesssim \sum_{k_1=0}^{\infty} \sum_{k_2=4}^{\infty} 2^{(s_1,s_2,s_3)\cdot(k_1,k_2,k_2)} \left\| \langle \xi, \eta \rangle^{-m} \mathcal{F}^{-1}(h_{(k_1,k_2,k_2)} \hat{\Gamma}) \right\|_{L^\infty},$$

$$\sum_{k_1=0}^{\infty} \sum_{(k_2,k_3) \in R_B} 2^{(s_1,s_2,s_3)\cdot(k_1,k_2,k_3)} \left\| \langle \xi, \eta \rangle^{-m} \mathcal{F}^{-1}(w_{(k_1,k_2,k_3)} \widehat{\sigma^{*1}}) \right\|_{L^\infty}$$
$$\lesssim \sum_{k_1=0}^{\infty} \sum_{k_3=4}^{\infty} \sum_{k_2=1}^{k_3-3} 2^{(s_1,s_2,s_3)\cdot(k_1,k_2,k_3)} \left\| \langle \xi, \eta \rangle^{-m} \mathcal{F}^{-1}(h_{(k_1,k_2,k_3)} \hat{\Gamma}) \right\|_{L^\infty},$$

$$\sum_{k_1=0}^{\infty} \sum_{(k_2,k_3) \in R_C} 2^{(s_1,s_2,s_3)\cdot(k_1,k_2,k_3)} \left\| \langle \xi, \eta \rangle^{-m} \mathcal{F}^{-1}(w_{(k_1,k_2,k_3)} \widehat{\sigma^{*1}}) \right\|_{L^\infty}$$

$$\lesssim \sum_{k_1=0}^{\infty}\sum_{k_2=1}^{\infty}\sum_{j=0}^{k_2+4} 2^{(s_1,s_2+s_3,0)\cdot(k_1,k_2,j)} \left\| \langle \xi,\eta\rangle^{-m}\mathcal{F}^{-1}(w_{(k_1,k_2,j)}\hat{\Gamma})\right\|_{L^\infty},$$

$$\sum_{k_1=0}^{\infty}\sum_{(k_2,k_3)\in R_1} 2^{(s_1,s_2,s_3)\cdot(k_1,k_2,k_3)} \left\| \langle \xi,\eta\rangle^{-m}\mathcal{F}^{-1}(w_{(k_1,k_2,k_3)}\widehat{\sigma^{*1}})\right\|_{L^\infty}$$

$$\lesssim \sum_{k_1=0}^{\infty}\sum_{k_2=3}^{\infty} 2^{(s_1,s_2,0)\cdot(k_1,k_2,k_2)} \left\| \langle \xi,\eta\rangle^{-m}\mathcal{F}^{-1}(h_{(k_1,k_2,k_2)}\hat{\Gamma})\right\|_{L^\infty},$$

$$\sum_{k_1=0}^{\infty}\sum_{(k_2,k_3)\in R_2} 2^{(s_1,s_2,s_3)\cdot(k_1,k_2,k_3)} \left\| \langle \xi,\eta\rangle^{-m}\mathcal{F}^{-1}(w_{(k_1,k_2,k_3)}\widehat{\sigma^{*1}})\right\|_{L^\infty}$$

$$\lesssim \sum_{k_1=0}^{\infty}\sum_{k_3=3}^{\infty} 2^{(s_1,s_2,s_3)\cdot(k_1,0,k_3)} \left\| \langle \xi,\eta\rangle^{-m}\mathcal{F}^{-1}(h_{(k_1,0,k_3)}\hat{\Gamma})\right\|_{L^\infty},$$

$$\sum_{k_1=0}^{\infty}\sum_{(k_2,k_3)\in R_3} 2^{(s_1,s_2,s_3)\cdot(k_1,k_2,k_3)} \left\| \langle \xi,\eta\rangle^{-m}\mathcal{F}^{-1}(w_{(k_1,k_2,k_3)}\widehat{\sigma^{*1}})\right\|_{L^\infty}$$

$$\lesssim \sum_{k_1=0}^{\infty}\sum_{k_2=0}^{3}\sum_{k_3=0}^{4} 2^{(s_1,s_2,s_3)\cdot(k_1,k_2,k_3)} \left\| \langle \xi,\eta\rangle^{-m}\mathcal{F}^{-1}(w_{(k_1,k_2,k_3)}\hat{\Gamma})\right\|_{L^\infty}.$$

The inequalities above imply that

$$\left\|\sigma^{*1}\right\|_{B^{s,m}_{\infty,1}} \lesssim \|\Gamma\|_{B^{(s_1,s_2+s_3,s_3),m}_{\infty,1}}. \tag{27}$$

We next estimate $\left\| \langle \xi,\eta\rangle^{-m}\mathcal{F}^{-1}(w_k\hat{\Gamma})\right\|_{L^\infty}$ for $k = (k_1,k_2,k_2) \in \mathbb{N}_0$ (the same computations apply to the factors with h instead of w.) Set $\tilde{w}_k(x,\xi,\eta) := \tilde{w}_{k_1}(x)\tilde{w}_{k_2}(\xi)\tilde{w}_{k_3}(\eta)$ and use part (b) of Lemma 3.1 to get

$$\left\| \langle \xi,\eta\rangle^{-m}\mathcal{F}^{-1}(w_k\hat{\Gamma})\right\|_{L^\infty}$$
$$\leq \left\| \langle 2^{-k_2}\xi, 2^{-k_3}\eta\rangle^{|m|}\mathcal{F}^{-1}(w(y)w(a)w(b)e^{-2\pi i 2^{k_1+k_2} a\cdot y})\right\|_{L^1} \left\| \langle \xi,\eta\rangle^{-m}\mathcal{F}^{-1}(\tilde{w}_k\hat{\sigma})\right\|_{L^\infty}, \tag{28}$$

where w should be replaced by w_0 in the corresponding cases when $k_1 = 0$ or $k_2 = 0$ or $k_3 = 0$. With $\lambda := 2^{-(k_1+k_2)}$, it follows that

$$\mathcal{F}^{-1}(w(y)w(a)w(b)e^{-2\pi i\lambda^{-1}a\cdot y})(x,\xi,\eta) = \check{w}(\eta)\lambda^n V_{\hat{w}}(w(\lambda\cdot))(x,-\lambda\xi),$$

where $V_{\hat{w}}(h)(x,\xi) := \int_{\mathbb{R}^n} h(a)\hat{w}(a-x)e^{-2\pi i\xi\cdot a}\,da$ is the short-time Fourier transform of h using the window $\hat{w}$. An estimate of the L^1 norm in (28) is based on a bound for the L^1 norm with respect to x and ξ of $(1+|2^{-k_2}\xi|)^{|m|}\lambda^n V_{\hat{w}}(w(\lambda\cdot))(x,-\lambda\xi)$. After a change of variables and using that $2^{k_1} \geq 1$ and $0 < \lambda < 1$, we obtain

$$\int_{\mathbb{R}^{2n}} (1+|2^{-k_2}\xi|)^{|m|}\lambda^n |V_{\hat{w}}(w(\lambda\cdot))(x,-\lambda\xi)|\,dxd\xi$$

$$\leq 2^{k_1|m|}\int_{\mathbb{R}^{2n}} (1+|\xi|)^{|m|}|V_{\hat{w}}(w(\lambda\cdot))(x,\xi)|\,dxd\xi \lesssim 2^{k_1|m|}\lambda^{-n} = 2^{k_1(|m|+n)}2^{k_2 n}.$$

The last inequality follows from the scaling properties of the modulation spaces $M^{1,1}_{\tau,0}(\mathbb{R}^n)$ with $\tau \in \mathbb{R}$ (see [5, Theorem 3.2]).

From (27) and the above computations we then have that

$$\left\|\sigma^{*1}\right\|_{B^{s,m}_{\infty,1}} \lesssim \|\Gamma\|_{B^{(s_1,s_2+s_3,s_3),m}_{\infty,1}} \lesssim \|\sigma\|_{B^{(s_1+|m|+n,s_2+s_3+n,s_3),m}_{\infty,1}}.$$

If we now assume the hypothesis that $\hat{\sigma}(y,a,b)$ has compact support in y and a uniformly in b, then the summations in k_1 and k_2 appearing in $\|\sigma\|_{B^{(s_1+|m|+n,s_2+s_3+n,s_3),m}_{\infty,1}}$ are finite and therefore $\|\sigma\|_{B^{(s_1+|m|+n,s_2+s_3+n,s_3),m}_{\infty,1}} \lesssim C_\sigma \|\sigma\|_{B^{(s_1,s_2,s_3),m}_{\infty,1}}$, where C_σ depends on the size of the support of σ in the variables y and a.

For the case $s \in \mathbb{R}^{3n}$ we can proceed in a similar way by splitting the summation in $(k_1,\dots,k_{3n}) \in \mathbb{N}_0^{3n}$ based on the regions R_A, R_B, R_C, R_1, R_2, R_3 for each pair (k_j,k_{n+j}), $j = n+1,\dots,2n$. □

Proof of Theorems 1.1 and 1.2 and Their Corollaries

In this section, we start by explaining how Corollaries 1.3 and 1.4 follow from Theorems 1.1 and 1.2, respectively. We then continue with the proofs of Theorems 1.1 and 1.2. Finally, we state some improvements of Theorem 1.1 corresponding to the case $1 \leq p < 2$.

Corollary 1.3 follows directly from Theorem 1.1 in view of the definition of $s(p)$ (see Definition 2.1) and the fact that $C_m^{[s(p)]+1}(\mathbb{R}^{3n})$ is continuously contained in $B^{s(p),m}_{\infty,1}(\mathbb{R}^{3n})$ (see (13)). As for Corollary 1.4, the assumptions (6) and (13) imply that $\sigma \in C_m^{(1,1,[\frac{n}{2}]+1)}(\mathbb{R}^{3n}) \subset B^{(s_1,s_2,\frac{n}{2})}_{\infty,1}(\mathbb{R}^{3n})$ for any $0 < s_1, s_2 < 1$; similarly, (7) and (13) give that $\sigma \in C_m^{(1,\dots,1)}(\mathbb{R}^{3n}) \subset B^{s,m}_{\infty,1}(\mathbb{R}^{3n})$ for any $s \in \mathbb{R}^{3n}$ with components in the interval $(0,1)$. Then part (a) of Corollary 1.4 follows from part (a) of Theorem 1.2. An analogous reasoning shows that part (b) of Corollary 1.4 follows from part (b) of Theorem 1.2.

We next proceed to prove Theorem 1.1. Recall that if $s = (s_1,s_2,s_3) \in \mathbb{R}^3$ or $s = (s_1,\dots,s_{2n}) \in \mathbb{R}^{3n}$, $\bar{s}$ denotes the vector $(s_2,s_3) \in \mathbb{R}^2$ or the vector $(s_{n+1},\dots,s_{3n}) \in \mathbb{R}^{2n}$, respectively. Moreover, an x-independent symbol σ belongs to $B^{s,m}_{\infty,q}(\mathbb{R}^{3n})$ if and only if $\sigma \in B^{\bar{s},m}_{\infty,q}(\mathbb{R}^{2n})$ (see (14) and the corresponding discussion).

Proof of Theorem 1.1. The boundedness results from $L^{p_1}(\mathbb{R}^n)\times L^{p_2}(\mathbb{R}^n)$ into $L^p(\mathbb{R}^n)$ for $2 \leq p_1, p_2 \leq \infty$ and $2 \leq p < \infty$ satisfying $\frac{1}{p_1}+\frac{1}{p_2}=\frac{1}{p}$ (region II of Fig. 2) and for all symbols in $B^{s,m}_{\infty,q}(\mathbb{R}^{3n})$ with $m < m(p_1,p_2)$ and s and q as in the hypothesis were proved in [8]. Boundedness corresponding to $p_1 = p_2 = p = \infty$ follows with a slight change in the proof of the latter case (see section "Boundedness from $L^\infty \times L^\infty$ into L^∞ for $B^{s(\infty),m}_{\infty,1}(\mathbb{R}^{3n})$, $m < m(\infty,\infty) = -n$").

We next prove the rest of the cases. By parts (a) and (c) of Proposition 2.1, it is enough to work with the spaces $B^{s(p),m}_{\infty,1}(\mathbb{R}^{3n})$ where $s(p)$ is in $\mathbb{R}^{3n}$ and $m < m(p_1,p_2)$.

If $1 \le p_1 \le 2$ and $\frac{1}{p_1} + \frac{1}{p_2} = 1$, then $(\frac{1}{p_1}, \frac{1}{p_2})$ is in the line segment joining $(1,0)$ and $(\frac{1}{2},\frac{1}{2})$ in Fig. 2 and $m(p_1,p_2) = -\frac{n}{p_1}$. We have $s(1) = (\frac{1}{2},\ldots,\frac{1}{2},1,\ldots,1,1,\ldots,1) \in \mathbb{R}^{3n}$ and $B^{s(1),m}_{\infty,1}(\mathbb{R}^{3n}) \subset B^{t,m}_{\infty,1}(\mathbb{R}^{3n})$ where $t \in \mathbb{R}^{3n}$ has its first $2n$ components equal to those of $s(1)$ and its last n components equal to $\frac{1}{2}$. Since $\overline{t^{*1}} = \overline{s(p_2)}$ (recall that $2 \le p_2 \le \infty$), then part (a) of Theorem 1.5 implies that $\sigma^{*1} \in B^{s(p_2),m}_{\infty,1}(\mathbb{R}^{3n})$ for any x-independent symbols $\sigma \in B^{t,m}_{\infty,1}(\mathbb{R}^{3n})$. Note that $(\frac{1}{\infty}, \frac{1}{p_2})$ is in the segment joining $(0,0)$ with $(0,\frac{1}{2})$ of Fig. 2 and $m(\infty,p_2) = -\frac{n}{p_2'} = -\frac{n}{p_1} = m(p_1,p_2)$. In view of the results corresponding to region II in Fig. 2, it follows

$$\|T_{\sigma^{*1}}(f,g)\|_{L^{p_2}} \lesssim \|\sigma^{*1}\|_{B^{s(p_2),m}_{\infty,1}} \|f\|_{L^\infty} \|g\|_{L^{p_2}}$$

for all $f,g \in \mathcal{S}(\mathbb{R}^n)$ and all x-independent symbols $\sigma \in B^{t,m}_{\infty,1}(\mathbb{R}^{3n})$ with $m < m(p_1,p_2)$. Duality and (10) imply

$$\|T_\sigma(f,g)\|_{L^1} \lesssim \|\sigma^{*1}\|_{B^{s(p_2),m}_{\infty,1}} \|f\|_{L^{p_1}} \|g\|_{L^{p_2}} \lesssim \|\sigma\|_{B^{t,m}_{\infty,1}} \|f\|_{L^{p_1}} \|g\|_{L^{p_2}},$$

for all $f,g \in \mathcal{S}(\mathbb{R}^n)$ and all x-independent symbols $\sigma \in B^{t,m}_{\infty,1}(\mathbb{R}^{3n})$ with $m < m(p_1,p_2)$. This holds in particular for all x-independent symbols in $B^{s(1),m}_{\infty,1}(\mathbb{R}^{3n})$ and for those symbols we have $\|\sigma\|_{B^{t,m}_{\infty,1}} \lesssim \|\sigma\|_{B^{s(1),m}_{\infty,1}}$. If $2 \le p_1 \le \infty$ and $\frac{1}{p_1} + \frac{1}{p_2} = 1$, then $(\frac{1}{p_1}, \frac{1}{p_2})$ is in the line segment joining $(0,1)$ and $(\frac{1}{2},\frac{1}{2})$ in Fig. 2 and $m(p_1,p_2) = -\frac{n}{p_2}$. We can proceed in a similar way as above, using σ^{*2} instead of σ^{*1}, to get the desired result.

We next consider $2 \le p_1, p_2 < \infty$ with $1 < p < 2$ given by $\frac{1}{p_1} + \frac{1}{p_2} = \frac{1}{p}$ (region I in Fig. 2). Let the trilinear operator T be defined as $T(\sigma,f,g) := T_\sigma(f,g)$. By the previously treated cases, T satisfies the following boundedness properties:

T is bounded from $B^{\overline{s(2)},m}_{\infty,1}(\mathbb{R}^{2n}) \times L^2(\mathbb{R}^n) \times L^\infty(\mathbb{R}^n)$ into $L^2(\mathbb{R}^n)$ for $m < -\frac{n}{2}$,

T is bounded from $B^{\overline{s(2)},m}_{\infty,1}(\mathbb{R}^{2n}) \times L^\infty(\mathbb{R}^n) \times L^2(\mathbb{R}^n)$ into $L^2(\mathbb{R}^n)$ for $m < -\frac{n}{2}$,

T is bounded from $B^{\overline{s(1)},m}_{\infty,1}(\mathbb{R}^{2n}) \times L^2(\mathbb{R}^n) \times L^2(\mathbb{R}^n)$ into $L^1(\mathbb{R}^n)$ for $m < -\frac{n}{2}$.

From the facts shown in section "Complex Interpolation of Besov Spaces of Product Type", it follows that

$$(B^{\overline{s(2)},m}_{\infty,1}(\mathbb{R}^{2n}), B^{\overline{s(1)},m}_{\infty,1}(\mathbb{R}^{2n}))_{[\theta]} = B^{\overline{s(p)},m}_{\infty,1}(\mathbb{R}^{2n}), \quad \frac{1}{p} = \frac{1-\theta}{2} + \theta,\ m \in \mathbb{R}.$$

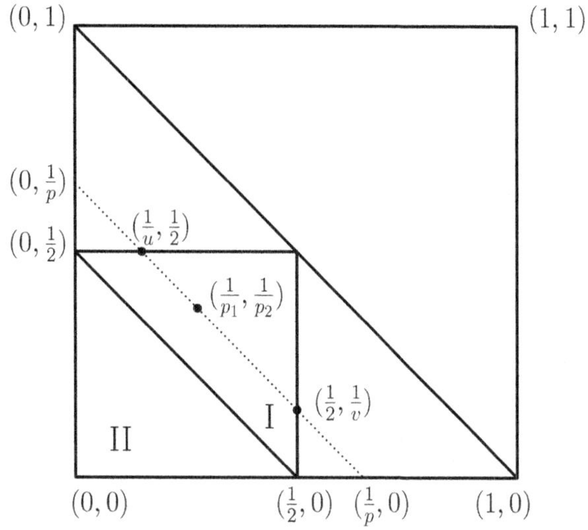

Fig. 2 Interpolation argument in the proof of Theorem 1.1

Using trilinear complex interpolation (see [4, Theorem 4.4.1]), we conclude that T is bounded from $B^{\overline{s(p)},m}_{\infty,1}(\mathbb{R}^{2n}) \times L^2(\mathbb{R}^n) \times L^{p_2}(\mathbb{R}^n)$ into $L^p(\mathbb{R}^n)$ for $m < -\frac{n}{2}$ and p_2, p such that $\frac{1}{2} + \frac{1}{p_2} = \frac{1}{p}$ and $(\frac{1}{2}, \frac{1}{p_2})$ is on the line segment joining $(\frac{1}{2}, 0)$ and $(\frac{1}{2}, \frac{1}{2})$ in Fig. 2. Similarly, we obtain the desired mapping properties for the indices corresponding to the segment joining $(0, \frac{1}{2})$ and $(\frac{1}{2}, \frac{1}{2})$ in Fig. 2. If $(\frac{1}{p_1}, \frac{1}{p_2})$ is a point in the interior of region I in Fig. 2, the boundedness result for operators with x-independent symbols in $B^{s(p),m}_{\infty,1}(\mathbb{R}^{3n})$, $m < m(p_1, p_2) = -\frac{n}{2}$, follows by complex interpolation and the boundedness corresponding to the indices given by the pairs $(\frac{1}{u}, \frac{1}{2})$ and $(\frac{1}{2}, \frac{1}{v})$ that satisfy $\frac{1}{u} + \frac{1}{2} = \frac{1}{2} + \frac{1}{v} = \frac{1}{p_1} + \frac{1}{p_2} = \frac{1}{p}$. □

Proof of Theorem 1.2. In view of Proposition 2.1 it is enough to consider $q = 1$. Also, $s \in \mathbb{R}^3$ will be assumed; a similar argument applies to the case $s \in \mathbb{R}^{3n}$.

We first observe that if $\sigma \in B^{s,m}_{\infty,1}(\mathbb{R}^{3n})$ is a symbol as in part (a) of Theorem 1.2, the assumption on the support of $\hat{\sigma}$ allows to conclude that $\sigma \in B^{\tilde{s},m}_{\infty,1}(\mathbb{R}^{3n})$ for any $\tilde{s} = (\tilde{s}_1, \tilde{s}_2, \frac{n}{2}) \in \mathbb{R}^3$. By taking $\tilde{s} = (\frac{n}{p}, \frac{n}{2}, \frac{n}{2})$ the boundedness corresponding to $2 \leq p_1, p_2, p \leq \infty$ is a particular case of Theorem 1.1. The same reasoning is valid for symbols as in part (b) of Theorem 1.2.

Fix now $1 \leq p_1, p \leq 2$ and $2 \leq p_2 \leq \infty$ such that $\frac{1}{p_1} + \frac{1}{p_2} = \frac{1}{p}$, $s = (s_1, s_2, \frac{n}{2}) \in \mathbb{R}^3$, $m < m(p_1, p_2)$ and let $\sigma \in B^{s,m}_{\infty,1}(\mathbb{R}^{3n})$ be a symbol as in part (a) of Theorem 1.2. From the assumptions on the support of $\hat{\sigma}$ it follows that $\sigma \in B^{\tilde{s},m}_{\infty,1}(\mathbb{R}^{3n})$ for $\tilde{s} = (\frac{n}{p_1'}, \frac{n}{2}, \frac{n}{2})$. By item (i) in part (b) of Theorem 1.5 we have that $\sigma^{*1} \in B^{\tilde{s},m}_{\infty,1}(\mathbb{R}^{3n})$; applying the mapping properties corresponding to region II in Fig. 2, it follows that $T_{\sigma^{*1}}$ is bounded from $L^{p'}(\mathbb{R}^n) \times L^{p_2}(\mathbb{R}^n)$ into $L^{p_1'}(\mathbb{R}^n)$. Duality then implies that T_σ is bounded from $L^{p_1}(\mathbb{R}^n) \times L^{p_2}(\mathbb{R}^n)$ into $L^p(\mathbb{R}^n)$ as desired.

Part (b) of Theorem 1.2 follows analogously through the use of σ^{*2}. □

An Improved Version of Theorem 1.1 for $1 \le p < 2$

We present an improvement of Theorem 1.1 in the sense that boundedness into $L^p(\mathbb{R}^n)$, $1 \le p < 2$, holds for operators with x-independent symbols belonging to certain Besov spaces that have regularity index below $s(p)$.

If s is a vector in $\mathbb{R}^2$ or $\mathbb{R}^{2n}$ and $\sigma = \sigma(\xi, \eta)$ is in $B^{s,m}_{\infty,1}(\mathbb{R}^{2n})$ for some $m \in \mathbb{R}$, we define $\|\sigma\|_{B^{s,m}_{\infty,1}} := \|\sigma\|_{B^{S,m}_{\infty,1}}$, where S is a vector in $\mathbb{R}^3$ or $\mathbb{R}^{3n}$, respectively, such that $\overline{S} = s$ (see (14) and the corresponding discussion).

For $1 \le p_1, p_2 \le \infty$ such that $\frac{1}{p_1} + \frac{1}{p_2} = 1$ define $s(p_1, p_2)$ as the vector in $\mathbb{R}^2$ given by

$$s(p_1,p_2) = \begin{cases} (n, \frac{n}{2}) & \text{if } 1 \le p_1 \le 2, \\ (\frac{n}{2}, n) & \text{if } 2 \le p_1 \le \infty, \end{cases}$$

or as the vector in $\mathbb{R}^{2n}$ given by

$$s(p_1,p_2) = \begin{cases} (\underbrace{1,\ldots,1}_{n}, \underbrace{\tfrac{1}{2},\ldots,\tfrac{1}{2}}_{n}) & \text{if } 1 \le p_1 \le 2, \\ (\underbrace{\tfrac{1}{2},\ldots,\tfrac{1}{2}}_{n}, \underbrace{1,\ldots,1}_{n}) & \text{if } 2 \le p_1 \le \infty. \end{cases}$$

Note that $s(2,2)$ receives two possible values in each dimension considered. The proof of Theorem 1.1 and part (c) of Proposition 2.1 give that

$$\|T_\sigma(f,g)\|_{L^1} \lesssim \|\sigma\|_{B^{s(p_1,p_2),m}_{\infty,1}} \|f\|_{L^{p_1}} \|g\|_{L^{p_2}},$$

for all $f, g \in \mathcal{S}(\mathbb{R}^n)$ and all symbols $\sigma \in B^{s(p_1,p_2),m}_{\infty,1}(\mathbb{R}^{2n})$ with $m < m(p_1,p_2)$ and where $s(p_1,p_2)$ is in $\mathbb{R}^2$ or in $\mathbb{R}^{2n}$. This is an improvement on the corresponding result in Theorem 1.1 since $B^{\overline{s(1)},m}_{\infty,1}(\mathbb{R}^{2n}) \subsetneq B^{s(p_1,p_2),m}_{\infty,1}(\mathbb{R}^{2n})$. Recall that $\overline{s(1)} = (n,n)$ or $\overline{s(1)} = (1,\ldots,1) \in \mathbb{R}^{2n}$, then the above states that boundedness still holds even if the symbol has $\frac{n}{2}$ fewer derivatives (in the sense of Besov spaces) with respect to one of its frequency variables.

For $2 \le p_1, p_2 \le \infty$ such that $\frac{1}{p_1} + \frac{1}{p_2} = \frac{1}{p}$ and $p_1 = 2$ or $p_2 = 2$ (segments joining points $(\frac{1}{2}, \frac{1}{2})$ and $(\frac{1}{2}, 0)$ and points $(\frac{1}{2}, \frac{1}{2})$ and $(0, \frac{1}{2})$, respectively, in Fig. 2) define $s(p_1,p_2)$ as the vector in $\mathbb{R}^2$ given by

$$s(p_1,p_2) = \begin{cases} (\frac{n}{2}, \frac{n}{p}) & \text{if } p_1 = 2, \\ (\frac{n}{p}, \frac{n}{2}) & \text{if } p_2 = 2, \end{cases}$$

or as the vector in $\mathbb{R}^{2n}$ given by

$$s(p_1,p_2) = \begin{cases} (\underbrace{\tfrac{1}{2},\dots,\tfrac{1}{2}}_{n},\underbrace{\tfrac{1}{p},\dots,\tfrac{1}{p}}_{n}) & \text{if } p_1 = 2, \\ (\underbrace{\tfrac{1}{p},\dots,\tfrac{1}{p}}_{n},\underbrace{\tfrac{1}{2},\dots,\tfrac{1}{2}}_{n}) & \text{if } p_2 = 2. \end{cases}$$

In this case we also have $B^{\overline{s(p)},m}_{\infty,1}(\mathbb{R}^{2n}) \subsetneq B^{s(p_1,p_2),m}_{\infty,1}(\mathbb{R}^{2n})$, if $(p_1,p_2) \neq (2,\infty)$ and $(p_1,p_2) \neq (\infty,2)$. Symbols in the latter class are allowed to have $\frac{n}{p} - \frac{n}{2} > 0$ fewer derivatives (in the sense of Besov spaces) than those in $B^{\overline{s(p)},m}_{\infty,1}(\mathbb{R}^{2n})$.

Finally, if p_1, p_2 are such that $\frac{1}{p_1} + \frac{1}{p_2} = \frac{1}{p}$ and $(\frac{1}{p_1}, \frac{1}{p_2})$ is in the interior of region I of Fig. 2, define $s(p_1,p_2) := (1-\theta)s(u,2) + \theta s(2,v)$ where $\theta \in (0,1)$ is such that $(\frac{1}{p_1}, \frac{1}{p_2}) = (1-\theta)(\frac{1}{u}, \frac{1}{2}) + \theta(\frac{1}{2}, \frac{1}{v})$ and u and v are as in Fig. 2. Once more, $B^{\overline{s(p)},m}_{\infty,1}(\mathbb{R}^{2n}) \subsetneq B^{s(p_1,p_2),m}_{\infty,1}(\mathbb{R}^{2n})$.

An analogous proof to that of Theorem 1.1 gives then the following:

Theorem 4.1. *Consider* $1 \le p_1, p_2 \le \infty$ *and* $p = 1$ *such that* $\frac{1}{p_1} + \frac{1}{p_2} = \frac{1}{p}$ *or,* $2 \le p_1, p_2 \le \infty$ *and* $1 \le p < 2$ *satisfying* $\frac{1}{p_1} + \frac{1}{p_2} = \frac{1}{p}$. *If* $m < m(p_1,p_2)$, $s(p_1,p_2)$ *is as defined above and* s *is a vector of the same dimension as* $s(p_1,p_2)$, *the following statements hold true:*

(a) If $0 < q \le 1$ *and* $s \ge s(p_1,p_2)$ *component-wise, then*

$$\|T_\sigma(f,g)\|_{L^p} \lesssim \|\sigma\|_{B^{s,m}_{\infty,q}} \|f\|_{L^{p_1}} \|g\|_{L^{p_2}},$$

for all $f,g \in \mathcal{S}(\mathbb{R}^n)$ *and all symbols* σ *in* $B^{s,m}_{\infty,q}(\mathbb{R}^{2n})$.

(b) If $1 < q \le \infty$ *and* $s > s(p_1,p_2)$ *component-wise, then*

$$\|T_\sigma(f,g)\|_{L^p} \lesssim \|\sigma\|_{B^{s,m}_{\infty,q}} \|f\|_{L^{p_1}} \|g\|_{L^{p_2}},$$

for all $f,g \in \mathcal{S}(\mathbb{R}^n)$ *and all symbols* σ *in* $B^{s,m}_{\infty,q}(\mathbb{R}^{2n})$.

Corresponding improvements of (4) and (5) in Corollary 1.3 follow from Theorem 4.1.

Complex Interpolation of Besov Spaces of Product Type

We refer the reader to [4, Chap. 4] regarding the method of complex interpolation and some of the notation used in this section. The following theorem is a particular case of [17, Theorem 1.3.3] (see also [9, Theorem 4]):

Theorem 5.A. *Let* $1 \le r_0, r_1, q_0, q_1 \le \infty$ *(excluding the cases* $r_0 = r_1 = \infty$ *and* $q_0 = q_1 = \infty$*),* $m_0, m_1 \in \mathbb{R}$, s_0, s_1 *both in* $\mathbb{R}^3$ *or both in* $\mathbb{R}^{3n}$. *If* $0 < \theta < 1$,

$$(B^{s_0,m_0}_{r_0,q_0}(\mathbb{R}^{3n}), B^{s_1,m_1}_{r_1,q_1}(\mathbb{R}^{3n}))_{[\theta]} = B^{s,m}_{r,q}(\mathbb{R}^{3n}), \tag{29}$$

$$(B^{\overline{s_0},m_0}_{r_0,q_0}(\mathbb{R}^{2n}), B^{\overline{s_1},m_1}_{r_1,q_1}(\mathbb{R}^{2n}))_{[\theta]} = B^{\overline{s},m}_{r,q}(\mathbb{R}^{2n}), \tag{30}$$

where $\frac{1}{r} = \frac{1-\theta}{r_0} + \frac{\theta}{r_1}$, $\frac{1}{q} = \frac{1-\theta}{q_0} + \frac{\theta}{q_1}$, $m = (1-\theta)m_0 + \theta m_1$ *and* $s = (1-\theta)s_0 + \theta s_1$.

In this section we will address the case $r_0 = r_1 = \infty$ and $m_0 = m_1$ and show that (29) and (30) also hold in such situation. This is, with the parameters as in the statement of Theorem 5.A and $m_0 = m_1 = m$,

$$(B^{s_0,m}_{\infty,q_0}(\mathbb{R}^{3n}), B^{s_1,m}_{\infty,q_1}(\mathbb{R}^{3n}))_{[\theta]} = B^{s,m}_{\infty,q}(\mathbb{R}^{3n}), \tag{31}$$

$$(B^{\overline{s_0},m}_{\infty,q_0}(\mathbb{R}^{2n}), B^{\overline{s_1},m}_{\infty,q_1}(\mathbb{R}^{2n}))_{[\theta]} = B^{\overline{s},m}_{\infty,q}(\mathbb{R}^{2n}). \tag{32}$$

For the sake of notation, only (31) will be treated; a completely analogous reasoning leads to (32).

We start by introducing some definitions and notations. Given a Banach space X, $1 \le q \le \infty$, and $s \in \mathbb{R}^L$, denote by $l^q_s(X)$ the Banach space of all sequences $x = \{x_k\}_{k\in\mathbb{N}_0^L}$ in X such that

$$\|x\|_{l^q_s(X)} := \left(\sum_{k\in\mathbb{N}_0^L} (2^{s\cdot k}\, \|x_k\|_X)^q \right)^{\frac{1}{q}} < \infty,$$

with the appropriate changes if $q = \infty$. If $m \in \mathbb{R}$, $s \in \mathbb{R}^L$ with $L = 3$ or $L = 3n$, and $1 \le q \le \infty$, then for $\sigma \in B^{s,m}_{\infty,q}(\mathbb{R}^{3n})$, we have that

$$\|\sigma\|_{B^{s,m}_{\infty,q}} = \left\| \{\mathcal{F}^{-1}(w_k\hat{\sigma})\}_{k\in\mathbb{N}_0^L} \right\|_{l^q_s(C^0_m(\mathbb{R}^{3n}))},$$

where $\{w_k\}_{k\in\mathbb{N}_0^L}$ is as in the definition of $B^{s,m}_{\infty,q}(\mathbb{R}^{3n})$. (Note that $\mathcal{F}^{-1}(w_k\hat{\sigma})$ is a continuous function in $\mathbb{R}^{3n}$ since it is analytic in $\mathbb{C}^{3n}$ as it is the inverse Fourier transform of a compactly supported tempered distribution in $\mathbb{R}^{3n}$).

A Banach space X is said to be a retract of a Banach space Y if there are linear bounded mappings $\mathscr{I} : X \to Y$ and $\mathscr{P} : Y \to X$ such that $\mathscr{P} \circ \mathscr{I} = Id$.

Lemma 5.1. *If* $m \in \mathbb{R}$, $s \in \mathbb{R}^L$ *with* $L = 3$ *or* $L = 3n$, *and* $1 \le q \le \infty$, *then* $B^{s,m}_{\infty,q}(\mathbb{R}^{3n})$ *is a retract of* $l^q_s(C^0_m(\mathbb{R}^{3n}))$.

Proof. Let $\{w_k\}_{k\in\mathbb{N}_0^L}$ be as in the definition of $B^{s,m}_{\infty,q}(\mathbb{R}^{3n})$ for $s \in \mathbb{R}^L$. Define $\mathscr{I} : B^{s,m}_{\infty,q}(\mathbb{R}^{3n}) \to l^q_s(C^0_m(\mathbb{R}^{3n}))$ as $\mathscr{I}(\sigma) := \{\mathcal{F}^{-1}(w_k\hat{\sigma})\}_{k\in\mathbb{N}_0^L}$. Then $\mathscr{I}$ is linear and $\|\mathscr{I}(\sigma)\|_{l^q_s(C^0_m(\mathbb{R}^{3n}))} = \|\sigma\|_{B^{s,m}_{\infty,q}}$.

Let $\{\tilde{w}_j\}_{j\in\mathbb{N}_0}$ be as in (12) and set $\tilde{w}_k(x,\xi,\eta) = \tilde{w}_{k_1}(x)\tilde{w}_{k_2}(\xi)\tilde{w}_{k_3}(\eta)$ for $k = (k_1,k_2,k_3) \in \mathbb{N}_0^3$ if $L = 3$, or $\tilde{w}_k(x,\xi,\eta) = \tilde{w}_{k_1}(x_1)\cdots\tilde{w}_{k_n}(x_n)\tilde{w}_{k_{n+1}}(\xi_1)\cdots\tilde{w}_{k_{3n}}(\eta_n)$

for $k = (k_1, \ldots, k_{3n}) \in \mathbb{N}_0^{3n}$ if $L = 3n$, then $\tilde{w}_k w_k \equiv w_k$ for all $k \in \mathbb{N}_0^L$. Note that $\sum_{|k| \le M} \mathcal{F}^{-1}(\tilde{w}_k \hat{x}_k)$ converges in $\mathcal{S}'(\mathbb{R}^{3n})$ as $M \to \infty$ if $\{x_k\}_{k \in \mathbb{N}_0^L} \in l_s^q(C_m^0(\mathbb{R}^{3n}))$.

Define $\mathscr{P} : l_s^q(C_m^0(\mathbb{R}^{3n})) \to B_{\infty,q}^{s,m}(\mathbb{R}^{3n})$ as $\mathscr{P}(\{x_k\}_{k \in \mathbb{N}_0^L}) := \sum_{k \in \mathbb{N}_0^L} \mathcal{F}^{-1}(\tilde{w}_k \hat{x}_k)$. The mapping $\mathscr{P}$ is bounded. Indeed, for $k = (k_1, \ldots, k_L) \in \mathbb{N}_0^L$ we have

$$\mathcal{F}^{-1}(w_k \mathcal{F}(\mathscr{P}(\{x_\nu\}_{\nu \in \mathbb{N}_0^L})) = \sum_{\substack{j=(j_1,\ldots,j_L) \in \mathbb{Z}^L \\ |j_\ell| \le 2,\, k_\ell + j_\ell \ge 0}} \mathcal{F}^{-1}(w_{k+j} \widehat{x_{k+j}}).$$

Therefore, by part (b) of Lemma 3.1,

$$\begin{aligned} \left\| \mathcal{F}^{-1}(w_k \mathcal{F}(\mathscr{P}(\{x_\nu\}_{\nu \in \mathbb{N}_0^L})) \right\|_{C_m^0} &= \sum_{\substack{j=(j_1,\ldots,j_L) \in \mathbb{N}_0^L \\ |j_\ell| \le 2,\, k_\ell + j_\ell \ge 0}} \left\| \mathcal{F}^{-1}(w_{k+j} \widehat{x_{k+j}}) \right\|_{C_m^0} \\ &\lesssim c(w, w_0) \sum_{\substack{j=(j_1,\ldots,j_L) \in \mathbb{N}_0^L \\ |j_\ell| \le 2,\, k_\ell + j_\ell \ge 0}} \left\| x_{k+j} \right\|_{C_m^0}, \end{aligned}$$

which implies that $\left\| \mathscr{P}(\{x_k\}_{k \in \mathbb{N}_0^L}) \right\|_{B_{\infty,1}^{s,m}} \lesssim \left\| \{x_k\}_{k \in \mathbb{N}_0^L} \right\|_{l_s^q(C_m^0(\mathbb{R}^{3n}))}$. Moreover, $\mathscr{P}$ is linear and we clearly have $\mathscr{P} \circ \mathscr{I} = Id$. □

Lemma 5.2. *Let $s_0, s_1 \in \mathbb{R}^L$, $1 \le q_0, q_1 \le \infty$ and $0 < \theta < 1$. If X_0 and X_1 are Banach spaces then, with equal norms,*

$$(l_{s_0}^{q_0}(X_0), l_{s_1}^{q_1}(X_1))_{[\theta]} = l_s^q(X),$$

where $s = (1-\theta)s_0 + \theta s_1$, $\frac{1}{q} = \frac{1-\theta}{q_0} + \frac{\theta}{q_1}$, and $X = (X_0, X_1)_{[\theta]}$.

Proof. This is a consequence, for instance, of Bergh and Löfström [4, Theorem 5.6.3], where the case corresponding to $L = 1$ is proved. □

Proof of (31). Note that $\mathscr{I}$ and $\mathscr{P}$ in the proof of Lemma 5.1 are independent of the parameters of the spaces involved. In particular $\mathscr{I}$ and $\mathscr{P}$ are linear mappings from $B_{\infty,q_0}^{s_0,m}(\mathbb{R}^{3n}) + B_{\infty,q_1}^{s_1,m}(\mathbb{R}^{3n})$ into $l_{s_0}^{q_0}(C_m^0(\mathbb{R}^{3n})) + l_{s_1}^{q_1}(C_m^0(\mathbb{R}^{3n}))$ and from $l_{s_0}^{q_0}(C_m^0(\mathbb{R}^{3n})) + l_{s_1}^{q_1}(C_m^0(\mathbb{R}^{3n}))$ into $B_{\infty,q_0}^{s_0,m}(\mathbb{R}^{3n}) + B_{\infty,q_1}^{s_1,m}(\mathbb{R}^{3n})$, respectively, such that $\mathscr{I}$ is linear and bounded from $B_{\infty,q_k}^{s_k,m}(\mathbb{R}^{3n})$ into $l_{s_k}^{q_k}(C_m^0(\mathbb{R}^{3n}))$ for $k = 0, 1$, $\mathscr{P}$ is linear and bounded from $l_{s_k}^{q_k}(C_m^0(\mathbb{R}^{3n}))$ into $B_{\infty,q_k}^{s_k,m}(\mathbb{R}^{3n})$ for $k = 0, 1$, and $\mathscr{P}(\mathscr{I}(\sigma)) = \sigma$ for all $\sigma \in B_{\infty,q_0}^{s_0,m}(\mathbb{R}^{3n}) + B_{\infty,q_1}^{s_1,m}(\mathbb{R}^{3n})$. By complex interpolation we then conclude that the space $(B_{\infty,q_0}^{s_0,m}(\mathbb{R}^{3n}), B_{\infty,q_1}^{s_1,m}(\mathbb{R}^{3n}))_{[\theta]}$ is a retract of $(l_{s_0}^{q_0}(C_m^0(\mathbb{R}^{3n})), l_{s_1}^{q_1}(C_m^0(\mathbb{R}^{3n})))_{[\theta]}$ with the mappings $\mathscr{I}$ and $\mathscr{P}$. This implies that

$$\mathscr{P}((l_{s_0}^{q_0}(C_m^0(\mathbb{R}^{3n})), l_{s_1}^{q_1}(C_m^0(\mathbb{R}^{3n})))_{[\theta]}) = (B_{\infty,q_0}^{s_0,m}(\mathbb{R}^{3n}), B_{\infty,q_1}^{s_1,m}(\mathbb{R}^{3n}))_{[\theta]}.$$

By Lemma 5.2 and the fact that $(C^0_m(\mathbb{R}^{3n}), C^0_m(\mathbb{R}^{3n}))_{[\theta]} = C^0_m(\mathbb{R}^{3n})$, we have

$$\mathscr{P}(l^q_s(C^0_m(\mathbb{R}^{3n}))_{[\theta]})) = (B^{s_0,m}_{\infty,q_0}(\mathbb{R}^{3n}), B^{s_1,m}_{\infty,q_1}(\mathbb{R}^{3n}))_{[\theta]}.$$

Since $\mathscr{P}(l^q_s(C^0_m(\mathbb{R}^{3n}))) = B^{s,m}_{\infty,q}(\mathbb{R}^{3n})$, (31) follows. The equality of the norms follows from the fact that $\mathscr{I}$ is an isometry and from the equality in norms given in Lemma 5.2. □

Remark 5.1. The above reasoning in the proof of (31) breaks down when $m_0 \neq m_1$. Indeed, assume without lost of generality that $m_0 < m_1$; then $(C^0_{m_0}(\mathbb{R}^{3n}), C^0_{m_1}(\mathbb{R}^{3n}))_{[\theta]} \neq C^0_m(\mathbb{R}^{3n})$ for $m = (1-\theta)m_0 + \theta m_1$ since $C^0_{m_0}(\mathbb{R}^{3n}) \cap C^0_{m_1}(\mathbb{R}^{3n}) = C^0_{m_0(\mathbb{R}^{3n})}$ is dense in $(C^0_{m_0}(\mathbb{R}^{3n}), C^0_{m_1}(\mathbb{R}^{3n}))_{[\theta]}$ while $C^0_{m_0}(\mathbb{R}^{3n})$ is not dense in $C^0_m(\mathbb{R}^{3n})$.

Boundedness from $L^\infty \times L^\infty$ into L^∞ for $B^{s(\infty),m}_{\infty,1}(\mathbb{R}^{3n})$, $m < m(\infty,\infty) = -n$

Boundedness from $L^{p_1}(\mathbb{R}^n) \times L^{p_2}(\mathbb{R}^n)$ into $L^p(\mathbb{R}^n)$ for operators associated to symbols in $B^{s(p),m}_{\infty,1}(\mathbb{R}^{3n})$ with $m < m(p_1,p_2)$, $2 \le p_1,p_2 \le \infty$, $2 \le p < \infty$ and $\frac{1}{p_1} + \frac{1}{p_2} = \frac{1}{p}$ was shown in [8]. The proof of such result can be adapted to obtain the corresponding mapping property for $p_1 = p_2 = p = \infty$ as explained in this section.

Given $f, g, h, \varphi, \psi, \theta \in \mathcal{S}(\mathbb{R}^n)$, define

$$V(f,g,h)(y,a,b) := \int_{\mathbb{R}^{3n}} e^{2\pi i(y\cdot x + a\cdot\xi + b\cdot\eta)} \hat{f}(\xi)\hat{g}(\eta) e^{2\pi i x\cdot(\xi+\eta)} \bar{h}(x)\, dx d\xi d\eta$$

$$= \int_{\mathbb{R}^n} e^{2\pi i x\cdot y} \bar{h}(x) f(x+a) g(x+b)\, dx, W(f,g,h,\varphi,\psi,\theta)(x,\xi,\eta) :$$

$$= \int_{\mathbb{R}^{3n}} e^{-2\pi i(x\cdot y + \xi\cdot a + \eta\cdot b)} \varphi(y)\psi(a)\theta(b) V(f,g,h)(y,a,b)\, dy da db.$$

The following estimate was shown in [8, Theorem 4] for $2 \le p_1,p_2 \le \infty$ and $2 \le p < \infty$ satisfying $\frac{1}{p_1} + \frac{1}{p_2} = \frac{1}{p}$:

$$\int_{\mathbb{R}^{3n}} \langle \xi,\eta\rangle^m |W(f,g,h,\varphi,\psi,\theta)(x,\xi,\eta)|\, dx d\xi d\eta$$

$$\lesssim \sum_{\alpha,\beta,\gamma \in \{0,1,2,3\}^n} \|\widehat{\partial^\alpha \psi}\|_{L^2} \|\|\widehat{\partial^\beta \theta}\|_{L^2} \|\widehat{\partial^\gamma \varphi}\|_{L^{p'}} \|f\|_{L^{p_1}} \|g\|_{L^{p_2}} \|h\|_{L^{p'}},$$

for all functions $f, g, h, \psi, \theta \in \mathcal{S}(\mathbb{R}^n)$ and φ of the form $\varphi(x) = \prod_{j=1}^n \varphi_j(x_j)$, where $x = (x_1, \ldots, x_n)$ and $\varphi_j \in \mathcal{S}(\mathbb{R})$, $j = 1, \ldots, n$.

We now prove that the above inequality also holds for $p = p_1 = p_2 = \infty$, this is

$$\int_{\mathbb{R}^{3n}} \langle \xi, \eta \rangle^m |W(f, g, h, \varphi, \psi, \theta)(x, \xi, \eta)| \, dx d\xi d\eta \tag{33}$$

$$\lesssim \sum_{\alpha,\beta,\gamma \in \{0,1,2,3\}^n} \|\widehat{\partial^\alpha \psi}\|_{L^2} \| \|\widehat{\partial^\beta \theta}\|_{L^2} \|\widehat{\partial^\gamma \varphi}\|_{L^1} \|f\|_{L^\infty} \|g\|_{L^\infty} \|h\|_{L^1}.$$

Once (33) is proved, the same ideas used in [8] give that

$$\|T_\sigma(f, g)\|_{L^\infty} \lesssim \|\sigma\|_{B^{s(\infty),m}_{\infty,1}} \|f\|_{L^\infty} \|g\|_{L^\infty}$$

for all $f, g \in \mathcal{S}(\mathbb{R}^n)$ and all $\sigma \in B^{s(\infty),m}_{\infty,1}(\mathbb{R}^{3n})$ with $m < m(\infty, \infty) = -n$ and $s(\infty)$ in $\mathbb{R}^3$ or in $\mathbb{R}^{3n}$. The corresponding statements in Theorem 1.1 for $q \neq 1$ and $s \geq s(\infty)$ follow by the embedding properties of Proposition 2.1.

Proof of (33). Defining

$$\mathcal{J}_k := \{\vec{j} = (j_1, \ldots, j_n) \in \{1, \ldots, n\}^n : j_l \neq j_{\tilde{l}} \text{ if } l \neq \tilde{l}, j_1 < \cdots < j_k, j_{k+1} < \cdots < j_n\}$$

for $k = 0, \ldots, n$, and denoting $W(f, g, h, \varphi, \psi, \theta)$ by W, the proof in [8, Theorem 4] shows that

$$W(x, \xi, \eta) = \sum_{\substack{k=0,\ldots,n \\ \vec{j} \in \mathcal{J}_k}} c_{k,\vec{j}} \, W_{k,\vec{j}}(x, \xi, \eta),$$

where for $k = 0, \ldots, n$ and $\vec{j} = (j_1, \ldots, j_n) \in \mathcal{J}_k$,

$$W_{k,\vec{j}}(x, \xi, \eta) := \int_{\mathbb{R}^{3n}} e^{-2\pi i(x\cdot(-\xi-\eta+\tau)+\xi\cdot a+\eta\cdot b)} \hat{\bar{h}}(\tau) f(a) g(b) \, S_{k,\vec{j}}(x, \tau, \xi, \eta, a, b) \, d\tau da db,$$

$$S_{k,\vec{j}}(x, \tau, \xi, \eta, a, b) := \int_{\mathbb{R}^n} e^{-2\pi i x\cdot y} \varphi_{k,\vec{j}}(\tau - \xi - \eta) \Phi_{k,\vec{j}}(y, \tau, \xi, \eta) \widehat{A_{a,b}}(y) \, dy,$$

$$A_{a,b}(t) := \psi(a + t)\theta(b + t),$$

$$\varphi_{k,\vec{j}}(\tau - \xi - \eta) := \prod_{l=1}^{k} \varphi_{j_l}(\tau_{j_l} - \xi_{j_l} - \eta_{j_l}), \quad (\varphi_{0,\vec{j}}(\tau - \xi - \eta) := 1),$$

$$\Phi_{k,\vec{j}}(y, \tau, \xi, \eta) := \prod_{l=k+1}^{n} y_{j_l} \int_0^1 \varphi^{(1)}_{j_l}(s_{j_l} y_{j_l} + \tau_{j_l} - \xi_{j_l} - \eta_{j_l}) \, ds_{j_l}$$

$$= y_{j_{k+1}} \cdots y_{j_n} \int_{[0,1]^{n-k}} \prod_{l=k+1}^{n} \varphi^{(1)}_{j_l}(s_{j_l} y_{j_l} + \tau_{j_l} - \xi_{j_l} - \eta_{j_l}) \, ds_{j_{k+1}} \ldots ds_{j_n},$$

with $\Phi_{n,\vec{j}} := 1$ and $\varphi^{(1)}_{j_l}$ denoting the first derivative of φ_{j_l}. It is then enough to prove the inequality (33) for each $W_{k,\vec{j}}$.

For the case $k = n$, define $F_x(a) := f(a)\psi(a-x)$, $G_x(b) := g(b)\theta(b-x)$, and $H_x(\tau) := \bar{h}(\tau)\hat{\varphi}(x-\tau)$, and obtain as in [8, Theorem 4] that

$$\int_{\mathbb{R}^{3n}} \langle \xi, \eta \rangle^m |W_{n,\vec{j}}(x,\xi,\eta)|\, dx d\xi d\eta$$

$$\lesssim \int_{\mathbb{R}^n} \|F_x\|_{L^2} \, \|G_x\|_{L^2} \left(\int_{\mathbb{R}^{2n}} \langle \xi, \eta \rangle^{2m} |\check{H}_x(\xi+\eta)|^2 d\xi d\eta \right)^{\frac{1}{2}} dx.$$

Then

$$\int_{\mathbb{R}^{3n}} \langle \xi, \eta \rangle^m |W_{n,\vec{j}}(x,\xi,\eta)|\, dx d\xi d\eta$$

$$\lesssim \sup_{x\in\mathbb{R}^n} (\|F_x\|_{L^2} \, \|G_x\|_{L^2}) \left\| \left(\int_{\mathbb{R}^{2n}} \langle \xi, \eta \rangle^{2m} |\check{H}_x(\xi+\eta)|^2 d\xi d\eta \right)^{\frac{1}{2}} \right\|_{L^1}$$

$$\lesssim \|f\|_{L^\infty} \, \|g\|_{L^\infty} \, \|\hat{\psi}\|_{L^2} \|\hat{\theta}\|_{L^2} \left\| \left(\int_{\mathbb{R}^{2n}} \langle \xi, \eta \rangle^{2m} |\check{H}_x(\xi+\eta)|^2 d\xi d\eta \right)^{\frac{1}{2}} \right\|_{L^1}.$$

Since $m(\infty,\infty) = -n$ and $m < m(\infty,\infty)$, we have $m = -n-\varepsilon$ for some $\varepsilon > 0$. Set $m_1 = m_2 := -\frac{n}{2} - \frac{\varepsilon}{2}$. The change of variable $\eta \to \eta - \xi$ and the fact that $\langle \xi, \eta - \xi \rangle^{2m} \leq (1+|\xi|)^{2m_1}(1+|\eta|)^{2m_2}$ imply

$$\left(\int_{\mathbb{R}^{2n}} \langle \xi, \eta \rangle^{2m} |\check{H}_x(\xi+\eta)|^2 d\xi d\eta \right)^{\frac{1}{2}}$$

$$\leq \left(\int_{\mathbb{R}^n} (1+|\xi|)^{2m_1}\, d\xi \right)^{\frac{1}{2}} \left(\int_{\mathbb{R}^n} (1+|\eta|)^{2m_2} |\check{H}_x(\eta)|^2\, d\eta \right)^{\frac{1}{2}}.$$

The integral in ξ is finite and

$$\left(\int_{\mathbb{R}^n} (1+|\eta|)^{2m_2} |\check{H}_x(\eta)|^2\, d\eta \right)^{\frac{1}{2}} = \left(\int_{\mathbb{R}^n} (1+|\eta|)^{-n-\varepsilon} |\check{H}_x(\eta)|^2\, d\eta \right)^{\frac{1}{2}}$$

$$\lesssim \left\| \check{H}_x \right\|_{L^\infty} \lesssim \|H_x\|_{L^1} = \left\| \bar{h}(\cdot)\hat{\varphi}(x-\cdot) \right\|_{L^1},$$

which implies

$$\left\| \left(\int_{\mathbb{R}^{2n}} \langle \xi, \eta \rangle^{2m} |\check{H}_x(\xi+\eta)|^2 d\xi d\eta \right)^{\frac{1}{2}} \right\|_{L^1} \lesssim \|h\|_{L^1} \, \|\hat{\varphi}\|_{L^1}.$$

We then obtain

$$\int_{\mathbb{R}^{3n}} \langle \xi,\eta\rangle^m |W_{n\vec{j}}(x,\xi,\eta)|\, dx d\xi d\eta \lesssim \|\hat{\psi}\|_{L^2}\|\hat{\theta}\|_{L^2}\, \|\hat{\varphi}\|_{L^1}\, \|f\|_{L^\infty}\, \|g\|_{L^\infty}\, \|h\|_{L^1}.$$

For the case $k \in \{0,\dots,n-1\}$, set $F_{t,\alpha_1,\gamma_1}(a) := f(a)(\partial^{\alpha_1+\gamma_1}\psi)(a+t)$, $G_{t,\alpha_2,\gamma_2}(b) := g(b)(\partial^{\alpha_2+\gamma_2}\theta)(b+t)$ and

$$H_{x,\bar{s}_k,\bar{y}_k}(\tau) := \bar{h}(\tau+x)\mathcal{F}^{-1}\left(\varphi_{k\vec{j}}(\cdot)\left(\prod_{l=k+1}^{n}\varphi_l^{(j_l)}(\cdot_l)s_l^{j_l-1}\right)\right)(\tau)\, e^{-2\pi i\sum_{l=k+1}^{n} s_l y_l \tau_l},$$

where $\bar{s}_k := (s_{k+1},\dots,s_n) \in [0,1]^{n-k}$, $\bar{y}_k := (y_{k+1},\dots,y_n) \in \mathbb{R}^{n-k}$, $x,\tau \in \mathbb{R}^n$ and ${\varphi_l}^{(j_l)}$ denotes the j_l derivative of φ_l with $j_l \le 3$. Proceeding as in [8, Theorem 4], it is enough to analyze

$$\begin{aligned} W_{k\vec{j},2}(x,\xi,\eta) = \int_{\mathbb{R}^{2n}}\int_{[0,1]^{n-k}} &\frac{e^{2\pi i x\cdot(\xi+\eta)}e^{-2\pi i(x+t)\cdot y}H(y)}{\prod_{j=1}^{n}(1+i(t_j+x_j))^2} \\ &\times \widehat{F_{t,\alpha_1,\gamma_1}}(\xi)\widehat{G_{t,\alpha_2,\gamma_2}}(\eta)\mathcal{F}^{-1}\left(H_{x,\bar{s}_k,\bar{y}_k}\right)(\xi+\eta)\, ds_{k+1}\dots ds_n\, dydt, \end{aligned}$$

where H is an appropriate fixed function on $L^1(\mathbb{R}^n)$ and α_i,γ_i, $i=1,2$, are multiindices with components at most equal to 3. We have

$$\begin{aligned} \int_{\mathbb{R}^{3n}} \langle \xi,\eta\rangle^m |W_{k\vec{j},2}(x,\xi,\eta)|\, d\xi d\eta dx \lesssim \int_{\mathbb{R}^{3n}} |H(y)| \int_{[0,1]^{n-k}} \frac{\left\|F_{t,\alpha_1,\gamma_1}\right\|_{L^2}\left\|G_{t,\alpha_2,\gamma_2}\right\|_{L^2}}{\prod_{j=1}^{n}(1+|t_j+x_j|^2)} \\ \times\left(\int_{\mathbb{R}^{2n}}\langle\xi,\eta\rangle^{2m}|\mathcal{F}^{-1}\left(H_{x,\bar{s}_k,\bar{y}_k}\right)(\xi+\eta)|^2\, d\xi d\eta\right)^{\frac{1}{2}} ds_{k+1}\dots ds_n\, dy\, dt\, dx, \end{aligned}$$

and therefore

$$\begin{aligned} \int_{\mathbb{R}^{3n}} \langle \xi,\eta\rangle^m |W_{k\vec{j},2}(x,\xi,\eta)|\, d\xi d\eta dx \lesssim \sup_t \left\|F_{t,\alpha_1,\gamma_1}\right\|_{L^2}\left\|G_{t,\alpha_2,\gamma_2}\right\|_{L^2}\int_{\mathbb{R}^n}|H(y)| \\ \times\int_{[0,1]^{n-k}}\left(\int_{\mathbb{R}^n}\left(\int_{\mathbb{R}^{2n}}\langle\xi,\eta\rangle^{2m}|\mathcal{F}^{-1}\left(H_{x,\bar{s}_k,\bar{y}_k}\right)(\xi+\eta)|^2\, d\xi d\eta\right)^{\frac{1}{2}} dx\right) ds_{k+1}\dots ds_n\, dy. \end{aligned} \tag{34}$$

The first factor in the right-hand side above satisfies

$$\sup_t \left\|F_{t,\alpha_1,\gamma_1}\right\|_{L^2}\left\|G_{t,\alpha_2,\gamma_2}\right\|_{L^2} \lesssim \|\widehat{\partial^{\alpha_1+\gamma_1}\psi}\|_{L^2}\|\widehat{\partial^{\alpha_2+\gamma_2}\theta}\|_{L^2}\|f\|_{L^\infty}\|g\|_{L^\infty}. \tag{35}$$

For the other factor, we proceed as in the case $k = n$ and recalling that $s_j \in [0, 1]$, it follows that

$$\left(\int_{\mathbb{R}^n}\left(\int_{\mathbb{R}^{2n}}\langle\xi,\eta\rangle^{2m}|\mathcal{F}^{-1}\left(H_{x,\bar{s}_k,\bar{y}_k}\right)(\xi+\eta)|^2\,d\xi d\eta\right)^{\frac{1}{2}}dx\right) \tag{36}$$

$$\lesssim\left(\int_{\mathbb{R}^{2n}}|H_{x,\bar{s}_k,\bar{y}_k}(\tau)|\,d\tau dx\right)\lesssim\|h\|_{L^1}\left\|\widehat{\varphi_{k,\vec{j}}}\prod_{j=k+1}^{n}\widehat{\varphi_l^{(j_l)}}\right\|_{L^1}.$$

Putting (34)–(36) together and using that $H \in L^1(\mathbb{R}^n)$, we get

$$\int_{\mathbb{R}^{3n}}\langle\xi,\eta\rangle^m|W_{k,\vec{j},2}(x,\xi,\eta)|\,d\xi d\eta dx$$

$$\lesssim\|\widehat{\partial^{\alpha_1+\gamma_1}\psi}\|_{L^2}\|\widehat{\partial^{\alpha_2+\gamma_2}\theta}\|_{L^2}\left\|\widehat{\varphi_{k,\vec{j}}}\prod_{j=k+1}^{n}\widehat{\varphi_l^{(j_l)}}\right\|_{L^1}\|f\|_{L^\infty}\|g\|_{L^\infty}\|h\|_{L^1}.$$

□

Acknowledgements The authors thank Stefania Marcantognini, Cristina Pereyra, Alex Stokolos, and Wilfredo Urbina for the invitation to contribute to this volume in honor of Cora Sadosky. The second author is also grateful to Laura De Carli, Alex Stokolos, and Wilfredo Urbina for inviting her to speak in the special session "Harmonic Analysis and Operator Theory, in memory of Cora Sadosky," held at the University of New Mexico (Albuquerque, NM) during the Spring 2014 Meeting of the Western Section of the American Mathematical Society. Virginia Naibo was supported by NSF under grant DMS 1500381.

References

1. Á. Bényi, R.H. Torres, Symbolic calculus and the transposes of bilinear pseudodifferential operators. Commun. Partial Differ. Equ. **28**(5–6), 1161–1181 (2003)
2. Á. Bényi, D. Maldonado, V. Naibo, R.H. Torres, On the Hörmander classes of bilinear pseudodifferential operators. Integr. Equ. Oper. Theory **67**(3), 341–364 (2010)
3. Á. Bényi, F. Bernicot, D. Maldonado, V. Naibo, R.H. Torres, On the Hörmander classes of bilinear pseudodifferential operators II. Indiana Univ. Math. J. **62**(6), 1733–1764 (2013)
4. J. Bergh, J. Löfström, *Interpolation Spaces: An Introduction*. Grundlehren der Mathematischen Wissenschaften, vol. 223 (Springer, Berlin, 1976)
5. E. Cordero, K. Okoudjou, Dilation properties for weighted modulation spaces. J. Func. Spaces Appl. Art. ID 145491, 29 (2012)
6. L. Grafakos, A. Miyachi, N. Tomita, On multilinear fourier multipliers of limited smoothness. Can. J. Math. **65**(2), 299–330 (2013)
7. J. Herbert, Boundedness properties of bilinear pseudodifferential operators. Ph.D. thesis, Kansas State University (2014)
8. J. Herbert, V. Naibo, Bilinear pseudodifferential operators with symbols in Besov spaces. J. Pseudo-Differ. Oper. Appl. **5**(2), 231–254 (2014)

9. J. Löfström, Interpolation of weighted spaces of differentiable functions on $\mathbf{R}^d$. Ann. Mat. Pura Appl. (4) **132**, 189–214 (1982/1983)
10. N. Michalowski, D. Rule, W. Staubach, Multilinear pseudodifferential operators beyond Calderón-Zygmund theory. J. Math. Anal. Appl. **414**(1), 149–165 (2014)
11. A. Miyachi, N. Tomita, Calderón-Vaillancourt-type theorem for bilinear operators. Indiana Univ. Math. J. **62**(4), 1165–1201 (2013)
12. A. Miyachi, N. Tomita, Minimal smoothness conditions for bilinear Fourier multipliers. Rev. Mat. Iberoam. **29**(2), 495–530 (2013)
13. A. Miyachi, N. Tomita, Boundedness criterion for bilinear Fourier multiplier operators. Tohoku Math. J. (2) **66**(1), 55–76 (2014)
14. V. Naibo, On the bilinear Hörmander classes in the scales of Triebel-Lizorkin and Besov spaces. J. Fourier Anal. Appl. **5**, 1077–1104 (2015)
15. V. Naibo, On the $L^\infty \times L^\infty \to BMO$ mapping property of certain bilinear pseudodifferential operators. Proc. Am. Math. Soc. **143**, 5323–5336 (2015)
16. S. Rodríguez-López, W. Staubach, Estimates for rough Fourier integral and pseudodifferential operators and applications to the boundedness of multilinear operators. J. Func. Anal. **264**(10), 2356–2385 (2013)
17. M. Sugimoto, L^p-boundedness of pseudodifferential operators satisfying Besov estimates. I. J. Math. Soc. Jpn. **40**(1), 105–122 (1988)
18. H. Triebel, *Theory of Function Spaces*. Monographs in Mathematics, vol. 78 (Birkhäuser, Basel, 1983)

Metric Characterizations of Some Classes of Banach Spaces

Mikhail Ostrovskii

Dedicated to the Memory of Cora Sadosky.

Abstract The main purpose of the paper is to present some recent results on metric characterizations of superreflexivity and the Radon–Nikodým property.

2010 Mathematics Subject Classification: Primary: 46B85; Secondary: 05C12, 20F67, 30L05, 46B07, 46B22

Introduction

By a *metric characterization* of a class of Banach spaces in the most general sense we mean a characterization which refers only to the metric structure of a Banach space and does not involve the linear structure. Some origins of the idea of a metric characterization can be seen in the classical theorem of Mazur and Ulam [75]: Two Banach spaces (over reals) are isometric as metric spaces if and only if they are linearly isometric as Banach spaces.

However study of metric characterizations became an active research direction only in mid-1980s, in the work of Bourgain [13] and Bourgain–Milman–Wolfson [16]. This study was motivated by the following result of Ribe [106].

Definition 1.1. Let X and Y be two Banach spaces. The space X is said to be *finitely representable* in Y if for any $\varepsilon > 0$ and any finite-dimensional subspace $F \subset X$ there exists a finite-dimensional subspace $G \subset Y$ such that $d(F, G) < 1 + \varepsilon$, where $d(F, G)$ is the Banach–Mazur distance.

The space X is said to be *crudely finitely representable* in Y if there exists $1 \le C < \infty$ such that for any finite-dimensional subspace $F \subset X$ there exists a finite-dimensional subspace $G \subset Y$ such that $d(F, G) \le C$.

M. Ostrovskii (✉)
Department of Mathematics and Computer Science, St. John's University, Queens, NY 11439, USA
e-mail: ostrovsm@stjohns.edu

M.C. Pereyra et al. (eds.), *Harmonic Analysis, Partial Differential Equations, Complex Analysis, Banach Spaces, and Operator Theory (Volume 1)*, Association for Women in Mathematics Series 4, DOI 10.1007/978-3-319-30961-3_15

Theorem 1.2 (Ribe [106]). *Let Z and Y be Banach spaces. If Z and Y are uniformly homeomorphic, then Z and Y are crudely finitely representable in each other.*

Three proofs of this theorem are known at the moment:

- The original proof of Ribe [106]. Some versions of it were presented in Enflo's survey [39] and the book by Benyamini and Lindenstrauss [11, pp. 222–224].
- The proof of Heinrich–Mankiewicz [52] based on ultraproduct techniques, also presented in [11].
- The proof of Bourgain [14] containing related quantitative estimates. This paper is a very difficult reading. The proof has been clarified and simplified by Giladi–Naor–Schechtman [44] (one of the steps was simplified earlier by Begun [8]). The presentation of [44] is easy to understand, but some of the ε-δ ends in it do not meet. I tried to fix this when I presented this result in my book [90, Sect. 9.2] (let me know if you find any problems with ε-δ choices there).

These three proofs develop a wide spectrum of methods of the nonlinear Banach space theory and are well worth studying.

The Ribe theorem implies stability under uniform homeomorphisms of each class $\mathcal{P}$ of Banach spaces satisfying the following condition LHI (local, hereditary, isomorphic): if $X \in \mathcal{P}$ and Y crudely finitely representable in X, then $Y \in \mathcal{P}$.

The following well-known classes have the described property:

- superreflexive spaces (see the definition and related results in section "Metric Characterizations of Superreflexivity" of this paper),
- spaces having cotype q, $q \in [2, \infty)$ (the definitions of type and cotype can be found, for example, in [90, Sect. 2.4]),
- spaces having cotype r for each $r > q$ where $q \in [2, \infty)$,
- spaces having type p, $p \in (1, 2]$,
- spaces having type r for each $r < p$ where $p \in (1, 2]$,
- Banach spaces isomorphic to q-convex spaces $q \in [2, \infty)$ (see Definition 2.7 below, more details can be found, for example, in [90, Sect. 8.4]),
- Banach spaces isomorphic to p-smooth spaces $p \in (1, 2]$ (see Definition 1.6 and [90, Sect. 8.4]),
- UMD (unconditional for martingale differences) spaces (recommended source for information on the UMD property is the forthcoming book [103]),
- Intersections of some collections of classes described above,
- One of such intersections is the class of spaces isomorphic to Hilbert spaces (by the Kwapień theorem [62], each Banach space having both type 2 and cotype 2 is isomorphic to a Hilbert space),
- Banach spaces isomorphic to subspaces of the space $L_p(\Omega, \Sigma, \mu)$ for some measure space (Ω, Σ, μ), $p \neq 2, \infty$ (for $p = \infty$ we get the class of all Banach spaces, for $p = 2$ we get the class of spaces isomorphic to Hilbert spaces).

Remark 1.3. This list seems to constitute the list of all classes of Banach spaces satisfying the condition LHI which were systematically studied.

By the Ribe Theorem (Theorem 1.2), one can expect that each class satisfying the condition LHI has a metric characterization. At this point metric characterizations are known for all classes listed above except p-smooth and UMD (and some of the intersections involving these classes). Here are the references:

- Superreflexivity—see section "Metric Characterizations of Superreflexivity" of this paper for a detailed account.
- Properties related to type—[76] (see [16, 40, 102] for previous important results in this direction, and [42] for some improvements).
- Properties related to cotype—[77], see [43] for some improvements.
- q-convexity—[78].
- Spaces isomorphic to subspaces of L_p ($p \neq 2, \infty$). Rabinovich noticed that one can generalize results of [71] and characterize the optimal distortion of embeddings of a finite metric space into L_p-space (see [74, Exercise 4 on p. 383 and comment of p. 380] and a detailed presentation in [90, Sect. 4.3]). Johnson, Mendel, and Schechtman (unpublished) found another characterization of the optimal distortion using a modification of the argument of Lindenstrauss and Pełczyński [70, Theorem 7.3]. These characterizations are very close to each other. They are not satisfactory in some respects.

Ribe Program

It should be mentioned that some of the metric characterizations (for example of the class of spaces having some type > 1) can be derived from the known 'linear' theory. Substantially nonlinear characterizations started with the paper of Bourgain [13] in which he characterized superreflexive Banach spaces in terms of binary trees.

This paper of Bourgain and the whole direction of metric characterizations was inspired by the unpublished paper of Joram Lindenstrauss with the tentative title "Topics in the geometry of metric spaces." This paper has never been published (and apparently has never been written, so it looks like it was just a *conversation*, and not a *paper*), but it had a significant impact on this direction of research. The unpublished paper of Lindenstrauss and the mentioned paper of Bourgain [13] initiated what is now known as the *Ribe program*.

Bourgain [13, p. 222] formulated it as the program of search for equivalent definitions of different LHI invariants in terms of metric structure with the next step consisting in studying these metrical concepts in general metric spaces in an attempt to develop an analogue of the linear theory.

Bourgain himself made several important contributions to the Ribe program, now it is a very deep and extensive research direction. In words of Ball [3]: "Within a decade or two the Ribe programme acquired an importance that would have been hard to predict at the outset." In this paper I am going to cover only a very small part of known results on this program. I refer interested people to the surveys of Ball [3] (short survey) and Naor [81] (extensive survey).

Many of the known metric characterizations use the following standard definitions:

Definition 1.4. Let $0 \leq C < \infty$. A map $f : (A, d_A) \to (Y, d_Y)$ between two metric spaces is called *C-Lipschitz* if

$$\forall u, v \in A \quad d_Y(f(u), f(v)) \leq C d_A(u, v).$$

A map f is called *Lipschitz* if it is C-Lipschitz for some $0 \leq C < \infty$.

Let $1 \leq C < \infty$. A map $f : A \to Y$ is called a *C-bilipschitz embedding* if there exists $r > 0$ such that

$$\forall u, v \in A \quad r d_A(u, v) \leq d_Y(f(u), f(v)) \leq r C d_A(u, v). \tag{1}$$

A *bilipschitz embedding* is an embedding which is C-bilipschitz for some $1 \leq C < \infty$. The smallest constant C for which there exist $r > 0$ such that (1) is satisfied is called the *distortion* of f.

There are at least two directions in which we can seek metric characterizations:

(1) We can try to characterize metric spaces which admit bilipschitz embeddings into some Banach spaces belonging to $\mathcal{P}$.
(2) We can try to find metric structures which are present in each Banach space $X \notin \mathcal{P}$.

Characterizations of type **(1)** would be much more interesting for applications. However, as far as I know such characterizations were found only in the following cases: (1) $\mathcal{P}$ = {the class of Banach spaces isomorphic to a Hilbert space} (it is the Linial–London–Rabinovich [71, Corollary 3.5] formula for distortion of embeddings of a finite metric space into ℓ_2). (2) $\mathcal{P}$ = {the class of Banach spaces isomorphic to a subspace of some L_p-space}, p is a fixed number $p \neq 2, \infty$, see the last paragraph preceding section "Ribe Program."

Local Properties for Which no Metric Characterization is Known

Problem 1.5. *Find a metric characterization of UMD.*

Here *UMD* stays for *unconditional for martingale differences*. The most comprehensive source of information on UMD is the forthcoming book of Pisier [103].

I have not found in the literature any traces of attempts to work on Problem 1.5.

Definition 1.6. A Banach space is called *p-smooth* if its modulus of smoothness satisfies $\rho(t) \leq Ct^p$ for $p \in (1, 2]$.

See [90, Sect. 8.4] for information on p-smooth spaces.

Problem 1.7. *Find a metric characterization of the class of Banach spaces isomorphic to p-smooth spaces* $p \in (1, 2]$.

This problem was posed and discussed in the paper by Mendel and Naor [78], where a similar problem is solved for q-convex spaces. Mendel and Naor wrote [78, p. 335]: "Trees are natural candidates for finite metric obstructions to q-convexity, but it is unclear what would be the possible finite metric witnesses to the "non-p-smoothness" of a metric space."

Metric Characterizations of Superreflexivity

Definition 2.1 (James [55, 56]). A Banach space X is called *superreflexive* if each Banach space which is finitely representable in X is reflexive.

It might look like a rather peculiar definition, but, as I understand, introducing it ($\approx$ 1967) James already had a feeling that it is a very natural and important definition. This feeling was shown to be completely justified when Enflo [38] completed the series of results of James by proving that each superreflexive space has an equivalent uniformly convex norm.

Definition 2.2. A Banach space is called *uniformly convex* if for every $\varepsilon > 0$ there is some $\delta > 0$ so that for any two vectors with $\|x\| \le 1$ and $\|y\| \le 1$, the inequality

$$1 - \left\| \frac{x+y}{2} \right\| < \delta$$

implies

$$\|x - y\| < \varepsilon.$$

Definition 2.3. Two norms $||\cdot||_1$ and $||\cdot||_2$ on a linear space X are called *equivalent* if there are constants $0 < c \le C < \infty$ such that

$$c||x||_1 \le ||x||_2 \le C||x||_1$$

for each $x \in X$.

After the pioneering results of James and Enflo numerous equivalent reformulations of superreflexivity were found and superreflexivity was used in many different contexts.

The metric characterizations of superreflexivity which we are going to present belong to the class of the so-called *test-space characterizations*.

Definition 2.4. Let $\mathcal{P}$ be a class of Banach spaces and let $T = \{T_\alpha\}_{\alpha \in A}$ be a set of metric spaces. We say that T is a set of *test-spaces* for $\mathcal{P}$ if the following two conditions are equivalent: **(1)** $X \notin \mathcal{P}$; **(2)** The spaces $\{T_\alpha\}_{\alpha \in A}$ admit bilipschitz embeddings into X with uniformly bounded distortions.

Characterization of Superreflexivity in Terms of Binary Trees

Definition 2.5. A *binary tree of depth n*, denoted T_n, is a finite graph in which each vertex is represented by a finite (possibly empty) sequence of 0 and 1, of length at most n. Two vertices in T_n are adjacent if the sequence corresponding to one of them is obtained from the sequence corresponding to the other by adding one term on the right. (For example, vertices corresponding to $(1, 1, 1, 0)$ and $(1, 1, 1, 0, 1)$ are adjacent.) A vertex corresponding to a sequence of length n in T_n is called a *leaf.*

An *infinite binary tree*, denoted T_∞, is an infinite graph in which each vertex is represented by a finite (possibly empty) sequence of 0 and 1. Two vertices in T_∞ are adjacent if the sequence corresponding to one of them is obtained from the sequence corresponding to the other by adding one term on the right.

Both for finite and infinite binary trees we use the following terminology. The vertex corresponding to the empty sequence is called a *root*. If a sequence τ is an initial segment of the sequence σ we say that σ is a *descendant* of τ and that τ is an *ancestor* of σ. If a descendant σ of τ is adjacent to τ, we say that σ is a *child* of τ and that τ is a *parent* of σ. Two children of the same parent are called *siblings*. Child of a child is called a *grandchild*. (It is clear that each vertex in T_∞ has exactly two children, the same is true for all vertices of T_n except leaves.)

Theorem 2.6 (Bourgain [13]). *A Banach space X is nonsuperreflexive if and only if it admits bilipschitz embeddings with uniformly bounded distortions of finite binary trees* $\{T_n\}_{n=1}^{\infty}$ *of all depths, see Fig. 1.*

In Bourgain's proof the difficult direction is the "if" direction, the "only if" is an easy consequence of the theory of superreflexive spaces. Recently Kloeckner [61] found a very simple proof of the "if" direction. I plan to describe the proofs of Bourgain and Kloeckner after recalling the results on superreflexivity which we need.

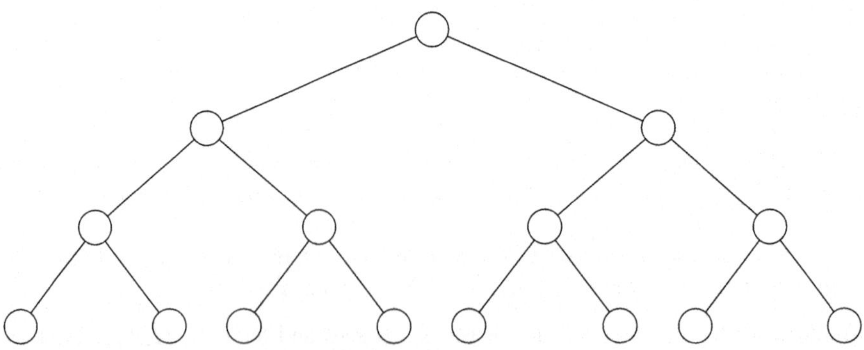

Fig. 1 The binary tree of depth 3, that is, T_3

Definition 2.7. The *modulus of* (uniform) *convexity* $\delta_X(\varepsilon)$ of a Banach space X with norm $\|\cdot\|$ is defined as

$$\inf\left\{1-\left\|\frac{x+y}{2}\right\| \,\middle|\, \|x\|=\|y\|=1 \text{ and } \|x-y\|\geqslant\varepsilon\right\}$$

for $\varepsilon\in(0,2]$. The space X or its norm is said to be *q-convex*, $q\in[2,\infty)$ if $\delta_X(\varepsilon)\geqslant c\varepsilon^q$ for some $c>0$.

Remark 2.8. It is easy to see that the definition of the uniform convexity given in Definition 2.2 is equivalent to: X is *uniformly convex* if and only if $\delta_X(\varepsilon)>0$ for each $\varepsilon\in(0,2]$.

Theorem 2.9 (Pisier [101]). *The following properties of a Banach space Y are equivalent:*

1. *Y is superreflexive.*
2. *Y has an equivalent q-convex norm for some* $q\in[2,\infty)$.

Using Theorem 2.9 we can prove the "if" part of Bourgain's characterization. Denote by $c_X(T_n)$ the infimum of distortions of embeddings of the binary tree T_n into a Banach space X.

By Theorem 2.9, for the "if" part of Bourgain's theorem it suffices to prove that if X is q-convex, then for some $c_1>0$ we have

$$c_X(T_n)\geq c_1(\log_2 n)^{\frac{1}{q}}.$$

Proof (Kloeckner [61]). Let F be the four-vertex tree with one root a_0 which has one child a_1 and two grandchildren a_2, a_2'. Sometimes such tree is called a *fork*, see Fig. 2. The following lemma is similar to the corresponding results in [73].

Lemma 2.10. *There is a constant* $K=K(X)$ *such that if* $\varphi: F\to X$ *is D-Lipschitz and distance non-decreasing, then either*

$$\|\varphi(a_0)-\varphi(a_2)\|\leqslant 2\left(D-\frac{K}{D^{q-1}}\right)$$

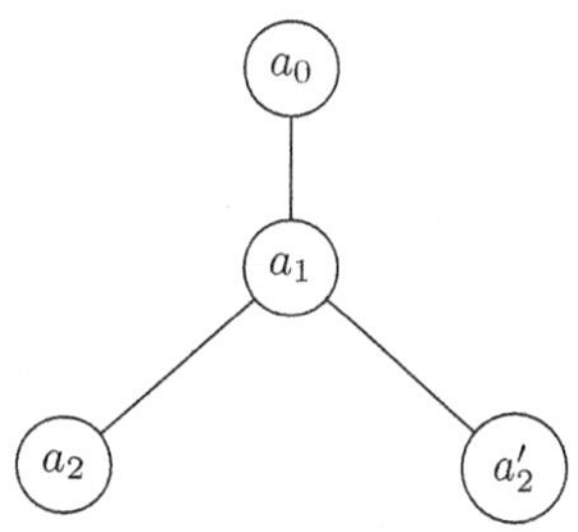

Fig. 2 A fork

or

$$\|\varphi(a_0) - \varphi(a_2')\| \leqslant 2\left(D - \frac{K}{D^{q-1}}\right)$$

First we finish the proof of $c_X(T_n) \geq c_1(\log_2 n)^{\frac{1}{q}}$ using Lemma 2.10. So let $\varphi : T_n \to X$ be a map of distortion D. Since X is a Banach space, we may assume that φ is D-Lipschitz, distance non-decreasing map, that is

$$d_{T_n}(u, v) \leq ||\varphi(u) - \varphi(v)|| \leq D d_{T_n}(u, v).$$

The main idea of the proof is to construct a less-distorted embedding of a smaller tree.

Given any vertex of T_n which is not a leaf, let us name arbitrarily one of its two children its *daughter*, and the other its *son*. We select two grandchildren of the root in the following way: we pick the grandchild mapped by φ closest to the root among its daughter's children and the grandchild mapped by φ closest to the root among its son's children (ties are resolved arbitrarily). Then we select inductively, in the same way, two grandchildren for all previously selected vertices up to generation $n - 2$.

The set of selected vertices, endowed with half the distance induced by the tree metric is isometric to $T_{\lfloor \frac{n}{2} \rfloor}$, and Lemma 2.10 implies that the restriction of φ to this set has distortion at most $f(D) = D - \frac{K}{D^{q-1}}$.

We can iterate such restrictions $\lfloor \log_2 n \rfloor$ times to get an embedding of T_1 whose distortion is at most

$$D - \lfloor \log_2 n \rfloor \frac{K}{D^{q-1}}$$

since each iteration improves the distortion by at least K/D^{q-1}. Since the distortion of any embedding is at least 1, we get the desired inequality. □

Remark 2.11. Kloeckner borrowed the approach based on "controlled improvement for embeddings of smaller parts" from the Johnson–Schechtman paper [58] in which it is used for diamond graphs (Kloeckner calls this approach a *self-improvement argument*). Arguments of this type are well known and widely used in the linear theory, where they go back at least to James [53]; but these two examples (Johnson–Schechtman [58] and Kloeckner [61]) seem to be the only two known results of this type in the nonlinear theory. It would be interesting to find further nonlinear arguments of this type.

Sketch of the Proof of Lemma 2.10. Assume $\varphi(a_0) = 0$ and let $x_1 = \varphi(a_1)$, $x_2 = \varphi(a_2)$ and $x_2' = \varphi(a_2')$. Recall that we assumed

$$d_{T_n}(u, v) \leq ||\varphi(u) - \varphi(v)|| \leq D d_{T_n}(u, v). \quad (*)$$

Consider the (easy) case where $\|x_2\| = 2D$ and $\|x_2'\| = 2D$ (that is, the distortion D is attained on these vectors). We claim that this implies that $x_2 = x_2'$. In fact, it is easy to check that this implies $\|x_1\| = D$, $\|x_2 - x_1\| = D$, and $\|x_2' - x_1\| = D$. Also $\left\|\frac{x_1+(x_2-x_1)}{2}\right\| = D$ and $\left\|\frac{x_1+(x_2'-x_1)}{2}\right\| = D$. By the uniform convexity we get $||x_1 - (x_2 - x_1)|| = 0$ and $||x_1 - (x_2' - x_1)|| = 0$. Hence $x_2 = x_2'$, and we get that the conditions $\|x_2\| = 2D$ and $\|x_2'\| = 2D$ cannot be satisfied simultaneously.

The proof of Lemma 2.10 goes as follows. We start by letting $\|x_2\| \geq 2(D - \eta)$ and $\|x_2'\| \geq 2(D - \eta)$ for some $\eta > 0$. Using a perturbed version of the argument just presented, the definition of the modulus of convexity, and our assumption $\delta_X(\varepsilon) \geq c\varepsilon^p$, we get an estimate of $||x_2 - x_2'||$ from above in terms of η. Comparing this estimate with the assumption $||x_2 - x_2'|| \geq 2$ (which follows from $d_{T_n}(u, v) \leq ||\varphi(u) - \varphi(v)||$), we get the desired estimate for η from below, see [61] for details. □

Remark 2.12. The approach of Kloeckner can be used for any uniformly convex space, it is not necessary to combine it with the Pisier Theorem (Theorem 2.9), see [103].

To prove the "only if" part of Bourgain's theorem we need the following characterization of superreflexivity, one of the most suitable sources for this characterization of superreflexivity is [103].

Theorem 2.13 (James [53, 55, 108]). *Let X be a Banach space. The following are equivalent:*

1. *X is not superreflexive*
2. *There exists $\alpha \in (0, 1]$ such that for each $m \in \mathbb{N}$ the unit ball of the space X contains a finite sequence $x_1, x_2, \ldots, x_m$ of vectors satisfying, for any $j \in \{1, \ldots, m-1\}$ and any real coefficients $a_1, \ldots, a_m$, the condition*

$$\left\|\sum_{i=1}^{m} a_i x_i\right\| \geq \alpha \left(\left|\sum_{i=1}^{j} a_i\right| + \left|\sum_{i=j+1}^{m} a_i\right|\right). \tag{2}$$

3. *For each $\alpha \in (0, 1)$ and each $m \in \mathbb{N}$ the unit ball of the space X contains a finite sequence $x_1, x_2, \ldots, x_m$ of vectors satisfying, for any $j \in \{1, \ldots, m-1\}$ and any real coefficients $a_1, \ldots, a_m$, the condition* (2).

Remark 2.14. It is worth mentioning that the proof of (1)⇒(3) in the case where $\alpha \in [\frac{1}{2}, 1)$ is much more difficult than in the case $\alpha \in (0, \frac{1}{2})$. A relatively easy proof in the case $\alpha \in [\frac{1}{2}, 1)$ was found by Brunel and Sucheston [20], see also its presentation in [103].

Remark 2.15. To prove the Bourgain's theorem it suffices to use (1)⇒(3) in the 'easy' case $\alpha \in (0, \frac{1}{2})$. The case $\alpha \in [\frac{1}{2}, 1)$ is needed only for "almost-isometric" embeddings of trees into nonsuperreflexive spaces.

Remark 2.16. The equivalence of (2)⇔(3) in Theorem 2.13 can be proved using a "self-improvement argument," but the proof of James is different. A proof of (2)⇔(3) using a "self-improvement argument" was obtained by Wenzel [114], it is based on the Ramsey theorem, so it requires a very lengthy sequence to get a better α. In [84] it was proved that to some extent the usage of 'very lengthy' sequences is necessary.

Proof of the "Only If" Part. There is a natural partial order on T_n: we say that $s < t$ ($s, t \in T_n$) if the sequence corresponding to s is the initial segment of the sequence corresponding to t.

An important observation of Bourgain is that there is a bijective mapping

$$\varphi : T_n \to [1, \dots, 2^{n+1} - 1]$$

such that φ maps two disjoint intervals of the ordering of T_n, starting at the same vertex and going "down" into disjoint intervals of $[1, \dots, 2^{n+1} - 1]$. The existence of φ can be seen from a suitably drawn picture of T_n (see Fig. 3), or using the expansion of numbers in base 2. To use the expansion of numbers, we observe that the map $\{\theta_i\}_{i=1}^n \to \{2\theta_i - 1\}_{i=1}^n$ maps a $0, 1-$sequence onto the corresponding $\pm 1-$sequence. Now we introduce a map $\psi : T_n \to [-1, 1]$ by letting $\psi(\emptyset) = 0$ and

$$\psi(\theta_1, \dots, \theta_n) = \sum_{i=1}^{n} 2^{-i} (2\theta_i - 1) .$$

To construct φ we relabel the range of ψ in the increasing order using numbers $[1, \dots, 2^{n+1} - 1]$.

Let $\{x_1, x_2, \dots, x_{2^{n+1}-1}\}$ be a sequence in a nonsuperreflexive Banach space X whose existence is guaranteed by Theorem 2.13 ((1)⇒(3)). We introduce an embedding $F_n : T_n \to X$ by

$$F_n(t) = \sum_{s \le t} x_{\varphi(s)},$$

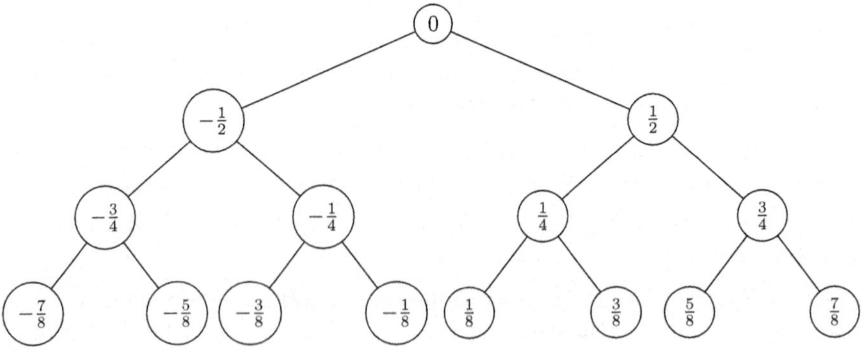

Fig. 3 The map of T_3 into $[-1, 1]$

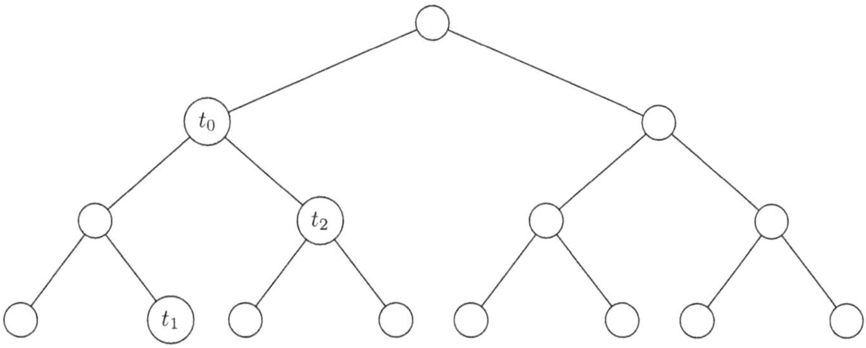

Fig. 4 t_0 is the closest common ancestor of t_1 and t_2

where $s \leq t$ for vertices of a binary tree means that s is the initial segment of the sequence t. Then $F_n(t_1) - F_n(t_2) = \sum_{t_0 < s \leq t_1} x_{\varphi(s)} - \sum_{t_0 < s \leq t_2} x_{\varphi(s)}$, where t_0 is the vertex of T_n corresponding to the largest initial common segment of t_1 and t_2, see Fig. 4. The condition in (2) and the choice of φ imply that

$$||F_n(t_1) - F_n(t_2)|| \geq \alpha(d_T(t_1, t_0) + d_T(t_2, t_0)) = \alpha d_{T_n}(t_1, t_2).$$

The estimate $||F_n(t_1) - F_n(t_2)||$ from above is straightforward. This completes the proof of bilipschitz embeddability of $\{T_n\}$ into any nonsuperreflexive Banach space with uniformly bounded distortions. □

Characterization of Superreflexivity in Terms of Diamond Graphs

Johnson and Schechtman [58] proved that there are some other sequences of graphs (with their graph metrics) which also can serve as test-spaces for superreflexivity. For example, binary trees in Bourgain's theorem can be replaced by the diamond graphs or by Laakso graphs.

Definition 2.17. Diamond graphs $\{D_n\}_{n=0}^{\infty}$ are defined as follows: The *diamond graph* of level 0 is denoted D_0. It has two vertices joined by an edge of length 1. The *diamond graph* D_n is obtained from D_{n-1} as follows. Given an edge $uv \in E(D_{n-1})$, it is replaced by a quadrilateral u, a, v, b, with edges ua, av, vb, bu. (See Fig. 5.)

Two different normalizations of the graphs $\{D_n\}_{n=1}^{\infty}$ are considered

- *Unweighted diamonds:* Each edge has length 1.
- *Weighted diamonds:* Each edge of D_n has length 2^{-n}

In both cases we endow vertex sets of $\{D_n\}_{n=0}^{\infty}$ with their shortest path metrics.

Fig. 5 Diamond D_2

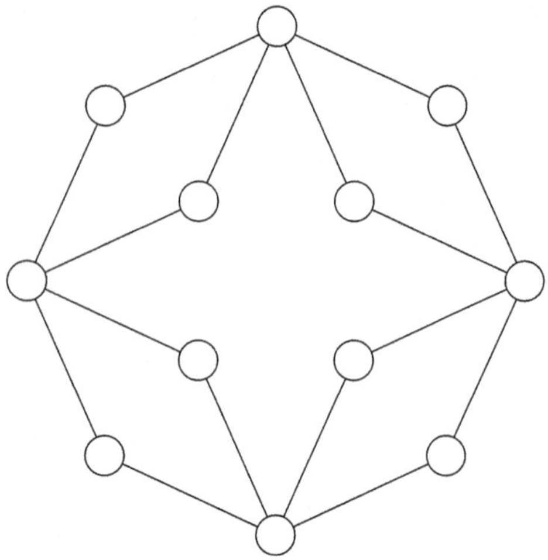

In the case of weighted diamonds the identity map $D_{n-1} \mapsto D_n$ is an isometry. In this case the union of D_n, endowed with its the metric induced from $\{D_n\}_{n=0}^{\infty}$ is called the *infinite diamond* and is denoted D_ω.

To the best of my knowledge the first paper in which diamond graphs $\{D_n\}_{n=0}^{\infty}$ were used in Metric Geometry is [50] (a conference version was published in 1999).

Definition 2.18. Laakso graphs $\{L_n\}_{n=0}^{\infty}$ are defined as follows: The *Laakso graph* of level 0 is denoted L_0. It has two vertices joined by an edge of length 1. The *Laakso graph* L_n is obtained from L_{n-1} as follows. Given an edge $uv \in E(L_{n-1})$, it is replaced by the graph L_1 shown in Fig. 6, the vertices u and v are identified with the vertices of degree 1 of L_1.

Two different normalizations of the graphs $\{L_n\}_{n=1}^{\infty}$ are considered

- *Unweighted Laakso graphs:* Each edge has length 1.
- *Weighted Laakso graphs:* Each edge of L_n has length 4^{-n}

In both cases we endow vertex sets of $\{L_n\}_{n=0}^{\infty}$ with their shortest path metrics.

In the case of weighted Laakso graphs the identity map $L_{n-1} \mapsto L_n$ is an isometry. In this case the union of L_n endowed with its the metric induced from $\{L_n\}_{n=0}^{\infty}$ is called the *Laakso space* and is denoted L_ω.

The Laakso graphs were introduced in [65], but they were inspired by the construction of Laakso in [63].

Theorem 2.19 (Johnson–Schechtman [58]). *A Banach space X is nonsuperreflexive if and only if it admits bilipschitz embeddings with uniformly bounded distortions of diamonds $\{D_n\}_{n=1}^{\infty}$ of all sizes.*

Theorem 2.20 (Johnson–Schechtman [58]). *A similar result holds for $\{L_n\}_{n=1}^{\infty}$.*

Fig. 6 Laakso graph L_1

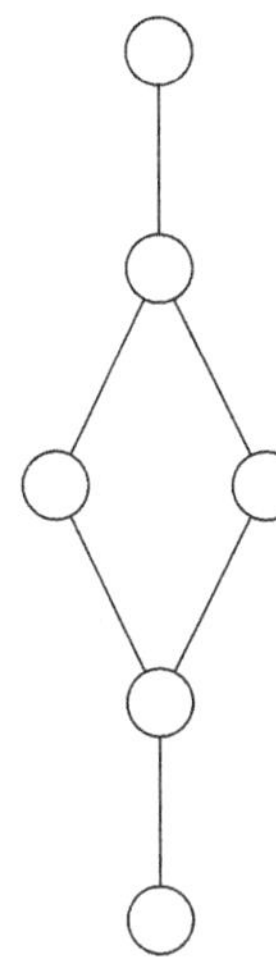

Without Proof. These results, whose original proofs (especially for diamond graphs) are elegant in both directions, are loved by expositors. Proof of Theorem 2.19 is presented in the lecture notes of Lancien [64], in the book of Pisier [103], and in my book [90]. □

Remark 2.21. In the "if" direction of Theorem 2.19, in addition to the original (controlled improvement for embeddings of smaller parts) argument of Johnson–Schechtman [58], there are two other arguments:

(1) The argument based on Markov convexity (see Definition 2.26). It is obtained by combining results of Lee–Naor–Peres [67] (each superreflexive Banach space is Markov p-convex for some $p \in [2,\infty)$) and Mendel–Naor [78] (Markov convexity constants of diamond graphs are not uniformly bounded from below, actually in [78] this statement is proved for Laakso graphs, but similar argument works for diamond graphs).
(2) The argument of [87, Sect. 3.1] showing that bilipschitz embeddability of diamond graphs with uniformly bounded distortions implies the finite tree property of the space, defined as follows:

Definition 2.22 (James [55]). Let $\delta > 0$. A *δ-tree* in a Banach space X is a subset $\{x_\tau\}_{\tau\in T_\infty}$ of X labelled by elements of the infinite binary tree T_∞, such that for each $\tau \in T_\infty$ we have

$$x_\tau = \frac{1}{2}(x_{\sigma_1} + x_{\sigma_2}) \quad \text{and } ||x_\tau - x_{\sigma_1}|| = ||x_\tau - x_{\sigma_2}|| \geq \delta, \tag{3}$$

where σ_1 and σ_2 are the children of τ. A Banach space X is said to have the *infinite tree property* if it contains a bounded δ-tree.

A *δ-tree of depth n* in a Banach space X is a finite subset $\{x_\tau\}_{\tau\in T_n}$ of X labelled by the binary tree T_n of depth n, such that the condition (3) is satisfied for each $\tau \in T_n$, which is not a leaf. A Banach space X has the *finite tree property* if for some $\delta > 0$ and each $n \in \mathbb{N}$ the unit ball of X contains a δ-tree of depth n.

Remark 2.23. In the "only if" direction of Theorem 2.19 there is a different (and more complicated) proof in [91, 96], which consists in a combination of the following two results:

(i) Existence of a bilipschitz embedding of the infinite diamond D_ω into any non-separable dual of a separable Banach space (using Stegall's [111] construction), see [91].
(ii) Finite subsets of a metric space which admits a bilipschitz embedding into any nonseparable dual of a separable Banach space, admit embeddings into any nonsuperreflexive Banach space with uniformly bounded distortions, see [96]. (The proof uses transfinite duals [9, 33, 34] and the results of Brunel–Sucheston [20, 21] and Perrott [100] on equal-signs-additive sequences.)

I would like to turn your attention to the fact that the Johnson–Schechtman Theorem 2.19 shows some obstacles on the way to a solution the (mentioned above) problem for superreflexivity:

Characterize metric spaces which admit bilipschitz embeddings into some superreflexive Banach spaces.

We need the following definitions and results. Let $\{M_n\}_{n=1}^\infty$ and $\{R_n\}_{n=1}^\infty$ be two sequences of metric spaces. We say that $\{M_n\}_{n=1}^\infty$ admits *uniformly bilipschitz embeddings* into $\{R_n\}_{n=1}^\infty$ if for each $n \in \mathbb{N}$ there is $m(n) \in \mathbb{N}$ and a bilipschitz map $f_n : M_n \to R_{m(n)}$ such that the distortions of $\{f_n\}_{n=1}^\infty$ are uniformly bounded.

Theorem 2.24 ([92]). *Binary trees $\{T_n\}_{n=1}^\infty$ do not admit uniformly bilipschitz embeddings into diamonds $\{D_n\}_{n=1}^\infty$.*

Without Proof. The proof is elementary, but rather lengthy combinatorial argument. □

There is also a non-embeddability in the other direction: The fact that diamonds $\{D_n\}$ do not admit uniformly bilipschitz embeddings into binary trees $\{T_n\}$ is well known, it follows immediately from the fact that D_n $(n \geq 1)$ contains a cycle of length 2^{n+1} isometrically, and the well-known observation of Rabinovich and Raz [105] stating that the distortion of any embedding of an m-cycle into any tree is $\geq \frac{m}{3} - 1$.

Remark 2.25. Mutual non-embeddability of Laakso graphs and binary trees is much simpler: (1) Laakso graphs are non-embeddable into trees because large Laakso graphs contain large cycles isometrically. (2) Binary trees are not embeddable into Laakso graphs because the Laakso graphs are uniformly doubling (see [51, p. 81] for the definition of a doubling metric space), but binary trees are not uniformly doubling.

Let us show that these results, in combination with some other known results, imply that it is impossible to find a sequence $\{C_n\}_{n=1}^{\infty}$ of finite metric spaces which admits uniformly bilipschitz embeddings into a metric space M if and only if M does not admit a bilipschitz embedding into a superreflexive Banach space. Assume the contrary: Such sequence $\{C_n\}_{n=1}^{\infty}$ exists. Then $\{C_n\}$ admits uniformly bilipschitz embeddings into the infinite binary tree. Therefore, by the result of Gupta [49], the spaces $\{C_n\}_{n=1}^{\infty}$ are uniformly bilipschitz-equivalent to weighted trees $\{W_n\}_{n=1}^{\infty}$. The trees $\{W_n\}_{n=1}^{\infty}$ should admit, by a result Lee–Naor–Peres [67] uniformly bilipschitz embeddings of increasing binary trees (these authors proved that $\{W_n\}_{n=1}^{\infty}$ would admit uniformly bilipschitz embeddings into ℓ_2 otherwise). Therefore, by Theorem 2.24 the spaces $\{C_n\}_{n=1}^{\infty}$ cannot be embeddable into diamonds with uniformly bounded distortion. Therefore they do not admit uniformly bilipschitz embeddings into D_ω (since the union of $\{D_i\}_{i=0}^{\infty}$ is dense in D_ω). On the other hand, Theorem 2.19 implies that D_ω does not admit a bilipschitz embedding into a superreflexive space, a contradiction.

One can try to find a characterization of metric spaces which are embeddable into superreflexive spaces in terms of some inequalities for distances. Some hope for such characterization was given by the already mentioned Markov convexity introduced by Lee–Naor–Peres [67], because it provides a reason for non-embeddability into superreflexive Banach spaces of both binary trees and diamonds (and many other trees and diamond-like spaces).

Definition 2.26 (Lee–Naor–Peres [67]). Let $\{X_t\}_{t\in\mathbb{Z}}$ be a Markov chain on a state space Ω. Given an integer $k \ge 0$, we denote by $\{\widetilde{X}_t(k)\}_{t\in\mathbb{Z}}$ the process which equals X_t for time $t \le k$, and evolves independently (with respect to the same transition probabilities) for time $t > k$. Fix $p > 0$. A metric space (X, d_X) is called *Markov p-convex with constant* Π if for every Markov chain $\{X_t\}_{t\in\mathbb{Z}}$ on a state space Ω, and every $f : \Omega \to X$,

$$\sum_{k=0}^{\infty}\sum_{t\in\mathbb{Z}} \frac{\mathbb{E}\left[d_X\left(f(X_t), f\left(\widetilde{X}_t\left(t-2^k\right)\right)\right)^p\right]}{2^{kp}} \le \Pi^p \cdot \sum_{t\in\mathbb{Z}} \mathbb{E}\left[d_X(f(X_t), f(X_{t-1}))^p\right]. \quad (4)$$

The least constant Π for which (4) holds for all Markov chains is called the *Markov p-convexity constant* of X, and is denoted $\Pi_p(X)$. We say that (X, d_X) is *Markov p-convex* if $\Pi_p(X) < \infty$.

Remark 2.27. The choice of the rather complicated left-hand side in (4) is inspired by the original Bourgain's proof [13] of the "if" part of Theorem 2.6.

Remark 2.28. It is unknown whether for general metric spaces Markov p-convexity implies Markov q-convexity for $q > p$. (This is known to be true for Banach spaces.)

Lee–Naor–Peres [67] showed that Definition 2.26 is important for the theory of metric embeddings by proving that each superreflexive space X is Markov q-convex for sufficiently large q. More precisely, it suffices to pick q such that X has an equivalent q-convex norm (see Definition 2.7), and by Theorem 2.9 of Pisier, such $q \in [2, \infty)$ exists for each superreflexive space.

On the other hand, Lee–Naor–Peres have shown that for any $0 < p < \infty$ the Markov p-convexity constants of binary trees $\{T_n\}$ are not uniformly bounded. Later Mendel and Naor [78] verified that the Markov p-convexity constants of Laakso graphs are not uniformly bounded. Similar proof works for diamonds $\{D_n\}$. See Theorem 3.11 and Remark 3.13 for a more general result.

Example 2.29 (Lee–Naor–Peres [67]). For every $m \in \mathbb{N}$, we have $\Pi_p(T_{2^m}) \geq 2^{1-\frac{2}{p}} \cdot m^{\frac{1}{p}}$.

Proof. Simplifying the description of the chain somewhat (precise description of Ω and the map f requires some formalities), we consider only times $t = 1, \ldots, 2^m$ and let $\{X_t\}_{t=0}^m$ be the downward random walk on T_{2^m} which is at the root at time $t = 0$ and X_{t+1} is obtained from X_t by moving down-left or down-right with probability $\frac{1}{2}$ each, see Fig. 7. We also assume that X_t is at the root with probability 1 if $t < 0$ (here more formal description of the chain is needed) and that for $t > 2^m$ we have $X_{t+1} = X_t$ (this is usually expressed by saying that *leaves are absorbing states*). Then

$$\sum_{t=1}^{2^m} \mathbb{E}\left[d_{T_{2^m}}(X_t, X_{t-1})^p\right] = 2^m.$$

Moreover, in the downward random walk, after splitting at time $r \leq 2^m$ with probability at least $\frac{1}{2}$ two independent walks will accumulate distance which is at least twice the number of steps (until a leaf is encountered). Thus

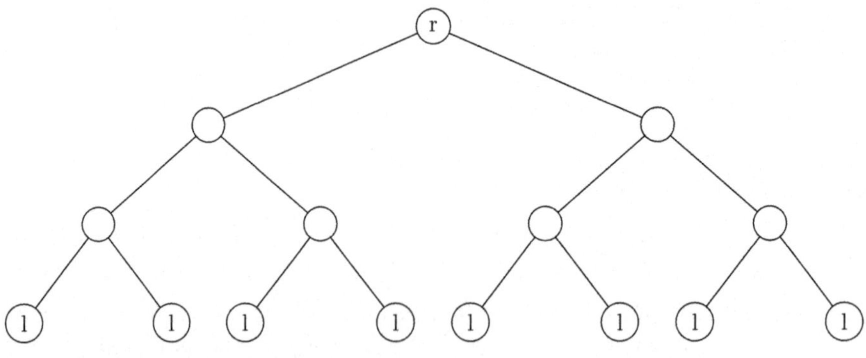

Fig. 7 T_3, with the root (r) and leaves (l) marked

$$\begin{aligned}\sum_{k=0}^{m}\sum_{t=1}^{2^m}\frac{\mathbb{E}\left[d_{T_{2^m}}\left(X_t,\widetilde{X}_t\left(t-2^k\right)\right)^p\right]}{2^{kp}} &\ge \sum_{k=0}^{m-1}\sum_{t=2^k}^{2^m}\frac{1}{2^{kp}}\cdot\frac{1}{2}\cdot 2^{(k+1)p}\\ &\ge m2^{m-1}2^{p-1}\\ &= 2^{p-2}\cdot m\cdot 2^m.\end{aligned}$$

The claim follows. □

For diamond graphs and Laakso graphs the argument is similar, but more complicated, because in such graphs the trajectories can come close after separation.

After uniting the reasons for non-embeddability for diamonds and trees one can hope to show that Markov convexity characterizes metric spaces which are embeddable into superreflexive spaces. It turns out that this is not the case. It was shown by Li [68, 69] that the Heisenberg group $\mathbb{H}(\mathbb{R})$ (see Definition 3.9) is Markov convex. On the other hand, it is known that the Heisenberg group does not admit a bilipschitz embedding into any superreflexive Banach space [26, 66]. (It is worth mentioning that in the present context we may consider the discrete Heisenberg group $\mathbb{H}(\mathbb{Z})$ consisting of the matrices with integer entries of the form shown in Definition 3.9 endowed with its word distance, see Definition 2.35.)

I suggest the following problem which is open as far as I know:

Problem 2.30. *Does there exist a test-space for superreflexivity which is Markov p-convex for some $0 < p < \infty$? (Or a sequence of test-spaces with uniformly bounded Markov p-convexity constants?)*

Remark 2.31. The Heisenberg group $H(\mathbb{Z})$ (with integer entries) has two properties needed for the test-space in Problem 2.30: it is not embeddable into any superreflexive space and is Markov convex. The only needed property which it does not have is: embeddability into each nonsuperreflexive space. Cheeger and Kleiner [28] proved that $H(\mathbb{Z})$ is not embeddable into some nonsuperreflexive Banach spaces, for example, into $L_1(0,1)$.

One more problem which I would like to mention here is

Problem 2.32 (Naor, July 2013). *Does there exist a sequence of finite metric spaces $\{M_i\}_{i=1}^{\infty}$ which is a sequence of test-spaces for superreflexivity with the following universality property: if $\{A_i\}_{i=1}^{\infty}$ is a sequence of test-spaces for superreflexivity, then there exist uniformly bilipschitz embeddings $E_i : A_i \to M_{n(i)}$, where $\{n(i)\}_{i=1}^{\infty}$ is some sequence of positive integers?*

Characterization of Superreflexivity in Terms of One Test-Space

Baudier [4] strengthened the "only if" direction of Bourgain's characterization and proved

Theorem 2.33 (Baudier [4]). *A Banach space X is nonsuperreflexive if and only if it admits bilipschitz embedding of the infinite binary tree T_∞.*

The following result hinted that possibly the Cayley graph of any nontrivially complicated hyperbolic group is the test-space for superreflexivity:

Theorem 2.34 (Buyalo–Dranishnikov–Schroeder [22]). *Every Gromov hyperbolic group admits a quasi-isometric embedding into the product of finitely many copies of the binary tree.*

Let us introduce notions used in this statement.

Definition 2.35. Let G be a group generated by a finite set S.

- The *Cayley graph* $\mathrm{Cay}(G, S)$ is defined as a graph whose vertex set is G and whose edge set is the set of all pairs of the form (g, sg), where $g \in G$, $s \in S$.
- In this context we consider each edge as a line segment of length 1 and endow $\mathrm{Cay}(G, S)$ with the shortest path distance. The restriction of this distance to G is called the *word distance*.
- Let u and v be two elements in a metric space (M, d_M). A *uv-geodesic* is a distance-preserving map $g : [0, d_M(u, v)] \to M$ such that $g(0) = u$ and $g(d_M(u, v)) = v$ (where $[0, d_M(u, v)]$ is an interval of the real line with the distance inherited from $\mathbb{R}$).
- A metric space M is *geodesic* if for any two points u and v in M, there is a uv-geodesic in M; $\mathrm{Cay}(G, S)$, with edges identified with line segments and with the shortest path distance is a geodesic metric space.
- A geodesic metric space M is called *δ-hyperbolic*, if for each triple $u, v, w \in M$ and any choice of a uv-geodesic, vw-geodesic, and wu-geodesics, each of these geodesics is in the δ-neighborhood of the union of the other two.
- A group is *word hyperbolic* or *Gromov hyperbolic* if $\mathrm{Cay}(G, S)$ is δ-hyperbolic for some $\delta < \infty$.

Remark 2.36.
- It might seem that the definition of hyperbolicity depends on the choice of the generating set S.
- It turns out that the value of δ depends on S, but its existence does not.
- The theory of hyperbolic groups was created by Gromov [46], although some related results were known before. The theory of hyperbolic groups plays an important role in group theory, geometry, and topology.
- Theory of hyperbolic groups is presented in many sources, see [1, 18].

Remark 2.37. It is worth mentioning that the identification of edges of $\mathrm{Cay}(G, S)$ with line segments is useful and important when we study geodesics and introduce the definition of hyperbolicity. In the theory of embeddings it is much more convenient to consider $\mathrm{Cay}(G, S)$ as a countable set (it is countable because we consider groups generated by finite sets), endowed with the shortest path distance (in the graph-theoretic sense), in this context it is called the *word distance*. See [89, 95] for relations between embeddability of graphs as vertex sets and as geodesic metric spaces.

It is worth mentioning that although different finite generating sets S_1 and S_2 in G lead to different word distances on G, the resulting metric spaces are bilipschitz equivalent: the identity map $(G, d_{S_1}) \to (G, d_{S_2})$ is bilipschitz, where d_{S_1} is the word distance corresponding to S_1 and d_{S_2} is the word distance corresponding to S_2.

We also need the following definitions used in [22]. A map $f : X \to Y$ between metric spaces (X, d_X) and (Y, d_Y) is called a *quasi-isometric embedding* if there are $a_1, a_2 > 0$ and $b \geq 0$, such that

$$a_1 d_X(u, v) - b \leq d_Y(f(u), f(v)) \leq a_2 d_X(u, v) + b \tag{5}$$

for all $u, v \in X$. By a *binary tree* the authors of [22] mean an infinite tree in which each vertex has degree 3. By a *product of trees*, denoted $(\oplus_{i=1}^n T(i))_1$, we mean their Cartesian product with the ℓ_1-metric, that is,

$$d(\{u_i\}, \{v_i\}) = \sum_{i=1}^n d_{T(i)}(u_i, v_i). \tag{6}$$

Observation 2.38. *The binary tree defined as an infinite tree in which each vertex has degree* 3 *is isometric to a subset of the product in the sense of* (6) *of three copies of* T_∞.

Therefore we may replace the infinite binary tree by T_∞ in the statement of Theorem 2.34. Hence the Buyalo–Dranishnikov–Schroeder Theorem 2.34 in combination with the Baudier Theorem 2.33 implies the existence of a quasi-isometric embedding of any Gromov hyperbolic group, which is embeddable into product of n copies of T_∞, into any Banach space containing an isomorphic copy of a direct sum of n nonsuperreflexive spaces. The fact that Buyalo–Dranishnikov–Schroeder consider quasi-isometric embeddings (which are weaker than bilipschitz) is not a problem. One can easily prove the following lemma. Recall that a metric space is called *locally finite* if all balls of finite radius in it have finite cardinality (a detailed proof of Lemma 2.39 can be found in [92, Lemma 2.3]).

Lemma 2.39. *If a locally finite metric space* M *admits a quasi-isometric embedding into an infinite-dimensional Banach space* X*, then* M *admits a bilipschitz embedding into* X.

Remark 2.40. One can easily construct a counterexample to a similar statement for general metric spaces.

Back to Embeddings However, we know from results of Gowers–Maurey [45] that there exist nonsuperreflexive spaces which do not contain isomorphically direct sums of any two infinite-dimensional Banach spaces, so we do not get immediately bilipschitz embeddability of hyperbolic groups into nonsuperreflexive Banach spaces.

Possibly this obstacle can be overcome by modifying Baudier's proof of Theorem 2.33 for the case of a product of several trees, but at this point a more general result is available. It can be stated as:

Embeddability of locally finite spaces into Banach spaces is finitely determined.

We need the following version of the result on finite determination (this statement also explains what we mean by "finite determination"):

Theorem 2.41 ([88]). *Let A be a locally finite metric space whose finite subsets admit bilipschitz embeddings into a Banach space X with uniformly bounded distortions. Then A admits a bilipschitz embedding into X.*

Remark 2.42. This result and its version for coarse embeddings have many predecessors: Baudier [4, 5], Baudier–Lancien [6], Brown–Guentner [19], and Ostrovskii [85, 86].

Now we return to embeddings of hyperbolic groups into nonsuperreflexive spaces. Recall that we consider finitely generated groups. It is easy to see that in this case $\mathrm{Cay}(G,S)$ is a locally finite metric space (recall that we consider $\mathrm{Cay}(G,S)$ as a countable set G with its word distance). By finite determination, it suffices to show only how to embed products of n finite binary trees into an arbitrary nonsuperreflexive Banach space with uniformly bounded distortions (the distortions are allowed to grow if we increase n, since for a fixed hyperbolic group the number n is fixed). This can be done using the embedding of a finite binary tree suggested by Bourgain (Theorem 2.6) and the standard techniques for constructions of basic sequences and finite-dimensional decompositions. These techniques (going back to Mazur) allow to show that for each n and N there exists a sequence of finite-dimensional spaces X_i such that X_i contains a 2-bilipschitz image of T_N and the direct sum $(\oplus_{i=1}^n X_i)_1$ is $C(n)$-isomorphic to their linear span in X ($C(n)$ is constant which depends on n, but not on N). See [92, pp. 157–158] for a detailed argument.

So we have proved that each Gromov hyperbolic group admits a bilipschitz embedding into any nonsuperreflexive Banach space. This proves the corresponding part of the following theorem:

Theorem 2.43. *Let G be a Gromov hyperbolic group which does not contain a cyclic group of finite index. Then the Cayley graph of G is a test-space for superreflexivity.*

The other direction follows by a combination of results of Bourgain [13], Benjamini and Schramm [10], and some basic theory of hyperbolic groups [18, 82], see [92, Remark 2.5] for details.

I find the following open problem interesting:

Problem 2.44. *Characterize finitely generated infinite groups whose Cayley graphs are test-spaces for superreflexivity.*

Possibly Problem 2.44 is very far from its solution and we should rather do the following. Given a group whose structure is reasonably well understood, check

(1) Whether it admits a bilipschitz embedding into an arbitrary nonsuperreflexive Banach space?

(2) Whether it admits bilipschitz embeddings into some superreflexive Banach spaces?

Remark 2.45. There are groups which do not admit bilipschitz embeddings into some nonsuperreflexive spaces, such as L_1. Examples which I know:

- Heisenberg group (Cheeger–Kleiner [28]).
- Gromov's random groups [48] containing expanders weakly are not even coarsely embeddable into L_1.
- Recently constructed groups of Osajda [83] with even stronger properties.

Remark 2.46. At the moment the only groups known to admit bilipschitz embeddings into superreflexive spaces are groups containing $\mathbb{Z}^n$ as a subgroup of finite index. de Cornulier–Tessera–Valette [35] conjectured that such groups are the only groups admitting a bilipschitz embedding into ℓ_2. This conjecture is still open. I asked about the superreflexive version of this conjecture on MathOverflow [97] (August 19, 2014) and de Cornulier commented on it as: "In the main two cases for which the conjecture is known to hold in the Hilbert case, the same argument also works for arbitrary uniformly convex Banach spaces."

Remark 2.47. Groups which are test-spaces for superreflexivity do not have to be hyperbolic. In fact, one can show that a direct product of finitely many hyperbolic groups is a test-space for superreflexivity provided at least one of them does not have a cyclic group as a subgroup of finite index. It is easy to check (using the definition) that such products are not Gromov hyperbolic unless all-except-one groups in the product are finite (the reason is that $\mathbb{Z}^2$ is not Gromov hyperbolic).

Now I would like to return to the title of this section: "Characterization of superreflexivity in terms of one test-space." This can actually be done using either the Bourgain or the Johnson–Schechtman characterization and the following elementary proposition (I published it [92], but I am sure that it was known to interested people):

Proposition 2.48 ([92, Sect. 5]).

(a) *Let $\{S_n\}_{n=1}^\infty$ be a sequence of finite test-spaces for some class $\mathcal{P}$ of Banach spaces containing all finite-dimensional Banach spaces. Then there is a metric space S which is a test-space for $\mathcal{P}$.*

(b) *If $\{S_n\}_{n=1}^\infty$ are*

- *unweighted graphs,*
- *trees,*
- *graphs with uniformly bounded degrees,*

then S also can be required to have the same property.

Sketch of the Proof. In all of the cases the constructed space S contains subspaces isometric to each of $\{S_n\}_{n=1}^\infty$. Therefore the only implication which is nontrivial is that the embeddability of $\{S_n\}_{n=1}^\infty$ implies the embeddability of S.

Each finite metric space can be considered as a weighted graph with its shortest path distance. We construct the space S as an infinite graph by joining S_n with S_{n+1} with a path P_n whose length is $\geq \max\{\text{diam}S_n, \text{diam}S_{n+1}\}$. To be more specific, we pick in each S_n a vertex O_n and let P_n be a path joining O_n with O_{n+1}. We endow the infinite graph S with its shortest path distance. It is clear that $\{S_n\}_{n=1}^{\infty}$ embed isometrically into S and all of the conditions in (b) are satisfied. It remains only to show that each infinite-dimensional Banach space X which admits bilipschitz embeddings of $\{S_n\}_{n=1}^{\infty}$ with uniformly bounded distortions, admits a bilipschitz embedding of S. This is done by embedding S_n into any hyperplane of X with uniformly bounded distortions. This is possible because the sets are finite, the space is infinite-dimensional, and all hyperplanes in a Banach space are isomorphic with the Banach–Mazur distances being $\leq$ some universal constant.

Now we consider in X parallel hyperplanes $\{H_n\}$ with the distance between H_n and H_{n+1} equal to the length of P_n and embed everything in the corresponding way. All computations are straightforward (see [92] for details). □

Non-local Properties

One can try to find metric characterizations of classes of Banach spaces which are not local. (We say that a class $\mathcal{P}$ of Banach spaces is *not local* if the conditions (1) $X \in \mathcal{P}$ and (2) Y is finitely representable in X, do not necessarily imply that $Y \in \mathcal{P}$). Apparently this study should not be considered as a part of the Ribe program, and this direction has developed much more slowly than the directions related to the Ribe program. It is clear that even if we restrict our attention to properties which are hereditary (inherited by closed subspaces) and isomorphic invariant, the class of non-local properties which have been already studied in the literature is huge. I found in the literature only four properties for which the problem of metric characterization was ever considered asymptotic uniform convexity and smoothness, Radon-Nykodým property, reflexivity, infinite tree property. The goal of this section is to survey the corresponding results.

Asymptotic Uniform Convexity and Smoothness

One of the first results of the described type is the following result of Baudier–Kalton–Lancien, where by T_∞^∞ we denote the tree defined similarly to the tree T_∞, but now we consider all possible finite sequences with terms in $\mathbb{N}$, and so degrees of all vertices of T_∞^∞ are infinite.

Theorem 3.1 ([7]). *Let X be a reflexive Banach space. The following assertions are equivalent:*

- *T_∞^∞ admits a bilipschitz embedding into X.*
- *X does not admit any equivalent asymptotically uniformly smooth norm* <u>*or*</u> *X does not admit any equivalent asymptotically uniformly convex norm.*
- *The Szlenk index of X is $> \omega$ or the Szlenk index of X^* is $> \omega$, where ω is the first infinite ordinal.*

It is worth mentioning that Dilworth et al. [37] found an interesting geometric description of the class of Banach spaces whose metric characterization is provided by Theorem 3.1.

Radon–Nikodým Property

The *Radon–Nikodým property* (RNP) is one of the most important isomorphic invariants of Banach spaces. This class also plays an important role in the theory of metric embeddings, this role is partially explained by the fact that for this class one can use differentiability to prove non-embeddability results.

There are many expository works presenting results on the RNP, we recommend the readers (depending on the taste and purpose) one of the following sources [11, Chap. 5], [12, 17, 36, 103, 113].

Equivalent Definitions of RNP

One of the reasons for the importance of the RNP is the possibility to characterize (define) the RNP in many different ways. I would like to remind some of them:

- Measure-theoretic definition (it gives the name to this property) $X \in \text{RNP} \Leftrightarrow$ The following analogue of the Radon–Nikodým theorem holds for X-valued measures.
  - Let (Ω, Σ, μ) be a positive finite real-valued measure, and (Ω, Σ, τ) be an X-valued measure on the same σ-algebra which is absolutely continuous with respect to μ (this means $\mu(A) = 0 \Rightarrow \tau(A) = 0$) and satisfies the condition $\tau(A)/\mu(A)$ is a uniformly bounded set of vectors over all $A \in \Sigma$ with $\mu(A) \neq 0$. Then there is an $f \in L_1(\mu, X)$ such that

$$\forall A \in \Sigma \quad \tau(A) = \int_A f(\omega) d\mu(\omega).$$

- Definition in terms of differentiability (goes back to Clarkson [31] and Gelfand [41]) $X \in \text{RNP} \Leftrightarrow X$-valued Lipschitz functions on $\mathbb{R}$ are differentiable almost everywhere.

- Probabilistic definition [24] $X \in$ RNP $\Leftrightarrow$ Bounded X-valued martingales converge.
  - In more detail: A Banach space X has the RNP if and only if each X-valued martingale $\{f_n\}$ on some probability space (Ω, Σ, μ), for which $\{||f_n(\omega)|| : n \in \mathbb{N}, \omega \in \Omega\}$ is a bounded set, converges in $L_1(\Omega, \Sigma, \mu, X)$.
- Geometric definition. $X \in$ RNP $\Leftrightarrow$ Each bounded closed convex set in X is dentable in the following sense:
  - A bounded closed convex subset C in a Banach space X is called *dentable* if for each $\varepsilon > 0$ there is a continuous linear functional f on X and $\alpha > 0$ such that the set

$$\{y \in C : f(y) \geq \sup\{f(x) : x \in C\} - \alpha\}$$

  has diameter $< \varepsilon$.
- **Examples** (these lists are far from being exhaustive):
  - RNP: Reflexive (for example, L_p, $1 < p < \infty$), separable dual spaces (for example, ℓ_1).
  - non-RNP: c_0, $L_1(0, 1)$, nonseparable duals of separable Banach spaces.

RNP and Metric Embeddings

Cheeger–Kleiner [26] and Lee–Naor [66] noticed that the observation of Semmes [109] on the result of Pansu [98] can be generalized to maps of the Heisenberg group into Banach spaces with the RNP. This implies that Heisenberg group with its sub-Riemannian metric (see Definition 3.9) does not admit a bilipschitz embedding into any space with the RNP.

Cheeger–Kleiner [27] generalized some part of differentiability theory of Cheeger [25] (see also [59, 60]) to maps of metric spaces into Banach spaces with the RNP. This theory implies some non-embeddability results, for example, it implies that the Laakso space does not admit a bilipschitz embedding into a Banach space with the RNP.

Metric Characterization of RNP

In 2009 Johnson [112, Problem 1.1] suggested the problem: Find a purely metric characterization of the Radon–Nikodým property (that is, find a characterization of the RNP which does not refer to the linear structure of the space). The main goal of the rest of section "Radon–Nikodým Property" is to present such characterization.

It turns out that the RNP can be characterized in terms of *thick families of geodesics* defined as follows (different versions of this definition appeared in [91, 93, 94], the following seems to be the most suitable definition).

Definition 3.2 ([91, 94]). A family T of uv-geodesics is called *thick* if there is $\alpha > 0$ such that for every $g \in T$ and for every finite collection of points $r_1, \ldots, r_n$ in the image of g, there is another uv-geodesic $\tilde{g} \in T$ satisfying the conditions:

(i) The image of $\tilde{g}$ also contains $r_1, \ldots, r_n$ (we call these points *control points*).
(ii) Possibly there are some more common points of g and $\tilde{g}$.
(iii) There is a sequence $0 = q_0 < s_1 < q_1 < s_2 < q_2 < \cdots < s_m < q_m = d_M(u, v)$, such that $g(q_i) = \tilde{g}(q_i)$ $(i = 0, \ldots, m)$ are common points containing $r_1, \ldots, r_n$; and $\sum_{i=1}^{m} d_M(g(s_i), \tilde{g}(s_i)) \geq \alpha$.
(iv) Furthermore, each geodesic which on some intervals between the points $0 = q_0 < q_1 < q_2 < \cdots < q_m = d_M(u, v)$ coincides with g and on others with $\tilde{g}$ is also in T.

Example 3.3. Interesting and important examples of spaces having thick families of geodesics are the infinite diamond D_ω and the Laakso space L_ω, but now we consider them not as unions of finite sets, but as unions of geodesic metric spaces obtained from weighted $\{D_n\}_{n=0}^{\infty}$ and $\{L_n\}_{n=0}^{\infty}$ in which edges are identified with line segments of lengths $\{2^{-n}\}_{n=0}^{\infty}$ and $\{4^{-n}\}_{n=0}^{\infty}$, respectively. Observe that for such graphs there are also natural (although non-unique) isometric embeddings of D_n into D_{n+1} and L_n into L_{n+1}, and therefore the unions are well-defined. It is easy to check that the families of all geodesics in D_ω and L_ω joining the vertices of D_0 and L_0, respectively, are thick.

Theorem 3.4 ([91, 94]). *A Banach space X does not have the RNP if and only if there exists a metric space M_X containing a thick family T_X of geodesics which admits a bilipschitz embedding into X.*

Remark 3.5. Theorem 3.4 implies the result of Cheeger and Kleiner [27] on nonexistence of bilipschitz embeddings of the Laakso space into Banach spaces with the RNP.

It turns out that the metric space M_X whose existence is established in Theorem 3.4 cannot be chosen independently of X, because the following result holds:

Theorem 3.6 ([91]). *For each metric space M containing a thick family of geodesics there exists a Banach space X which does not have the RNP and does not admit a bilipschitz embedding of M.*

Because of Theorem 3.6 the following is an open problem:

Problem 3.7. *Can we characterize the RNP using test-spaces?*

Also I would like to mention the problem of the metric characterization of the RNP can have many different (correct) answers, so it is natural to try to find metric characterizations of the RNP in some other terms.

- Proof of Theorem 3.4 (in both directions) is based on the characterization of the RNP in terms of martingales. It will be presented in section "Proof of Theorem 3.4".
- It is not true that each Banach space without RNP contains a thick family of geodesics, because Banach spaces without RNP can have the uniqueness of geodesics property (consider a strictly convex renorming of a separable Banach space without RNP), so the words 'bilipschitz embedding' in Theorem 3.4 cannot be replaced by 'isometric embedding.'
- Proof of Theorem 3.6 is based on the construction of Bourgain and Rosenthal [15] of 'small' subspaces of $L_1(0,1)$ which still do not have the Radon–Nikodým property.

- Studying metric characterizations of the RNP, it would be much more useful and interesting to get a characterization of all metric spaces which do not admit bilipschitz embeddings into Banach spaces with the RNP.
- In view of Theorem 3.4 it is natural to ask: whether the presence of bilipschitz images of thick families of geodesics characterizes metric spaces which do not admit bilipschitz embeddings into Banach spaces with the RNP?
- It is clear that the answer to this question in full generality is negative: we may just consider a dense subset of a Banach space without the RNP which does not contain any continuous curves.
- So we restrict our attention to spaces containing sufficiently large collections of continuous curves. Our next result is a negative answer even in the case of geodesic metric spaces. Recall a metric space is called *geodesic* if any two points in it are joined by a geodesic.

Theorem 3.8 ([93]). *There exist geodesic metric spaces which satisfy the following two conditions simultaneously:*

- *Do not contain bilipschitz images of thick families of geodesics.*
- *Do not admit bilipschitz embeddings into Banach spaces with the Radon–Nikodým property.*

In [93] it was shown that the Heisenberg group with its sub-Riemannian metric is an example of such metric space. Let us recall the corresponding definitions.

Definition 3.9. The *Heisenberg group* $\mathbb{H}(\mathbb{R})$ can be defined as the group of real upper-triangular matrices with 1's on the diagonal:

$$\begin{bmatrix} 1 & x & z \\ 0 & 1 & y \\ 0 & 0 & 1 \end{bmatrix}.$$

One of the ways to introduce the *sub-Riemannian metric on* $\mathbb{H}(\mathbb{R})$ is to find the tangent vectors of the curves produced by left translations in x and in y directions, that is,

$$\frac{d}{d\varepsilon}\bigg|_{\varepsilon=0} \begin{bmatrix} 1 & \varepsilon & 0 \\ 0 & 1 & 0 \\ 0 & 0 & 1 \end{bmatrix} \begin{bmatrix} 1 & x & z \\ 0 & 1 & y \\ 0 & 0 & 1 \end{bmatrix} = \begin{bmatrix} 0 & 1 & y \\ 0 & 0 & 0 \\ 0 & 0 & 0 \end{bmatrix}$$

$$\frac{d}{d\varepsilon}\bigg|_{\varepsilon=0} \begin{bmatrix} 1 & 0 & 0 \\ 0 & 1 & \varepsilon \\ 0 & 0 & 1 \end{bmatrix} \begin{bmatrix} 1 & x & z \\ 0 & 1 & y \\ 0 & 0 & 1 \end{bmatrix} = \begin{bmatrix} 0 & 0 & 0 \\ 0 & 0 & 1 \\ 0 & 0 & 0 \end{bmatrix}$$

All elements of $\mathbb{H}(\mathbb{R})$ can be regarded as elements of $\mathbb{R}^3$ with coordinates x, y, z. For $u, v \in \mathbb{H}(\mathbb{R})$ we consider the set of all differentiable curves joining u and v with the restriction that the tangent vector at each point of the curve is a linear combination of the two tangent vectors computed above. Finally we introduce the distance between u and v as the infimum of lengths (in the usual Euclidean sense) of projections onto xy-plane of all such curves.

This metric has been systematically studied (see [23, 47, 80]), it has very interesting geometric properties. The Heisenberg group $\mathbb{H}(\mathbb{R})$ with its subriemannian metric is a very important example for Metric Geometry and its applications to Computer Science. One of the reasons for this is its poor embeddability into many classes of Banach spaces. As we already mentioned, Cheeger–Kleiner [26] and Lee–Naor [66] proved that the Heisenberg group does not admit a bilipschitz embedding into a Banach space with the RNP. It remains to show that it does not admit a bilipschitz embedding of a thick family of geodesics.

Remark 3.10. It is not needed for our argument, but is worth mentioning that

- Cheeger–Kleiner [28] proved that $\mathbb{H}(\mathbb{R})$ does not admit a bilipschitz embedding into $L_1(0, 1)$.
- Cheeger–Kleiner–Naor [29] found quantitative versions of the previous result for embeddings of finite subsets of $\mathbb{H}(\mathbb{R})$ into $L_1(0, 1)$. These quantitative results are important for Theoretical Computer Science.

We finish the proof of Theorem 3.8 by using the notion of Markov convexity (Definition 2.26), proving

Theorem 3.11 ([93]). *A metric space with a thick family of geodesics is not Markov p-convex for any $p \in (0, \infty)$.*

and combining it with the following result:

Theorem 3.12 ([68, 69]). *The Heisenberg group $\mathbb{H}(\mathbb{R})$ is Markov* 4*-convex.*

Remark 3.13. Since the infinite diamond D_ω and the Laakso space L_ω contain thick families of geodesics, they are not Markov p-convex for any $p \in (0, \infty)$. Since the unions of $\{D_n\}_{n=0}^{\infty}$ and $\{L_n\}_{n=0}^{\infty}$ (considered as finite sets) are dense in D_ω and L_ω, respectively; we conclude that Markov p-convexity constants of diamond graphs and Laakso graphs are not uniformly bounded for any $p \in (0, \infty)$.

Remark 3.14. It is worth mentioning that the discrete Heisenberg group $\mathbb{H}(\mathbb{Z})$ embeds into a Banach space with the RNP. Since $\mathbb{H}(\mathbb{Z})$ is locally finite, this follows by combining the well-known observation of Fréchet on isometric embeddability of any n-element set into ℓ_∞^n (see [90, p. 6]) with the finite determination (Theorem 2.41, actually the earlier result of [6] suffices here). In fact, these results imply bilipschitz embeddability of $\mathbb{H}(\mathbb{Z})$ into the direct sum $(\oplus_{n=1}^\infty \ell_\infty^n)_2$, which has the RNP because it is reflexive.

Proof of Theorem 3.4

First we prove: **No RNP $\Rightarrow$ bilipschitz embeddability of a thick family of geodesics.**

We need to define a more general structure than that of a δ-tree (see Definition 2.22), in which each element is not a midpoint of a line segment, but a convex combination.

Definition 3.15. Let Z be a Banach space and let $\delta > 0$. A set of vectors $\{z_{n,j}\}_{n=0,j=1}^{\infty\ m_n}$ in Z is called a *δ-bush* if $m_0 = 1$ and for every $n \geq 1$ there is a partition $\{A_k^n\}_{k=1}^{m_{n-1}}$ of $\{1, \ldots, m_n\}$ such that

$$||z_{n,j} - z_{n-1,k}|| \geq \delta \tag{7}$$

for every $n \geq 1$ and for every $j \in A_k^n$, and

$$z_{n-1,k} = \sum_{j \in A_k^n} \lambda_{n,j} z_{n,j} \tag{8}$$

for some $\lambda_{n,j} \geq 0$, $\sum_{j \in A_k^n} \lambda_{n,j} = 1$.

Theorem 3.16. *A Banach space Z does not have the RNP if and only if it contains a bounded δ-bush for some $\delta > 0$.*

Remark 3.17. Theorem 3.16 can be derived from Chatterji's result [24]. Apparently Theorem 3.16 was first proved by James, possibly even before Chatterji, see [57].

- We construct a suitable thick family of geodesics using a bounded δ-bush in a Banach space without the RNP.
- It is not difficult to see (for example, using the Clarkson–Gelfand characterization) that a subspace of codimension 1 (hyperplane) in a Banach space without the RNP does not have the RNP.
- Let X be a non-RNP Banach space. We pick a norm-one vector $x \in X$, then a norm-one functional x^* on X satisfying $x^*(x) = 1$. Then (by the previous remark) we find a bounded δ-bush (for some $\delta > 0$) in the kernel $\ker x^*$. We shift this bush adding x to each of its elements, and get a (still bounded) δ-bush $\{x_{i,j}\}$ satisfying the condition $x^*(x_{i,j}) = 1$ for each i and j.

- Now we change the norm of X to equivalent. The purpose of this step is to get a norm for which we are able to construct the thick family of geodesics in X, and there will be no need in a bilipschitz embedding. (One can easily see that this would be sufficient to prove the theorem.)
- The unit ball of the norm $||\cdot||_1$ is defined as the closed convex hull of the unit ball in the original norm and the set of vectors $\{\pm x_{i,j}\}$ (recall that $\{x_{i,j}\}$ form a bush in the hyperplane $\{x : x^*(x) = 1\}$). It is easy to check that in this new norm the set $\{x_{i,j}\}$ is a bounded δ-bush (possibly with a somewhat smaller $\delta > 0$, but we keep the same notation). Also in the new norm we have $||x_{i,j}||_1 = 1$ for all i and j. For simplicity of notation we shall use $||\cdot||$ to denote the new norm.
- We are going to use this δ-bush to construct a thick family T_X of geodesics in X joining 0 and $x_{0,1}$. First we construct a subset of the desired set of geodesics, this subset will be constructed as the set of limits of certain broken lines in X joining 0 and $x_{0,1}$. The constructed broken lines are also geodesics (but they do not necessarily belong to the family T_X).
- The mentioned above broken lines will be constructed using representations of the form $x_{0,1} = \sum_{i=1}^m z_i$, where z_i are such that $||x_{0,1}|| = \sum_{i=1}^m ||z_i||$. The broken line represented by such finite sequence $z_1, \ldots, z_m$ is obtained by letting $z_0 = 0$ and joining $\sum_{i=0}^k z_i$ with $\sum_{i=0}^{k+1} z_i$ with a line segment for $k = 0, 1, \ldots, m-1$. Vectors $\sum_{i=0}^k z_i$, $k = 0, 1, \ldots, m$ will be called *vertices* of the broken line.
- The infinite set of broken lines which we construct is labelled by vertices of the infinite binary tree T_∞ in which each vertex is represented by a finite (possibly empty) sequence of 0 and 1.
- The broken line corresponding to the empty sequence $\emptyset$ is represented by the one-element sequence $x_{0,1}$, so it is just a line segment joining 0 and $x_{0,1}$.
- We have

$$x_{0,1} = \lambda_{1,1}x_{1,1} + \cdots + \lambda_{1,m_1}x_{1,m_1},$$

where $||x_{1,j} - x_{0,1}|| \geq \delta$. We introduce the vectors

$$y_{1,j} = \frac{1}{2}(x_{1,j} + x_{0,1}).$$

- For these vectors we have

$$x_{0,1} = \lambda_{1,1}y_{1,1} + \cdots + \lambda_{1,m_1}y_{1,m_1},$$

$||y_{1,j} - x_{1,j}|| = ||y_{1,j} - x_{0,1}|| \geq \frac{\delta}{2}$, and $||y_{1,j}|| = 1$.

- As a preliminary step to the construction of the broken lines corresponding to one-element sequences (0) and (1) we form a broken line represented by the points

$$\lambda_{1,1}y_{1,1}, \ldots, \lambda_{1,m_1}y_{1,m_1}. \tag{9}$$

We label the broken line represented by (9) by $\overline{\emptyset}$.

- The broken line corresponding to the one-element sequence (0) is represented by the sequence obtained from (9) if we replace each term $\lambda_{1,j}y_{1,j}$ by a two-element sequence

$$\frac{\lambda_{1,j}}{2}x_{0,1}, \frac{\lambda_{1,j}}{2}x_{1,j}. \tag{10}$$

- The broken line corresponding to the one-element sequence (1) is represented by the sequence obtained from (9) if we replace each term $\lambda_{1,j}y_{1,j}$ by a two-element sequence

$$\frac{\lambda_{1,j}}{2}x_{1,j}, \frac{\lambda_{1,j}}{2}x_{0,1}. \tag{11}$$

- At this point one can see where are we going to get the thickness property from.
- In fact, one of the inequalities above is $||x_{1,j} - x_{0,1}|| \geq \delta$. Therefore

$$\left\| \frac{\lambda_{1,j}}{2}x_{1,j} - \frac{\lambda_{1,j}}{2}x_{0,1} \right\| \geq \frac{\lambda_{1,j}}{2}\delta.$$

Summing over all j, we get that the total sum of deviations is $\geq \frac{\delta}{2}$.
- In the obtained broken lines each line segment corresponds either to a multiple of $x_{0,1}$ or to a multiple of some $x_{1,j}$. In the next step we replace each such line segment by a broken line. Now we describe how we do this.
- Broken lines corresponding to 2-element sequences are also formed in two steps. To get the broken lines labelled by $(0,0)$ and $(0,1)$ we apply the described procedure to the geodesic labelled (0), to get the broken lines labelled by $(1,0)$ and $(1,1)$ we apply the described procedure to the geodesic labelled (1).
- In the first step we replace each term of the form $\frac{\lambda_{1,k}}{2}x_{0,1}$ by a multiplied by $\frac{\lambda_{1,k}}{2}$ sequence $\lambda_{1,1}y_{1,1}, \ldots, \lambda_{1,m_1}y_{1,m_1}$, and we replace a term of the form $\frac{\lambda_{1,k}}{2}x_{1,k}$ by the multiplied by $\frac{\lambda_{1,k}}{2}$ sequence

$$\{\lambda_{2,j}y_{2,j}\}_{j\in A_k^2}, \tag{12}$$

ordered arbitrarily, where $y_{2,j} = \frac{x_{1,k}+x_{2,j}}{2}$ and $\lambda_{2,j}$, $x_{2,j}$, and A_k^2 are as in the definition of the δ-bush (it is easy to check that in the new norm we have $||y_{2,j}|| = 1$). We label the obtained broken lines by $\overline{(0)}$ and $\overline{(1)}$, respectively.
- To get the sequence representing the broken line labelled by $(0,0)$ we do the following operation with the preliminary sequence labelled $\overline{(0)}$.
  - Replace each multiple $\lambda y_{1,j}$ present in the sequence by the two-element sequence

$$\lambda\frac{x_{0,1}}{2}, \lambda\frac{x_{1,j}}{2}. \tag{13}$$

– Replace each multiple $\lambda y_{2,j}$, with $j \in A_k^2$, present in the sequence by the two-element sequence

$$\lambda \frac{x_{1,k}}{2}, \lambda \frac{x_{2,j}}{2}. \tag{14}$$

- To get the sequence representing the broken line labelled by $(0, 1)$ we do the same but changing the order of terms in (13) and (14). To get the sequences representing the broken lines labelled by $(1, 0)$ and $(1, 1)$, we apply the same procedure to the broken line labelled $\overline{(1)}$.
- We continue in an "obvious" way and get broken lines for all vertices of the infinite binary tree T_∞. It is not difficult to see that vertices of a broken line corresponding to some vertex $(\theta_1, \dots, \theta_n)$ are contained in the broken line corresponding to any extension $(\theta_1, \dots, \theta_m)$ of $(\theta_1, \dots, \theta_n)$ $(m > n)$.
- This implies that broken lines corresponding to any ray (that is, a path infinite in one direction) in T_∞ has a limit (which is not necessarily a broken line, but is a geodesic), and limits corresponding to two different infinite paths have common points according to the number of common $(\theta_1, \dots, \theta_n)$ in the vertices of those paths.
- A thick family of geodesics is obtained by pasting pieces of these geodesics in all "reasonable" ways. All verifications are straightforward; see the details in [94].

It remains to prove:

Bilipschitz embeddability of a thick family of geodesics ⇒ No RNP.

Proof. We assume that a metric space (M, d) with a thick family of geodesics admits a bilipschitz embedding $f : M \to X$ into a Banach space X and show that there exists a bounded divergent martingale $\{M_i\}_{i=0}^{\infty}$ on $(0, 1]$ with values in X. We assume that

$$\ell d(x, y) \le ||f(x) - f(y)||_X \le d(x, y) \tag{15}$$

for some $\ell > 0$. We assume that the thick family consists of uv-geodesics for some $u, v \in M$ and that $d(u, v) = 1$ (dividing all distances in M by $d(u, v)$, if necessary).

Each function in the martingale $\{M_i\}_{i=0}^{\infty}$ will be obtained in the following way. We consider some finite sequence $V = \{v_i\}_{i=0}^{m}$ of points on any uv-geodesic, satisfying $v_0 = u$, $v_m = v$ and $d(u, v_{k+1}) \ge d(u, v_k)$. We define M_V as the function on $(0, 1]$ whose value on the interval $(d(u, v_k), d(u, v_{k+1})]$ is equal to

$$\frac{f(v_{k+1}) - f(v_k)}{d(v_k, v_{k+1})}.$$

It is clear that the bilipschitz condition (15) implies that $||M_V(t)|| \le 1$ for any collection V and any $t \in (0, 1]$. Since $\{v_i\}$ are on a geodesic, is clear that an infinite collection of such functions $\{M_{V(k)}\}_{k=0}^{\infty}$ forms a martingale if for each $k \in \mathbb{N}$ the sequence $V(k)$ contains $V(k-1)$ as a subsequence. So it remains to find a collection of sequences $\{V(k)\}_{k=0}^{\infty}$ for which the martingale $\{M_{V(k)}\}_{k=0}^{\infty}$ diverges. We denote $M_{V(k)}$ by M_k.

Now we describe some of the ideas of the construction.

- It suffices to have differences $||M_k - M_{k-1}||$ to be bounded away from zero for some infinite set of values of k.
- On steps for which we achieve such estimates from below we add exactly one new point z'_j into $V(k)$ between any two consequent points w_{j-1} and w_j of $V(k-1)$. In such a case it suffices to make the choice of points in such a way that the values of M_k on the intervals corresponding to pairs (w_{j-1}, z'_j) and (z'_j, w_j) are 'far' from each other, and thus from the value of M_{k-1} corresponding to (w_{j-1}, w_j). Actually we do not need this condition for each pair (w_{j-1}, w_j), but only "on average."
- Using the definition of a thick family of geodesics and the bilipschitz condition, we can achieve this goal. A detailed description follows.

We let $V(0) = \{u, v\}$ and so M_0 is a constant function on $(0, 1]$ taking value $f(v)-f(u)$. In the next step we apply the condition of the definition of a thick family to control points $\{u, v\}$ and any geodesic g of the family. We get another geodesic $\tilde{g}$, the corresponding sequence of common points $\{w_i\}_{i=0}^m$ and the corresponding pair of sufficiently well-separated sequences $\{z_i, \tilde{z}_i\}_{i=1}^m$ on the geodesics g and $\tilde{g}$. The separation condition is $\sum_{i=1}^m d(z_i, \tilde{z}_i) \geq \alpha$.

We let $V(1) = \{w_i\}_{i=0}^m$. Observe that in this step we cannot claim any nontrivial estimates for $||M_1 - M_0||_{L_1(X)}$ from below because we have not made any nontrivial assumptions on this step of the construction (it can even happen that $M_1 = M_0$). Lower estimates for martingale differences in our argument are obtained only for differences of the form $||M_{2k} - M_{2k-1}||_{L_1(X)}$.

We choose $V(2)$ to be of the form

$$w_0, z'_1, w_1, z'_2, w_n, \ldots, z'_m, w_m, \tag{16}$$

where each z'_i is either z_i or $\tilde{z}_i$ depending on the behavior of the mapping f. We describe this dependence below. Observe that since z_i or $\tilde{z}_i$ are images of the same point in $[0, 1]$, the corresponding partition of the interval $(0, 1]$ does not depend on our choice.

To make the choice of z'_i we consider the quadruple $w_{i-1}, z_i, w_i, \tilde{z}_i$. The bilipschitz condition (15) implies $||f(z_i) - f(\tilde{z}_i)|| \geq \ell d(z_i, \tilde{z}_i)$. Consider two pairs of vectors corresponding to two different choices of z'_i:

Pair 1: $f(w_i) - f(z_i), f(z_i) - f(w_{i-1})$. **Pair 2:** $f(w_i) - f(\tilde{z}_i), f(\tilde{z}_i) - f(w_{i-1})$.

The inequality $||f(z_i)-f(\tilde{z}_i)|| \geq \ell d(z_i, \tilde{z}_i)$ implies that at least one of the following is true:

$$\left\| \frac{f(w_i) - f(z_i)}{d(w_i, z_i)} - \frac{f(z_i) - f(w_{i-1})}{d(z_i, w_{i-1})} \right\| \geq \frac{\ell}{2} d(z_i, \tilde{z}_i) \left(\frac{1}{d(w_i, z_i)} + \frac{1}{d(z_i, w_{i-1})} \right) \tag{17}$$

or

$$\left\| \frac{f(w_i) - f(\tilde{z}_i)}{d(w_i, \tilde{z}_i)} - \frac{f(\tilde{z}_i) - f(w_{i-1})}{d(\tilde{z}_i, w_{i-1})} \right\| \geq \frac{\ell}{2} d(z_i, \tilde{z}_i) \left(\frac{1}{d(w_i, \tilde{z}_i)} + \frac{1}{d(\tilde{z}_i, w_{i-1})} \right). \tag{18}$$

Since in the definition of a thick family of geodesics we have $\sum_i d(z_i, \tilde{z}_i) \geq \alpha$, these inequalities show that if we choose z'_i to be the one of z_i and $\tilde{z}_i$, for which

$\frac{f(w_i)-f(z_i')}{d(w_i,z_i')}$ and $\frac{f(z_i')-f(w_{i-1})}{d(z_i',w_{i-1})}$ are more distant from each other, we have a chance to get the desired condition. (This is what we verify below.)

We pick z_i' to be z_i if the left-hand side of (17) is larger than the left-hand side of (18), and pick $z_i' = \tilde{z}_i$ otherwise.

Let us estimate $||M_2 - M_1||_1$. First we estimate the part of this difference corresponding to the interval $(d(w_0, w_{i-1}), d(w_0, w_i)]$. Since the restriction of M_2 to the interval $(d(w_0, w_{i-1}), d(w_0, w_i)]$ is a two-valued function, and M_1 is constant on the interval, the integral

$$\int_{d(w_0,w_{i-1})}^{d(w_0,w_i)} ||M_2 - M_1||dt \tag{19}$$

can be estimated from below in the following way. Denote the value of M_2 on the first part of the interval by x, the value on the second by y, the value of M_1 on the whole interval by z, the length of the first interval by A and of the second by B. We have: the desired integral is equal to $A||x-z|| + B||y-z||$ and therefore can be estimated in the following way:

$$\begin{aligned} A||x-z|| + B||y-z|| &\geq \max\{||x-z||, ||y-z||\} \cdot \min\{A, B\} \\ &\geq \frac{1}{2}||x-y|| \min\{A, B\}. \end{aligned}$$

Therefore, assuming without loss of generality that the left-hand side of (17) is larger than the left-hand side of (18), we can estimate the integral in (19) from below by

$$\begin{aligned} &\frac{1}{2}\left\|\frac{(f(w_i)-f(z_i))}{d(w_i,z_i)} - \frac{(f(z_i)-f(w_{i-1}))}{d(z_i,w_{i-1})}\right\| \cdot \min\{d(w_i,z_i), d(z_i,w_{i-1})\} \\ &\geq \frac{1}{4}\ell\, d(z_i,\tilde{z}_i)\left(\frac{1}{d(w_i,z_i)} + \frac{1}{d(z_i,w_{i-1})}\right) \cdot \min\{d(w_i,z_i), d(z_i,w_{i-1})\} \\ &\geq \frac{1}{4}\ell\, d(z_i,\tilde{z}_i). \end{aligned}$$

Summing over all intervals and using the condition $\sum_{i=1}^{m} d(z_i,\tilde{z}_i) \geq \alpha$, we get $||M_2 - M_1|| \geq \frac{1}{4}\ell\alpha$.

Now we recall that the last condition of the definition of a thick family of geodesics implies that

$$w_0, z_1', w_1, z_2', w_2, \ldots, z_m', w_m, \tag{20}$$

where each z_i' is either z_i or $\tilde{z}_i$ depending on the choice made above belongs to some geodesic in the family.

We use all of points in (20) as control points and find new sequence $\{w_i^2\}_{i=0}^{m_2}$ of common points and a new sequence of pairs $\{z_i^2, \tilde{z}_i^2\}_{i=1}^{m_2}$ with substantial separation: $\sum_{i=1}^{m_2} d(z_i^2, \tilde{z}_i^2) \geq \alpha$.

We use $\{w_i^2\}_{i=0}^{m_2}$ to construct M_3 and the suitably selected sequence

$$w_0^2, z_1'^2, w_1^2, z_2'^2, w_2^2, \ldots, z_{m_2}'^2, w_{m_2}^2$$

to construct M_4. We continue in an obvious way. The inequalities $||M_{2k} - M_{2k-1}|| \geq \frac{1}{4}\ell\alpha$ imply that the martingale is divergent. □

Reflexivity

Problem 3.18. *Is it possible to characterize the class of reflexive spaces using test-spaces?*

Some comments on this problem:

Remark 3.19. It is worth mentioning that a metric space (or spaces) characterizing in the described sense reflexivity or the Radon–Nikodým property cannot be uniformly discrete (that is, cannot satisfy $\inf_{u\neq v} d(u, v) > 0$). This statement follows by combining the example of Ribe [107] of Banach spaces belonging to these classes which are uniformly homeomorphic to Banach spaces which do not belong to the classes, and the well-known fact (Corson–Klee [32]) that uniformly continuous maps are Lipschitz for (nontrivially) "large" distances.

I noticed that combining two of the well-known characterizations of reflexivity (Pták [104]—Singer [110]—Pełczyński [99]—James [54]—Milman and Milman [79]) and some differentiation theory (Mankiewicz [72]—Christensen [30]—Aronszajn [2], see also presentation in [11]) we get a purely metric characterization of reflexivity. This characterization can be described as a submetric test-space characterization of reflexivity:

Definition 3.20. A *submetric test-space* for a class $\mathcal{P}$ of Banach spaces is defined as a metric space T with a marked subset $S \subset T \times T$ such that the following conditions are equivalent for a Banach space X:

1. $X \notin \mathcal{P}$.
2. There exist a constant $0 < C < \infty$ and an embedding $f : T \to X$ satisfying the condition

$$\forall (x, y) \in S \quad d_T(x, y) \leq ||f(x) - f(y)|| \leq Cd_T(x, y). \tag{21}$$

An embedding satisfying (21) is called a *partially bilipschitz* embedding. Pairs (x, y) belonging to S are called *active*.

Let $\Delta \geq 1$. The submetric space X_Δ is the space ℓ_1 with its usual metric. The only thing which makes it different from ℓ_1 is the set of active pairs S_Δ: A pair $(x, y) \in X_\Delta \times X_\Delta$ is active if and only if

$$||x-y||_1 \leq \Delta ||x-y||_s, \tag{22}$$

where $||\cdot||_s$ is the summing norm, that is,

$$||\{a_i\}_{i=1}^{\infty}||_s = \sup_k \left|\sum_{i=1}^{k} a_i\right|.$$

Theorem 3.21 ([91]). *X_Δ, $\Delta \geq 2$ is a submetric test-space for reflexivity.*

The proof goes as follows. Let Z be a nonreflexive space. If you know the characterization of reflexivity which I meant, you see immediately that it implies that the space ℓ_1 admits a partially bilipschitz embedding into Z with the set of active pairs described as above.

The other direction. If ℓ_1 admits a partially bilipschitz embedding with the described set of active pairs, then the embedding is Lipschitz on ℓ_1, because each vector in ℓ_1 is a difference of two positive vectors.

Now, if Z does not have the Radon–Nikodým property (RNP), we are done (Z is nonreflexive). If Z has the RNP, we use the result of Mankiewicz–Christensen–Aronszajn and find a point of Gâteaux differentiability of this embedding. The Gâteaux derivative is a bounded linear operator which is "bounded below in certain directions." Using this we can get a sequence in Z which, after application of the non-reflexivity criterion (due to Pták—Singer—Pełczyński—James—D.&V. Milman), implies non-reflexivity of Z; see [91] for details.

Infinite Tree Property

See Definition 2.22 for the definition of the infinite tree property. Using a bounded δ-tree in a Banach space X one can easily construct a bounded divergent X-valued martingale. Hence the infinite tree property implies non-RNP. For some time it was an open problem whether the infinite tree property coincides with non-RNP. A counterexample was constructed by Bourgain and Rosenthal [15] in the paper mentioned above. The infinite tree property admits the following metric characterization.

Theorem 3.22 ([91]). *The class of Banach spaces with the infinite tree property admits a submetric characterization in terms of the metric space D_ω with the set S_ω of active pairs defined as follows: a pair is active if and only if it is a pair of vertices of a quadrilateral introduced in one of the steps.*

It would be interesting to answer the following open problem:

Problem 3.23. *Whether the infinite diamond D_ω is a test-space for the infinite tree property?*

Remark 3.24. It is worth mentioning that if we restrict our attention to **dual Banach spaces**, the following three properties are equivalent:

(1) Non-RNP.
(2) Infinite tree property.
(3) Bilipschitz embeddability of D_ω.

The implication (1) $\Rightarrow$ (2) is due to Stegall [111]. The implication (2) $\Rightarrow$ (1) follows from Chatterji [24]. The equivalence of (1) and (3) was proved in [91].

Acknowledgements The author gratefully acknowledges the support by NSF DMS-1201269.

The author would like to thank Gilles Lancien for the invitation to give a series of five lectures at "Autumn School on Nonlinear Geometry of Banach Spaces and Applications" (Métabief, France, October 2014). This paper was prepared on the basis of notes of that lecture series. The same lecture series was also presented at the seminar at St. John's University. The author would like to thank the participants of both the school and the seminar for the numerous questions, corrections, and comments, many of which were incorporated into the final version of the paper. Also the author would like to thank Gideon Schechtman and Hannes Thiel for very useful comments on the first draft of this paper.

References

1. J.M. Alonso, T. Brady, D. Cooper, V. Ferlini, M. Lustig, M. Mihalik, M. Shapiro, H. Short, Notes on word hyperbolic groups. Edited by Short, in *Group Theory from a Geometrical Viewpoint* (Trieste, 1990) (World Scientific, River Edge, NJ, 1991), pp. 3–63
2. N. Aronszajn, Differentiability of Lipschitzian mappings between Banach spaces. Stud. Math. **57**(2), 147–190 (1976)
3. K. Ball, The Ribe programme, in *Séminaire Bourbaki*, vol. 2011/2012. Exposés 1043–1058. Astérisque, vol. 352 (2013). Exp. No. 1047, viii, 147–159
4. F. Baudier, Metrical characterization of super-reflexivity and linear type of Banach spaces. Archiv. Math. **89**(5), 419–429 (2007)
5. F. Baudier, Embeddings of proper metric spaces into Banach spaces. Houst. J. Math. **38**(1), 209–223 (2012)
6. F. Baudier, G. Lancien, Embeddings of locally finite metric spaces into Banach spaces. Proc. Am. Math. Soc. **136**, 1029–1033 (2008)
7. F. Baudier, N.J. Kalton, G. Lancien, A new metric invariant for Banach spaces. Stud. Math. **199**(1), 73–94 (2010)
8. B. Begun, A remark on almost extensions of Lipschitz functions. Isr. J. Math. **109**, 151–155 (1999)
9. S.F. Bellenot, Transfinite duals of quasireflexive Banach spaces. Trans. Am. Math. Soc. **273**(2), 551–577 (1982)
10. I. Benjamini, O. Schramm, Every graph with a positive Cheeger constant contains a tree with a positive Cheeger constant. Geom. Funct. Anal. **7**(3), 403–419 (1997)

11. Y. Benyamini, J. Lindenstrauss, *Geometric Nonlinear Functional Analysis*, vol. 1. American Mathematical Society Colloquium Publications, vol. 48 (American Mathematical Society, Providence, RI, 2000)
12. R.D. Bourgin, *Geometric Aspects of Convex Sets with the Radon-Nikodým Property*. Lecture Notes in Mathematics, vol. 993 (Springer, Berlin, 1983)
13. J. Bourgain, The metrical interpretation of superreflexivity in Banach spaces. Isr. J. Math. **56**(2), 222–230 (1986)
14. J. Bourgain, Remarks on the extension of Lipschitz maps defined on discrete sets and uniform homeomorphisms, in *Geometrical Aspects of Functional Analysis* (1985/86). Lecture Notes in Mathematics, vol. 1267 (Springer, Berlin, 1987), pp. 157–167
15. J. Bourgain, H.P. Rosenthal, Martingales valued in certain subspaces of L^1. Isr. J. Math. **37**(1–2), 54–75 (1980)
16. J. Bourgain, V. Milman, H. Wolfson, On type of metric spaces. Trans. Am. Math. Soc. **294**(1), 295–317 (1986)
17. J. Bourgain, La propriété de Radon-Nikodym Publications mathématiques de l'Université Pierre et Marie Curie, Paris **36**, 104 (1979)
18. M.R. Bridson, A. Haefliger, *Metric Spaces of Non-positive Curvature*. Grundlehren der Mathematischen Wissenschaften, vol. 319 (Springer, Berlin, 1999)
19. N. Brown, E. Guentner, Uniform embeddings of bounded geometry spaces into reflexive Banach space. Proc. Am. Math. Soc. **133**(7), 2045–2050 (2005)
20. A. Brunel, L. Sucheston, On J-convexity and some ergodic super-properties of Banach spaces. Trans. Am. Math. Soc. **204**, 79–90 (1975)
21. A. Brunel, L. Sucheston, Equal signs additive sequences in Banach spaces. J. Funct. Anal. **21**(3), 286–304 (1976)
22. S. Buyalo, A. Dranishnikov, V. Schroeder, Embedding of hyperbolic groups into products of binary trees. Invent. Math. **169**(1), 153–192 (2007)
23. L. Capogna, D. Danielli, S.D. Pauls, J.T. Tyson, *An Introduction to the Heisenberg Group and the Sub-Riemannian Isoperimetric Problem*. Progress in Mathematics, vol. 259 (Birkhäuser, Basel, 2007)
24. S.D. Chatterji, Martingale convergence and the Radon-Nikodym theorem in Banach spaces. Math. Scand. **22**, 21–41 (1968)
25. J. Cheeger, Differentiability of Lipschitz functions on metric measure spaces. Geom. Funct. Anal. **9**(3), 428–517 (1999)
26. J. Cheeger, B. Kleiner, On the differentiability of Lipschitz maps from metric measure spaces to Banach spaces, in *Inspired by S.S. Chern*. Nankai Tracts in Mathematics, vol. 11 (World Scientific, Hackensack, NJ, 2006), 129–152
27. J. Cheeger, B. Kleiner, Differentiability of Lipschitz maps from metric measure spaces to Banach spaces with the Radon-Nikodým property. Geom. Funct. Anal. **19**(4), 1017–1028 (2009)
28. J. Cheeger, B. Kleiner, Differentiating maps into L^1, and the geometry of BV functions. Ann. Math. (2) **171**(2), 1347–1385 (2010). `arXiv:math.MG/0611954`
29. J. Cheeger, B. Kleiner, A. Naor, Compression bounds for Lipschitz maps from the Heisenberg group to L_1. Acta Math. **207**, 291–373 (2011)
30. J.P.R. Christensen, Measure theoretic zero sets in infinite dimensional spaces and applications to differentiability of Lipschitz mappings. Publ. Dép. Math. (Lyon) **10**(2), 29–39 (1973). Actes du Deuxième Colloque d'Analyse Fonctionnelle de Bordeaux (Univ. Bordeaux, 1973), I, pp. 29–39.
31. J.A. Clarkson, Uniformly convex spaces. Trans. Am. Math. Soc. **40**(3), 396–414 (1936)
32. H. Corson, V. Klee, Topological classification of convex sets, in *Proceedings of Symposia in Pure Mathematics*, vol. VII, pp. 37–51 (American Mathematical Society, Providence, RI, 1963)
33. W.J. Davis, J. Lindenstrauss, The ℓ_1^n problem and degrees of non-reflexivity. II. Stud. Math. **58**(2), 179–196 (1976)

34. W.J. Davis, W.B. Johnson, J. Lindenstrauss, The ℓ_1^n problem and degrees of non-reflexivity. Stud. Math. **55**(2), 123–139 (1976)
35. Y. de Cornulier, R. Tessera, A. Valette, Isometric group actions on Hilbert spaces: growth of cocycles. Geom. Funct. Anal. **17**(3), 770–792 (2007)
36. J. Diestel, J.J. Uhl, Jr. Vector measures, in *With a Foreword by B.J. Pettis*. Mathematical Surveys, vol. 15 (American Mathematical Society, Providence, RI, 1977)
37. S.J. Dilworth, D. Kutzarova, G. Lancien, N.L. Randrianarivony, Asymptotic geometry of Banach spaces and uniform quotient maps. Proc. Am. Math. Soc. **142**(8), 2747–2762 (2014)
38. P. Enflo, Banach spaces which can be given an equivalent uniformly convex norm. Isr. J. Math. **13**, 281–288 (1972)
39. P. Enflo, Uniform homeomorphisms between Banach spaces. *Séminaire d'Analyse Fonctionnelle* (1975–1976), Exp. No. 18, 6 pp., École Polytechnique, Palaiseau, 1976
40. P. Enflo, On infinite-dimensional topological groups. Séminaire sur la Géométrie des Espaces de Banach (1977–1978), Exp. No. 10–11, 11 pp., École Polytechnique, Palaiseau, 1978
41. I. Gelfand, Abstrakte Funktionen und lineare Operatoren. Mat. Sb. **4 (46)**(2), 235–286 (1938). Available at http://www.mathnet.ru/
42. O. Giladi, A. Naor, Improved bounds in the scaled Enflo type inequality for Banach spaces. Extracta Math. **25**(2), 151–164 (2010)
43. O. Giladi, M. Mendel, A. Naor, Improved bounds in the metric cotype inequality for Banach spaces. J. Funct. Anal. **260**, 164–194 (2011)
44. O. Giladi, A. Naor, G. Schechtman, Bourgain's discretization theorem. Annales Mathematiques de la faculte des sciences de Toulouse, **XXI**(4), 817–837 (2012)
45. W.T. Gowers, B. Maurey, The unconditional basic sequence problem. J. Am. Math. Soc. **6**(4), 851–874 (1993)
46. M. Gromov, Hyperbolic groups, in *Essays in Group Theory*. Mathematical Sciences Research Institute Publications, vol. 8 (Springer, New York, 1987), pp. 75–263
47. M. Gromov, Carnot-Carathéodory spaces seen from within, in *Sub-Riemannian Geometry*, ed. by A. Bellaïche, J.-J. Risler. Progress in Mathematics, vol. 144 (Birkhäuser, Basel, 1996), pp. 79–323
48. M. Gromov, Random walk in random groups. Geom. Funct. Anal. **13**(1), 73–146 (2003)
49. A. Gupta, Steiner points in tree metrics don't (really) help, in *Proceedings of the Twelfth Annual ACM-SIAM Symposium on Discrete Algorithms*, Washington, DC, 2001 (SIAM, Philadelphia, PA, 2001), pp. 220–227
50. A. Gupta, I. Newman, Y. Rabinovich, A. Sinclair, Cuts, trees and ℓ_1-embeddings of graphs. Combinatorica **24**, 233–269 (2004). Conference version in 40th Annual IEEE Symposium on Foundations of Computer Science, 1999, pp. 399–408
51. J.M. Heinonen, *Lectures on Analysis on Metric Spaces*. Universitext (Springer, New York, 2001)
52. S. Heinrich, P. Mankiewicz, Applications of ultrapowers to the uniform and Lipschitz classification of Banach spaces. Stud. Math. **73**(3), 225–251 (1982)
53. R.C. James, Uniformly non-square Banach spaces. Ann. Math. **80**, 542–550 (1964)
54. R.C. James, Weak compactness and reflexivity. Isr. J. Math. **2**, 101–119 (1964)
55. R.C. James, Some self-dual properties of normed linear spaces, in *Symposium on Infinite-Dimensional Topology*, Louisiana State University, Baton Rouge, LA, 1967). Annals of Mathematics Studies, vol. 69 (Princeton University Press, Princeton, NJ, 1972), pp. 159–175
56. R.C. James, Super-reflexive Banach spaces. Can. J. Math. **24**, 896–904 (1972)
57. R.C. James, Structure of Banach spaces: Radon-Nikodým and other properties, in *General Topology and Modern Analysis*. Proceedings of Conference, University of California, Riverside, CA, 1980 (Academic Press, New York, London, 1981), pp. 347–363
58. W.B. Johnson, G. Schechtman, Diamond graphs and super-reflexivity. J. Topol. Anal. **1**(2), 177–189 (2009)
59. S. Keith, A differentiable structure for metric measure spaces. Adv. Math. **183**(2), 271–315 (2004)

60. B. Kleiner, J. Mackay, Differentiable structures on metric measure spaces: a primer. arXiv:1108.1324
61. B. Kloeckner, Yet another short proof of the Bourgain's distortion estimate for embedding of trees into uniformly convex Banach spaces. Isr. J. Math. **200**(1), 419–422 (2014). doi:10.1007/s11856-014-0024-4
62. S. Kwapień, Isomorphic characterizations of inner product spaces by orthogonal series with vector valued coefficients. Stud. Math. **44**, 583–595 (1972)
63. T.J. Laakso, Ahlfors Q-regular spaces with arbitrary $Q > 1$ admitting weak Poincare inequality. Geom. Funct. Anal. **10**(1), 111–123 (2000)
64. G. Lancien, A short course on nonlinear geometry of Banach spaces, in *Topics in Functional and Harmonic Analysis*. Theta Series in Advanced Mathematics, vol. 14 (Theta, Bucharest, 2013), pp. 77–101
65. U. Lang, C. Plaut, Bilipschitz embeddings of metric spaces into space forms. Geom. Dedicata **87**, 285–307 (2001)
66. J.R. Lee, A. Naor, L_p metrics on the Heisenberg group and the Goemans-Linial conjecture, in *Proceedings of the 47th Annual IEEE Symposium on Foundations of Computer Science* (IEEE, Piscataway, NJ, 2006), pp. 99–108
67. J.R. Lee, A. Naor, Y. Peres, Trees and Markov convexity. Geom. Funct. Anal. **18**, 1609–1659 (2009). Conference version: Proceedings of the Seventeenth Annual ACM-SIAM Symposium on Discrete Algorithms, pp. 1028–1037 (ACM, New York, 2006), arXiv:0706.0545
68. S. Li, Coarse differentiation and quantitative nonembeddability for Carnot groups. J. Funct. Anal. **266**, 4616–4704 (2014)
69. S. Li, Markov convexity and nonembeddability of the Heisenberg group. Ann. Inst. Fourier, to appear, arXiv: 1404.6751
70. J. Lindenstrauss, A. Pełczyński, Absolutely summing operators in $\mathcal{L}_p$-spaces and their applications. Studia Math. **29**, 275–326 (1968)
71. N. Linial, E. London, Y. Rabinovich, The geometry of graphs and some of its algorithmic applications. Combinatorica **15**(2), 215–245 (1995)
72. P. Mankiewicz, On the differentiability of Lipschitz mappings in Fréchet spaces. Stud. Math. **45**, 15–29 (1973)
73. J. Matoušek, On embedding trees into uniformly convex Banach spaces. Isr. J. Math. **114**, 221–237 (1999)
74. J. Matoušek, *Lectures on Discrete Geometry*. Graduate Texts in Mathematics, vol. 212 (Springer, New York, 2002)
75. S. Mazur, S. Ulam, Sur les transformationes isométriques d'espaces vectoriels normés. C. R. Acad. Sci. Paris **194**, 946–948 (1932)
76. M. Mendel, A. Naor, Scaled Enflo type is equivalent to Rademacher type. Bull. Lond. Math. Soc. **39**, 493–498 (2007)
77. M. Mendel, A. Naor, Metric cotype. Ann. Math. **168**, 247–298 (2008)
78. M. Mendel, A. Naor, Markov convexity and local rigidity of distorted metrics. J. Eur. Math. Soc. **15**(1), 287–337 (2013). Conference version: Computational geometry (SCG'08) (ACM, New York, 2008), pp. 49–58
79. D.P. Milman, V.D. Milman, The geometry of nested families with empty intersection. Structure of the unit sphere of a nonreflexive space (Russian). Matem. Sbornik **66**(1), 109–118 (1965). English transl.: Am. Math. Soc. Transl. (2) **85**, 233–243 (1969)
80. R. Montgomery, *A Tour of Subriemannian Geometries, Their Geodesics and Applications*. Mathematical Surveys and Monographs, vol. 91 (American Mathematical Society, Providence, RI, 2002)
81. A. Naor, An introduction to the Ribe program. Jpn. J. Math. **7**(2), 167–233 (2012)
82. P.W. Nowak, G. Yu, *Large Scale Geometry*. EMS Textbooks in Mathematics (European Mathematical Society, Zürich, 2012)
83. D. Osajda, Small cancellation labellings of some infinite graphs and applications. arXiv:1406.5015

84. M.I. Ostrovskii, Finite dimensional characteristics related to superreflexivity of Banach spaces. Bull. Aust. Math. Soc. **69**, 289–295 (2004)
85. M.I. Ostrovskii, Coarse embeddings of locally finite metric spaces into Banach spaces without cotype. C. R. Acad. Bulg. Sci. **59**(11), 1113–1116 (2006)
86. M.I. Ostrovskii, Coarse embeddability into Banach spaces, Topol. Proc. **33**, 163–183 (2009)
87. M.I. Ostrovskii, On metric characterizations of some classes of Banach spaces. C. R. Acad. Bulg. Sci. **64**(6), 775–784 (2011)
88. M.I. Ostrovskii, Embeddability of locally finite metric spaces into Banach spaces is finitely determined. Proc. Am. Math. Soc. **140**, 2721–2730 (2012)
89. M.I. Ostrovskii, Different forms of metric characterizations of classes of Banach spaces. Houst. J. Math. **39**(3), 889–906 (2013)
90. M.I. Ostrovskii, *Metric Embeddings: Bilipschitz and Coarse Embeddings into Banach Spaces*. de Gruyter Studies in Mathematics, vol. 49 (Walter de Gruyter, Berlin, 2013)
91. M.I. Ostrovskii, On metric characterizations of the Radon-Nikodým and related properties of Banach spaces. J. Topol. Anal. **6**(3), 441–464 (2014)
92. M.I. Ostrovskii, Metric characterizations of superreflexivity in terms of word hyperbolic groups and finite graphs. Anal. Geom. Metr. Spaces **2**, 154–168 (2014)
93. M.I. Ostrovskii, Metric spaces nonembeddable into Banach spaces with the Radon-Nikodým property and thick families of geodesics. Fundam. Math. **227**, 85–95 (2014)
94. M.I. Ostrovskii, Radon-Nikodým property and thick families of geodesics. J. Math. Anal. Appl. **409**(2), 906–910 (2014)
95. M.I. Ostrovskii, Test-space characterizations of some classes of Banach spaces, in *Algebraic Methods in Functional Analysis* ed. by I.G. Todorov, L. Turowska. The Victor Shulman Anniversary Volume. Operator Theory: Advances and Applications, vol. 233 (Birkhäuser, Basel, 2014), pp. 103–126
96. M.I. Ostrovskii, Connections between metric characterizations of superreflexivity and the Radon-Nikodým property for dual Banach spaces. Can. Math. Bull. **58**(1), 150–157 (2015). `arXiv:1406.0904`; http://dx.doi.org/10.4153/CMB-2014-049-9
97. M.I. Ostrovskii, Embeddings of finitely generated groups into uniformly convex Banach spaces. http://mathoverflow.net/questions/178898/
98. P. Pansu, Métriques de Carnot-Carathéodory et quasiisométries des espaces symétriques de rang un. Ann. Math. (2) **129**(1), 1–60 (1989)
99. A. Pełczyński, A note on the paper of I. Singer "Basic sequences and reflexivity of Banach spaces". Stud. Math. **21**, 371–374 (1961/1962)
100. J.C.B. Perrott, Transfinite duals of Banach spaces and ergodic super-properties equivalent to super-reflexivity. Q. J. Math. Oxford Ser. (2) **30**(117), 99–111 (1979)
101. G. Pisier, Martingales with values in uniformly convex spaces. Isr. J. Math. **20**, 326–350 (1975)
102. G. Pisier, Probabilistic methods in the geometry of Banach spaces, in *Probability and Analysis (Varenna, 1985)*. Lecture Notes in Mathematics, vol. 1206 (Springer, Berlin, 1986), pp. 167–241
103. G. Pisier, Martingales in Banach spaces. Cambridge Studies in Advanced Mathematics, vol. 155 (Cambridge University Press, Cambridge, 2016)
104. V. Pták, Biorthogonal systems and reflexivity of Banach spaces. Czechoslov. Math. J. **9**, 319–326 (1959)
105. Y. Rabinovich, R. Raz, Lower bounds on the distortion of embedding finite metric spaces in graphs. Discret. Comput. Geom. **19**(1), 79–94 (1998)
106. M. Ribe, On uniformly homeomorphic normed spaces. Ark. Mat. **14**(2), 237–244 (1976)
107. M. Ribe, Existence of separable uniformly homeomorphic nonisomorphic Banach spaces. Isr. J. Math. **48**(2–3), 139–147 (1984)
108. J.J. Schäffer, K. Sundaresan, Reflexivity and the girth of spheres. Math. Ann. **184**, 163–168 (1969/1970)
109. S. Semmes, On the nonexistence of bi-Lipschitz parameterizations and geometric problems about A_∞-weights. Rev. Mat. Iberoamericana **12**(2), 337–410 (1996)

110. I. Singer, Basic sequences and reflexivity of Banach spaces. Stud. Math. **21**, 351–369 (1961/1962)
111. C. Stegall, The Radon-Nikodým property in conjugate Banach spaces. Trans. Am. Math. Soc. **206**, 213–223 (1975)
112. Texas 2009 nonlinear problems, open problems suggested at the seminar "Nonlinear geometry of Banach spaces", Texas A & M University, August 2009. Available http://facpub.stjohns.edu/ostrovsm/Texas2009nonlinearproblems.pdf
113. D. van Dulst, The geometry of Banach spaces with the Radon-Nikodým property. Rend. Circ. Mat. Palermo (2) Suppl. 7, ix+81 pp (1985).
114. J. Wenzel, Superreflexivity and J-convexity of Banach spaces. Acta Math. Univ. Comenian. (N.S.) **66**, 135–147 (1997)

On the IVP for the k-Generalized Benjamin–Ono Equation

Gustavo Ponce

Dedicado a Cora Sadosky con infinita gratitud

Abstract We shall study special properties of solutions to the IVP associated to the k-generalized Benjamin–Ono equation. We shall compare them with those for the k-generalized Korteweg-de Vries equation and for the k-generalized dispersive Benjamin–Ono equation. Also we shall discuss some open questions appearing in this subject.

Mathematics Subject Classification: Primary: 35Q53; Secondary: 35B05

Introduction

This work is concerned with the initial value problem (IVP) associated to the k-generalized Benjamin–Ono (k-gBO) equation

$$\begin{cases} \partial_t u - \partial_x^2 \mathscr{H} u + u^k \, \partial_x u = 0, & x, t \in \mathbb{R}, \ k \in \mathbb{Z}^+, \\ u(x,0) = u_0(x), \end{cases} \tag{1}$$

where $\mathscr{H}$ denotes the Hilbert transform,

$$\begin{aligned} \mathscr{H} f(x) &= \frac{1}{\pi} \text{p.v.}\big(\frac{1}{x} * f\big)(x) \\ &= \frac{1}{\pi} \lim_{\epsilon \downarrow 0} \int\limits_{|y| \geq \epsilon} \frac{f(x-y)}{y} \, dy = (-i \operatorname{sgn}(\xi) \hat{f}(\xi))^{\vee}(x). \end{aligned} \tag{2}$$

G. Ponce (✉)
Department of Mathematics, University of California, Santa Barbara, CA 93106, USA
e-mail: ponce@math.ucsb.edu

M.C. Pereyra et al. (eds.), *Harmonic Analysis, Partial Differential Equations, Complex Analysis, Banach Spaces, and Operator Theory (Volume 1)*, Association for Women in Mathematics Series 4, DOI 10.1007/978-3-319-30961-3_16

The case $k = 1$ in (1)

$$\partial_t u - \partial_x^2 \mathcal{H} u + u\,\partial_x u = 0, \quad x, t \in \mathbb{R}, \tag{3}$$

corresponds to the Benjamin–Ono (BO) equation. This was first deduced by Benjamin [5] and Ono [44] as a model for long internal gravity waves in deep stratified fluids. Later, it was also shown to be a completely integrable system (see [2, 11], and the references therein). Thus, the BO equation has followed the same historical pattern of the celebrated Korteweg-de Vries (KdV) equation

$$\partial_t u + \partial_x^3 u + u\,\partial_x u = 0. \tag{4}$$

One of our goals here will be to compare the mathematical results for the IVP (1) with those for the IVP associated to the k-generalized Korteweg-de Vries (k-gKdV) equation

$$\begin{cases} \partial_t u + \partial_x^3 u + u^k \partial_x u = 0, & x, t \in \mathbb{R},\ k \in \mathbb{Z}^+, \\ u(x, 0) = u_0(x). \end{cases} \tag{5}$$

In addition, we shall mention some striking differences as well as some open problems (these will be labelled and numbered with a $Q\ldots$). Also, we shall discuss the extension of some known results to the IVP (1) with $k = 1$ to the case of general $k \in \mathbb{Z}^+$.

Another of our goals here will be to illustrate how well-known results in harmonic analysis have provided the key arguments in the proof of some crucial results describing the behavior of the solutions to the IVP (1).

To understand the relationship between the dispersion and the nonlinearity it is convenient to consider the so-called k-generalized dispersive Benjamin–Ono equation

$$\partial_t u - D_x^{1+\alpha} \partial_x u + u^k \partial_x u = 0, \quad x, t \in \mathbb{R},\ \alpha \in [0, 1],\ k \in \mathbb{Z}^+. \tag{6}$$

We notice that in (6) the case $\alpha = 0$ corresponds to the k-gBO equation in (1) and the case $\alpha = 1$ to the k-gKdV equation in (5).

Well-Posedness Results

We shall say (see [26]) that the IVP (1) is locally well-posed (LWP) in the function space X if given any datum $u_0 \in X$ there exist $T > 0$ and a unique solution

$$u \in C([-T, T] : X) \cap \ldots.$$

of the IVP (3) with the map data $\to$ solution, $u_0 \mapsto u$, being continuous. If T can be taken arbitrarily large, one says that the IVP (3) is globally well-posed (GWP) in X.

Well-posedness in Sobolev spaces

We observe that $e^{i|\xi|\xi}$ is not a multiplier in L^p for $p \neq 2$ and that the family of operators $\{U(t) \,:\, t \in \mathbb{R}\}$

$$U(t)u_0(x) = e^{\partial_x^2 \mathscr{H} t}u_0(x) = (e^{4\pi^2 it|\xi|\xi}\widehat{u_0}(\xi))^{\vee}(x), \tag{7}$$

defines a group of isometries in

$$H^s(\mathbb{R}) = \left(1 - \partial_x^2\right)^{-s/2} L^2(\mathbb{R}), \qquad s \in \mathbb{R}. \tag{8}$$

Therefore these classical Sobolev spaces (8) are the natural functional setting to study well-posedness of the IVP (1).

The problem of finding the minimal regularity property, measured in the classical Sobolev scale $H^s(\mathbb{R})$, $s \in \mathbb{R}$, required to guarantee that the IVP (1) is LWP or GWP in $H^s(\mathbb{R})$ has been extensively studied.

First, one has the following scaling invariant argument : if $u(x,t)$ is solution of the IVP (1) in the time interval $[-T, T]$, then for any $\lambda > 0$

$$u_\lambda(x,t) = \lambda^{1/k}u(\lambda x, \lambda^2 t), \tag{9}$$

is also solution of the equation in (1) in the time interval $[-\lambda^{-2}T, \lambda^{-2}T]$ with initial data

$$u_\lambda(x,0) = \lambda^{1/k}u_0(\lambda x),$$

such that

$$\| u_\lambda \|_{\dot{H}^s} = \|D_x^s u_\lambda\|_2 = \lambda^{s+1/k-1/2}\| u_0\|_{\dot{H}^s}, \tag{10}$$

where $D_x = \mathscr{H}\partial_x$. When

$$s = s_k = 1/2 - 1/k, \tag{11}$$

one cannot have that $T = T(\| u_0\|_{\dot{H}^s}) > 0$, since by changing λ the time interval of existence expands (or contracts) arbitrarily. Hence, $s = s_k = 1/2 - 1/k$ is the critical exponent for well-posedness suggested by the above scaling argument.

Second, due to the integrability, real solutions of the IVP (1) with $k = 1$ satisfy infinitely many conservation laws. These quantities provide an a priori estimate for the $H^{n/2}$-norm, $n \in \mathbb{Z}^+$, of the solution $u = u(x,t)$ for (3). Here, we shall only consider real- valued solutions of the IVP (1). For general $k \in \mathbb{Z}^+$ (real) solutions of (1) satisfy the following three conservation laws (time invariant quantities):

$$\begin{aligned} I_1(u) &= \int_{-\infty}^{\infty} u(x,t)dx, \qquad I_2(u) = \int_{-\infty}^{\infty} u^2(x,t)dx, \\ I_3(u) &= \int_{-\infty}^{\infty}\left(|D_x^{1/2}u|^{1/2} - \frac{2}{(k+1)(k+2)}u^{k+2}\right)(x,t)dx. \end{aligned} \tag{12}$$

Under appropriate assumptions these conservation laws provide an a priori estimate of the H^s-norm of the solution which combined with a LWP with $T = T(\|u_0\|_s) > 0$ yields a GWP result.

Returning to the well-posedness results for the IVP (1) in the classical Sobolev spaces (8) first we shall concentrate on the case of the BO equation (3). In this regards, one finds the following string of results : in [1, 22] LWP was obtained for $s > 3/2$, in [46] GWP was proven for $s \geq 3/2$, LWP was established in [34] for $s > 5/4$ and in [28] for $s > 9/8$, in [51] GWP was shown for $s \geq 1$, in [8] LWP was found to hold for $s > 1/4$, and in [21] GWP was demonstrated in $H^0(\mathbb{R}) = L^2(\mathbb{R})$, (for further details and results regarding the well-posedness of the IVP associated to the BO equation (3) in $H^s(\mathbb{R})$ see [38]).

It is important to mention that all the above results have been obtained by "compactness methods." In particular, one can only show that the map data $\rightarrow$ solution is just locally continuous.

In fact, in [41] it was established that for any $s \in \mathbb{R}$ the flow map $u_0 \rightarrow u$ from $H^s(\mathbb{R})$ to $C([-T, T] : H^s(\mathbb{R}))$ is not locally of class C^2 (see also [35]).

We recall that if LWP in X can be obtained using just the contraction principle applied to the corresponding integral equation, then the implicit function theorem guarantees that the flow map $u_0 \rightarrow u$ from X to $C([-T, T] : X)$ is (locally) as smooth as the nonlinearity, (see, for example, [36]).

For the IVP (1) one has the corresponding integral equation

$$u(t) = U(t)u_0 - \int_0^t U(t-t')u^k \partial_x u(t')dt' \tag{13}$$

$[U(t)$ as in (7)$]$ where the nonlinearity is a polynomial (analytic).

Hence, as a consequence of the result in [41] one sees that the contraction principle cannot be used by itself to established the LWP of the IVP (1) with $k = 1$.

Next, we consider the IVP (1) with $k \geq 2$. In the case $k = 2$ in [29] LWP has been proven in $H^s(\mathbb{R}),\ s \geq 1/2$. In [53] LWP was obtained for the IVP (1) for $k = 3$ in $H^s(\mathbb{R}),\ s > 1/3$ and for $k \geq 4$ in $H^s(\mathbb{R}),\ s \geq s_k = 1/2 - 1/k$, (see also [40]).

These results were also obtained by compactness methods and seem to be optimal. More precisely : in [29] it was shown that for the case $k = 2$ the map data $\rightarrow$ solution is not uniformly continuous in $H^s(\mathbb{R}),\ s < 1/2$ (in [39] it was proven not to be C^3), in [52] it was shown that in the case $k = 3$ the map data $\rightarrow$ solution is not C^4 in $H^s(\mathbb{R}),\ s < 1/3$. In [6] it was shown that the map data $\rightarrow$ solution is not uniformly continuous in $H^{s_k}(\mathbb{R}),\ s_k = 1/2 - 1/k,\ k \geq 1$ (and in [39] it was shown that the map data $\rightarrow$ solution is not C^{k+1} in $H^{s_k}(\mathbb{R}),\ s_k = 1/2 - 1/k,\ k \geq 4$).

We observe that for $k \geq 3$ these well-posedness results agree with the values suggested by the scaling argument $s_k = 1/2 - 1/k$ [see (11)], being larger for $k = 1,\ s \geq 0$ with $s_1 = -1/2$, for $k = 2,\ s \geq 1/2$ with $s_2 = 0$ and for $k = 3,\ s > 1/3$ with $s_3 > 1/6$.

This is somehow similar to the case of the IVP (5) for the k-gKdV for which the values suggested by the scaling argument $\tilde{s}_k = 1/2 - 2/k$ have been achieved for the LWP results only for the powers $k \geq 3$, being larger the case $k = 1,\ s \geq -3/4$ with

$\tilde{s}_1 = -3/2$ and for $k = 2$, $s \geq 1/4$ with $\tilde{s}_2 = -1/2$. All these results for the IVP (5) have been established by an argument based on the contraction principle applied to the corresponding integral equation. This is the first main difference between the IVP's (1) and (5).

Also, one notices that from the well-posedness point of view the worst power is $k = 2$ for both IVP's (1) and (5) at least for $1 \leq k \leq 8$.

In addition to the above well-posedness and ill-posedness results, one has that in [33, 39] LWP and GWP have been established for the IVP (1) with $k \geq 2$ for small data in $H^s(\mathbb{R})$ with $s > \alpha_k$ solely by a contraction principle argument. So in this case the map (small) data $\to$ solution is analytic. Since the result for the BO equation (3) found in [41] does not extend to higher power $k \geq 2$ in (1), the next question presents itself:

Q 1: Given any $k \in \mathbb{Z}^+$, $k \geq 2$ which of following two statements holds:

(i) there exists $a_k > 0$ such that the IVP (1) can be proved to be LWP in the whole space $H^s(\mathbb{R})$ with $s > a_k$ by using a contraction principle in the corresponding integral equation (13)

or

(ii) for any $s > 0$ the map data $\to$ solution for the IVP (1) defined in $H^s(\mathbb{R})$ is not (locally) smooth.

It is interesting to notice that the ill-posedness result in [41] applies to all equations in (6) with $k = 1$, i.e., in [41] it was shown that for any $s \in \mathbb{R}$ the map data $\to$ solution associated to the IVP for Eq. (6) with $\alpha \in [0, 1)$ and $k = 1$ is not locally of class C^2. Thus, only for the KdV, (6) with $\alpha = 1$ and $k = 1$, the dispersive relation (modeled by a third-order operator) is strong enough to allow a proof based on the contraction principle.

In this regard, one has that the techniques and arguments presented in [32] can be used to establish LWP for the IVP associated to Eq. (6) with $\alpha \in (0, 1)$ and $k \geq 2$ with an argument based on the contraction principle. In the case $\alpha \in (0, 1)$ and $k = 1$ GWP for the IVP (6) in $H^s(\mathbb{R})$ for $s \geq 0$ was obtained in [19] by compactness methods.

Next, we consider the problem of extending LWP results to global ones. One sees that due to the infinitely many conservation laws satisfied by solutions of the BO equation (3) this follows directly. In fact, the GWP results in [21] only requires the use of I_2 in (12). In general, combining the inequality

$$\|u\|_{k+2}^{k+2} \leq c_k \|D_x^{k/2(k+2)} u\|_2^{k+2} \leq c_k \|D_x^{1/2} u\|_2^k \, \|u\|_2^2, \tag{14}$$

and the conservation law I_3 in (12)

$$I_3(u) = I_3(u_0) \leq \|D_x^{1/2} u(t)\|_2^2 - c_k \|D_x^{1/2} u(t)\|_2^k \, \|u_0\|_2^2 \tag{15}$$

one obtains an a priori estimate for the solutions corresponding for small enough data in L^2 in the case $k = 2$, see [29], and for small enough data in $H^{1/2}$ in the case $k \geq 3$.

In fact, the existence of global smooth solutions for the IVP (1) with $k \geq 2$ and data in $H^{1/2}(\mathbb{R})$ is an open problem, (Q 2).

For the k-gKdV equation [see (5)] one has that for $k = 1, 2, 3$ local solutions corresponding to initial data in $u_0 \in H^1(\mathbb{R})$ extend globally in time. In [37] it was shown that there exist data $u_0 \in H^1(\mathbb{R})$ for which the corresponding local solution $u = u(x,t)$ of the IVP (5) with $k = 4$ blows-up in finite time, i.e., $\exists\, T > 0$ such that

$$\lim_{t \uparrow T} \|\partial_x u(t)\|_2 = \infty.$$

A similar result for higher powers $k \geq 5$ in (5) remains as an open problem, (Q 3).

To complete this section we shall briefly comment some of the techniques used in the proof of the well-posedness results for the BO equation (3) commented before.

By using the commutator estimate established in [27] (as an application of results in [10])

$$\|J^s(fg) - fJ^s g\|_p \leq c(\|\partial f\|_\infty \|J^{s-1} g\|_p + \|J^s f\|_p \|g\|_\infty), \qquad 1 < p < \infty,\ s > 0,$$

with $J = (1 - \partial_x^2)^{1/2}$, one formally has that solutions of the BO equation (3) satisfy

$$\frac{d}{dt}\|J^s u(t)\|_2 \leq c\|\partial_x u(t)\|_\infty \|J^s u(t)\|_2, \qquad s > 0. \tag{16}$$

Hence, if one controls the quantity

$$\int_{-T}^{T} \|\partial_x u(t)\|_\infty \, dt, \tag{17}$$

then the solution, in the time interval $[-T, T]$, is as regular in $H^s(\mathbb{R})$ as the data.

Thus, the result in [1, 22] ($s > 3/2$) follows by using Sobolev embedding in (16). The result in [46] combined the Strichartz estimates [50] and the Kato local smoothing effect [26]. In the case of the associated linear problem [see (7)], the Strichartz estimates are (mainly) given by the inequality:

$$\Big(\int_{-\infty}^{\infty} \|U(t)u_0\|_p^q dt\Big)^{1/q} \leq c\|u_0\|_2, \quad 2/q = 1/2 - 1/p, \quad 2 \leq p \leq \infty \tag{18}$$

and the Kato local smoothing effect by

$$\Big(\int_{-\infty}^{\infty} |D_x^{1/2} U(t) u_0(x)|^2 \, dt\Big)^{1/2} = \frac{\sqrt{2\pi}}{2} \|u_0\|_2, \qquad \forall\, x \in \mathbb{R}. \tag{19}$$

and

$$\sup_{x \in \mathbb{R}} \|\partial_x \int_0^t U(t - t')F(\cdot, t')dt'\|_{L_t^2} \leq c \int_{-\infty}^{\infty} \|F(x, \cdot)\|_{L_t^2} dx. \tag{20}$$

Roughly, by using the estimate (20), established in [30], one gains one derivative which compensates the loss introduced by the nonlinear term in (13). We observe that the proof of the identity (19) is a direct consequence of the Plancherel theorem and a change of variable. However, since we are working with the differential equation only a weaker form of (19) can be proved, i.e., if the data $u_0 \in H^{3/2}(\mathbb{R})$, then the corresponding solution of the BO equation (3) satisfies

$$\sup_{x_0,R} \Big(\int_{-T}^{T} \int_{x_0-R}^{x_0+R} |\partial_x^2 u(x,t)|^2 dx\, dt \Big)^{1/2} \leq c(T; R; \|u_0\|_{3/2,2}),$$

and using Stichartz estimate in (18) with $(p,q) = (\infty, 4)$ one gets that

$$\Big(\int_{-\infty}^{\infty} \|\partial_x u(\cdot,t)\|_{\infty}^4 dt \Big)^{1/4} \leq c(T; R; \|u_0\|_{3/2,2}),$$

which basically yields the result in [46]. The above techniques were refined in [28, 34] to obtain the improvement in the LWP mentioned before.

The ill-posedness result established in [41] is a consequence of the low-high frequency interaction emerging from the nonlinear term. In [51] a gauge transformation (reminiscence the Cole-Hopf transformation) which involves projection in the high positive frequency was introduced to obtain the GWP result in $H^1(\mathbb{R})$. In [8, 21] the previous idea was combined with the so-called Bourgain spaces $X_{s,b}$, see [7], to obtain the improvement mentioned above. In the context of the BO equation (3) the spaces $X_{s,b}$ $s,\ b \in \mathbb{R}$ can be defined as the $f \in \mathcal{S}'(\mathbb{R}^2)$ such that

$$\|f\|_{X_{s,b}} = \Big(\int_{-\infty}^{\infty} \int_{-\infty}^{\infty} (1 + |\tau - \xi|\xi||)^{2b} (1 + |\xi|)^{2s} |\hat{f}(\xi,\tau)|^2 d\xi d\tau \Big)^{1/2} < \infty.$$

Well-posedness in weighted Sobolev spaces

In [26] it was shown that the solution flow associated to the IVP (5) preserves the Schwartz class. However, it was first established by Iorio [22, 23] that, in general, decay of polynomial type is not preserved by the solution flow of the BO equation. The results in [22, 23] have been extended to fractional order weighted Sobolev spaces and have shown to be optimal in [15, 16].

This is the second main difference concerning the behavior of solutions of the IVP's (1) and (5). This difference in the decay is already reflected in the explicit form of the traveling wave solutions $\phi_{\{\cdot\}}$ of these equations. For the BO equation (3) one has if

$$\phi_{BO}(x) = \frac{4}{1+x^2}, \tag{21}$$

(the unique positive, even, decreasing for $x > 0$ and tending to zero as $x \to \pm\infty$ solution of $-\varphi - D_x\varphi + \varphi^2/2 = 0$, whose uniqueness was proved in [3]), then

$$u(x,t) = c\phi_{BO}(c(x - ct)), \qquad c > 0,$$

describes the family of traveling wave solution. For the KdV equation (4)

$$\phi_{KdV}(x) = \frac{3}{2}\operatorname{sech}^2\left(\frac{x}{2}\right).$$

describes the family of traveling wave solutions

$$u(x,t) = c\phi_{KdV}(c^2(x-c^2t)), \qquad c>0.$$

To make the above statements precise, we introduce the weighted Sobolev spaces

$$Z_{s,r} = H^s(\mathbb{R}) \cap L^2(|x|^{2r}dx), \;\; s,r\in\mathbb{R}, \tag{22}$$

and

$$\dot{Z}_{s,r} = \{f\in H^s(\mathbb{R})\cap L^2(|x|^{2r}dx) \,:\, \hat{f}(0)=0\}, \qquad s,\, r\in\mathbb{R}. \tag{23}$$

Notice that the conservation law for solutions of (3), see (12),

$$I_1(u_0) = \int_{-\infty}^{\infty} u(x,t)dx = \int_{-\infty}^{\infty} u_0(x)dx,$$

guarantees that the property $\hat{f}(0) = 0$ is preserved by the solution flow.

The following well-posedness results for the IVP associated to the BO equation, (1) with $k=1$, in weighted Sobolev spaces $Z_{s,r}$ were found in [15]:

Theorem 2.1.

(i) *If $s > 9/8$ ($s \geq 3/2$), $r\in[0,s]$, and $r<5/2$, then IVP associated to the BO equation* (3) *is LWP (GWP resp.) in $Z_{s,r}$.*
(ii) *If $r\in[5/2,7/2)$ and $r\leq s$, then the IVP associated to Eq.* (3) *is GWP in $\dot{Z}_{s,r}$.*

Theorem 2.2. *Let $u\in C([-T,T]:Z_{2,2})$ be a solution of the IVP for Eq.* (3). *If there exist two different times $t_1,t_2\in[-T,T]$ such that*

$$u(\cdot,t_j)\in Z_{5/2,5/2},\; j=1,2, \;\text{then}\;\; \hat{u}_0(0)=0, \;\; (\text{so}\;\; u(\cdot,t)\in \dot{Z}_{5/2,5/2}). \tag{24}$$

Theorem 2.3. *Let $u\in C([-T,T]:\dot{Z}_{3,3})$ be a solution of the IVP for Eq.* (3). *If there exist three different times $t_1,t_2,t_3\in[-T,T]$ such that*

$$u(\cdot,t_j)\in \dot{Z}_{7/2,7/2},\;\; j=1,2,3, \;\;\text{then}\;\;\; u(x,t)\equiv 0. \tag{25}$$

Remarks.

(a) Iorio's results [22, 23] correspond to the indexes $s\geq r=2$ in Theorem 2.1 part (i), $s\geq r=3$ in Theorem 2.1 part (ii) and $s\geq r=4$ in Theorem 2.3.

(b) Theorem 2.2 shows that the condition $\hat{u}_0(0) = 0$ is necessary to have persistence property of the solution in $Z_{s,5/2}$, with $s \geq 5/2$, so in that regard Theorem 2.1 part (i) is sharp.
(c) Theorem 2.3 affirms that there is an upper limit of the spatial L^2-decay rate of the solution (i.e., $|x|^{7/2}u(\cdot,t) \notin L^\infty([0,T] : L^2(\mathbb{R}))$, for any $T > 0$) regardless of the decay and regularity of the non-zero initial data u_0. In particular, Theorem 2.3 shows that Theorem 2.1 part (ii) is sharp.

From Theorem 2.2 it is natural to ask if the assumption (25) in Theorem 2.3 can be reduced from three times to a two different times $t_1 < t_2$.

Surprisingly, the next result found in [16] shows that this is not the case, the condition involving three different times in Theorem 2.3 is necessary:

Theorem 2.4. *For any $u_0 \in \dot{Z}_{5,4}$ such that*

$$\int_{-\infty}^{\infty} x\, u_0(x)dx \neq 0, \tag{26}$$

the corresponding solution $u \in C(\mathbb{R} : \dot{Z}_{5,7/2-})$ of the IVP for Eq. (3) *provided by Theorem 2.1 part (ii) satisfies that*

$$u(\cdot,t^*) \in \dot{Z}_{4,4}, \tag{27}$$

with

$$t^* = -\frac{4}{\|u_0\|_2^2}\int_{-\infty}^{\infty} x\, u_0(x)dx. \tag{28}$$

Remarks.

(a) The result in Theorem 2.4 is due to the inter play between the dispersive relation and the nonlinearity of the BO equation (3). In particular, one can see that if $u_0 \in \dot{Z}_{5,4}$ verifying (26), then the solution $U(t)u_0(x)$ of the associated linear IVP

$$\partial_t u - \mathscr{H}\partial_x^2 u = 0, \qquad u(x,0) = u_0(x),$$

satisfies

$$U(t)u_0(x) = c(e^{4\pi^2 it|\xi|\xi}\widehat{u_0}(\xi))^\vee \in L^2(|x|^{7-}) - L^2(|x|^7), \qquad \forall\, t \neq 0.$$

However, for the same data u_0 one has that the solution $u(x,t)$ of Eq. (3) satisfies

$$u(\cdot,0),\ u(\cdot,t^*) \in L^2(|x|^8dx), \ \text{ and } \ u(\cdot,t) \in L^2(|x|^{7-}) - L^2(|x|^7), \ \ \forall\, t \notin \{0,t^*\}.$$

(b) The value of t^* in (28) can be motivated as follows, the identity

$$\frac{d}{dt}\int_{-\infty}^{\infty} xu(x,t)dx = \frac{1}{2}\|u(\cdot,t)\|_2^2 = \frac{1}{2}\|u_0\|_2^2,$$

[using the second conservation law $I_2(u)$ in (12)] describes the time evolution of the first momentum of the solution

$$\int_{-\infty}^{\infty} xu(x,t)dx = \int_{-\infty}^{\infty} xu_0(x)dx + \frac{t}{2}\|u_0\|_2^2.$$

So assuming that

$$\int_{-\infty}^{\infty} x\,u_0(x)dx \neq 0, \tag{29}$$

one looks for the times where the average of the first momentum of the solution vanishes, i.e., for t such that

$$\int_0^t \int_{-\infty}^{\infty} x\,u(x,t)dxdt = \int_0^t \Big(\int_{-\infty}^{\infty} x\,u_0(x)dx + \frac{t'}{2}\|u_0\|_2^2\Big)dt' = 0,$$

which under the assumption (29) has a unique solution $t = t^*$ given by the formula in (28).

(c) To prove Theorem 2.4 we shall work with the integral equation version of the problem (3). Roughly, from the result in [41] one cannot regard the nonlinear term as a perturbation of the linear one. So one needs to rely on an argument similar to that in [23]. This is based on the special structure of the equation and allows us to reduce the contribution of two terms in the integral equation to just one. Also the use of the integral equation in the proof and the result in [41] explains our assumption $u_0 \in \dot{Z}_{5,4}$ instead of the expected one from the differential equation point of view $u_0 \in \dot{Z}_{4,4}$.

(d) In view of the above results one may ask if it is possible to reduce the hypothesis in Theorem 2.3 involving three times to one for two different times by strengthening the hypothesis on the decay. More precisely, (*Q 4*) can one find $r > 4$ such that the following statement holds ? :

if a solution $u \in C([-T,T] : \dot{Z}_{s,7/2^-})$ $s \gg 1$ of the BO equation satisfies that there exist $t_1,\ t_2 \in [-T,T],\ t_1 \neq t_2$ such that $u(t_j) \in \dot{Z}_{s,r},\ j = 1,2$, then $u \equiv 0$.

In [14] it was shown that this is not the case at least for $r \in (4, 11/2)$:

Theorem 2.5 ([14]). *Given $r \in (4, 11/2)$ there exists $u_0 \in \dot{Z}_{10,r}$ with*

$$\int_{-\infty}^{\infty} x\,u_0(x)dx \neq 0, \tag{30}$$

such that the corresponding solution $u \in C(\mathbb{R} : \dot{Z}_{10,7/2^-})$ of the IVP for Eq. (3) provided by Theorem 2.1 part (ii) satisfies that

$$u(\cdot, t^*) \in \dot{Z}_{10,r}, \tag{31}$$

with

$$t^* = -\frac{4}{\|u_0\|_2^2}\int_{-\infty}^{\infty} x\,u_0(x)dx. \tag{32}$$

Thus, question *Q 4* remains open for $r \geq 11/2$.

One of the main ideas in [14] is the use of the time evolution of the second moment of the solution

$$\int_{-\infty}^{\infty} x^2 u(x,t)dx = \int_{-\infty}^{\infty} x^2 u_0(x)dx + t\int_{-\infty}^{\infty} xu_0^2(x)dx + t^2 I_3(u_0).$$

with I_3 as in (12). Notice that Theorem 2.3 suggests that the third moment of the solution

$$\int_{-\infty}^{\infty} x^3 u(x,t)dx,$$

is not defined.

Next, we discuss the possible extensions of Theorems 2.1–2.4 to higher powers $k \geq 2$ in (1). From the method of proof used in [15] it is clear that similar results to those in Theorems 2.1 and 2.2 hold, locally in time, for the IVP (1) with $k \geq 2$. Theorem 2.3 extends to all the odd powers k. The issue in this case is the following (see [16]): let

$$u \in C([-T,T] : \dot{Z}_{10,7/2-}) \cap \ldots. \tag{33}$$

be a solution of the IVP (1) with $k \geq 2$, then for each $t^* \in (-T,T)$ such that

$$\int_0^{t^*}\left(\int_{-\infty}^{\infty} xu_0(x)dx + \frac{1}{k+1}\int_0^t\int_{-\infty}^{\infty} u^{k+1}(x,\tau)dx\,d\tau\right)dt = 0, \tag{34}$$

one has that

$$u(\cdot,t^*) \in \dot{Z}_{10,4}. \tag{35}$$

In the case $k = 1$, from the conservation law I_2 [see (12)] and the fact that the solution extends globally in time, one has that if $\int xu_0(x)dx \neq 0$, then there exists exactly one time $t^* \neq 0$, see (32), such that (35) holds.

If $k \geq 3$ is odd and $\int xu_0(x)dx \neq 0$, then Eq. (34) has at most one solution $t^* \neq 0$; however, we do not know (*Q 5*) if this solution t^* exists it belongs to the time interval of existence $[-T,T]$. Notice that a scaling argument suggests that

$$T = O(\|u_0\|_{s,2}^{-k}), \quad \text{as} \quad \|u_0\|_{s,2} \downarrow 0 \quad s > s_k \quad \text{defined in (11).}$$

In this case, $k \geq 3$ and odd, Theorem 2.3 holds and Theorem 2.4 also holds in the conditional manner described before.

In the case where k is even, we do not know (*Q 6*) given a solution u as in (33) how to determinate the exact number of times t_j's solutions of (34).

Above we have worked with the space $Z_{s,r}$ with indices $s \leq r$. This is due to the fact that, in general, for dispersive equations, the solution flow preserves the L^2 polynomial decay only for smooth enough, in H^s, solutions. This has been established in [24, 43] for solution of the IVP associated to the nonlinear Schrödinger equation and the k-gKdV, respectively. However, the equivalent result for solution of the IVP (1) is not available. More precisely, for simplicity let us consider the case $k = 1$ in (1): let $u \in C(\mathbb{R} : L^2(\mathbb{R})) \cap \ldots$ be a solution of the IVP for Eq. (3) found in [21] (*Q 7*). If there exist two different times t_1, t_2 such that

$$u(\cdot, t_j) \in L^2(|x|^{\alpha}dx),\, j = 1, 2 \;\; \alpha \in (0, 7/2^-), \text{then} \;\; u \in C(\mathbb{R} : Z_{\alpha,\alpha}).$$

Working with the differential equation one may need to assume that $u \in C([-T, T] : H^{9/8+}(\mathbb{R})) \cap \ldots$ so that the quantity in (17) is finite, see [28].

It was already mentioned that a result in [41] shows that the IVP associated to Eq. (6) with $\alpha \in [0, 1)$ and $k = 1$ cannot be solved by a contraction principle argument in $H^s(\mathbb{R})$ for any $s \in \mathbb{R}$. For $\alpha \in (0, 1)$ and $k \geq 2$ this is not the case, i.e., the IVP associated to Eq. (6) with $\alpha \in (0, 1)$ and $k \geq 2$ can be solved in $H^s(\mathbb{R})$ with $s \geq a_k$ for some $a_k \in \mathbb{R}$ by the contraction principle. The proof in this case follows by a simple modification of the arguments given in [32]. Also, one has that a slight variation of the method found in [31] confirms that the IVP associated to Eq. (6) with $\alpha \in (0, 1)$ and $k = 1$ can be solved in the weighted Sobolev spaces $Z_{s,r}$ (22), (for appropriate values of s, r) by an argument based on the contraction principle. For the remaining case $\alpha = 0$ and $k = 1$, i.e., the BO equation, one has that a similar argument to that given in [31] shows that the contraction principle in the weighted Sobolev spaces $Z_{s,r}$ (for appropriate values of s, r) works, at least, for small data. Thus, (*Q 8*) for the IVP associated to the BO equation (6) with $\alpha = 0$ and $k = 1$, can one find s, $r \in \mathbb{R}$ such that the contraction principle provides LWP in $Z_{s,r}$ or can one extend the result in [41] to show that for any s, $r \in \mathbb{R}$ the map data $\to$ solution from $Z_{s,r}$ to $C([-T, T] : Z_{s,r})$ is not smooth? Next, we briefly discuss some of the techniques used in the proof of Theorems 2.1–2.4.

Let us first recall the definition of the A_p condition. We shall restrict here to the cases $p \in (1, \infty)$ and the one-dimensional case $\mathbb{R}$, (see [42]).

Definition 2.6. A nonnegative function $w \in L^1_{loc}(\mathbb{R})$ satisfies the A_p inequality with $1 < p < \infty$ if

$$\sup_{Q \text{ interval}} \left(\frac{1}{|Q|}\int_Q w\right)\left(\frac{1}{|Q|}\int_Q w^{1-p'}\right)^{p-1} = c(w) < \infty, \tag{36}$$

where $1/p + 1/p' = 1$.

Theorem 2.7 ([20]). *The condition* (36) *is necessary and sufficient for the boundedness of the Hilbert transform* $\mathcal{H}$ *in* $L^p(w(x)dx)$, *i.e.*,

$$\Big(\int_{-\infty}^{\infty} |\mathcal{H}f|^p w(x)dx\Big)^{1/p} \leq c^* \Big(\int_{-\infty}^{\infty} |f|^p w(x)dx\Big)^{1/p}. \tag{37}$$

In the case $p = 2$, a previous characterization of w in (37) was found in [18] (for further references and comments we refer to [13, 17, 47, 49]). In particular, one has that in $\mathbb{R}$

$$|x|^{\alpha} \in A_p \quad \Leftrightarrow \quad \alpha \in (-1, p-1). \tag{38}$$

In order to justify some of the arguments in the proof of Theorem 2.1 we need some further continuity properties of the Hilbert transform. More precisely, our proof requires the constant c^* in (37) to depend only on $c(w)$ in (36) and on p (it is only required for the case $p = 2$). This fact is implicit in the proof given in [20], see also [13, 17]. However, we recall the following optimal result:

Theorem 2.8 ([45]). *For $p \in [2, \infty)$ the inequality* (37) *holds with $c^* \leq c(p)c(w)$, with $c(p)$ depending only on p and $c(w)$ as in* (36). *Moreover, for $p = 2$ this estimate is sharp.*

Next, we define the truncated weights $w_N(x)$ using the notation $\langle x \rangle = (1 + x^2)^{1/2}$ as

$$w_N(x) = \begin{cases} \langle x \rangle & \text{if } |x| \leq N, \\ 2N & \text{if } |x| \geq 3N, \end{cases} \tag{39}$$

$w_N(x)$ are smooth and non-decreasing in $|x|$ with $w_N'(x) \leq 1$ for all $x \geq 0$.

Proposition 2.9. *For any $\theta \in (-1, 1)$ and any $N \in \mathbb{Z}^+$, $w_N^{\theta}(x)$ satisfies the A_2 inequality* (36). *Moreover, the Hilbert transform $\mathscr{H}$ is bounded in $L^2(w_N^{\theta}(x)dx)$ with a constant depending on θ but independent of $N \in \mathbb{Z}^+$.*

The proof of Proposition 2.9 follows by combining the fact that for a fixed $\theta \in (-1, 1)$ the family of weights $w_N^{\theta}(x)$, $N \in \mathbb{Z}^+$ satisfies the A_2 inequality in (36) with a constant c independent of N and Theorem 2.8.

Proposition 2.9 is used in weighted energy estimates with a bounded weight for each $N \in \mathbb{Z}^+$ (so all the quantities involved are finite) and with constants independent of N. This allows to pass to the limit to get the desired result.

Next, we have the following generalization of the Calderón commutator estimate given in [4] (for a different proof see [12]):

Theorem 2.10 ([4, 9]). *For any $p \in (1, \infty)$ and l, $m \in \mathbb{Z}^+ \cup \{0\}$, $l + m \geq 1$ there exists $c = c(p; l; m) > 0$ such that*

$$\|\partial_x^l [\mathscr{H}; a] \partial_x^m f\|_p \leq c \|\partial_x^{l+m} a\|_{\infty} \|f\|_p. \tag{40}$$

Calderón's result [9] corresponds to the case $l + m = 1$.

We shall also use the pointwise identities

$$[\mathscr{H}; x]\partial_x f = [\mathscr{H}; x^2]\partial_x^2 f = 0,$$

and more general

$$[\mathscr{H};x]f = 0 \quad \text{if and only if} \quad \int f dx = 0.$$

We recall the following characterization of the $L^p_s(\mathbb{R}^n) = (1-\Delta)^{-s/2}L^p(\mathbb{R}^n)$ spaces given in [48].

Theorem 2.11 ([48]). *Let $b \in (0,1)$ and $2n/(n+2b) < p < \infty$. Then $f \in L^p_b(\mathbb{R}^n)$ if and only if*

$$(a)\ f \in L^p(\mathbb{R}^n),$$

$$(b) \quad \mathscr{D}^b f(x) = \Big(\int_{\mathbb{R}^n} \frac{|f(x)-f(y)|^2}{|x-y|^{n+2b}} dy\Big)^{1/2} \in L^p(\mathbb{R}^n), \tag{41}$$

with

$$\begin{aligned} \|f\|_{b,p} \equiv \|(1-\Delta)^{b/2} f\|_p = \|J^b f\|_p &\simeq \|f\|_p + \|D^b f\|_p \\ &\simeq \|f\|_p + \|\mathscr{D}^b f\|_p. \end{aligned} \tag{42}$$

Above we have used the notation: for $s \in \mathbb{R}$

$$D^s = (-\Delta)^{s/2} \quad \text{with} \quad D^s = (\mathscr{H}\,\partial_x)^s, \quad \text{if} \quad n = 1.$$

For the proof of this theorem we refer the reader to [48].

We observe that from (41) for $p = 2$ and $b \in (0,1)$ one has

$$\|\mathscr{D}^b(fg)\|_2 \le \|f\,\mathscr{D}^b g\|_2 + \|g\,\mathscr{D}^b f\|_2, \tag{43}$$

which is a stronger version of the standard Leibniz rule for fractional derivatives whose right-hand side involved the product of appropriate norms of f and g. So it is natural to ask (Q 9) whether or not the inequality (43) holds for $p \neq 2$ and with D instead of $\mathscr{D}$ for $p = 2$ and for $p \neq 2$.

These estimates are essential in the proof of Theorem 2.3. In particular, one has the following applications of Theorem 2.11 given in [43] :

Proposition 2.12. *Let $b \in (0,1)$. For any $t > 0$*

$$\mathscr{D}^b(e^{-itx|x|}) \le c(|t|^{b/2} + |t|^b|x|^b). \tag{44}$$

Lemma 2.13. *Let $a,\, b > 0$. Assume that $J^a f = (1-\Delta)^{a/2} f \in L^2(\mathbb{R})$ and $\langle x\rangle^b f = (1+|x|^2)^{b/2} f \in L^2(\mathbb{R})$. Then for any $\theta \in (0,1)$*

$$\|J^{\theta a}(\langle x\rangle^{(1-\theta)b} f)\|_2 \le c\|\langle x\rangle^b f\|_2^{1-\theta}\,\|J^a f\|_2^{\theta}. \tag{45}$$

Moreover, the inequality (45) *is still valid with $w_N(x)$ in* (39) *instead of $\langle x\rangle$ with a constant c independent of N.*

Propagation of Regularities

To simplify our exposition we shall restrict ourselves to the case of the BO equation, i.e., $k = 1$ in (1). We shall work with the following kind of solutions whose existence was established in [46] .

Theorem 3.1. *If $u_0 \in H^s(\mathbb{R})$ with $s \geq 3/2$, then there exists a unique solution $u = u(x,t)$ of the IVP for Eq.* (3) *such that for any $T > 0$*

$$
\begin{aligned}
&\text{(i)} \quad u \in C(\mathbb{R} : H^s(\mathbb{R})) \cap L^\infty(\mathbb{R} : H^s(\mathbb{R})),\\
&\text{(ii)} \quad \partial_x u \in L^4([-T,T] : L^\infty(\mathbb{R})), \quad \textit{(Strichartz)},\\
&\text{(iii)} \quad \int_{-T}^{T}\int_{-R}^{R} (|\partial_x D u|^2 + |\partial_x^2 u|^2)(x,t)|^2 dx dt \leq c_0,\\
&\text{(iv)} \quad \int_{-T}^{T}\int_{-R}^{R} |\partial_x D^{r-1/2} u(x,t)|^2 dx dt \leq c_1, \quad r \in [1,s],
\end{aligned}
\tag{46}
$$

with $c_0 = c_0(R, T, \|u_0\|_{3/2,2})$ and $c_1 = c_1(R, T, \|u_0\|_{s,2})$.

Remark. From our previous comments it is clear that the results in Theorem 3.1 still holds for the IVP (1) locally in time for $k \geq 2$. Indeed, one can lower the requirement $s \geq 3/2$ to a value between $[1, 3/2]$, depending on the k considered, such that well-posedness including the estimate (46) (ii) still holds.

To state the results found in [25] we define a two parameter family a, b with $a < b$ functions $\chi_{\epsilon,b} \in C^\infty(\mathbb{R})$ non-decreasing such that $\chi'_{a,b}(x) \geq 0$ with

$$
\chi_{a,b}(x) = \begin{cases} 0, & x \leq a, \\ 1, & x \geq b, \end{cases} \tag{47}
$$

satisfying some appropriate estimates, see a more precise definition below and in [25].

Theorem 3.2. *Let $u_0 \in H^{3/2}(\mathbb{R})$ and $u = u(x,t)$ be the solution of the IVP for Eq.* (3) *provided by Theorem 3.1. If for some $x_0 \in \mathbb{R}$ and for some $m \in \mathbb{Z}^+$, $m \geq 2$,*

$$
\int_{x_0}^{\infty} (\partial_x^m u_0)^2(x)\, dx < \infty, \tag{48}
$$

then for any $v > 0$, $T > 0$, $a > 0$, $b > a$

$$\sup_{0\le t\le T}\int (\partial_x^m u(x,t))^2\,\chi_{a,b}(x+x_0-vt)\,dx + \int_0^T\int (D^{1/2}(\partial_x^m u(x,t))^2\chi'_{a,b}(x+x_0-vt)\,dxdt < c = c(T;a;b;v). \tag{49}$$

If in addition to (48) *there exists* $x_0 \in \mathbb{R}$ *such that any* $a > 0,\ b > a$

$$D^{1/2}(\partial_x^m u_0\,\chi_{a,b}(\cdot + x_0)) \in L^2(\mathbb{R}), \tag{50}$$

then

$$\sup_{0\le t\le T}\int (D^{1/2}(\partial_x^m u(x,t)\,\chi_{a,b}(x+x_0-vt)))^2\,dx + \int_0^T\int (\partial_x^{m+1}u(x,t))^2\,\chi'_{a,b}\,\chi_{a,b}(x+x_0-vt)\,dxdt < c, \tag{51}$$

with $c = c(T;a;b;v)$.

Theorem 3.3. *With the same hypotheses the results in Theorem 3.2 apply locally in time to solutions of the IVP* (1) *with* $k \ge 2$.

Remark. From our comments above and our proof of Theorem 3.2 it will be clear that the requirement $u_0 \in H^{3/2}(\mathbb{R})$ in Theorem 3.2 can be lower to $u_0 \in H^{9/8^+}(\mathbb{R})$ by considering the problem in a finite time interval (see [28]).

Next, we discuss some outcome of Theorems 3.2 and 3.3. In order to simplify the exposition we shall state them only for solutions of the IVP for the BO equation (3). First, as a direct consequence of Theorem 3.2 and the time reversible character of Eq. (3) one has

Corollary 3.4. *Let* $u \in C(\mathbb{R} : H^{3/2}(\mathbb{R}))$ *be the solution of the IVP for the BO equation* (3) *provided by Theorem 3.1. If there exist* $m \in \mathbb{Z}^+,\ m \ge 2,\ \hat{t} \in \mathbb{R}$, $a \in \mathbb{R}$ *such that*

$$\partial_x^m u(\cdot,\hat{t}) \notin L^2((a,\infty)),$$

then for any $t \in (-\infty,\hat{t})$ *and any* $\beta \in \mathbb{R}$

$$\partial_x^m u(\cdot,t) \notin L^2((\beta,\infty)).$$

Next, one has that for an appropriate class of data singularities of the corresponding solutions travel with infinite speed to the left as time evolves.

Corollary 3.5. *Let* $u \in C(\mathbb{R} : H^{3/2}(\mathbb{R}))$ *be the solution of the IVP* (3) *provided by Theorem 3.1. If for* $r \in \mathbb{Z}^+$

$$\int_{x_0}^{\infty} |\partial_x^r u_0(x)|^2 dx < \infty \quad but \quad \partial_x^r u_0 \notin L^2((x_0 - 1, \infty)),$$

then for any $t \in (0, \infty)$ *and any* $v > 0$ *and* $\epsilon > 0$

$$\int_{x_0+\epsilon-vt}^{\infty} |\partial_x^r u(x,t)|^2 dx < \infty,$$

and for any $t \in (-\infty, 0)$ *and any* $\alpha \in \mathbb{R}$

$$\int_{\alpha}^{\infty} |\partial_x^r u(x,t)|^2 dx = \infty.$$

Remark. Notice that (49) implies: for any $\epsilon > 0$, $v > 0$, $T > 0$

$$\sup_{0 \le t \le T} \int_{x_0+\epsilon-vt}^{\infty} (\partial_x^r u(x,t))^2 \, dx \le c = c(\epsilon, v, T). \tag{52}$$

This tells us that the local regularity of the initial datum u_0 described in (48) propagates with infinite speed to its left as time evolves.

In [24] similar results concerning the IVP for the k-generalized Korteweg-de Vries equation (5) were obtained. However, the proof for the BO equation is quite more involved. First, it includes a non-local operator, the Hilbert transform (2). Second, in the case of the k-gKdV the Kato local smoothing effect produces a gain of one derivative (see the argument below) which allows to pass to the next step in the inductive process. However, in the case of the BO equation the gain of the local smoothing is just $1/2$-derivative (see the argument below) so the iterative argument has to be carried out into two steps, one for positive integers m and another for $m + 1/2$. Also the explicit identity obtained in [26] describing the local smoothing effect in solutions of the KdV equation is not available for the BO equation. In this case, to establish the local smoothing one has to rely on several commutator estimates. The main one is the extension of the Calderón first commutator estimate for the Hilbert transform [9] given in [4], see Theorem 2.10.

As it was already mentioned Theorem 3.2 describes the propagation of regularities in solutions of the k-gBO equation ($\alpha = 0$ in (6)) and the corresponding result for the k-gKdV equation ($\alpha = 1$ in (6)) was proved in [24]. However, a similar result for the equations in (6) with $\alpha \in (0, 1)$ is unknown (*Q 10*).

To illustrate a key argument in these proofs and the difficulties in extending it to the case $\alpha \in (0, 1)$ we consider the linear IVP associated to generalized dispersive BO equation

$$\begin{cases} \partial_t w_\alpha - D_x^{1+\alpha}\partial_x w_\alpha = 0, & \alpha \in [0,1], \quad x, t \in \mathbb{R}, \\ w_\alpha(x,0) = w \in L^2(\mathbb{R}), & w|_{(0,\infty)} \in H^1((0,\infty)). \end{cases} \tag{53}$$

We define the following family of functions : for each $a, b \in \mathbb{R}$, $a < b$ we define $\chi_{a,b} \in C^\infty$ as in (47) such that if $b - a = 4\epsilon$ one has $\chi'_{a,b}(x) \geq 0,\ x \in \mathbb{R}$ and $\chi'_{a,b}(x) = (b-a)/2,\ x \in [a+\epsilon, b-\epsilon]$.

First, we consider the case $\alpha = 1$ in (53). Following the argument in [26] for any $v > 0$ one formally has that

$$\begin{aligned} &\frac{1}{2}\frac{d}{dt}\int w_1^2(x,t)\chi(x+vt)dx + \frac{3}{2}\int (\partial_x w_1)^2(x,t)\chi'(x+vt)dx \\ &\quad -\frac{1}{2}\int w_1^2(x,t)\chi^{(3)}(x+vt)dx - v\int w_1^2(x,t)\chi'(x+vt)dx = 0. \end{aligned} \tag{54}$$

Thus, integrating in the time interval $[0, T]$ one gets that

$$\begin{aligned} &\sup_{0\le t\le T}\int w_1^2(x,t)\chi(x+vt)dx + \int_0^T\int (\partial_x w_1)^2(x,t)\chi'(x+vt)dxdt \\ &\quad \le c = c(\|w\|_2; T; v; a; b). \end{aligned} \tag{55}$$

Taking derivatives in the equation in (53) with $\alpha = 1$, reapplying the above argument with $a > 0$ and using (55) one gets that

$$\begin{aligned} &\sup_{0\le t\le T}\int (\partial_x w_1)^2\chi(x+vt)dx + \int_0^T\int (\partial_x^2 w_1)^2\chi'(x+vt)dxdt \\ &\quad \le c = c(\|w\|_2; \|\partial_x w\|_{L^2((0,\infty))}; T; v; a; b), \end{aligned} \tag{56}$$

which very roughly explains one of the main ideas given in [24] to prove the propagation of regularities in solutions of the k-gKdV.

Next, we consider the case $\alpha = 0$ in (53). A similar argument yields the identity

$$\begin{aligned} &\frac{1}{2}\frac{d}{dt}\int w_0^2(x,t)\chi(x+vt)dx + \int \mathcal{H}\partial_x w_0\,\partial_x w_0\,\chi(x+vt)dx \\ &\quad + \int \mathcal{H}\partial_x w_0\, w_0\,\chi'(x+vt)dx - v\int w_0^2\chi'(x+vt)dx = 0. \end{aligned} \tag{57}$$

We observe that

$$\begin{aligned} A(t) &= \int \mathcal{H}\partial_x w_0\,\partial_x w_0\,\chi(x+vt)dx \\ &= -\int \partial_x w_0\,\mathcal{H}(\partial_x w_0\,\chi(x+vt))dx \\ &= -\int \partial_x w_0\,\mathcal{H}\partial_x w_0\,\chi(x+vt)dx - \int \partial_x w_0\,[\mathcal{H};\chi]\partial_x w_0(x,t)dx \\ &= -A(t) + \int w_0\,\partial_x[\mathcal{H};\chi]\partial_x w_0 dx. \end{aligned} \tag{58}$$

Therefore, combining Theorem 2.10 and the fact that the L^2-norm of the solution is preserved it follows that

$$A(t) = \frac{1}{2}\int w_0\, \partial_x[\mathscr{H};\chi]\partial_x w_0 dx = O(\|w\|_2^2), \qquad t \in \mathbb{R}.$$

Also, one sees that

$$\begin{aligned} B(t) &= \int \mathscr{H}\partial_x w_0 w_0 \chi'(x+vt)dx = \int D_x w_0 w_0 \chi'(x+vt)dx \\ &= -\int D^{1/2} w_0\, D_x^{1/2}(w_0\chi'(x+vt))dx \\ &= \int (D_x^{1/2}w_0)^2\, \chi'(x+vt)dx + \int D_x^{1/2}w_0\, [D_x^{1/2};\chi']w_0(x,t)dx \\ &= \int (D_x^{1/2}w_0)^2\, \chi'(x+vt)dx + \int w_0\, D_x^{1/2}[D_x^{1/2};\chi']w_0(x,t)dx, \end{aligned} \tag{59}$$

where the second term in the right-hand side of (59) can be estimated using

$$\|D_x^{1/2}[D_x^{1/2};\chi']f\|_2 \le c_\chi \|f\|_2. \tag{60}$$

Inserting the above estimates in (57) one concludes that

$$\begin{aligned} &\sup_{0\le t\le T} \int (w_0(x,t))^2\chi(x+vt)dx + \int_0^T \int (D_x^{1/2}w_0)^2\chi'(x+vt)dxdt \\ &\quad \le c = c(\|w\|_2; T; a; b). \end{aligned} \tag{61}$$

Repeating the above argument, after applying the operator $D_x^{1/2}$ to the equation in (53) with $\alpha = 0$, one gets

$$\begin{aligned} &\frac{1}{2}\frac{d}{dt}\int (D_x^{1/2}w_0)^2\chi(x+vt)dx \\ &\quad + \int \mathscr{H}\partial_x D_x^{1/2}w_0\, \partial_x D_x^{1/2}w_0 \chi(x+vt)dx \\ &\quad + \int \mathscr{H}\partial_x D_x^{1/2}w_0\, D_x^{1/2}w_0\chi'(x+vt)dx \\ &\quad - v\int (D_x^{1/2}w_0)^2\chi'(x+vt)dx = 0. \end{aligned} \tag{62}$$

From (61) after integration in time the last term in (62) is bounded. To control the second term in (62) we write

$$
\begin{aligned}
A_1(t) &\equiv \int \mathscr{H}\partial_x D_x^{1/2}w_0\,\partial_x D_x^{1/2}w_0\chi(x+vt)dx\\
&= -\int \partial_x D_x^{1/2}w_0\,\mathscr{H}(\partial_x D_x^{1/2}w_0\chi(x+vt))dx\\
&= -\int \partial_x D_x^{1/2}w_0\,\mathscr{H}\partial_x D_x^{1/2}w_0\,\chi(x+vt)dx\\
&\quad -\int \partial_x D_x^{1/2}w_0\,[\mathscr{H};\chi]\partial_x D_x^{1/2}w_0(x,t)dx\\
&= -A_1(t) - \int w_0\,D_x^{1/2}\partial_x[\mathscr{H};\chi]\partial_x D_x^{1/2}w_0dx.
\end{aligned} \tag{63}
$$

Hence, from Theorem 2.10 it follows that

$$
A_1(t) = -\frac{1}{2}\int w_0\,D_x^{1/2}\partial_x[\mathscr{H};\chi]\partial_x D_x^{1/2}w_0dx = O(\|w\|_2^2), \qquad t\in\mathbb{R}.
$$

Finally, to bound the third term in (62) one defines $\rho(\cdot)$ as

$$
\chi'(x) = \rho^2(x),
$$

then

$$
\begin{aligned}
B_1(t) &\equiv \int \mathscr{H}\partial_x D_x^{1/2}w_0\rho(x+vt)D_x^{1/2}w_0\rho(x+vt)dx\\
&= \int \mathscr{H}(\partial_x D^{1/2}w_0\rho(x+vt))\,D_x^{1/2}w_0\rho(x+vt)dx\\
&\quad - \int [\mathscr{H};\rho]\partial_x D_x^{1/2}w_0\,D_x^{1/2}w_0\rho(x+vt)dx\\
&= \int \mathscr{H}\partial_x(D_x^{1/2}w_0\rho(x+vt))\,D_x^{1/2}w_0\rho(x+vt)dx\\
&\quad - \int \mathscr{H}(D_x^{1/2}w_0\rho'(x+vt))\,D_x^{1/2}w_0\rho(x+vt)dx - B_2(t)\\
&= \int (D_x^{1/2}(D_x^{1/2}w_0\rho(x+vt)))^2dx - B_3(t) - B_2(t).
\end{aligned} \tag{64}
$$

with

$$
B_3(t) = \int \mathscr{H}(D_x^{1/2}w_0\rho'(x+vt))\,D_x^{1/2}w_0\rho(x+vt)dx,
$$

which after integration in time is basically bounded by (61), and

$$
\begin{aligned}
B_2(t) &= \int [\mathcal{H}; \rho]\partial_x D_x^{1/2} w_0 \, D_x^{1/2} w_0 \rho(x+vt)dx \\
&= \int [\mathcal{H}; \rho]\partial_x D_x^{1/2} w_0 \, D_x^{1/2} (w_0 \rho(x+vt))dx \\
&\quad - \int [\mathcal{H}; \rho]\partial_x D_x^{1/2} w_0 \, [D_x^{1/2}; \rho] w_0 dx,
\end{aligned}
\tag{65}
$$

whose boundedness follows by combining Theorem 2.10 and (60). Gathering the above information one sees that

$$
\begin{aligned}
&\sup_{0\le t\le T} \int (D_x^{1/2} w_0)^2 \chi(x+vt)dx \\
&\quad + \int_0^T \int (D_x^{1/2}(D_x^{1/2} w_0 \rho(x+vt)))^2 dxdt \\
&\quad \le c = c(\|w\|_2; \|\partial_x w\|_{L^2((0,\infty))}; T; v; a; b),
\end{aligned}
\tag{66}
$$

with $\rho^2(x) = \chi'(x)$.

Reapplying the above argument with the appropriate modifications and using the estimate (66) one can conclude that

$$
\begin{aligned}
&\sup_{0\le t\le T} \int (\partial_x w_0)^2 \chi(x+vt)dx + \int_0^T \int (D_x^{1/2}(\partial_x w_0 \rho(x+vt)))^2 dxdt \\
&\quad \le c = c(\|w\|_2; \|\partial_x w\|_{L^2((0,\infty))}; T; v; a; b),
\end{aligned}
\tag{67}
$$

with $\eta^2 = \chi$.

As it was mentioned above the result in (56) for $\alpha = 1$ were extended in [24] to solutions of the IVP (5) and those in (67) for $\alpha = 0$ were extended in [25] to solutions of the IVP (1). As it was mentioned before, a similar result for the solutions of the IVP associated to the equations in (6) with $\alpha \in (0,1)$ is unknown (*Q 10*).

Acknowledgements The author would like to thank German Fonseca, Felipe Linares, and Jean-Claude Saut for fruitful conversations concerning with this work.

References

1. L. Abdelouhab, J.L. Bona, M. Felland, J.-C. Saut, Nonlocal models for nonlinear dispersive waves. Physica D **40**, 360–392 (1989)
2. M.J. Ablowitz, A.S. Fokas, The inverse scattering transform for the Benjamin-Ono equation, a pivot for multidimensional problems. Stud. Appl. Math. **68**, 1–10 (1983)
3. C.J. Amick, J.F. Toland, Uniqueness and related analytic properties for the Benjamin-Ono equation - a nonlinear Neumann problem in the plane. Acta Math. **167**, 107–126 (1991)

4. B. Bajsanski, R.R. Coifman, *Singular Integrals*. Proceedings of Symposia in Pure Mathematics (American Mathematical Society, Providence, RI, 1966), pp. 1–17
5. T.B. Benjamin, Internal waves of permanent form in fluids of great depth. J. Fluid Mech. **29**, 559–592 (1967)
6. H.A. Biagioni, F. Linares, Ill-posedness for the derivative Schrödinger and generalized Benjamin-Ono equations. Trans. Am. Math. Soc. **353**, 3649–3659 (2001)
7. J. Bourgain, Fourier restriction phenomena for certain lattice subsets and applications to nonlinear evolution equations. Geom. Funct. Anal. **3**, 107–156/209–262 (1993)
8. N. Burq, F. Planchon, On the well-posedness of the Benjamin-Ono equation. Math. Ann. **340**, 497–542 (2008)
9. A.P. Calderón, Commutators of singular integral operators. Proc. Nat. Acad. Sci. USA **53**, 1092–1099 (1965)
10. R.R. Coifman, Y. Meyer, Au delá des opérateurs pseudodifférentiels, Asterisque **57**, Société Mathemátique de France (1973)
11. R.R. Coifman, M. Wickerhauser, The scattering transform for the Benjamin-Ono equation. Inverse Prob. **6**, 825–860 (1990)
12. L. Dawson, H. McGahagan, G. Ponce, On the decay properties of solutions to a class of Schrödinger equations. Proc. Am. Math. Soc. **136**, 2081–2090 (2008)
13. J. Duoandikoetxea, *Fourier Analysis*. Graduate Studies in Mathematics, vol. 29 (American Mathematical Society, Providence, RI, 2000)
14. C. Flores, On decay properties of solutions to the IVP for the Benjamin-Ono equation. J. Dyn. Differ. Equ. **25**, 907–923 (2013)
15. G. Fonseca, G. Ponce, The IVP for the Benjamin-Ono equation in weighted Sobolev spaces. J. Funct. Anal. **260**, 436–459 (2010)
16. G. Fonseca, F. Linares, G. Ponce, The IVP for the Benjamin-Ono equation in weighted Sobolev spaces II. J. Funct. Anal. **262**, 2031–2049 (2012)
17. J. Garcia-Cuerva, J.L. Rubio de Francia, *Weighted Norm Inequalities and Related topics* (North Holland, Amsterdam, 1985)
18. H. Helson, G. Szegö, A problem in prediction theory. Ann. Math. Pure Appl. **51**, 107–138 (1960)
19. S. Herr, A. Ionescu, C.E. Kenig, H. Koch, A para-differential renormalization technique for nonlinear dispersive equations. Commun. Partial Differ. Equ. **35**, 1827–1875 (2010)
20. R. Hunt, B. Muckenhoupt, R. Wheeden, Weighted norm inequalities for the conjugate function and Hilbert transform. Trans. Am. Math. Soc. **176**, 227–251 (1973)
21. A.D. Ionescu, C.E. Kenig, Global well-posedness of the Benjamin-Ono equation on low-regularity spaces. J. Am. Math. Soc. **20**(3), 753–798 (2007)
22. R.J. Iorio, On the Cauchy problem for the Benjamin-Ono equation. Commun. Partial Differ. Equ. **11**, 1031–1081 (1986)
23. R.J. Iorio, Unique continuation principle for the Benjamin-Ono equation. Differ. Integr. Equ. **16**, 1281–1291 (2003)
24. P. Isaza, F. Linares, G. Ponce, On the propagation of regularity and decay of solutions to the k-generalized Korteweg-de Vries equation. Commun Partial Differ. Equ. **40**, 1336–1364 (2015)
25. P. Isaza, F. Linares, G. Ponce, On the propagation of regularity in solutions to the Benjamin-Ono equation. J. Funct. Anal. **270**, 976–1000 (2016)
26. T. Kato, On the Cauchy problem for the (generalized) Korteweg-de Vries equation. Adv. Math. Suppl. Stud. Stud. Appl. Math. **8**, 93–128 (1983)
27. T. Kato, G. Ponce, Commutator estimates and the Euler and Navier-Stokes equations. Commun. Pure Appl. Math. **41**, 891–907 (1988)
28. C.E. Kenig, K.D. Koenig, On the local well-posedness of the Benjamin-Ono and modified Benjamin-Ono equations. Math. Res. Lett. **10**, 879–895 (2003)
29. C.E. Kenig, H. Takaoka, Global well-posedness of the modified Benjamin-Ono equation with initial data in $H^{1/2}$. Int. Math. Res. Not. **9**, 1–44 (2006). Art. ID 95702
30. C.E. Kenig, G. Ponce, L. Vega, Oscillatory integrals and regularity of dispersive equations. Indiana Univ. Math. J. **40**, 33–69 (1991)

31. C.E. Kenig, G. Ponce, L. Vega, Well-posedness of the initial value problem for the Korteweg-de Vries equation. J. Am. Math. Soc. **4**, 323–347 (1991)
32. C.E. Kenig, G. Ponce, L. Vega, Well-posedness and scattering results for the generalized Korteweg-de Vries equation via contraction principle. Commun. Pure Appl. Math. **46**, 527–620 (1993)
33. C.E. Kenig, G. Ponce, L. Vega, On the generalized Benjamin-Ono equation. Trans. Am. Math. Soc. **342**, 155–172 (1994)
34. H. Koch, N. Tzvetkov, On the local well-posedness of the Benjamin-Ono equation on $H^s(\mathbb{R})$. Int. Math. Res. Not. **26**, 1449–1464 (2003)
35. H. Koch, N. Tzvetkov, Nonlinear wave interactions for the Benjamin-Ono equation. Int. Math. Res. Not. **30**, 1833–1847 (2005)
36. F. Linares, G. Ponce, *Introduction to Nonlinear Dispersive Equations* (Springer, New York, 2009)
37. Y. Martel, F. Merle, Blow up in finite time and dynamics of blow up solutions for the L^2-critical generalized KdV equation. J. Am. Math. Soc. **15**, 617–664 (2002)
38. L. Molinet, D. Pilod, The Cauchy problem for the Benjamin-Ono equation in L^2 revisited. Anal. PDE **5**, 365–395 (2012)
39. L. Molinet, F. Ribaud, Well-posedness results for the generalized Benjamin-Ono equation with small initial data. J. Math. Pures et Appl. **83**, 277–311 (2004)
40. L. Molinet, F. Ribaud, Well-posedness results for the Benjamin-Ono equation with arbitrary large initial data. Int. Math. Res. Not. **70**, 3757–3795 (2004)
41. L. Molinet, J.C. Saut, N. Tzvetkov, Ill- posedness issues for the Benjamin-Ono and related equations. SIAM J. Math. Anal. **33**, 982–988 (2001)
42. B. Muckenhoupt, Weighted norm inequalities for the Hardy maximal function. Trans. Am. Math. Soc. **165**, 207–226 (1972)
43. J. Nahas, G. Ponce, On the persistent properties of solutions to semi-linear Schrodinger equation. Commun. Partial Differ. Equ. **34**, 1–20 (2009)
44. H. Ono, Algebraic solitary waves on stratified fluids. J. Phys. Soc. Jpn. **39**, 1082–1091 (1975)
45. S. Petermichl, The sharp bound for the Hilbert transform on weighted Lebesgue spaces in terms of the classical A_p characteristic. Am. J. Math. **129**, 1355–1375 (2007)
46. G. Ponce, On the global well-posedness of the Benjamin-Ono equation. Differ. Integr. Equ. **4**, 527–542 (1991)
47. C. Sadosky, *Interpolation of Operators and Singular Integrals* (Marcel Dekker Inc, New York, 1979)
48. E.M. Stein, The characterization of functions arising as potentials. Bull. Am. Math. Soc. **67**, 102–104 (1961)
49. E.M. Stein, *Harmonic Analysis* (Princeton University Press, Princeton, 1993)
50. R.S. Strichartz, Restriction of Fourier transforms to quadratic surface and decay of solutions of wave equations. Duke Math. J. **44**, 705–714 (1977)
51. T. Tao, Global well-posedness of the Benjamin-Ono equation on H^1. J. Hyp. Differ. Equ. **1**, 27–49 (2004)
52. S. Vento, Sharp well-posedness results for the generalized Benjamin-Ono equations with higher nonlinearity. Differ. Integr. Equ. **22**, 425–446 (2009)
53. S. Vento, Well-posedness of the generalized Benjamin-Ono equations with arbitrary large initial data in the critical space. Int. Math. Res. Not. **2**, 297–319 (2010)

CPSIA information can be obtained
at www.ICGtesting.com
Printed in the USA
BVOW10*0240220916
462961BV00001B/4/P

9 783319 309590